W9-BKV-397

The History of Mathematics from Antiquity to the Present

BIBLIOGRAPHIES OF THE HISTORY
OF SCIENCE AND TECHNOLOGY
(Vol. 6)

GARLAND REFERENCE LIBRARY
OF THE HUMANITIES
(Vol. 313)

Bibliographies of the History of Science and Technology

Editors

Robert Multhauf, Smithsonian Institution, Washington, D.C.
Ellen Wells, Smithsonian Institution, Washington, D.C.

The History of Mathematics from Antiquity to the Present
A Selective Bibliography

Joseph W. Dauben

GARLAND PUBLISHING, INC. • NEW YORK & LONDON
1985

Library of Congress Cataloging in Publication Data

Dauben, Joseph Warren, 1944–
The history of mathematics from antiquity to the present.

(Bibliographies of the history of science and technology ; v. 6) (Garland reference library of the humanities ; v. 313)
Includes index.
1. Mathematics—History—Bibliography. I. Title. II. Series. III. Series: Garland reference library of the humanities ; v. 313.
Z6651.D38 1985 [QA21] 016.510'9 81-43364
ISBN 0-8240-9284-8 (alk. paper)

Printed on acid-free, 250-year-life paper
Manufactured in the United States of America

GENERAL INTRODUCTION

This bibliography is one of a series designed to guide the reader into the history of science and technology. Anyone interested in any of the components of this vast subject area is part of our intended audience, not only the student, but also the scientist interested in the history of his own field (or faced with the necessity of writing an "historical introduction") and the historian, amateur or professional. The latter will not find the bibliographies "exhaustive," although in some fields he may find them the only existing bibliographies. He will in any case not find one of those endless lists in which the important is lumped with the trivial, but rather a "critical" bibliography, largely annotated, and indexed to lead the reader quickly to the most important (or only existing) literature.

Inasmuch as everyone treasures bibliographies, it is surprising how few there are in this field. George Sarton's *Guide to the History of Science* (Waltham, Mass., 1952; 316 pp.), Eugene S. Ferguson's *Bibliography of the History of Technology* (Cambridge, Mass., 1968; 347 pp.), François Russo's *Histoire des Sciences et des Techniques. Bibliographie* (Paris, 2nd ed., 1969; 214 pp.) are justifiably treasured but they are, of necessity, limited in their coverage and need to be updated.

For various reasons, mostly bad, the average scholar prefers adding to the literature to sorting it out. The editors are indebted to the scholars represented in this series for their willingness to expend the time and effort required to pursue the latter objective. Our aim, and that of the publisher, has been to give the series enough uniformity to give some consistency to the series, but otherwise to leave the format and contents to the author/compiler. We have urged that introductions be used for essays on "the state of the field," and that selectivity be exercised to limit the length of each volume. Since the historical literature ranged from very large (e.g., medicine) to very small (e.g., chemical technology), some bibliographies will be limited to truly important writings while others will include modest "contributions" and even primary sources. The problem is intelligible guidance into a particular field—or subfield—and its solution is largely left to the author/compiler. In general, topical volumes (e.g., chemistry) will deal with the subject since about 1700, leaving earlier literature to area or chronological volumes (e.g., medieval science); but here, too, the volumes will vary according to the judgment of the author. The volumes are international (except for two, *Science and Technology in the United States* and *Science and Technology in Eastern Asia*) but the literature covered depends, of course, on the linguistic equipment of the author and his access to "exotic" literatures.

Robert Multhauf
Ellen Wells

Smithsonian Institution
Washington, D.C.

CONTENTS

PREFACE

The general philosophy of the Garland series of annotated bibliographies in the history of science is to provide an introduction to all major areas and approaches to a given subject—in this case, the history of mathematics. This has been accomplished here through the efforts of 49 scholars on five continents, who have drawn upon their individual expertise in specific aspects of the history of mathematics to provide as wide a coverage of topics and time periods as possible. Considering the breadth of this undertaking—which attempts to encompass more than 4000 years of mathematics and almost as many years in the writing of its history—no individual scholar could have written the extensive critical annotations that make this bibliography unique, and above all, indispensable to anyone with a serious interest in the history of mathematics.

Appropriately, this bibliography is dedicated to the memory of Kenneth O. May, whose own commitment to the history of mathematics and especially to its bibliography did much to encourage scholarship in the subject. He would have been the first to applaud the kind of international cooperation that this annotated bibliography reflects on behalf of the history of mathematics.

Joseph W. Dauben
New York City
June 1984

CONTRIBUTORS

Kirsti Andersen
University of Aarhus
Denmark

E.J. Ashworth
University of Waterloo
Waterloo, Ont., Canada

J.L. Berggren
Simon Fraser University
Burnaby, B.C., Canada

H.J.M. Bos
University of Utrecht
The Netherlands

Stephen G. Brush
University of Maryland
College Park, Md., U.S.A.

Robert Bunn
University of British Columbia
Vancouver, B.C., Canada

H.L.L. Busard
Venlo
The Netherlands

Paul J. Campbell
Beloit College
Beloit, Wis., U.S.A.

James J. Cross
University of Melbourne
Victoria, Australia

Joseph W. Dauben
City University
New York, N.Y., U.S.A.

Steven B. Engelsman
Museum Boerhaave
Leiden, the Netherlands

Philip Enros
University of Toronto
Toronto, Ont., Canada

Menso Folkerts
Universität München
Munich, Germany

Jöran Friberg
Chalmers Tekniska Högskola
Göteborg, Sweden

Bernard R. Goldstein
University of Pittsburgh
Pittsburgh, Pa., U.S.A.

Ivor Grattan-Guinness
Middlesex Polytechnic
England

Jeremy Gray
The Open University
England

Judy Green
Rutgers University
Camden, N.J., U.S.A.

Takao Hayashi
Shiozawamachi
Niigataken, Japan

R.W. Home
University of Melbourne
Australia

Shuntaro Ito
University of Tokyo
Japan

S.A. Jayawardene
Science Museum
South Kensington, England

Dale Johnson
Hampden-Sydney College
Hampden-Sydney, Va., U.S.A.

Phillip S. Jones
University of Michigan
Ann Arbor, Mich., U.S.A.

E.S. Kennedy
The American University
Beirut, Lebanon

Eberhard Knobloch
Technische Universität
Berlin, Germany

Jeanne LaDuke
DePaul University
Chicago, Ill., U.S.A.

Lam Lay-Yong
National University
Singapore

Albert C. Lewis
McMaster University
Hamilton, Ont., Canada

Richard Lorch
Bayerische Akademie der Wissenschaften
Munich, Germany

Wolfgang Mathis
Technische Universität
Braunschweig, Germany

G.P. Matvievskaya
Tashkent
U.S.S.R.

Herbert Mehrtens
Technische Universität
Berlin, Germany

Gregory H. Moore
University of Toronto
Toronto, Ont., Canada

Jon V. Pepper
N.E. London Polytechnic
England

Teri H. Perl
The Learning Company
Portola Valley, Cal., U.S.A.

Helena M. Pycior
University of Wisconsin
Milwaukee, Wis., U.S.A.

N.L. Rabinovitch
Jews' College
London, England

B. Randell
University of Newcastle upon Tyne
England

V. Frederick Rickey
Bowling Green State University
Bowling Green, Ohio, U.S.A.

Leo Rogers
Roehampton Institute
London, England

B.A. Rosenfeld
Institute of the History of Science and Technology
Moscow, U.S.S.R.

David E. Rowe
Pace University
Pleasantville, N.Y., U.S.A.

Christoph Scriba
University of Hamburg
Germany

Robert H. Silliman
Emory University
Atlanta, Ga., U.S.A.

Lynn Arthur Steen
St. Olaf College
Northfield, Minn., U.S.A.

Edith Sylla
North Carolina State University
Raleigh, U.S.A.

Garry J. Tee
University of Auckland
New Zealand

Claudia Zaslavsky
New York, N.Y., U.S.A.

INTRODUCTION

> Bibliographic extravagance is a sin rather than a virtue, a real perversity.
>
> —George Sarton

George Sarton liked to describe the history of science as a "secret history," and the history of mathematics as the secret within the secret, for while most scholars might know something of the history of science in general, few mathematicians, scientists, or even specialists in the history of science could be expected to know much about the history of mathematics. In part, this was because of the theoretical and often abstruse nature of mathematics, but also in part because so little had been done to make the history of mathematics accessible to those who might have need of it, or to others who might simply be interested in learning more about it. A major goal of this bibliography in the history of mathematics is, in fact, to furnish a reference work that will help to open the doors to this *secreta secretorum*, and to provide something of an Ariadne's thread through the labyrinth of this increasingly specialized and often difficult domain of human knowledge.

The History of Mathematics

Mathematics has a history that begins in *pre*-history. The earliest archaeological and anthropological evidence we possess makes it clear that even as *homo sapiens* began to speak, he began to count as well. Much before there was any way to express even the simplest numbers in written form, man was counting and calculating. By the time neolithic cultures had emerged along the rivers of the Tigris-Euphrates, Nile, Hwang Ho, and Indus, mathematics was already developed and in some cases to a high and sophisticated degree. This was especially true in Mesopotamia, where the Babylonians developed algebra to a considerable extent and devised a very powerful sexagesimal (base-sixty) arithmetic which was well suited to their complex but correspondingly effective mathematical astronomy. For the most part, however, ancient knowledge of arithmetic, algebra, and geometry was imperfect or only approximate and was reserved for the small numbers of royal scribes and priests who became adept at solving basic problems in accounting, surveying, and astronomy, often with mystical and religious overtones. The Egyptian *Book of the Dead* makes this quite clear in describing a Pharaoh who knows a mnemonic device, learned in the form of a rhymed chant, to remember how to count from one to ten on his fingers. His ability to do so is considered sufficient to establish that he is indeed a king with full knowledge of things reserved for the gods, thereby convincing the ferryman to transport him across the river of forgetfulness to the divine world of eternal life.

Centuries later, the Greeks of the archaic period began to "demythologize" their world (to borrow Henri Frankfort's phrase) and adopted both a more secular and objective perspective on nature than did their predecessors, the Egyptians or Babylonians. But they also succeeded in elevating mathematics to the idealized realm of perfect, eternal knowledge, whose truths were associated with the loftiest philosophical ideals. The Greeks, having learned their mathematics from the Egyptians and Babylonians, soon transformed it into a powerful epistemological tool. From Thales to Euclid numerous discoveries were made in geometry and arithmetic. Equally important, methods were developed to *prove* the validity of an assertion or mathematical theorem. This, of course, culminated in two major achievements—Aristotle's formal logic of the syllogism and Euclid's axiomatic geometry. By this time the practical importance and theoretical significance of mathematics was also clear not only to philosophers and mathematicians, but to shopkeepers, mariners, generals, and government administrators. Moreover, mathematics had found its first true historian (of whom any record survives)—Eudemos of Rhodes, a disciple of Aristotle who wrote a summary of Greek mathematics in the early 4th century (fl. 335 B.C.).

For the rest of antiquity the Greek advances remained the norm, and for over 900 years few advances were made. Despite nearly a millennium of the so-called Dark Ages, mathematics eventually experienced a *second* miracle, as George Sarton called it, that is, the interest of Arab scholars first in discovering and preserving the mathematics of the ancients and then in augmenting and transmitting that knowledge. In so doing, they transformed and invigorated much of the received mathematics. They also developed the Hindu-Arabic decimal system, a seminal discovery that eventually was transmitted to the Latin West.

Both the Arabic and Latin traditions were particularly strong, and much richer than is usually thought, largely because until recently little in the way of systematic research to produce viable texts and informative commentaries was available to historians of science. When Gustav Eneström began publishing corrections in the 1890s in *Bibliotheca Mathematica* (item 31) to the monumental *Geschichte der Mathematik* (item 74) of Moritz Cantor, the areas most often found wanting were those for ancient, Islamic, and medieval mathematics. Since 1900, these areas have received considerable attention, and recent research reflects strong interest now in each of these periods. Such activity is well represented in this bibliography, with substantial sections devoted to Egyptian, Babylonian, Greek, and medieval mathematics of the Indian, Islamic, Hebrew, and Western Latin cultures.

The most important stage in the advance of European mathematics coincided with the Italian and Northern Renaissance of the fourteenth and fifteenth centuries, when the progress of mathematics and the sciences generally was as rapid as the social and economic development of Western Europe. Beginning with 1600, this bibliography proceeds by roughly 100-year intervals (although a certain amount of overlap between centuries is unavoidable) as it surveys the development of mathematics from the Renaissance through the 17th, 18th, 19th, and 20th centuries.

In the last hundred years, mathematics has developed into a highly abstract and esoteric body of knowledge which, nevertheless, has extraordinary versatility in applications that permeate almost every aspect of life in any modern society. Its past history, as already indicated, is enormous, from its beginnings evident in the notched bones of caveman tally sticks to, more recently, the sublimities of group theory, abstract algebra, and the powerful applications of differential equations and high-speed computers. In fact, the last two major divisions of this bibliography investigate the history of mathematics by 22 subject areas, ranging from the philosophy and sociology of mathematics to mathematics in Africa and the Orient, as well as the subject of women in mathematics.

One way to bring order to this long, diverse, and complicated history of mathematics, however, is through the guidance of a good critical, annotated bibliography.

The Role and Significance of Bibliographies

The art of compiling bibliographies is nearly as old as the history of written documents themselves.[1] It is of interest to historians of science that the first bibliographies of which any record is known were drawn up by the Roman physician Galen in the second century A.D. They are his *De libris propriis liber*, followed by a second version, the *De ordine librorum suorum liber* (which survives only in a fragment), both of which were intended to authenticate his own works and distinguish them from the many spurious writings attributed to him. Later bibliographies in antiquity and the early medieval period, like those of St. Jerome and the Venerable Bede, fall into the tradition of compiling lists of ecclesiastical authors and their works.

The first bibliography encompassing printed works rather than manuscript material was compiled in the 15th century by Johann Tritheim, whose *Liber de scriptoribus ecclesiasticis* (1494) continued in the tradition of Jerome and Bede. None of these works, however, whether of books or manuscripts, was actually called a bibliography—the words used most often were those like "bibliotheca," "catalogus," "repertorium," "inventarium," or "index." The word "bibliography" was actually used first in France, it seems, by Gabriel Naudé, secretary and librarian to Cardinal Richelieu, for his *Bibliographia politica*.[2]

Bibliography actually came into its own in France, although slowly. The word is absent from the first edition of the *Dictionnaire de l'Académie Françoise*, and was still missing in 1751. Nor does it occur in Diderot and D'Alembert's *Encyclopédie* (although the term "bibliographer" does, but only in the sense of one skilled in the use of ancient manuscripts, e.g., a paleographer; there is no reference to catalogues or lists). In the fourth edition of the Académie's *Dictionnaire*, however, "bibliography" in the modern sense, finally appears.[3]

It was the French Revolution, however, that marked the real turning point in the history of bibliography. In fact, the subject became a matter of considerable urgency and was the special subject of a "Rapport sur la bibliographie" issued on 22 Germinal of the Year II of the Revolution (April 11, 1794). Not only was this the first official document of a government on the subject of bibliography, submitted by Henri Grégoire (1750–1831), the constitutional Bishop of Blois and a Deputy of the Convention, but it addressed directly the problem of cataloguing the mass of books confiscated from religious organizations and emigres, all of which subsequently became the property of the French nation. It was also in France that the subject was first institutionalized, at the Ecole des Chartes, where a professorship was established for bibliography in 1869. Courses were regularly offered, chiefly on the classification of archives and libraries.

As for bibliographies of interest to the history of mathematics, one of the first is attributed to Cornelius à Beughem, a bookseller and publisher who lived for a long period in Emmerich, Westphalia, and who produced a number of specialized bibliographies, including one in 1688 entitled *Bibliographia mathematica et artificiosa novissima*, which ran to nearly 500 pages and included the works of nearly 2000 writers.[4]

Not until the 19th century, however, was any real impetus given to the systematic production of bibliographic resources. These were naturally stimulated by the progress of public education and the proliferation of universities, learned societies, and related institutions. Stabilization of the book trade, the growth of periodical presses, and the establishment of the first great public archives and libraries all provoked an obvious need for systematic bibliography. For the first time, rather than trying primarily to record, note, or save from oblivion the works of the past, the role of bibliography advanced to one of dissemination, calling to the attention of scholars and interested readers the most current advances in learning. Here, not surprisingly, leadership came first from Germany. The organization of universities and their emphasis upon careful scholarship made the creation of research libraries essential, and these in turn both depended upon and stimulated the subject of bibliography. If one surveys the major bibliographies produced in this period and, especially those of interest to the historian of mathematics, an instructive pattern emerges:

1830. I. Rogg. *Bibliotheca Mathematica* (Tübingen).
1839. J.O. Halliwell. *Rara Mathematica* (London).
1847. A. de Morgan. *Arithmetical Books from the Invention of Printing* (London).
1854. L.A. Sohncke. *Bibliotheca Mathematica* (Leipzig).
1863. J.C. Poggendorff, ed. *Biographisch-literarisches Handwörterbuch* (Leipzig).
1868–1887. B. Boncompagni. *Bulletino di Biblographia e Storia delle Scienze matematiche e fisiche* (Rome).
1873–1928. P. Riccardi. *Biblioteca matematica italiana* (Modena).
1873. A. Erlecke. *Bibliotheca Mathematica* (Halle).
1884–1914. G. Eneström, ed. *Bibliotheca Mathematica* (Stockholm).

Among these works, the most significant, even today, remains the seminal contribution of J.C. Poggendorff, an historian, biographer, and bibliographer. At age 27, he became editor of the *Annalen der Physik und Chemie* (founded in 1790) and, as a result, initiated a correspondence with the leading scientists of his day. This eventually prompted a project that was naturally suited to his historical interests (also expressed in his lectures and in a book on the history of physics), namely, the *Biographisch-literarisches Handwörterbuch zur Geschichte der exakten Naturwissenschaften*. In 1863 it consisted of only two volumes, comprised primarily of brief biographical information and bibliographic references for 8400 scientists up to 1858. By 1974, however, the continuation of this series had grown to 18 volumes.

Mention should also be made of Baldassarre Boncompagni, who even established his own private printing plant in order to ensure the high standards of his publications related to the history of science. Especially important is the *Bullettino Boncom-*

pagni, which George Sarton once described as "a very rich collection, a model of its kind."[5] Another of those dedicated to bibliography and the history of mathematics before this century was Gustav Eneström, who began his career as a librarian. Although little is known of his private life, one biographer has called him "very original and eccentric."[6] Perhaps no greater tribute could be made to Eneström's contributions to the history of mathematics than one paid by George Sarton in 1922: "Personal notes for living scholars have been thus far avoided in *Isis*, but an exception must be made in favor of Gustav Eneström, than whom no one has ever done more for the sound development of our studies."[7]

More recently, George Sarton's guides to the history of science set a new generation on a more interdisciplinary and historical course. Of special interest to users of this bibliography is Sarton's *The Study of the History of Mathematics* (item 48), published in 1937 and devoted, in part, to a discussion and critique of bibliographic resources on the subject. Sarton's work was followed a decade later, in 1946, by Gino Loria's *Guida allo studio della storia delle matematiche* (item 47). This too provided excellent critical discussion of the history of mathematics, with the advantage that it was able to profit from the appearance of Sarton's guide as well.

Shortly after the appearance of Loria's guide, George Sarton published *Horus*, a major effort on his part not only to provide a rationale for the purposes and meaning of the history of science (Part I of *Horus*), but a bibliographic summary as well, meant "to provide a kind of *vade mecum* for students" (Part II). "The first part is meant to be read," Sarton wrote, "the second to be used as a tool."[8] In his introduction to Part II, Sarton asserted that "nothing is more instructive than a good bibliography. . . . Every bibliography must begin with a bibliography, and it must end with a better bibliography."[9]

The most recent of all the bibliographic guides concerned with the *history* of mathematics, however, is Kenneth May's (item 17). Its aim was "to assist mathematicians, users of mathematics, and historians in finding and communicating information required for research, applications, teaching, exposition and policy decisions."[10] May's bibliography was inspired, in part, by the rampant growth of literature related to mathematics, which he estimated in 1966 as already consisting of half-a-million titles and growing at the rate of 15,000 new items annually. May was particularly concerned that "this enormous collection is not indexed. No one knows its nature or contents. Preliminary studies suggest that there is a vast amount of duplication, and that the important information is contained in perhaps as little as 10% of the titles."[11] In order to reduce this problem and obtain what he called "effective entry to this storehouse so as to find information, orientation, and enlightenment," May's bibliography offered 31,000 entries under 3700 topics, with an appendix listing about 3000 periodicals in which papers on mathematics and its history are published.

Bibliographies and Information Retrieval

Recently, some scholars have sensed an impending crisis in the domain of information retrieval. Within the past decade, most major libraries, universities, and publishers have begun to rely on computerized indexes and reference systems to facilitate data management. Some fear that soon we shall be so inundated with the printed word that even attempting to do basic research on a given subject will be a bibliographical nightmare. Automation may indeed help to solve some bibliographical problems, but it may also contribute to the problem by facilitating even more rapid production of the written word and of information in general.

Not all agree, however, and among the most persuasive of the dissenters is Yehoshua Bar-Hillel, who more than twenty years ago posed the question, "Is information retrieval approaching a crisis?" In Bar-Hillel's opinion, there is no crisis. Simply because the growth of mathematical information may seem to be pathologically out of control (to adopt a point of view made popular by Derek de Solla Price) does not mean that researchers will wallow in a hopeless miasma of printed material, making it impossible for anyone to keep abreast of the latest books, monographs, papers, and articles on a given subject. Bar-Hillel observes that:

> Scientists did not spend on the average in 1961 more time on reading than they did 12 years ago, though printed scientific output has indeed almost doubled during this period. There must

> therefore have been a way out between the horns of the dilemma. What is it? Everybody knows it: *Specialization.*[12]

Nonetheless, as research becomes increasingly focused, bibliographies become all the more important, especially for those who choose to move from their own areas of specialization into others that may be related, or even wholly new. Here bibliographies play a crucial role in helping to orient new readers to a field in which they may be less than proficient.

There is another service bibliographies perform, and this is related to facilitating research and helping to reduce duplication of efforts. Not long ago it was suggested that hundreds of thousands of dollars (estimates ranged from $200,000 to $250,000) had been wasted because a Russian paper on Boolean matrix algebra and relay contact networks was not known to American researchers. According to an article in *Science*, the Russian paper had appeared in an "important, readily available Soviet journal" and "simply reposed on a library shelf waiting to be noticed."[13] This example had been noted as early as 1956, when William Locke wrote in *Scientific American*, "groups of people in several companies in the United States did, in fact, work for five frustrating years on the very points cleared up by this paper before discovering it."[14] The paper in question was written by A.G. Lunts (also transliterated as Lunc) in 1950: "The Application of Boolean Matrix Algebra to the Analysis and Synthesis of Relay Contact Networks," but published in Russian. The waste of both time and money suggested by this example was even brought to the floor of Congress as an example of costly duplication and inefficiency of research efforts due to lack of proper information dissemination.[15]

The thrust of Locke's article in *Scientific American* was simple: an inexpensive means of translating material from Russian to English should be made available so that such a waste of time and money could be avoided in the future. Ironically, as a letter published in the March issue of *Scientific American* noted, the Lunts paper should have been known already, since it had been abstracted in *Mathematical Reviews*.[16] As R.P. Boas, Jr., also pointed out in a letter to *Science* (Boas was then executive editor of *Mathematical Reviews*): "If people are unable or unwilling to use the bibliographic aids that are already provided, there is little point in supplying them with even more in the form of translations and so on."[17] E.H. Cutler posed an equally germane question. Because the article by Lunts had been annotated in *Mathematical Reviews*, he asked, "Does not the case suggest more the need for the application of machines to the bibliographic problems of cross-referencing publications rather than the more fascinating application to the problem of translation?"[18] Locke agreed. On the same page of the March issue of *Scientific American*, he noted that "My example does show the need for machines to keep the bibliographies of specialized fields up to date."

Whether or not machines are used to store, catalogue, and retrieve bibliographic information, there are many ways in which bibliographies, whatever their form, are essential resources for serious research. Not only does a good bibliography eliminate the need to read everything on a given subject—clearly impossible with the overwhelming number of publications produced annually in mathematics, not to mention the burgeoning growth of literature in the *history* of mathematics—but it ought to provide a simple way of orienting the user to the most significant works in print on a particular topic.

Another and often unappreciated value of bibliographies is the extent to which they are historical and sociological resources in their own right. This, in fact, was the point made in 1952 by Victor Zoltowski in an article entitled "Les cycles de la création intellectuelle et artistique" in *Année sociologique*, and quoted by Louise Malclès to the effect that "the gifts of bibliography are capable of leading to the discovery of cycles of intellectual and artistic creation." As Malclès says, "Just as the demographer inventories populations, and studies their movements, without knowing each citizen of the country in question, the bibliographer, without having read all books, follows their creation, their purport, and distribution."[19]

What sets this annotated bibliography apart is that all entries *have* been read critically and annotated by experts. This means that readers have at their disposal those works regarded as essential by specialists in a large number of different areas of the history of mathematics. From these, it is a far simpler task for researchers to compile increasingly detailed bibliographies of their own, with the knowledge that the most essential material on a given topic appears here as a starting point.

Insofar as trends or cycles of intellectual interest might be discerned from reading this bibliography, it may be of interest to contrast it briefly with the only other major bibliographic reference work to appear for the history of mathematics in the last decade, namely, Kenneth O. May's *Bibliography and Research Manual of the History of Mathematics* (item 17). As early as 1966 May had written a short article for *Science*, in which he discussed the problems raised by the fact that the annual list of publications in mathematics was growing rapidly.[20] May's solution was a project he did not complete until 1973—his *Bibliography and Research Manual* (and even then, it was not really finished, for he continued to update it until his death in 1977). May's *Bibliography* remains an important reference work for the history of mathematics in all periods and subjects. What sets it apart from this bibliography, however, is the fact that it lacks annotations and is now more than ten years out of date. The number of articles which have appeared in new journals like *Historia Mathematica* (founded in 1974), and the number of serious book-length biographies and special histories of specific branches of the history of mathematics (not to mention the stimulus given to all areas of the history of science by the monumental *Dictionary of Scientific Biography*, item 10), show the extent to which the history of mathematics has become professionalized in its own right as a subject for serious study in the last ten years.

Special Aspects of This Bibliography: Caveats and Comments

Ralph Waldo Emerson once quipped that consistency was the hobgoblin of small minds. Users of this bibliography should probably keep that in mind—along with George Sarton's equally sardonic statement that bibliography was "a sin . . . a real perversity." By this he meant that some researchers elevate bibliography to the point where it appears to take precedence over everything else, including the subject matter. However, Sarton was very much aware of the need and crucial contribution that good bibliographies make to professionals and amateurs alike, for Sarton was a consummate if idiosyncratic bibliographer. But he disdained, above all, more lists of titles, however long, as ends in themselves. Such lists he termed "bewildering" (Sarton, item 48, p. 26). Bibliographies took on greater value when annotated, constructed in such a way that expert understanding might guide newcomers to a particular subject.

Each section of this bibliography reflects a considered judgment as to what works are absolutely essential on any given topic or period, accompanied with critical descriptions of those works. Emphasis has been given to the most useful and authoritative secondary sources and, when appropriate, to texts, manuscripts, correspondence, and other varieties of primary sources. Reviews of major items have also been included, especially when they warn readers of special quirks, problems, or prejudices in a given item, or provide substantive additional information relevant to a given subject. Older standard works that have established themselves as a continuing part of the history of mathematics are also, of course, included. The major European languages, especially French and German, are essential tools for the historian of mathematics, and no attempt has been made to minimize the frequency of their appearance in this bibliography. References in other languages, however, are included only when thought essential, but titles in Italian, Spanish, Russian, Chinese, and Japanese have been listed whenever appropriate. Where English translations or their equivalents exist, these have been noted. Finally, truly obscure studies, unless thought crucial, have been minimized in favor of more easily available sources, but not always.

Although most entries are annotated, some are not, if in the opinion of the reviewer the title accurately reflects the content of the reference. In other cases, where an item has appeared elsewhere in the bibliography with sufficient annotation, only a cross-reference has been given. However, if a work is important for differing reasons to more than one section, it has been repeated (again with cross-references), but with a separate annotation tailored specifically to the relevance of the work for the section in which it appears.

Ultimately, in editing this volume I have adopted as pragmatic a position as possible; the final consideration has always been to make this as useful a reference work as possible. Because of the large number of individual contributors, however, it has not always been possible to attain complete consistency in the format and amount of information supplied for each title. Although format guidelines were issued to all contributors, not everyone chose to follow them exactly or consistently. Where practicable, the editor has brought as much uniformity of format to references as possible, but there were limits to which additional information could be retrieved. In most cases what discrepancies remain are of little consequence; the one exception is in the case of different editions of a given work. Some contributors have tried to be as inclusive as possible, indicating the significance between editions, or at least noting the number of different editions of a given work and their dates. Some have preferred to supply the first, others only the most recent edition. Readers should therefore be aware that it is always wise to make their own bibliographic searches to discover which editions or versions of a work may be available to them, as well as the number and differences among various reprintings and editions of a given work. (Usually this can be done most easily by referring to the National Union Catalog of the Library of Congress.)

A major orthographic dilemma facing any bibliographer dealing with material in multiple languages is, of course, the different possible spellings and transliterations of a title or author's name. This is most acute for citations in Russian, Japanese, Chinese, and Arabic. One notable example in this bibliography is that of A.P. Youschkevitch, whose name variously appears in some references as Juschkewitsch, Juškevič, Juschkevic, Youskevich, or Youschkevitch. The practice that has been followed in all cases here is to present a given name *as it is spelled on the title page of the work in question.* Thus, one will find item 355 listing "Youschkevitch," with item 467 as "Juschkewitsch," one reflecting a French, the other a German translation. In the indexes at the end of the book, however, all are cross-referenced to the canonical spelling that has been otherwise adopted in this bibliography, namely, "Youschkevitch." Other but less complicated cases of variant transliterations include "Bashmakova," "Kowalevskaya," and "Lobachevsky."

Similar variations occur in the case of Greek names. Here the practice has been followed of giving names in terms of transliterations of Greek spellings, e.g., Diophantos. However, in all citations, names are always given as they appear on *title* pages, which often follow earlier conventon and use the Latinized version of Greek names, e.g., Diophantus.

With but two exceptions, all entries within a given section of the bibliography are arranged alphabetically. As explained in the introduction to the section on Babylonian mathematics, titles there are arranged in chronological order. This makes it possible to follow at a glance the changing currents of research, and for readers to follow sequentially the extent to which understanding of Babylonian mathematics has depended greatly upon the primary sources available at any given time. Thus, the more recent material on Babylonian mathematics tends to be more authoritative in terms of current knowledge, even though earlier works still remain important for their essential texts and other information they contain.

The 19th-century bibliography (pages 176–203) also departs from a strictly alphabetical arrangement, as certain topics within a particular subject area are grouped together and then annotated with a single common entry.

Conclusion

Why would anyone wish to undertake a project such as this one? My own reasons have been both professional and personal, but before saying more, I would like to dismiss one motive for undertaking bibliographic research that has been attributed to George Sarton. Marc De Mey, writing in the 1984 Sarton memorial volume of *Isis*, interprets Sarton's penchant for bibliography in very psychological terms. What interested De Mey was that Sarton's interest in bibliography seems to have antedated his interest in the history of science. Based upon a letter that Sarton sent to the Chief Librarian of the University of Ghent, dated November 4, 1902, De Mey conjectures as follows:

> If, as in classical embryology, the order of formation is taken as an indication of primacy (here intellectual), it is obvious that Sarton's interest in bibliographies is more basic than his interest in

> science to which he, only several years later, applied this attitude in so masterly a fashion. Confronted with such a vigorous need, one feels compelled to take certain psychoanalytic claims seriously and search for the sources of this attitude in early childhood. It seems almost a textbook case. The lonely child George Sarton, losing his mother at a very early age, starves, in May Sarton's words, for the tenderness that vanished with her. Deep insecurity could derive from lesser causes. The collector's attitude is considered a classical response to such insecurity: if life, on the whole, is uncertain, establishing and controlling a well-organized and complete collection of items belonging to a specific domain provides solidity and certainty for one subrealm at least. That attitude is later enthusiastically extended to science as the most solid domain. This is one plausible suggestion, but there might be others. In any case, if one were to engage in a study of Sarton along the lines that Erik Erikson applied to Martin Luther, Sarton's bibliographical bias should play a pivotal role, since in its fervor it comes close to the "obsessive compensation" characteristic of many great achievements.[21]

I doubt that any of the contributors to this annotated bibliography would wish to see it in terms of "obsessive compensation"! In fact, this should remind us of Sarton's *own* phraseology, as well as his caveat, that bibliography was "a sin . . . a real perversity." Surely this puts the lie to De Mey's interesting if arcane speculation, because Sarton, in the best sense of this bibliography, saw such guides as practical sources of information.

Despite the valiant efforts of the copy editor, it has not always been possible to fill certain gaps. Sometimes publishers' names could not be found, or issue numbers of periodicals within a given volume could not be identified. From time to time the National Union Catalog of the Library of Congress was inaccurate, or its information incomplete. Although I have attempted to bring an overall uniformity to the format of the citations, and to check them all for completeness, I have also tried to resist letting this become a bibliographic obsession. On the other hand, what *has* been an uncompromised goal is to make certain that each citation provides the essential information necessary to retrieve it from libraries or book sellers with relative ease.

My own reasons, as just indicated, for undertaking this project have been to some degree personal, and to a larger extent professional. Kenneth May was both a friend and a moving force in my own interest in the history of mathematics. This bibliography, for me, honors the efforts he made, especially through *Historia Mathematica*, to promote the subject in the most professional and international way possible. But it was primarily because I felt a strong need for such a bibliography that I was ultimately persuaded to undertake this project.

Actually, when Robert Multhauf and Ellen Wells first approached me about editing a volume on the history of mathematics for their Bibliographies on the History of Science and Technology series, I was doubtful whether any single individual was capable of surveying the entire history of mathematics from antiquity to the present in any sort of authoritative way. I agreed, however, with their basic premise: authoritative, annotated bibliographies in the history of science would be of great utility to the scholarly community. With this in mind, it seemed reasonable to suggest that a collaborative effort might be the perfect solution, involving a dozen or so experts who might reasonably be expected to cull the best forty or fifty titles in a given area, and provide annotations within a few months. As the editor of *Historia Mathematica* I was in close contact with leading authorities on virtually every aspect of the history of mathematics. Given assurances that this would be fine, I wrote to several dozen colleagues and modestly proposed (with what in hindsight was too much optimism) that if all were willing to draw up basic lists of essential works in their special fields, a preliminary draft of the bibliography might be possible within six months, with a completed annotated version in print within a year.

My initial letter was answered with a variety of responses. The majority of those to whom I wrote, I am happy to say, responded positively, even enthusiastically. Most recognized a definite need for such a critical bibliography to serve the interests of the history of mathematics and, furthermore, they were willing to take on the job without remuneration, for the sake of the subject and its future. Most agreed that they could comply with my request for preliminary lists within six months. This would allow time to check all of the proposed bibliographies for duplication, cross-reference them as needed, and thereby prevent unnecessary duplication.

The present volume is the result of an extraordinary amount of effort. It could never have been accomplished by a single person, but required the combined efforts of all those contributors who worked together in the best cooperative spirit of scholarly collaboration. This would have pleased Kenneth May greatly, and it is both fitting—and a reflection of the magnanimity of the many scholars who have contributed to this bibliography—that all royalties accruing from its publication will be used to establish a fund in his memory. This fund, in the names of all contributors to this volume, is to be administered by the International Commission for the History of Mathematics, and will be designated specifically to help promote the history of mathematics internationally.

I especially want to acknowledge the continuing help and moral support of three individuals in particular: Robert Multhauf for his persistent encouragement along the way; David Rowe for his diligence in helping to complete and proofread the final version of the bibliography; and Rita Quintas for her care in seeing this volume through the last editorial stages of its production. Moreover, I am happy to express my indebtedness and gratitude to all of the contributors to this volume, not only for their care in producing each of the individual sections, but for their patience owing to the time it has taken to cement the myriad pieces into a coherent and useful whole.

The ultimate goal of this bibliography has been to make the *secreta secretorum* of the history of mathematics much less a secret history than it may seem to many at present. I am grateful to all who have given so generously of their time and energy in order to make this reference work possible.

JWD

Notes

1. The information contained in this section of the Introduction draws heavily from Archer Taylor, *A History of Bibliographies of Bibliographies* (New Brunswick, N.J.: The Scarecrow Press, 1955); Louise N. Malclès, *Bibliography*, trans. T.C. Hines (New York: The Scarecrow Press, 1961); Georg Schneider, *Theory and History of Bibliography*

(New York: Columbia University Press, 1934), especially pp. 3–24; John Thornton and R.I.J. Tully, *Scientific Books, Libraries and Collectors* (London: Library Association, 1954); and Theodore Besterman, *The Beginnings of Systematic Bibliography* (London: Humphrey Milford for Oxford University Press, 1935).
2. G. Naudé, *Bibliographia Politica* (Venice: F. Baba, 1633); also issued in French as *La bibliographie politique* (Paris: G. Pelé, 1642). Strictly speaking, however, the word "bibliography" was also used in 1645 by Lewis Jacob de Saint Charles, but in a different sense from that currently understood. Jacob used the word to signify the mechanical writing and transcription of books. See Philip H. Vitale, *Bibliography, Historical and Bibliothecal* (Chicago: Loyola University Press, 1971), p. 14.
3. *Le Dictionnaire de l'Académie françoise* (Paris: J.B. Coignard, 1694; reprinted Lille: L. Danel, 1901; fourth edition, 1762).
4. Cornelius à Beughem, *Bibliographia mathematica et artificiosa novissima* (Amsterdam: J. à Waĕsberge, 1688).
5. *Isis* 2 (1914), 133.
6. W. Lorey, "Gustav Eneström (1852–1923)," *Isis* 8 (1925), 314.
7. G. Sarton, "For Gustav Eneström's 71st Anniversary," *Isis* 5 (1922), 421.
8. G. Sarton, *Horus* (Waltham, Mass.: Chronica Botanica, 1952), p. ix.
9. *Ibid.*, p. 71.
10. K.O. May, item 17, p. iii.
11. K.O. May, "Growth and Quality of the Mathematical Literature," *Science* 154 (1966), 1672–1673. See also his article with the same title in *Isis* 59 (1968), 363–371.
12. Y. Bar-Hillel, "Is Information Retrieval Approaching a Crisis?" Chapter 20 of *Language and Information. Selected Essays on Their Theory and Application* (Reading, Mass.: Addison-Wesley, 1964), p. 365.
13. Ralph E. O'Dette, "Russian Translation," *Science* 125 (March 29, 1957), 579–585; especially p. 580. At the time, R.E. O'Dette was Program Director of the Foreign Science Information Program of the National Science Foundation.
14. William N. Locke, "Translation by Machine," *Scientific American* 194 (January 1956), 29–33, especially p. 29. Locke was working at MIT in the Department of Modern Languages when he wrote this article.
15. See A.G. Oettinger, "An Essay in Information Retrieval or the Birth of a Myth," *Information and Control* 8 (1965), 64–79. Oettinger argues that the deleterious effects of the so-called information explosion are greatly exaggerated, and that the incident involving the Lunts paper is really a "comedy of errors" occasioned by "over-

zealous proponents in the area of information retrieval," as characterized by A.J. Lohwater in his review of Oettinger's essay in *Mathematical Reviews* 29 (6) (June 1965), #6960.

16. *Mathematical Reviews* 11 (September 1950), 574.
17. *Science* 125 (June 21, 1957), 1260.
18. *Scientific American* 194 (March 1956), 6.
19. Malclès, ibid., p. 8.
20. See above, note 11.
21. Marc De Mey, "Sarton's Earliest Ambitions at the University of Ghent," *Isis* 75 (276) (1984), 42.

ACKNOWLEDGMENTS

Gratitude is hereby expressed to the following for permission to use works or illustrations as indicated: Parts of the section on the History of Computing were previously published in the *Annals of the History of Computing*. I am grateful to the American Federation of Information Processing Societies, Inc., for allowing material from Volume 1 (2) (October 1979), pp. 101–207, to be used here. The following illustrations appear in this bibliography by permission of the individuals, publishers, and libraries as indicated:

Figure 1. Courtesy of the Egyptian National Library and David A. King, New York.

Figure 2. From the frontispiece, *Abhandlungen zur Geschichte der Mathematik* 9 (1899), courtesy of the Columbia University Science Library.

Figure 3. From A.B. Chace, *The Rhind Mathematical Papyrus* (Oberlin, Ohio: Mathematical Association of America, 1927–1929), Vol. II, Plate 58.

Figure 4. Courtesy of Professor Jöran Friberg.

Figure 5. Reproduced from Otto Neugebauer, *A History of Ancient Astronomy*, Part III (Berlin: Springer Verlag, 1975), Plate VIII, by permission of Springer Verlag.

Figure 6. From the first printed edition of Euclid's *Elements* (1482), courtesy of the Bern D. Dibner Library and the Smithsonian Institution Libraries, Washington, D.C.

Figure 7. L. Pacioli, *Summa de arithmetica, geometria, proportioni, et proportionalita* (1494), courtesy of the Bern D. Dibner Library and the Smithsonian Institution Libraries, Washington, D.C.

Figure 8. R. Descartes, *La Géométrie* (1637). Title page reproduced from James R. Neuman, *The World of Mathematics* (New York: Simon and Schuster), by permission of the publisher.

Figure 9. Title page of Leibniz's "Nova Methodus . . .," from Dirk J. Struik, *A Source Book in Mathematics, 1200–1800* (Cambridge, Mass.: Harvard University Press, 1969), p. 273, by permission of the publisher.

Figure 10. Illustration of Pascal's calculator, from *Machines et inventions* . . ., Vol. 4 (Paris, 1735); reproduced from I. Bernard Cohen, *Album of Science* (New York: Charles Scribner's Sons, 1980), p. 80, Illustration 106, by kind permission of the Houghton Library, Harvard University, and Charles Scribner's Sons.

Figure 11. From *Recueil de planches sur les sciences, les arts liberaux, et les arts mécaniques*, Vol. 22 (Paris, 1767). Courtesy of the Graduate Center Library of the City University of New York.

Figure 12. From A. Koyré and I.B. Cohen, *Variorum Edition of Isaac Newton's Philosophia Naturalis Principia Mathematica* (Cambridge, Mass.: Harvard University Press, 1972), Vol. I, p. 89, by permission of Harvard University Press.

Figure 13. From Gregorius Reisch, *Margarita philosophica* (Freiberg, 1503), reproduced from I. Bernard Cohen, *Album of Science* (New York: Charles Scribner's Sons, 1980), p. 74, Illustration 99, by kind permission of the Houghton Library, Harvard University, and Charles Scribner's Sons.

Figure 14. Title page of P.S. Laplace, *Théorie analytique des probabilités* (3rd ed., Paris: Courcier, 1820), taken from Vol. 7 of the *Oeuvres complètes* (Paris: Courcier, 1886), courtesy of the Graduate Center Library of the City University of New York.

Figure 15. From Constance Reid, *Hilbert* (New York: Springer Verlag, 1970), p. 288, by permission of Springer Verlag.

Figure 16. Reproduced from Joseph Needham and Wang Ling, *Science and Civilisation in China*, Vol. III (Cambridge: Cambridge University Press, 1959), p. 136, by permission of Cambridge University Press.

Contributors to Individual Parts of the Bibliography, by Section

GENERAL REFERENCE WORKS: J. Dauben, A. Lewis; Source Materials: S. Jayawardene; General Histories: J. Dauben, A. Lewis.

CHRONOLOGICAL PERIODS: Egyptian: P. Campbell; Babylonian: J. Friberg; Greek: L. Berggren: Islamic: R. Lorch, with E.S. Kennedy, G. Matvievskaya, B. Rosenfeld; Indian: T. Hayashi; Latin West: H. Busard, M. Folkerts, E. Sylla; Hebrew: B. Goldstein, N. Rabinovitch; Renaissance: J. Dauben, S. Jayawardene; 17th Century: K. Andersen, H. Bos, C. Scriba; 18th Century: S. Engelsman; 19th Century: J. Gray, H. Pycior; 20th Century: G. Moore, L. Steen.

SUBJECT AREAS: Algebra: E. Knobloch, H. Pycior; Analysis: D. Rowe; Computing: G.J. Tee, B. Randell; Differential Equations: S. Engelsman; Electricity and Magnetism: R.W. Home; Geometry: D. Rowe; Logic: I. Grattan-Guinness, J. Ashworth, V.F. Rickey; Mathematical Physics: J. Cross; Navigation: J. Pepper; Number Theory: J. Dauben; Optics: R.H. Silliman; Potential Theory and Mechanics: J. Cross; Probability and Statistics: D. Rowe; Quantum Mechanics: W. Mathis; Relativity: D. Rowe; Set Theory: J. Dauben; Thermodynamics: S. Brush; Topology: D. Johnson.

SELECTED TOPICS: Education: P. Jones, L. Rogers; Institutions: P. Enros; Sociology of Mathematics: H. Bos, H. Mehrtens; Philosophy of Mathematics: R. Bunn, G. Moore; African: C. Zaslavsky; Chinese: Lam Lay-Yong; Japanese: S. Ito; Women in Mathematics: J. Green, J. LaDuke, T. Perl.

The History of Mathematics from Antiquity to the Present

I. GENERAL REFERENCE WORKS

The following works include general reference sources of particular utility for historians of mathematics. Readers with specific interests are referred to the separate sections of this bibliography dealing with individual time periods or topics. Moreover, it should also be noted that the reference works annotated here have been selected for their specific utility in the *history of mathematics*, and by no means exhaust the tools available for scholarly research. Many additional reference works are also of great use, including general guides to archives and other library resources, encyclopedias, historical dictionaries, dissertation abstracts, and government publications. Although such materials are not included here, they should always be kept in mind and are listed in the following guides.

1. Sheehy, Eugene P., ed. *Guide to Reference Books*. 9th ed. Chicago: American Library Association, 1976; Supplement, 1980.

2. Walford, A.J. *Walford's Guide to Reference Material*. Vol. 1: *Science and Technology*. 4th ed. London: The Library Association, 1980.

DICTIONARIES, ENCYCLOPEDIAS, AND BIBLIOGRAPHIES

3. Archibald, Raymond Clare. "Bibliographia de mathematicis." *Scripta Mathematica* 1 (1932-1933), 173-181, 265-274, 346-362; 2 (1933-1934), 75-85, 181-187, 282-292, 363-373; 3 (1935), 83-92, 179-190, 266-276, 348-354; 4 (1936-1937), 82-87, 176-188, 273-282, 317-330.

   Bio-bibliographical notices of some 200 mathematicians, mostly of the twentieth century. Indexed.

4. Bynum, W.F., E.J. Browne, and R. Porter, eds. *Dictionary of the History of Science*. London: The Macmillan, Press, 1981.

   Unlike dictionaries of scientific biography the organization of this dictionary is conceptual. It includes more than 700 entries explaining the origins, meaning, and significance of major theories and ideas in the history of science, including mathematics. The emphasis is on Western science over the last 500 years, with particular attention to current historiography and philosophy of science. Each of the major entries has its own bibliography. There is also an index of all scientists mentioned and extensive cross-referencing.

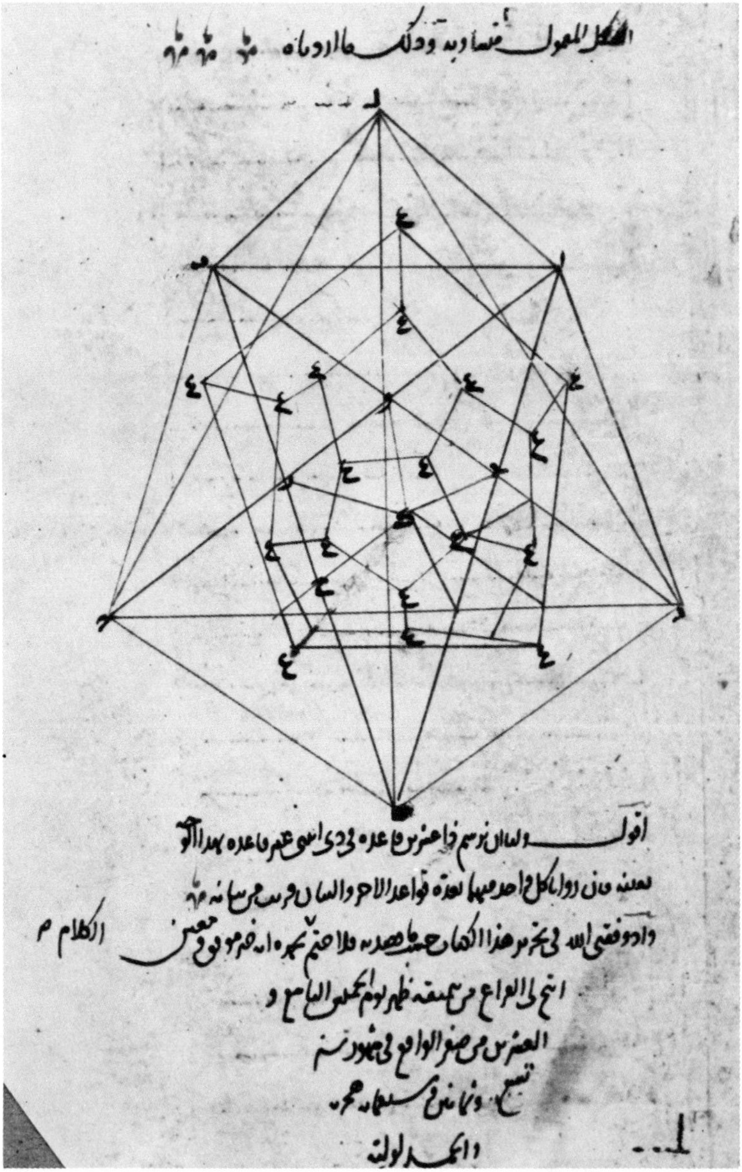

Figure 1. The last page and colophon from a copy of Nasīr al-Dīn al-Ṭūsī's recension of Euclid's *Elements*. Taken from MS Cairo Ṭal^{c}at riyāḍa 107, fol. 135r, copied in 1387.

5. *Encyklopädie der mathematischen Wissenschaften mit Einschluss ihrer Anwendungen*. Leipzig: B.G. Teubner, 1898-1935. 6 vols. in 23 parts. Parts of a second edition appeared as *Enzyklopädie* ..., 1939-1958.

   Long articles by specialists with full bibliographic notes. A French edition of the original set (Paris: Gauthier-Villars, 1904-1916, 7 vols.) was never completed, but included some revisions, especially historical footnotes that are often very useful. See also items 967, 1424.

6. Eneström, G. "Bio-bibliographie der 1881-1900 verstorbenen Mathematiker." *Bibliotheca Mathematica* 2 (1901), 326-350.

   List of obituary notices of some three hundred mathematicians. Many are not recorded in Poggendorff, items 20, 21, nor in Kenneth May's *Bibliography*, item 17.

7. Fang, J. *A Guide to the Literature of Mathematics Today*. Hauppauge, N.Y.: Paideia, 1972.

   Provides a guide to international congresses, mathematical societies, a comparison of major topics in mathematics in 1900 versus 1970, and lists of major series (colloquiums, memoirs, proceedings, translations). A section is devoted to collected works.

8. Forsythe, G.E. *Bibliography of Russian Mathematics Books*. New York: Chelsea, 1956.

   Provides a general survey of books in Russian (with English translations given for titles) in all areas of mathematics. The subject index guides readers to sources for history, including bibliographic materials and collected papers.

9. Gaffney, M.P., and L.A. Steen. *Annotated Bibliography of Expository Writing in the Mathematical Sciences*. Washington, D.C.: The Mathematical Association of America, 1976.

   Has sections devoted to history and biography. The lists are highly selective, and only a portion are annotated, but they do include articles as well as books.

10. Gillispie, Charles Coulston, editor in chief. *Dictionary of Scientific Biography*. New York: Scribners, 1970-1980. 16 vols.

    Probably the single most important reference work in the history of science today, with signed bio-bibliographical articles on non-living scientists who were deemed to have made an "identifiable difference to the profession or community of knowledge." Volume 16 contains a name and subject index and a listing of scientists by field.

11. Grattan-Guinness, Ivor. "History of Mathematics." *Uses of Mathematical Literature*. Edited by A.R. Dorling. London: Butterworth, 1977, 60-77.

    A basic introduction to all aspects of the literature, followed by a discussion of the state and nature of the art of writing the history of mathematics.

12. *International Catalogue of Scientific Literature*. Section A: *Mathematics*; Section B: *Mechanics*. London: Harrison, 1902-1918. Reprinted New York: Johnson Reprint Corp., 1968.

Grew out of the Royal Society's *Catalogue of Scientific Papers*, and was intended to cover literature after 1900. The 28 volumes for sections A and B cover 1901 to 1914. Each volume contains schedules and indexes in four languages, bibliography arranged by author, and bibliography arranged by subject (including history and biography).

13. *Isis Cumulative Bibliography: A Bibliography of the History of Science Formed from Isis Critical Bibliographies 1-90, 1913-1965*. London: Mansell, 1971-1976.

Volumes 1 and 2 (1971) are indexed by individuals and institutions; volume 3 (1976) is indexed by subject. As no item is indexed under more than one heading, it may be necessary to refer to all the volumes.
Review: Isis, 70 (1979), 160-163.

A second cumulation is in progress: John Neu, editor, *Isis Cumulative Bibliography ... Formed from Isis Critical Bibliographies 91-100*. London: Mansell, 1980-. Volume 1, *Personalities and Institutions*, has been published.

14. *Isis Critical Bibliography of the History of Science and its Cultural Influences*.

Issued annually as a supplemental issue of the journal *Isis* surveying both books and articles. A detailed table of contents and an index to personal names (and institutions when authors or subjects) are provided.

15. Jayawardene, S.A. "Mathematical Sciences." In *Information Sources for the History of Science and Medicine*. Edited by P. Corsi and P. Weindling. London: Butterworth, to appear.

Contains bibliographic essays on various topics in the history of science and medicine. The chapter on mathematics is devoted to an extensive essay on mathematics since the Renaissance.

16. Karpinski, Louis C. *Bibliography of Mathematical Works Printed in America through 1850*. Ann Arbor: University of Michigan Press; London: Oxford University Press, 1940.

Provides basic information, some commentary, and frequent illustrations of books, pamphlets, and broadsides from the sixteenth century to 1850, including encyclopedias, journals, and newspapers with mathematical articles. An appendix deals with American mathematical developments. Indexed by topic. See also item 962.

17. May, Kenneth O. *Bibliography and Research Manual of the History of Mathematics*. Toronto: University of Toronto Press, 1973.

Aims to include all secondary literature in the field of the history of mathematics for the period 1868-1965. Arrangement is alphabetical under the following sections: biography, mathematical topics, epi-mathematical topics (e.g., the abacus, women, and

Zeno's paradoxes), historical classifications (time periods, countries, cities, organizations), and information retrieval (bibliographies, historiography, information systems, libraries, manuscripts, museums, monuments, exhibits). This is an essential starting point for any beginning historian of mathematics who wishes a speedy introduction to the relevant literature on a given topic or individual.

The bibliographies and abstracting journals incorporated by May are noted as such in the present bibliography. For Oriental studies May should be supplemented by other works such as Suter, item 28, Sezgin, item 27, and Pingree, item 19. For Chinese mathematics, there is Joseph Needham, *Science and Civilization in China*, Vol. 3 (Cambridge University Press, 1959). May is reviewed in *Historia Mathematica* 1 (1974), 192-194. See also item 138.

18. Müller, Felix. *Führer durch die mathematische Literatur, mit besonderer Berücksichtigung der historisch wichtigen Schriften*. (Abhandlungen zur Geschichte der mathematischen Wissenschaften, 27.) Leipzig: Teubner, 1909.

Provides annotations in the form of running commentary on the literature of mathematics and its history. Classified by subject with indexes to subjects and names. Still useful for works in pure mathematics before 1868 and a guide to the historical literature before 1909.

19. Pingree, David. "Census of the Exact Sciences in Sanskrit." *Memoirs of the American Philosophical Society*, Series A, 81 (1970), 1-60; 86 (1971), 1-147; 111 (1976), 1-208; 146 (1981), 1-447.

Provides a "preliminary exploration and organization of the vast mass of Sanskrit and Sanskrit-influenced literature devoted to the exact sciences (including astronomy, mathematics, astrology and divination)." Both a bibliography of published literature and a survey of manuscripts. Includes a bio-bibliography of some 1,500 scientists. Reviewed in *Historia Mathematica* 4 (1977), 226-227. See also item 383.

20. Poggendorff, Johann Christian, ed. *Biographisch-literarisches Handwörterbuch zur Geschichte der exakten Naturwissenschaften*. Leipzig: J.A. Barth, 1863-1919?; Berlin: Verlag Chemie, 1925-1940. 6 vols. in 11.

21. ———. *Biographisch-literarisches Handwörterbuch der exakten Naturwissenschaften, unter Mitwirkung der Akademien der Wissenschaften zu Berlin, Göttingen, Heidelberg, München, und Wien*. Berlin: Akademie Verlag, 1955-.

Contains short biographical information followed by a list of books and articles by and about each scientist. Entries were updated from volume to volume and thus all chronological divisions may need to be consulted for complete information on a given personality. Some entries have been provided by the scientists themselves. Though weak on pre-1800 scientists, it is indispensable for the nineteenth century. Historians of science are included from volume 6 on. See also item 963.

22. Read, Cecil B., and James K. Bidwell. "Selected Articles Dealing with the History of Elementary Mathematics. (Periodical Articles Dealing with the History of Advanced Mathematics)." *School Science and Mathematics* 76 (1976), 477-483, 581-598, 687-703.

Arranges some 2,000 articles under 28 subject headings, with coverage beginning in 1934.

23. Riccardi, Pietro. *Biblioteca matematica italiana dalla origine della stampa ai primi anni del secolo XIX*. Modena, 1873-1928. Reprinted Milan: Goerlich, 1952.

A bibliography of some 10,000 books on mathematics and its applications by Italian authors who lived before 1811. Access is by author and chronological under 262 subject headings. List of about 800 works cited in the annotations.

24. Rogers, Leo F. *Finding Out in the History of Mathematics. A Resource File of Bibliographic, Audio-visual and Other Materials ... for Students and Teachers*. Leicester: Leapfrogs, 1979. Typescript.

Restricted to works in English which are easily available through libraries. There are about 450 titles (approximately 200 annotated) under fourteen subject headings; includes works by and about some sixty mathematicians. Frequently updated and revised in the light of criticism and experience.

25. Royal Society of London. *Catalogue of Scientific Papers, 1800-1900*. London: C.J. Clay, 1867-1902; Cambridge: University Press, 1914-1925. Reprinted New York: Johnson Reprint, 1965. 19 vols. *Subject Index (Pure Mathematics, Mechanics, Physics)*. Cambridge: University Press, 1908-1914. 3 vols. in 4.

Intended to contain every scientific memoir (including history of science) which appeared in transactions and proceedings of societies and in journals. The index volumes for pure mathematics and mechanics contain separate sections for historical material and serve to supplement May's *Bibliography*, item 17. See also item 964.

26. Russo, F. *Histoire des sciences et des techniques. Bibliographie*. Paris: Hermann, 1954. Supplement, 1955.

An essential reference book for the historian of science, whatever his speciality. While Sarton's guides (items 13, 14, and 48) do not deal for the most part with primary sources, Russo devotes about half of his book to them. The volume opens with a brief survey of materials devoted to societies and institutes for the history of science, congresses, and biographies of historians of science, then discusses libraries, archives, museums, periodicals, general works, works by time period, and works for the various sciences. Nearly ten pages are devoted to mathematics. Well indexed.

27. Sezgin, Fuat. *Geschichte des arabischen Schrifttums*. Vol. 5: *Mathematik bis ca. 430 H*. Leiden: Brill, 1974.

Surveys bio-bibliographical sources of the history of Arabic mathematics, including Greek and Indian sources. The principal collections of manuscripts are listed in volumes 1, 3, 5, and 6. Volume 5 is reviewed in *Archives internationales d'histoire des sciences* 28 (1978), 325-329. See also items 204, 272.

28. Suter, Heinrich. "Die Mathematiker und Astronomen der Araber und ihre Werke." *Abhandlungen zur Geschichte der mathematischen Wissenschaften* 10 (1900), 1-278; 14 (1902), 157-185.

Bio-bibliographical survey of Arabic mathematicians and astronomers, arranged chronologically. Locations of manuscripts given. Name and title index. Additions and corrections by H.P.J. Renaud in *Isis* 18 (1932), 166-183. Reviewed in *Biblioteca Mathematica* 2 (1901), 161-164. See also item 273.

PERIODICALS

The *Isis Guide to the History of Science 1980* (published by the History of Science Society) includes a listing of over 100 journals that relate to the history of science, many of them including the history of mathematics. The *Archive for History of Exact Sciences* (Berlin: Springer Verlag, 1960-) publishes a substantial number of history of mathematics papers. The only journals devoted exclusively to the history of mathematics are *Historia Mathematica* (New York: Academic Press, 1974-), which has also published the *World Directory of Historians of Mathematics* (Toronto: Historia Mathematica, 1978), *Bollettino di Storia della Scienze Matematiche* (Florence: Istituto di Matematica, to appear), and *Ganita Bharati, Bulletin of the Indian Society for History of Mathematics*, edited by R.C. Gupta (Ranchi, India, 1979-).

Useful journals which have ceased publication are listed below:

29. *Abhandlungen zur Geschichte der mathematischen Wissenschaften mit Einschluss ihrer Anwendungen*. 1877-1913. 30 vols. Volumes 1-10 issued as supplement to *Zeitschrift für Mathematik und Physik*. Leipzig, 1877-1913.

30. *Archiv für die Geschichte der Naturwissenschaften und der Technik*. Leipzig, 1900-1931. Continued as *Archiv für Geschichte der Mathematik*, and then as *Quellen und Studien zur Geschichte der Naturwissenschaften und der Medicin*. See item 35.

31. *Bibliotheca Mathematica* 1 (1884)-3 (1886); Neue Folge, 1 (1887)-13 (1899); Dritte Folge, 1 (1900)-14 (1913-1914). Stockholm, 1887-1899; Leipzig, 1900-1914.

The first series indexed newly published books, articles, and papers in the area of pure mathematics, giving occasional brief historical notes. The second series emphasized the history of mathematics exclusively. The third series, in addition to listing newly published works, included substantial historical articles treating subjects from antiquity to the most current mathematical topics.

32. *Bollettino di Bibliografia e Storia delle Scienze Matematiche.* Edited by G. Loria. Turin, 1898-1919.

    Places great emphasis on book reviews and announcements covering literature in all languages.

33. *Bullettino di Bibliografia e di Storia delle Scienze Matematiche e Fisiche*. Rome: Tipografia delle Scienze Matematiche e Fisiche, 1868-1887.

    Volume 20 contains an index by author for the 20 volumes of the *Bulletino*. Authors include Boncompagni, M. Cantor, Favaro, Genocchi, Sédillot, Steinschneider. A list of edited documents also appears, from Abū-l-Wafā' and the Arabic text of his *Kitab al Mobarek* to a note by Zucchetti in the Gonzaga archives in Mantua.

34. *Osiris; Studies on the History and Philosophy of Science, and on the History of Learning and Culture*. Bruges, 1936-.

35. *Quellen und Studien zur Geschichte der Mathematik, Astronomie und Physik*. Berlin: J. Springer, 1930-1936. 4 vols. Series A is devoted to sources. Title varies slightly. See item 30.

36. *Rete: Strukturgeschichte der Naturwissenschaften*. Hildesheim, 1971-1975.

37. *Scripta Mathematica: A Quarterly Journal Devoted to the Philosophy, History, and Expository Treatment of Mathematics*. New York, 1932-1973?

38. *Zeitschrift für Mathematik und Physik. Historisch-literarische Abtheilung*. Leipzig, 1865-1900.

ABSTRACTING JOURNALS

39. *Bulletin signalétique*. 522. *Histoire des sciences et des techniques*. Paris: Centre de Documentation Sciences Humaines, 1962-.

    Covers a wide range of journals to produce a bibliography of articles and book reviews, arranged by subject and period and briefly annotated. There are annual cumulative subject and author indexes.

40. *Historia Mathematica*. New York: Academic Press, 1974-.

    Articles, books, and other media, in all languages, in the field of the history of mathematics are abstracted in each quarterly issue with an index by author or main entry in the last issue of each year.

41. *Jahrbuch über die Fortschritte der Mathematik*. 1868-1942. Berlin: de Gruyter, 1871-1942. 68 vols.

Classified annotated bibliography of articles in contemporaneous journals with a section for history and philosophy. Indexed by author. See also item 973.

42. *Mathematical Reviews*. Providence, R.I.: American Mathematical Society, 1940-.

Abstracts and reviews books and articles in nearly 1,200 journals. A special effort is made to cover material published in Russian. Reviews, written by experts in each field, are critical in nature. *Mathematical Reviews* appears monthly and includes a separate section for history and biography. Historical items for the years 1940-1966 are included in May's *Bibliography*, item 17.

43. *Referativny Zhurnal*. Moscow: Akademiia Nauk SSSR, 1953-.

All reviews are in Russian. The scope is comparable to *Mathematical Reviews*, item 42, and the *Zentralblatt*, item 45, but with greater emphasis given to pedagogy.

44. *Revue semestrielle des publications mathématiques*. 1893-1932.

Half-yearly index of journal articles. The five-year cumulative indexes have a history section and an index of biographies. Historical items are included in May's *Bibliography*, item 17.

45. *Zentralblatt für Mathematik und ihre Grenzgebiete (reine und angewandte Mathematik, theoretische Physik, Astrophysik, Geophysik)*. Berlin: Springer, 1931-.

Covers a range of mathematical literature similar to *Mathematical Reviews*, item 42, but relies upon author abstracts, and consequently is not a source for critical or evaluative reviews. In general, abstracts are published within ten weeks after their receipt.

HANDBOOKS

46. Jayawardene, S.A. *Reference Books for the Historian of Science: A Handlist*. London: Science Museum Library, 1982.

Consists of some 1,000 titles arranged in 44 chapters. There are detailed subject and author/title indexes. Sections I, 3; II, 4; VI, 1, b; VI, 3, b; and VII, 2 are specifically devoted to mathematics. Special sections deal with libraries, archives, manuscripts, theses, and dissertations.

47. Loria, Gino. *Guida allo studio della storia delle matematiche; generalità, didattica, bibliografia*. 2nd ed. Milan: Hoepli, 1946.

A series of bibliographical essays in which some 1,600 titles are enumerated and commented on. In spite of superfluous detail in some chapters, and many errors in the index, this first bibliographical handbook for the historian of mathematics is still informative. Personal names are indexed, but not subjects.

48. Sarton, George. *The Study of the History of Mathematics*. Cambridge, Mass.: Harvard University Press, 1937. Reprinted New York: Dover, 1957.

Consists of a text on the meaning of the history of mathematics, and an annotated bibliography (25 pages) intended for beginners. Although the work is outdated in many respects, Sarton's annotations are still of interest. As a companion volume should be mentioned Sarton's *A Guide to the History of Science (Horus)* (Waltham, Mass.: Chronica Botanica Co., 1952), which includes a chapter on mathematics. No single guide since *Horus* has covered all the topics treated in it and many of these are of general interest to historians of mathematics.

INDEXES

49. Jayawardene, S.A., and Jennifer Lawes. "Biographical Notices of Historians of Science: A Checklist." *Annals of Science* 36 (1979), 315-394.

Includes 128 historians of mathematics among the 800 historians of science and 3,000 notices. Bibliographies and portraits are indicated.

50. Wallis, P.J. *An Index of British Mathematicians. A Check-list.*

Copies are available from the author (£1.50) at the School of Education, University of Newcastle-upon-Tyne, England. A second part (1701-1760) was issued in 1976.

II. SOURCE MATERIALS

SOURCE BOOKS

51. Bellman, R. *A Collection of Modern Mathematical Classics*. New York: Dover, 1961.

Photographically reproduces thirteen classical papers in eighteenth- and nineteenth-century analysis ("the field we have arbitrarily selected for this initial collection"--no more were published) by each of the following: C. Hermite; G.H. Hardy and J.E. Littlewood; P.L. Chebyshev; L. Fejér; I. Fredholm; L. Fuchs; A. Hurwitz; H. Weyl; B. van der Pol; G.D. Birkhoff; G.D. Birkhoff and O.D. Kellogg; J. von Neumann; D. Hilbert. Each paper is preceded by a short one- or two-page introduction.

52. Birkhoff, Garrett, ed. with the assistance of Uta Merzbach. *A Source Book in Classical Analysis*. Cambridge, Mass.: Harvard University Press, 1973.

Presents English translations, with brief historical introductions and explanatory footnotes, of nineteenth-century treatises on analysis. The selection is of "some of the most influential writings of the greatest mathematicians of a period in which, truly, (classical) analysis dominated mathematics." A two-page bibliography and two-page index are appended. See also item 966.

53. Midonick, Henrietta. *The Treasury of Mathematics. A Collection of Source Material in Mathematics ... with Introductory Biographical and Historical Sketches*. London and New York, 1965. Reprinted Harmondsworth, England: Penguin Books, 1968.

Translates into English, some for the first time, excerpts from published works. Covers antiquity (including Eastern) up to C.S. Peirce, G. Cantor, and G. Frege.

54. Moritz, Robert E. *Memorabilia Mathematica or The Philomath's Quotation Book*. New York: Macmillan, 1914. Reprinted 1958.

Over 1,000 quotations arranged by subject (e.g., the value of mathematics, mathematics and logic, algebra, paradoxes and curiosities) in English translation with index.

55. Newman, James R. *The World of Mathematics. A Small Library of the Literature of Mathematics from A'h-mosé the Scribe to Albert Einstein, Presented with Commentaries and Notes*. New York: Simon and Schuster, 1956. 4 vols.

One of the broadest anthologies in scope. Volume One, Part II, is devoted to history and biography from the Rhind Papyrus to

B. Russell and A.N. Whitehead, though there are a number of historical articles elsewhere. General index in Volume Four.

56. Smith, David Eugene, ed. *A Source Book in Mathematics*. 1st ed. New York: McGraw-Hill, 1929. Reprinted New York: Dover, 1959. 2 vols.

 Excerpts, generally five or six pages in length, from Renaissance and post-Renaissance treatises. All are translated into English and many of the translators are well-known modern specialists in the relevant mathematical subject (e.g., "Gauss on the Third Proof of the Law of Quadratic Reciprocity" is translated from the Latin by D.H. Lehmer). Short introductions and some modern explanations (carefully put in square-bracketed footnotes) are given. See also items 968, 1925.

57. Speiser, Andreas. *Klassische Stücke der Mathematik*. Zurich, Leipzig: Füssli, 1925.

 German translations of twenty-five selections, generally expositions about mathematics, from Euclid to Einstein. Short pieces are included from Dante, da Vinci, Kepler, Goethe, Rousseau, Euler, and Sylvester.

58. Struik, Dirk J. *A Source Book in Mathematics: 1200-1800*. Cambridge, Mass.: Harvard University Press, 1969.

 Selections in English translation from works in the Latin world (i.e., none from Arabic or Oriental authors unless a much used Latin translation was available). Confined to pure mathematics and those fields of applied mathematics having a direct bearing on the development of pure mathematics. Extensive historical introductions and commentaries. Arranged by subject: arithmetic; algebra; goemetry; analysis before Newton and Leibniz; Newton, Leibniz, and their schools.

59. Wieleitner, Heinrich. *Mathematische Quellenbücher*. Berlin: Verlag Salle, 1927-1928.

 Issued in four parts as volumes 3, 11, 19, and 24 in a larger series devoted to science. Excerpts from the literature are given under the headings: numerical calculations and algebra (Roman numerals and Egyptian fractions to Descartes and Euler); geometry and trigonometry (Hippocrates, Pappos, and Euclid to De Moivre); analytic and synthetic geometry; and infinitesimal calculus (Axiom of Archimedes, results due to Torricelli, Fermat, Pascal, Newton, and Leibniz). Each volume is indexed and contains explanatory and historical notes.

See also items 205, 206, 403, 511, 1619, 2272.

COLLECTED PAPERS AND CORRESPONDENCE

Following is a list of mathematicians who were active before World War I and whose correspondence (at least in part) or collected works have been published or are in preparation. The bibliographical details may be found in Poggendorff, item 20, the British Library catalogue of printed books, the *National Union Catalogue*, the John Crerar Library's

Author-Title Catalog (150 titles of mathematical collected works are listed in their subject catalog), and Gillispie's *Dictionary of Scientific Biography*, item 10. More recent works will be found listed in abstracting journals.

Annotations: W = collected works published; W* = more than one edition of collected works; C = correspondence published; L = details in Loria, *Guida*, item 47; S = details in Sarton, *Study*, item 48.

Abbe, Ernst (1840-1905). W
Abel, Niels Henrik (1802-1829). W*, C, L
Adams, John Couch (1819-1892). W
Alembert, Jean Baptiste le Rond d' (1717-1783). W*, C, L
Apollonius of Perga (ca. 262-ca. 190 B.C.). W*
Archimedes (ca. 287-212 B.C.). W*
Aristotle (384-322 B.C.). W*
Arnauld, Antoine (1612-1694). W
Ball, Sir Robert Stawell (1840-1913). S, C
Barrow, Isaac (1630-1677). W
Beltrami, Eugenio (1835-1899). W
Bernoulli, Jakob (1654-1705). W, C, L
Bernoulli, Johann, I (1667-1748). W, C, L
Bernstein, Sergei Natanovich (1880-1968). W
Bessel, Friedrich Wilhelm (1784-1846). W, C, S
Betti, Enrico (1823-1892). W
Bezout, Etienne (1739-1783). W
Bianchi, Luigi (1856-1928). W
Birkhoff, George David (1884-1944). W
Bohr, Harald August (1887-1951). W
Boltzmann, Ludwig (1844-1906). W
Bolyai, Farkas Wolfgang (1775-1856). W, C, L, S
Bolzano, Bernard (1781-1848). W*, C
Boole, George (1815-1864). W
Borchardt, Carl Wilhelm (1817-1880). W
Bordini, Francesco (1535-1591). W
Borel, Emile (1871-1956). W
Bradwardine, Thomas (1290?-1349). W
Brahe, Tycho (1546-1601). C, L
Brauer, Richard (1901-1977). W
Brendel, Johann Gottfried (1712-1758). W
Brioschi, Francesco (1824-1897). W
Brouwer, Luitzen Egbertus Jan (1881-1966). W
Burgatti, Pietro (1868-1938). W
Camus, Charles Etienne Louis (1699-1768). W
Cantelli, Francesco Paolo (1875-1966). W
Cantor, Georg (1845-1918). W, C, S
Caporali, Ettore (1855-1886). W
Carathéodory, Constantin (1873-1950). W
Cardano, Girolamo (1501-1576). W
Carnot, Lazare (1753-1823). W, C, S
Cartan, Elie Joseph (1869-1951). W
Casorati, Felice (1835-1890). W
Castelnuovo, Guido (1864-1952). W
Catalan, Eugene Charles (1814-1894). W
Cataldi, Pietro Antonio (1552-1626). W
Cauchy, Augustin (1789-1857). W

Cayley, Arthur (1821-1895). W, C, S
Cesaro, Ernesto (1859-1906). W
Ceva, T. (1648-1737). C, L
Chaplygin, Sergei Aleksandrovich (1869-1942). W
Chebyshev, Pafnutii L'vovich (1821-1894). W
Christoffel, Elvin Bruno (1829-1900). W
Ciani, Edgardo (1864-1942). W*
Clavius, Christoph (1536-1612). W
Clebsch, Alfred (1833-1872). W
Clifford, William Kingdon (1845-1879). W
Collins, J. (1625-1683). C, L
Condorcet, Jean Antoine Nicolas de (1743-1794). W
Cossali, Pietro (1748-1815). W
Cotes, Roger (1682-1716). W, C, L
Cremona, Luigi (1830-1903). W
Darboux, Gaston (1842-1917). W
Darwin, Sir George Howard (1845-1912). W
Dedekind, Richard (1831-1916). W, C, S
De Moivre, A. (1667-1754). C, L
De Morgan, A. (1806-1871). C, S
Denjoy, Arnaud (1884-1974). W
Deargues, Girard (1591-1662). W
Desargues, Girard (1591-1662). W
Descartes, René (1596-1650). W, C
Dini, Ulisse (1845-1918). W
Diophantus (fl. ca. 250 A.D.). W*
Dirichlet, Gustav Lejeune (1805-1859). W, C, S
Dolbnia, Ivan P. (1853-1912). W
Eisenstein, Gotthold (1823-1852). W, C, S
Ellis, Robert Leslie (1817-1859). W
Encke, Johann Franz (1791-1865). W
Enriques, Federigo (1871-1946). W
Euclid (fl. ca. 300 B.C.). W*
Euler, Leonhard (1707-1783). W*, C, L
Fagnano, Giulio Carlo (Marchese de'Toschi) (1682-1766). W
Faulhaber, Johann (1580-1635). W
Fejér, Leopold (1880-1959). W
Fermat, Pierre (1601-1665). W
Ferrari, L. (1522-1565). C
Ferraris, Galileo (1847-1897). W
Fibonacci, Leonardo, of Pisa (ca. 1180-ca. 1250). W
Fine, Oronce (1494-1555). W
Fontenelle, Bernard Le Bovier de (1657-1757). W
Fourier, Joseph (1768-1830). W
Fredholm, Eric Ivar (1866-1927). W
Frege, Gottlob (1848-1925). W*
Frénicle de Bessy, Bernard (1605-1675). W
Fresnel, Augustin (1788-1827). W
Frisi, Paolo (1728-1784). W
Frobenius, Ferdinand Georg (1849-1917). W
Fubini, Guido (1879-1943). W
Fuchs, Immanuel Lazarus (1833-1902). W
Fuss, P.H. (1797-1855). C, L, S
Galerkin, Boris Grigorievich (1871-1945). W
Galois, Evariste (1811-1832). W*

Gauss, Carl Friedrich (1777-1855). W, C, L, S
Gerbert, Pope Sylvester II (ca. 945-1003). W
Gerling, C.L. (1788-1864). C, L
Ghetaldi, Marino (1566-1626). W
Gibbs, Josiah Willard (1839-1903). W
Glaisher, James Whitbread Lee (1848-1928). W
Grandi, G. (1671-1742). C, L
Grassmann, Hermann G. (1809-1877). W
Grave, Dmitri Aleksandrovich (1863-1939). W
Gravesande, Willem Jacob van's (1688-1742). W
Green, George (1793-1841). W
Gregory, Duncan Farquharson (1813-1844). W
Gregory, James (1638-1675). W
Guglielmini, Domenico (1655-1710). W
Haag, Jules (1882-1953). W
Haar, Alfred (1885-1933). W
Hadamard, Jacques (1865-1963). W
Halley, Edmund (1656-1741). W, C
Halphen, Georges Henri (1844-1889). W
Hamilton, Sir William Rowan (1805-1865). W
Hardy, Godfrey Harold (1877-1947). W
Hausdorff, Felix (1868-1942). W
Hecke, Erich (1887-1947). W
Helmholtz, Hermann Ludwig Ferdinand von (1821-1894). W
Hennert, Johann Friedrich (1733-1813). W*
Herglotz, Gustav (1881-1953). W
Hermann, Jakob (1678-1733). C
Hermite, Charles (1822-1901). W, C, L, S
Hero, of Alexandria (ca. 75 A.D.). W
Hertz, Heinrich Rudolph (1857-1894). W*, C, S
Hesse, Ludwig Otto (1811-1874). W
Hilbert, David (1862-1943). W
Hill, George William (1838-1914). W
Hoëné-Wroński, Josef Maria (1776-1853). W
Hopkinson, Bertram (1874-1918). W
Hudde, J. (1633-1704). C
Humbert, Georges (1859-1921). W
Hurwitz, Adolph (1859-1919). W
Hutton, Charles (1737-1823). W
Jacobi, Carl Gustav Jacob (1804-1851). W*, C, S
Janiszewski, Zygmunt (1888-1920). W
Jerrard, George Birch (1804-1863). W
Jones, Sir William (1675-1749). C
Jordan, Camille (1838-1922). W
Kaestner, Abraham Gotthelf (1719-1801). w
Karman, Theodore von (1881-1963). W
Karsten, Wenceslaus Johann Gustav (1732-1787). W*
al-Kashi (d. 1429). W
Keckermann, Bartholomew (d. 1609). W
Kelvin, Lord (Sir William Thomson, first Baron Kelvin) (1824-1907). W
al-Khayyami (Omar Khayyam) (ca. 1050-1130). W
Kirchhoff, Gustav (1824-1887). W
Kitao, Diro (1851-1907). W
Klein, Felix (1849-1925). W, C, L, S
Kochanski, A.A. (17th century). C, L

Korkin, Aleksandr Nikolaevich (1837-1908). W
Kovalewski, S. (1850-1891). C
Krasovskii, Theodosii Nikolaevich (1878-1948). W
Kronecker, Leopold (1823-1891). W, C, L, S
Krylov, Aleksei Nikolaevich (1863-1945). W
Kummer, Ernst Eduard (1810-1893). W, C, S
Lagrange, Joseph Louis de (1736-1813). W, C
Laguerre, Edmond (1834-1886). W
Lambert, Johann Heinrich (1728-1777). W*
Landen, John (1719-1790). W
Laplace, Pierre Simon (1749-1827). W*
Larmor, Sir Joseph (1857-1942). W
Lebesgue, Henri (1875-1941). W
Lefschetz, Solomon (1884-1972). W
Legendre, A.M. (1752-1833). C, L, S
Leibenzon, Leonid Samuilovich (1879-1951). W
Leibniz, Gottfried Wilhelm (1646-1716). W*, C, L
Levi, Eugenio Elia (1883-1917). W
Levi-Civita, Tullio (1873-1941). W
Levy, Paul (1886-1971). W
L'Hospital, G.F. de (1661-1704). C, L
Lie, Sophus (1842-1899). W
Liouville, J. (1809-1882). C, L, S
Lobachevskii, Nikolai Ivanovich (1793-1856). W
Lorentz, Hendrik Antoon (1853-1928). W
Lorenz, Ludwig Valentin (1829-1891). W
Luzin, Nikolai Nikolaievich (1883-1950). W
Lyapunov, Aleksandr Mikhailovich (1857-1918). W
MacCullagh, James (1809-1847). W
MacMahon, Percy Alexander (1854-1929). W
Maggi, Giovan Antonio (1856-1937). W
Magini, Giovanni Antonio (1555-1617). C, L
Malebranche, Nicolas (1638-1715). W
Malfatti, Gianfrancesco (Giovanni Francesco Giuseppe) (1731-1807). C, L
Manfredi, Eustachio (1674-1739). W
Mansfield, Jared (1759-1830). W
Mansion, Paul (1844-1919). W
Markov, Andrei Andreivich (1856-1922). W
Maupertuis, Pierre Louis Moreau de (1698-1759). W
Maurolico, Francesco (1494-1575). W
Maxwell, James Clerk (1831-1879). W
Mayer, Johann Tobias (1723-1762). W
Mersenne, Marin (1588-1648). C, L
Miller, George Abram (1863-1951). W
Minkowski, Hermann (1864-1909). W, C
Mises, Richard von (1883-1953). W
Moebius, August Ferdinand (1790-1868). W
Mossotti, Ottaviano Fabrizio (1791-1863). W
Napier, John (1550-1617). W
Nekrasov, Aleksandr Ivanovich (1883-1957). W
Neumann, Franz (1798-1895). W
Newton, Sir Isaac (1642-1727). W*, C, L
Nicolai, B. (1793-1846). C, L
Nicolas of Cusa (1401-1464). W
Nunez, Pedro (1492-1577). W
Olbers, Wilhelm (1758-1840). W, C, S

Oldenburg, H. (1618-1677). C
Ostrogradskii, Mikhail Vasilevich (1801-1862). W*
Oughtred, William (1575-1660). W
Ozanam, Jacques (1640-1717). W
Painlevé, Paul (1863-1933).
Pappus of Alexandria (fl. ca. 320 A.D.). W*
Pascal, Blaise (1621-1662). W*
Peano, Giuseppe (1858-1932). W*
Peirce, Charles Sanders (1839-1914). W
Pfaff, J.F. (1765-1825). C, L
Piazzolla, Margherita (Beloch) (1879-). W
Picard, Charles Emile (1856-1941). W
Pincherle, Salvatore (1853-1936). W
Playfair, John (1748-1819). W
Pluecker, Julius (1801-1868). W
Poincaré, Henri (1854-1912). W, C, S
Pol, Balthasar van der (1889-1959). W
Poleni, Giovanni (1683-1761). C
Pólya, George (1887-). W
Pompeiu, Dimitrie (1873-1954). W
Ptolemy, Claudius (ca. 85?-ca. 165? A.D.). W
Ramanujan, Srinivasa (1887-1920). W*
Rankine, William John Macquorn (1820-1872). W
Rayleigh, Lord (John William Strutt, 3rd Baron Rayleigh) (1842-1919). W
Regiomontanus (1436-1476). W, C
Riccati, Jacopo Francesco (1676-1754). W
Ricci, Matteo (1552-1610). C
Ricci-Curbastro, Gregorio (1853-1925). W
Richard of Wallingford (1292-1336). W
Riemann, Bernhard (1826-1866). W*
Riesz, Frigyes (Frederic) (1880-1956). W
Rigaud, Stephen Peter (1774-1839). C, L
Roberval, Giles Personne de (1602-1675). W, C
Robin, Victor Gustave (1855-1897). W
Robins, Benjamin (1707-1751). W
Robinson, Abraham (1918-1974). W
Rohault, Jacques (1620-1675). W
Rohde, Johann Philipp von (1759-1834). W
Ruffini, Paolo (1765-1822). W, C, L, S
Russell, Bertrand (1872-1970). W
Sauri, Abbé (1741-1785). W
Schering, Ernst Christian Julius (1833-1897). W
Schiaparelli, Giovanni Virginio (1835-1910). W
Schläfli, Ludwig (1814-1895). W, C, L, S
Schlömilch, Oskar Xavier (1823-1901). W
Schoute, Pieter Hendrik (1846-1923). W
Schubert, Hermann Cäsar Hannibal (1848-1911). W
Schumacher, H.C. (1780-1850). C, L, S
Schur, Issai (1875-1941). W
Schwarz, Hermann Amandus (1843-1921). W
Scorza, Gaetano (1876-1939). W
Segner, Janos Andras (1704-1777). W
Segre, Corrado (1863-1924). W
Serenus, Antinoensis (4th century A.D.). W
Severi, Francesco (1879-1961). W
Siacci, Uge Aldo de Francesco (1839-1907). W

Sierpínski, Waclaw (1882-1969). W
Simson, Robert (1687-1768). W, C, L
Sluse, René François Walter de (1622-1685). C, L, S
Smith, Henry John Stephen (1826-1883). W
Somigliana, Carlo (1860-1955). W
Steiner, Jakob (1796-1863). W, C, S
Stern, Maritz Abraham (1807-1894). C, L
Stevin, Simon (1548-1620). W*
Stewart, Matthew (1717-1785). C, L
Stieltjes, Thomas (1856-1894). W, C, L, S
Stirling, James (1692-1770). C, L
Stokes, Sir George Gabriel (1819-1903). W, C, S
Sturm, Charles François (1803-1855). W
Sylow, Ludvig (1832-1918). W
Sylvester, James Joseph (1814-1897). W, C, S
Szász, Otto (1884-1952). W
Tacquet, André (1612-1660). W, C
Tait, Peter Guthrie (1831-1901). W
Takagi, Teiji (1875-1960). W
Tannery, Paul (1843-1904). W
Tartaglia, Niccolo (1505-1557). W, C
Taylor, Brook (1685-1731). C, L
Tedone, Orazio (1870-1922). W
Teixeira, Francisco Gomez (1851-1933). W
Theon, of Smyrna (fl. early 2nd century A.D.). W
Tonelli, Leonida (1885-1946). W
Torricelli, Evangelista (1608-1647). W
Tschirnhaus, Ehrenfried Walter von (1651-1708). C
Vailati, Giovanni (1863-1909). W
Viète, François (1540-1603). W
Vojtech, Jan (1879-1953). W
Volterra, Vito (1860-1940). W
Voronoi, Georgi Fedosievich (1868-1908). W
Wallis, John (1616-1703). W
Weber, Heinrich (1842-1913). W
Weierstrass, Karl (1815-1897). W, C
Weyl, Hermann (1885-1955). W*
Witt, Jan de (1625-1672). C
Wolff, Christian (1679-1754). W
Ximines, Leonardo (1716-1786). C, L
Zanotti, Francesco Maria (1692-1777). W
Zhukovskii, Nikolai Egorovich (1847-1921). W*
Zolotarev, Egor Ivanovich (1847-1878). W

ARCHIVAL COLLECTIONS

60. Merzbach, Uta. *Guide to Mathematical Papers in United States Repositories*. Wilmington, Del.: Scholarly Resources, to appear.

Provides brief descriptions of manuscript collections with itemization of some major collections. Collections can include correspondence, lecture and class notes, research notes and other types of manuscripts, and memorabilia. The introduction includes information on the preservation, importance, and use of mathematical papers. Reference is made to holdings of papers of some American mathematicians in repositories abroad. Dr. Merzbach is Curator of Mathematics, National Museum of American History, Smithsonian Institution, Washington, D.C. 20560.

Information on mathematical archives in the United States may also be obtained from the Archives for American Mathematics and Archives for History of Statistics, Humanities Research Center, University of Texas at Austin, Box 7219, Austin, Texas 78712.

Mathematical archives, including those in other countries, will also be listed in general guides to archives described in the reference works by Sheehy, item 1, and Walford, item 2, which are cited in the introduction to section I of this bibliography.

See also item 46.

III. GENERAL HISTORIES OF MATHEMATICS

61. Archibald, Raymond Clare. *Outline of the History of Mathematics*. Oberlin, Ohio: Mathematical Association of America, 1949. 6th ed. Supplement to *American Mathematical Monthly* 56 (1) (January 1949).

    Originally two lectures given in 1931, numerous additions and changes were made in subsequent editions. See also item 199.

62. Ball, W.W.R. *Short Account of the History of Mathematics*. New York: Dover, 1960. Reprinted from the fourth edition of 1908.

    First published in 1888, and based, for facts concerning pre-1800 mathematics, on Cantor's *Vorlesungen*, item 74. Approximately ten percent of the volume is devoted to the nineteenth century. Often unreliable and needs to be used with considerable caution.

63. Becker, Oscar. *Grundlagen der Mathematik in geschichtlicher Entwicklung*. Freiburg: Alber, 1954.

    With extensive quotations from mathematical and philosophical texts (translated into German), analyzes "the historical development of the problems of the foundations of mathematics from its beginnings to today." See also item 2224.

64. ———, and J.E. Hofmann. *Geschichte der Mathematik*. Bonn: Athenäum-Verlag, 1951.

    Described by the authors as a sketch, for those already knowledgeable in the subject, from ancient times to about 1900. H.W. Turnbull, in *Mathematical Reviews* 14 (1953), 341, calls attention to the treatment of Japanese mathematics and the excellent bibliography.

65. Bell, Eric T. *Men of Mathematics*. New York: Simon and Schuster, 1937.

    A popular book, still in print, whose author's highly readable style has to be balanced against an often over-imaginative content. This work, like the following item 66, should be used with caution, for Bell's facts are often unreliable, and his conclusions either exaggerated or now out of date. The fact that his works are often the best known among general readers interested in the history of mathematics is all the more reason to verify any assertions that Bell makes. The fact that he rarely offers any documentation often makes it difficult, but all the more necessary, to check the veracity of his assertions. See also item 983.

66. ———. *The Development of Mathematics*. 2nd ed. New York: McGraw-Hill, 1945.

"Not a history of the traditional kind, but a narrative of the decisive epochs in the development of mathematics." Bell gives his analysis of "why certain things continue to interest mathematicians, technologists, and scientists, while others are ignored or dismissed as being no longer vital." See also item 982.

67. Bochner, S.F. *The Role of Mathematics in the Rise of Science*. Princeton: Princeton University Press, 1966.

Misleading title. Collects together essays which give a personal summary of the history of mathematics from its beginnings to the early twentieth century. The main concern, "the role of mathematics in the rise and unfolding of Western intellectuality," is treated on a higher mathematical level and generally with a less popular appeal than, for example, Morris Kline's *Mathematics in Western Culture*, item 81. Bochner often gives his evaluations of other commentators and historians. Part II consists of about seventy pages of biographical sketches. Indexed.

68. Bolgarskii, B.V. *Ocherki po istorii matematiki*. 2nd ed. Minsk: Vysheishaia Shkola, 1979.

A concise textbook providing many portraits of mathematicians while emphasizing the history of problems.

69. Bourbaki, Nicolas. *Eléments d'histoire des mathématiques*. (Histoire de la pensée IV.) Paris: Hermann, 1960.

Gathers together, without substantial change, most of the historical notes accompanying Bourbaki's *Eléments des mathématiques*. Consequently, it provides neither a continuous nor a complete history of mathematics. What it does provide, with emphasis given to modern mathematics, are historical accounts of the evolution of algebra, quadratic forms, topological spaces, real numbers, n-dimensional spaces, complex numbers, metric spaces, infinitesimal calculus, the gamma function, functional spaces, topological vector spaces, and integration, among other topics. However, such important areas as differential geometry, algebraic geometry, and the calculus of variations are not included. Virtually no biographical information is given, but in those areas of mathematics that are considered, special attention is given to formative ideas and how they developed and interacted with other leading mathematical ideas. A bibliography of nearly 300 items is also given. See also items 984, 1103.

70. Boyer, Carl B. *A History of Mathematics*. New York: John Wiley, 1968.

A college-level textbook, with chapter exercises; more extensive in scope than H. Eves, *An Introduction*, item 78, and more historically oriented than Eves or M. Kline, *Mathematical Thought*, item 82. Includes a chapter on Chinese and Indian mathematics. Though there is a chapter on aspects of 20th-century mathematics, consistent detail is give only up to about 1900. Bibliography and 12-page chronological table. A revised edition, giving special emphasis to the 19th and 20th centuries, is currently being prepared by U. Merzbach.

71. Bunt, L.N.H., P.S. Jones, and J.D. Bedient. *The Historical Roots of Elementary Mathematics*. Englewood Cliffs, N.J.: Prentice-Hall, 1976.

Contains a history of arithmetic, algebra, geometry, and number systems emphasizing the beginnings in Egyptian, Babylonian, and Greek civilizations. Intended as an elementary introductory textbook for prospective teachers as well as for students with a high school background.

72. Cajori, Florian. *A History of Mathematics*. 1st ed. New York: Macmillan, 1893. New York: Chelsea, 1980.

Brings many detailed references up to 1919, the date of the last substantially revised edition. Anonymous editor's additions and a few references and refinements, such as marking names of mathematicians in the index whose collected papers exist, have been added to the 1980 edition.

73. ———. *A History of Mathematical Notations*. Chicago: Open Court, 1928-1929. 2 vols.

Aims to give "not only the first appearance of a symbol and its origin (whenever possible), but also to indicate the competition encountered and the spread of the symbol among writers in different countries." This is perhaps a unique approach in the literature of the history of mathematics. There is a detailed table of contents and index. See also items 643, 986, 1821.

74. Cantor, Moritz B. *Vorlesungen über Geschichte der Mathematik*. Leipzig: Teubner, 1880-1908. Reprinted New York: Johnson Reprint Corporation, 1965. 4 vols.

This mammoth effort, totalling nearly 3,000 pages in four volumes, is still useful although now somewhat outdated. Volume I (antiquity to 1200 A.D.) emphasizes Greek mathematics, but devotes chapters to the Romans, Hindus, Chinese, and Arabs, and ends with medieval mathematics, including a section on Gerbert. Volume II (1200-1668) is divided into time periods, either by century or half-century, and ends with the work of Kepler, Cavalieri, Descartes, Fermat, Roberval, Torricelli, Wallis, Pascal, Sluse, Hudde, et al. Volume III (1668-1699) is concerned primarily with the invention and development of the calculus by Newton, Leibniz, and their various adherents, although there are sections on combinatorial analysis, infinite series, algebra, number theory, analytic geometry, and the enormous contributions of Euler. Volume IV (1759-1799) was written with the collaboration of a number of Cantor's contemporaries, including Cajori, Netto, Loria, and Veronese. It covers various topics in separate chapters by given authors (for example, Vivanti did the longest section, on "Infinitesimalrechnung"), and ends with a 20-page general overview of mathematics between 1758 and 1799 by M. Cantor. There are two pages of corrections, and an index (both name and subject). See also commentaries and annotations by Eneström in

Figure 2. Moritz Cantor. Author of the monumental four-volume *Vorlesungen über Geschichte der Mathematik* (item 74). This photograph is taken from the frontispiece of a *Festschrift* in Cantor's honor, *Abhandlungen zur Geschichte der Mathematik* 9 (1899), simultaneously issued as a supplement to the *Zeitschrift für Mathematik und Physik.*

Bibliotheca Mathematica, item 31, as well as items 243, 327, 466, 622.

75. Dantzig, T. *Number: The Language of Science*. 1st ed. New York, 1930. Paperback edition, New York: The Free Press, 1967.

Concerned not with the mathematical techniques of using numbers, but with the idea of number itself, this work is especially interested in symbols and forms underlying the number concept. In discussing the evolution of number in various cultural contexts, from prehistory to the present, the book examines the individuals and the epochs that produced significant advances, beginning with finger counting and use of devices like the abacus, and ending with recent developments, including the views of Kronecker, Cantor, Dedekind, Poincaré, Brouwer, et al. A very brief bibliography, as well as subject and name indexes, are included. See also item 1822.

76. Dedron, P., and J. Itard. *Mathématiques et mathématiciens*. Paris: Magnard, 1959. English translation, *Mathematics and Mathematicians*, published by Richard Sadler Ltd., 1973. Paperback edition by Open University Press, England, 1978. 2 vols.

With extensive quotations from original sources, provides a modern historical survey of classical Greek and European Renaissance (up to 1800) mathematics. Examples are at the elementary level in geometry and algebra. The first volume tries "to show some of the intentions, the hesitations and the doubts in the minds of the greatest thinkers" of the pre-Renaissance time.

77. Dubbey, John M. *Development of Modern Mathematics*. London: Butterworth, 1970.

Aims "to present the basic facts in the development of mathematics from earliest times to the present day." Eighteen references are collected at the end.

78. Eves, Howard. *An Introduction to the History of Mathematics*. 4th ed. New York: Rinehart, 1976.

Intended as a "textbook for a one-semester undergraduate course which meets three hours a week." The mathematical level is only up to the beginnings of calculus. Problems (with solutions) are presented in each section along with bibliographies, and an 8-page chronological table and an index are appended.

79. Hofmann, Joseph E. *Geschichte der Mathematik*. (Sammlung Goeschen, volumes 226, 875, 882.) Berlin: Walter de Gruyter, 1953-1957.

Ancient mathematics is covered quickly in order to give greater detail to less well-known areas of medieval and Renaissance mathematics. Volume I covers antiquity through Fermat and Descartes; volume II is devoted to the creation of new methods and developments of the calculus; volume III carries the history up to the French Revolution. Short bibliographies are given, as well as name and subject indexes.

80. Iseki, Kiyoshi, and M. Kondo. (Modern mathematics: the development and the problems). Tokyo: Nihon Hyoron, 1977. In Japanese.

Gives brief biographical and bibliographical information on prominent twentieth-century mathematicians, such as the Fields medalists, often accompanied by a photographic portrait. Divided into subject groupings (e.g., abstract algebra, topology, and functional analysis), its purpose is to provide a biographical history of modern mathematics.

81. Kline, Morris. *Mathematics in Western Culture*. New York: Oxford University Press, 1953.

On an elementary level advances the thesis "that mathematics has been a major cultural force in Western civilization" by emphasizing mathematical applications in a range of subjects including painting, music, and relativity theory. Illustrated.

82. ———. *Mathematical Thought from Ancient to Modern Times*. New York: Oxford University Press, 1972.

"Emphasizes the leading mathematical themes rather than the men" and excludes Chinese, Japanese, and Mayan civilizations. This is the most complete and the most modern in coverage of the general histories of mathematics. It gives more attention than Boyer's history, item 70, for example, to the first few decades of the twentieth century. Chapter bibliographies and an index are provided. See also items 867, 990, 1106, 1317, 1571, 2145.

83. Kowalewski, Gerhard W.H. *Grosse Mathematiker*. Munich: J.F. Lehmanns Verlag, 1938.

A meandering, as the book describes itself, through the history of mathematics from antiquity to the time of the book. While often anecdotal, Kowalewski gives considerable emphasis to the technical side of mathematics.

84. Meschkowski, H. *Problemgeschichte der Mathematik*. Mannheim: Bibliographisches Institut-Wissenschaftsverlag, 1978-1981. 3 vols.

Discusses the "fundamental problems of each epoch" at an undergraduate level, from Chinese and Pythagorean number mysticism to the decidability problem (up to 1950). Roughly evenly divided between quotations from original works and explanations by the author.

85. Smith, David Eugene. *History of Mathematics*. New York: Dover, 1958. Reprinted from the Boston, Ginn and Company, edition of 1951, first published in 1923-1925. 2 vols.

Although out of date in many respects, still useful as a source for names, illustrations, and chronology if used in conjunction with more recently written histories. Covers ancient times up to the late nineteenth century. See also item 629.

86. Struik, Dirk J. *A Concise History of Mathematics*. 3rd rev. ed. New York: Dover, 1967, 195 pp.

Though "concise," devotes two of the eight chapters to Oriental mathematics. Covers up to the end of the nineteenth century and is probably the only modern history of comparable size and reliability. Generously illustrated. Index and chapter bibliographies are provided. See also item 994.

87. Wilder, Raymond L. *Evolution of Mathematical Concepts: An Elementary Study*. New York: John Wiley, 1968. Reprinted by Open University Press, 1973.

Studies the "mathematical subculture from the standpoint of an anthropologist, rather than that of a mathematician" (though the author is also a mathematician). Principal concern is with the development of the concepts of number and geometry, but, as the author adds in a special preface to the Open University paperback edition, this is not a purely historical work. These historical examples are "secondary to furnishing a means for detecting and formulating the nature of the cultural forces affecting the evolution of mathematics." See also item 88.

88. ———. *Mathematics as a Cultural System*. Oxford: Pergamon, 1981.

"Describes the nature of mathematics and its relations to society from the standpoint of cultural anthropology." The author intends this as a "more mature treatment" than his *Evolution of Mathematical Concepts*, item 87, "in that citations to mathematical theory are not restricted to number and geometry," and his concepts of hereditary stress and consolidation are related more closely to mathematical developments.

89. Young, Laurence. *Mathematicians and Their Times*. Amsterdam: North-Holland, 1981.

Anecdotal lectures. Most valuable to the historian are probably the personal accounts of the author, professor at the University of Wisconsin, Madison, and Past Fellow of Trinity College, Cambridge. There is an index to persons, but no bibliography and no sources for the quotations are given.

IV. THE HISTORY OF MATHEMATICS: CHRONOLOGICAL PERIODS

ANCIENT MATHEMATICS

EGYPTIAN MATHEMATICS

The main surviving mathematical documents from ancient Egyptian culture date from about 1700 B.C., although the mathematics in them may have been known as early 3500 B.C. Of chief interest are the Egyptian use of distinct unit fractions to express the result of division, formulas for areas and volumes, and calendar-oriented astronomy.

Documents

Survey of Extant Sources

90. Chace, Arnold Buffum, et al., eds. "Chronological List of Documents." In item 95 below, 1979, 67.

    Lists all known Egyptian mathematical papyri with approximate dates and current locations.

91. Neugebauer, Otto. *The Exact Sciences in Antiquity*. 1st ed. Copenhagen: E. Munksgaard, and Princeton: Princeton University Press, 1951; also Princeton, 1952. 2nd ed. Providence, R.I.: Brown University, 1957. Reprinted New York: Harper, 1962.

    Cites publications of the most important papyri, pp. 86-87. See also items 182, 245.

The Rhind Papyrus

Texts

92. Eisenlohr, A. *Ein mathematisches Handbuch der alten Ägypter (Papyrus Rhind des British Museum) übersetzt und erklärt*. Leipzig, 1877. Reprinted Wiesbaden: Sändig, 1972. 2 vols.

93. British Museum Department of Oriental Antiquities. *Facsimile of the Rhind Mathematical Papyrus*. 1898.

94. Peet, T.E. *The Rhind Mathematical Papyrus, British Museum 10057 and 10058*. London, 1923. Reprinted 1977.

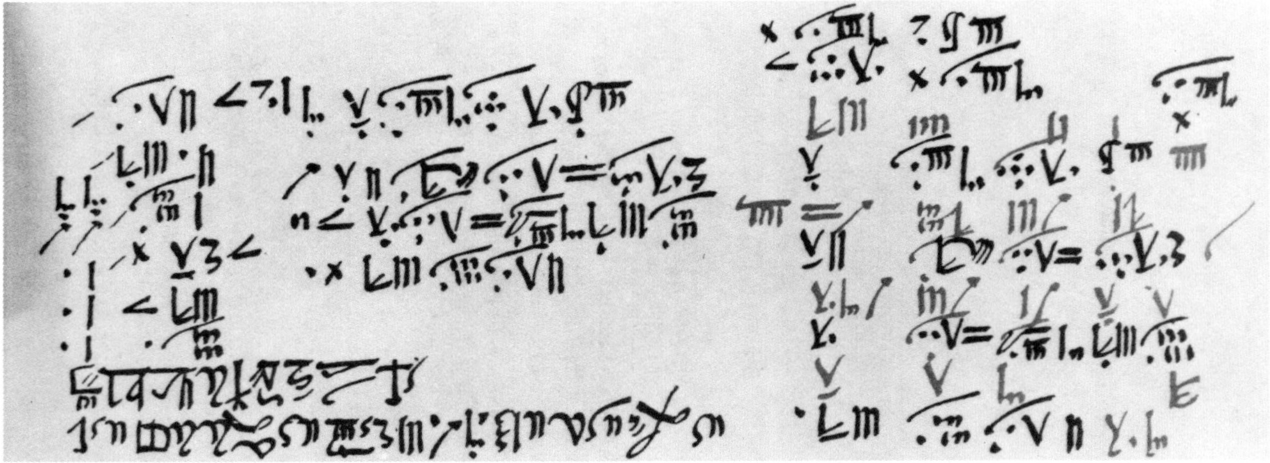

Figure 3. Problem 36 of the Rhind Papyrus translates as "I have gone in three times, my ⅓ and ⅕ have been added to me, and I return having filled the measure. What is the quantity that says this?" It is particularly noteworthy that in providing the solution to this problem, the mathematician uses the only example in which fractional expressions are applied to a particular number for the purpose of dividing.

95. Chace, Arnold Buffum, et al., eds. *The Rhind Mathematical Papyrus.* Oberlin, Ohio: Mathematical Association of America, 1927-1929. Abridged reprint, Reston, Va.: National Council of Teachers of Mathematics, 1979. 2 vols.

This edition is the most elaborate reproduction of the papyrus, with two-color plates and abundant background and introduction. There is also a comprehensive, annotated, chronological bibliography of commentary on Egyptian and Babylonian mathematics, Volume I, pp. 119-192, with supplement, Volume II, (unpaged). This bibliography is not reproduced in the abridged reprint. But see item 90 above.

96. Guggenbuhl, Laura. "The New York Fragments of the Rhind Mathematical Papyrus." *Mathematics Teacher* 57 (1964), 406-410.

Commentary

97. Gillain, O. *La science égyptienne: L'arithmétique au Moyen Empire.* Brussels: Edition de la Fondation Egyptologique Reine Elisabeth, 1927.

98. Vogel, Kurt. *Die Grundlagen der Ägyptishen Arithmetik in ihrem Zusammenhang mit der 2:n Tabelle des Papyrus Rhind.* Dissertation. University of Munich, 1929. Reprinted Wiesbaden: Sändig, 1970.

See also entries under Egyptian Fractions, items 112-126, especially Gillings, item 116.

Moscow Papyrus

Text

99. Struve, W.W. "Mathematischer Papyrus des Staatlichen Museums der schönen Künste in Moskau." *Quellen und Studien zur Geschichte der Mathematik*, Abteilung A: Quellen 1. Berlin: Springer, 1930, 12 + 198 pp. + 10 pls.
Reviews: Archibald, R.C., *Isis* 16 (1931), 148-155.
Peet, T.E., *Journal of Egyptian Archaeology* 17 (1931), 100-106.

Commentary

100. Nims, Charles F. "The Bread and Beer Problems of the Moscow Mathematical Papyrus." *Journal of Egyptian Archaeology* 44 (1958), 56-65.

101. Gillings, Richard J. "The Area of the Curved Surface of a Hemisphere in Ancient Egypt." *Australian Journal of Science* 30 (1967-1968), 113-116.

Concerns interpretation of the text of the geometric Problem 10, in which there are several breaks. The review surveys previous interpretations but concludes that the meaning of the problem is

still uncertain.
Review: Allen, E.B. *Mathematical Reviews* 42 (1971), 1043-1044, #5751.

Egyptian Mathematical Leather Roll

Texts

102. Glanville, S.R.K. "The Mathematical Leather Roll in the British Museum." *Journal of Egyptian Archaeology* 13 (1927), 232-239, Plates LVIII-LXII.

103. Scott, A., and H.R. Hall. "Egyptian Leather Roll of the 17th Century B.C." *British Museum Quarterly* 2 (1927), 56-57, plate.

104. Glanville, S.R.K. "The Mathematical Leather Roll in the British Museum." *The Rhind Mathematical Papyrus*, Vol. II (unpaged). Edited by Arnold Buffum Chace et al. Item 95.

See also Gillings, item 116, 85-103.

Commentary

105. Vogel, Kurt. "Erweitert die Lederolle unserer Kenntniss Ägyptischer Mathematik?" *Archiv für Geschichte der Mathematik der Naturwissenschaften und der Technik* 2 (1929), 386-407.

106. Neugebauer, Otto. "Zur Ägyptischen Bruchrechnung." *Zeitschrift für Ägyptische Sprache* 64 (1929), 44-48.

107. Gillings, Richard J. "What Is the Relation between EMLR and the RMP Recto?" *Archive for History of Exact Sciences* 14 (3) (1975), 159-167.

108. ———. "The Recto of the Rhind Mathematical Papyrus and the Egyptian Mathematical Leather Roll." *Historia Mathematica* 6 (1979), 442-447.

See item 117.

109. ———. "The Egyptian Mathematical Leather Role--Line 8. How Did the Scribe Do It?" *Historia Mathematica* 8 (1981), 456-457.

A follow-up to Gillings's earlier article, item 108. Here he offers two possible methods scribes may have devised in producing Number 8 of the 26 inequalities in the Egyptian Mathematical Leather Roll. See also item 117.

Demotic Papyri

Text

110. Parker, Richard A. *Demotic Mathematical Papyri*. Providence, R.I.: Brown University Press, 1972.
Review: Vogel, Kurt. *Historia Mathematica* 1 (1974), 195-199.

Commentary

111. Vogel, Kurt. "Ein arithmetisches Problem aus dem Mittleren Reich in einem demotischen Papyrus." *Enchoria: Zeitschrift für Demotistik und Koptologie* 4 (1974), 67-70.

Egyptian Fractions

112. Neugebauer, Otto. *Die Grundlagen der Ägyptischen Bruchrechnung.* Berlin: J. Springer, 1926.
Review: Jahrbuch über die Fortschritte der Mathematik, 52 (1926), 4. This review also lists *other* reviews of the book up to 1929; this volume of the *Fortschritte* was not actually published until 1935.

113. Van der Waerden, Bartel Leendert. "Die Entstehungsgeschichte der ägyptischen Bruchrechnung." *Quellen und Studien zur Geschichte der Mathematik*, Abteilung B: Studien 4 (1937-1938), 359-382.

114. Bruins, Evert M. "Ancient Egyptian Arithmetic: 2/N." *Koninklijke Nederlandse Akademie van Wetenschappen, Proceedings, Series A, Mathematical Sciences* (Amsterdam) 55 (1952), 83-93.

Neugebauer and van der Waerden hold that the 2/n table of the Rhind Mathematical Papyrus was developed gradually--perhaps over centuries. Bruins disagrees, claiming that the table might have been fashioned in an afternoon. He suggests that the author in treating primes confined himself to "direct" decompositions, those in which all but one of the denominators is a multiple of the prime; and that the decompositions could have been effected by means of the method of multipliers, that is, multiplying numerator and denominator by an integer abundant in factors.

115. ———. Platon et la table égyptienne 2/n." *Janus* 46 (1957), 253-263.

116. Gillings, Richard J. *Mathematics in the Time of the Pharaohs.* Cambridge, Mass.: MIT Press, 1972.

Suggests five precepts guided the author of the 2/n table of the Rhind Mathematical Papyrus in choosing decompositions:

1. Smaller denominators are preferred, and all are smaller than 1000.
2. Shorter decompositions are preferred, and all are no more than 4 terms long.
3. Smaller denominators precede larger in a decomposition, and never is a denominator repeated.
4. Smallness of the smallest denominator is the main consideration, but a slightly larger smallest denominator is acceptable if it greatly reduces the largest denominator.
5. Even numbers are preferred to odd numbers, even if large, or if the preference lengthens the decomposition.

Gillings regards the decomposition of 2/35 as giving a clue to a general method used by the table's author wherever possible, though some decompositions required specialized treatment.
Reviews: Guggenbuhl, L., *Isis* 64 (1973), 533-534; *Mathematical Reviews* 57 (1979), 1228, #9396.
Raĭk, Anna Eremeeva, *Historia Mathematica* 1 (1974), 464-468.

117. ———. "The Recto of the Rhind Mathematical Papyrus: How Did the Ancient Egyptian Scribe Prepare It?" *Archive for History of Exact Sciences* 12 (1974), 291-298.

See items 108 and 109.

118. Rising, Gerald R. "The Egyptian Use of Unit Fractions for Equitable Distribution." *Historia Mathematica* 1 (1974), 73-74; response by R.J. Gillings, 74.

Rising points out (and Gillings agrees) that Ahmose's uses of 2/3 along with unit fractions provide counterexamples to Gillings's observation that "Egyptian fractions allow equal distribution not only of quantity but also in size and number of pieces."
Review: Guggenbuhl, L. *Mathematical Reviews* 58 (1979), 2359, #15803.

119. Bruins, Evert M. "The Part in Ancient Egyptian Mathematics." *Centaurus* 19 (1975), 241-251.

Restates Bruins's 1944 research results recounted in his 1952 article, item 114. Here he refers to the "beautiful multipliers"--doublings of 2 and 3 and the tenfolds--which can account for all of the entries in the 2/n table except for n = 43, 59, and 97.
Review: Guggenbuhl, L. *Mathematical Reviews* 56 (1978), 695, #5123.

120. Bruckheimer, M., and Y. Salomon. "Some Comments on R.J. Gillings' Analysis of the 2/n Table in the Rhind Papyrus." *Historia Mathematica* 4 (1977), 445-452; response by Gillings, 5 (1978), 221-227.

Bruckheimer and Salomon criticize Gillings's book, item 116: (1) for carelessness in counting Egyptian fraction decompositions of various kinds in the Rhind Papyrus; (2) for failing to account for all possible decompositions of 2/n; (3) for partiality toward the scribe's choices of decompositions, in the formulations of Gillings's "precepts" for choosing a decomposition; (4) various other errors. Gillings defends his precepts as deductions from the data, focusing in detail on the cases of 2/35 and 2/95 ("a solitary error" by the scribe) and citing other commentators.
Review: Guggenbuhl, L. *Mathematical Reviews* 58 (1979), 3951, #26694.

121. Campbell, Paul J. "A 'Practical' Approach to Egyptian Fractions." *Journal of Recreational Mathematics* 10 (1977-1978), 81-86.

Explores feasibility of producing an Egyptian fraction decomposition by the method of multipliers. Other algorithms less closely related to the procedures of the papyri are also discussed.

122. Gillings, Richard J. "The Mathematics of Ancient Egypt." *Dictionary of Scientific Biography* XV, Supplement 1. Edited by Charles C. Gillispie. New York: Scribners, 1978, 681-705.

See item 10.

123. Raïk, Anna Eremeeva. "On the Theory of Egyptian Fractions." *Istoriko-Matematicheskie Issledovaniia*, No. 23 (1978), 181-191, 358. In Russian.

Supports Gillings's claim that all Egyptian decompositions of 2/n can be obtained by transformation of 2/101 = 1/101 + 1/202 + 1/303 + 1/606.
Review: Bruins, E.M. *Mathematical Reviews* 58 (1979), 3957, #26743.

124. Campbell, Paul J. "Bibliography of Algorithms for Egyptian Fractions." 1979, 21 pp.; revised 1981. Available from the author.

Indexed bibliography of articles dealing with mathematical methods of producing Egyptian fraction decompositions.

125. Van der Waerden, Bartel Leendert. "The (2:n) Table in the Rhind Papyrus." *Centaurus* 23 (1980), 259-274.

Contends that "quite satisfactory" answers to how the 2:n table was obtained were given years ago by Vetter, Neugebauer, and van der Waerden himself. Presents condensed version of van der Waerden's earlier paper, tracing the table entries mainly to Egyptian division and the method of multipliers. Van der Waerden formerly thought the table was the culmination of a long historical process, but here is inclined to attribute it to a single ingenious individual.

126. Knorr, Wilbur R. "Techniques of Fractions in Ancient Egypt and Greece." *Historia Mathematica* 9 (1982), 133-171.

Recent writers have debated the alleged esthetic criteria underlying the Egyptian computations; by contrast, van der Waerden, items 113, 125, and 129, has stressed the purely technical character of these computations. Although the Greek evidence is rarely brought to bear on this question, the present study seeks to show that this evidence strongly upholds van der Waerden's position.

See also items 97, 108, 109 for further material on Egyptian fractions.

Egyptian Geometry

Commentary

127. Wheeler, Noel F. "Pyramids and Their Purpose." *Antiquity* 9 (1935), 5-21, 161-189, 292-304.

128. Bruins, Evert M. "Over de Benandering van $\pi/4$ in de Aegyptische Meetkunde." *Koninklijke Nederlandse Akademie van Wetenschappen, Proceedings, Series A, Mathematical Sciences* (Amsterdam) 48 (1945), 206-210.

129. Van der Waerden, Bartel Leendert. *Science Awakening*. Groningen: Noordhoff, 1954, 31-35.

    See items 189, 247.

130. Gillings, Richard J. *Mathematics in the Time of the Pharaohs*. Item 116 above, 137-153, 185-201.

    Problems on areas and volumes, pyramids and truncated pyramids, and areas of semicylinder and hemisphere, from various papyri.

131. Engels, Hermann. "Quadrature of the Circle in Ancient Egypt." *Historia Mathematica* 4 (1977), 137-140.

    Offers a simple construction to explain the Egyptian approximation of $(16/9)^2$ for π.

Egyptian Astronomy

Text

132. Parker, R.A., and Otto Neugebauer. *Egyptian Astronomical Texts*. London: Lund Humphreys, 1960-1969. 3 vols.
    Reviews: Vol. I: Bruins, E.M. *Isis* 53 (1962), 523-525.
    Vol. II: Pingree, David. *Isis* 57 (1966), 136-137.

Commentary

133. Bruins, Evert M. "Egyptian Astronomy." *Janus* 52 (1965), 161-180.

    Attempts to show that the Egyptian system of decans was more precise than indicated by Parker and Neugebauer.
    Review: Toomer, G.J. *Mathematical Reviews* 39 (1970), 3, #12.

134. Van der Waerden, Bartel Leendert. *Science Awakening II: The Birth of Astronomy*. New York: Oxford University Press, 1974, 8-45.

135. Neugebauer, Otto. *A History of Ancient Mathematical Astronomy. Book III: Egyptian Astronomy*. Berlin, Heidelberg, New York: Springer-Verlag, 1975.

    "Egypt has no place in a work on the history of mathematical astronomy." See also items 246 and 1806.
    Review: Swerdlow, N.M. *Historia Mathematica* 6 (1979), 76-85.

Guide to Further Literature

136. Archibald, Raymond Clare. "Bibliography of Egyptian Mathematics." *The Rhind Mathematical Papyrus*, Vol. I. "Supplement," in Vol. II. Edited by Arnold Buffum Chace et al. Item 95.

    Splendid comprehensive annotated bibliography through 1929.

137. Guggenbuhl, Laura. "Mathematics in Ancient Egypt: A Checklist (1930-1965)." *Mathematics Teacher* 58 (1965), 630-634.

Most references are to general works on Egyptian culture or mathematics. There are no annotations.

138. May, Kenneth O. *Bibliography and Research Manual of the History of Mathematics*. Toronto: University of Toronto Press, 1973, 611-615.

Indexes articles reviewed in the major review journals through 1965-1966. Relevant headings under "Historical Classifications: Egyptian" are "General," "Fractions," "Moscow papyrus," and "Rhind papyrus." Article titles are omitted. See also items 17, 618.

BABYLONIAN MATHEMATICS

The matter of chronology is of the greatest importance in any bibliography devoted to works on (the history of) Babylonian mathematics, because the state of our knowledge about the subject is strongly dependent on the number of mathematical cuneiform texts that have been excavated and published at any given point in time. Therefore the following bibliography is arranged with the titles of books and articles in chronological rather than in alphabetical order.

Source materials (S): The majority of the more interesting mathematical cuneiform texts have been published or republished in a number of comprehensive books and catalogues. These have been included in the general bibliography below, but are indicated by the letter *S* in front of the name of the author or the authors.

General histories of Babylonian mathematics (G): Similarly, books and articles setting out to give a general historical or topological survey of a smaller or larger part of Babylonian mathematics are indicated by the letter *G*.

General Reference

139. Archibald, Raymond Clare. "Bibliography of Egyptian and Babylonian Mathematics (Supplement)." In *The Rhind Mathematical Papyrus*. Vol. II. Edited by A.B. Chase et al. Item 95, 12.

Contains a carefully annotated, although by no means complete, bibliography of works on Babylonian mathematics, covering the years from 1854 to 1930.

Chronological Periods

Pre-Babylonian Mathematics and Metrology in the Middle East

There are clear indications that the "Babylonian" mathematicians inherited many of their mathematical methods and traditions from their Sumerian (and other) predecessors in Mesopotamia and surrounding regions. However, very few mathematical texts in the true sense of the word have been preserved from the pre-Babylonian period. Therefore even commercial or economical texts from those remote times have to be taken into account when one is trying to study the very early development of mathematical concepts and ideas. In particular, it turns out that it is impossible to understand certain central aspects of the history of Babylonian mathematics without paying due attention to Babylonian and pre-Babylonian metrology. In fact, the history of the Babylonian sexagesimal system, for instance, is intimately associated with the history of the Sumero-Babylonian metrological number systems for length, area, capacity, and weight.

140. (*S*) Hilprecht, Hermann Vollrath. *Mathematical, Metrological and Chronological Tablets from the Temple Library of Nippur.* (Babylonian Expedition of the University of Pennsylvania, Series A, XX:1.) Philadelphia: Dept. of Archaeology, University of Pennsylvania, 1906, 70 pp., 30 + 15 pls.

A pioneering work in the field, in which two groups of mathematical and metrological table texts are published and analyzed in detail; the older is dated to the time of the first dynasty of Isin, i.e., the first subperiod of the Old Babylonian period c. 1900 B.C. Among the tablets are 20 multiplication tables, 3 tables of reciprocals (called "division tables"), one table of squares and one of square roots, and 15 "metrological lists" describing the structure of the current metrological systems and the forms of the corresponding number signs in all possible combinations. Older than all the other tablets is the text CBM 10201, which is in part correctly interpreted by the author, but which also gives rise to fantastic speculations about the role in Babylonian mathematics of the "Number of Plato," simply because the author failed to take into account the already then-known fact that sexagesimal fractions are written without the use of any "sexagesimal point" in Babylonian sexagesimal notation.

141. Allotte de la Fuÿe, François-Maurice. "Mesures agraires et formules d'arpentage à l'époque présargonique." *Revue d'Assyriologie* 12 (1915), 117-146.

A study of the structure of the Sumerian metrological systems for length and area, with departure from eight texts from Telloh. In some of these texts appear the earliest known application of the approximative "agrimensor formula" for the area of general quadrilaterals.

142. Neugebauer, Otto. "Zur Entstehung des Sexagesimalsystems." *Abhandlungen der Gesellschaft der Wissenschaften zu Göttingen, Math.-Phys. Klasse*, Neue Folge XIII (1). Berlin: Weidmannsche Buchhandlung, 1927, 55 pp.

The most interesting of several books and articles from about this time, by a number of authors, with various conjectures about the origin of the Babylonian sexagesimal system. However, since the proto-Sumerian Jemdet Nasr tablets had not yet been published when this book was written, the discussion in the book is based mainly on the Fara texts, several centuries younger, which means that it is now outdated in several respects. The main thesis is still valid, namely, that the history of the Babylonian sexagesimal system is intimately associated with the history of the Sumero-Babylonian metrological systems. (Cf. Thureau-Dangin, "Sketch of a History of the Sexagesimal System," *Osiris* 7 [1939], 95-141.)

143. *(S)* Langdon, Stephen Herbert. *Pictographic Inscriptions from Jemdet Nasr.* (Oxford Editions of Cuneiform Texts, VII.) London: Oxford University Press, 1928, VIII + 72 pp., 41 pls.

Contains the largest published collection of account tablets from the proto-Sumerian Jemdet Nasr period c. 3000 B.C. In particular, the sign list offers a discussion of the various types of number signs (numerational or metrological) in these early texts, with a preliminary assignment of values to the individual metrological units.

144. Allotte de la Fuÿe, François-Maurice. "Mesures agraires et calcul des superficies dans les textes pictographiques de Djemdet-Nasr." *Revue d'Assyriologie* 27 (1930), 65-71.

Shows that essentially the same metrological number system for area was in use during the Jemdet Nasr period as in, for instance, the much later Old Babylonian period.

145. *(S)* Scheil, Vincent. *Textes de comptabilité proto-élamites.* (Mémoires de la Mission Archéologique de Perse, XXVI.) Paris: Libraire E. Leroux, 1935, 14 pp., 65 pls.

One of several volumes in this series of memoirs published by Scheil and devoted to the publication of large collections of account tablets from the "proto-Elamite" period in Iran, which was contemporary with the Jemdet Nasr period in Mesopotamia. (The first proto-Elamite tablet, of considerable mathematical interest, was published by Scheil in the same series, *MAP* 2 [1900], without any attempt at that time, of course, to interpret the inscription.) This volume includes a discussion of an important "exercise tablet" concerned with the summing of a long sequence of very complicated numbers expressed in the proto-Elamite metrological system for capacity (grain measure).

146. Van der Meer, Petrus E. "Dix-sept tablettes semi-pictographiques." *Revue d'Assyriologie* 33 (1936), 185-187, 3 pls.

Gives an incorrect "proof" of the value 300 *sila* for one of the units in the metrological number system for capacity in the texts from the Jemdet Nasr period. For several decades, until the mistake finally was corrected, it was impossible to understand the mathematical contents of all proto-Sumerian and proto-Elamite texts in which the capacity system is involved, i.e., nearly all texts of mathematical interest.

147. *(S)* Falkenstein, Adam. *Archaische Texte aus Uruk.* (Ausgrabungen der Deutschen Forschungsgemeinschaft in Uruk-Warka, 2.) Berlin: Kommissions-Verlag O. Harrassowitz, 1936, 76 + 216 pp., 71 pls.

Contains the largest published collection of account tablets from the proto-Sumerian Uruk IV period, which preceded the Jemdet Nasr period by a few hundred years. The sign list offers a discussion of the numerational and metrological number signs used in these texts, essentially identical with the number signs of the Jemdet Nasr type texts (of which a nice collection is published in this volume, too).

148. ———. "Archaische Texte des Iraq-Museums in Bagdad." *Orientalistische Literatur-Zeitung* 40 (1937), 402-410.

Offers a thorough discussion of a proto-Sumerian "bread-and-beer-text" from the Jemdet Nasr period, which contains a quite complicated series of computations involving both the proto-Sumerian "proto-sexagesimal" system and the proto-Sumerian metrological system for capacity. Although Falkenstein's conclusions concerning the meaning and the structure of the text are correct, his interpretations of the computations are based on van der Meer's mistaken analysis of the capacity system and therefore quite misleading. (Cf. Friberg, item 158.)

149. Vogel, Kurt. "Ist die babylonische Mathematik sumerisch oder akkadisch?" *Mathematische Nachrichten* 18 (1958), 377-382.

Discusses the evidence supporting the very reasonable conjecture that much of the mathematical corpus of the Old Babylonian period was inherited from the Sumero-Akkadian predecessors of the Babylonians, in spite of the fact that very few Sumerian or Akkadian mathematical problem texts or mathematical tables have been preserved to our time.

150. *(G)* ———. *Vorgriechische Mathematik*, II: *Die Mathematik der Babylonier*. Hannover: H. Schroedel, 1959, 93 pp.

A comprehensive and clear account of all known aspects of Sumero-Babylonian mathematics, in some parts based on the author's own contributions in a number of smaller papers.

151. *(G)* Vaĭman, A.A. *Šumero-vavilonskaya matematika III-I tysyačeletiya do n.e.* (Sumero-Babylonian mathematics of the third to first millennia B.C.). Moscow: Izdatel'stvo Vostočnoĭ Literatury, 1961, 279 pp.

Another comprehensive and clear account of Sumero-Babylonian mathematics. Contains, in particular, two chapters on Sumero-Babylonian number systems and metrology. An interesting Appendix presents an important text from the Hermitage suggesting that the Babylonian mathematicians were familiar with "Heronic triangles." The Appendix also discusses some important fragments of mathematical tables in Sachs 1955 (cf. Aaboe 1965, item 192).

152. Edzard, Dietz Otto. "Eine altsumerische Rechentafel (OIP 14,70)." *Lišān mithurti, Festschrift W. Freiherr von Soden zum 19. VI. 1968 gewidmet*. Edited by W. Röllig. (Alter Orient und Altes Testament, 1.) Neukirchen-Vluyn: Neukirchener Verlag (Butzon und Kevelaer), 1969, 101-104.

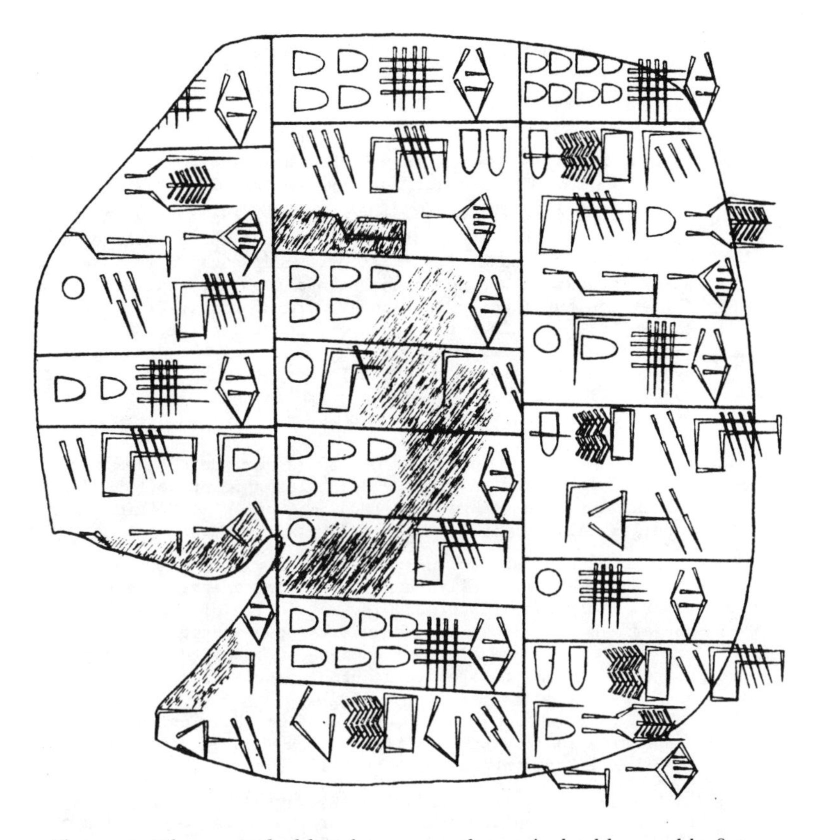

Figure 4. The second oldest known mathematical table, a table for the areas of small squares, from about 2400 B.C., and published by D.O. Edzard in 1969 (item 152). The text, a small clay tablet, belongs to the Oriental Institute in Chicago. The table starts with a "1 cubit square = 1 SA mana 15 gin," and goes on to "10 cubits square = ⅔ SAR 2 gin − 1 SA mana," where the areas of the squares are expressed in the complex Sumerian system of area measures with, in particular, $(12 \text{ cubits})^2$ = 1 SAR = 60 gin; 1 gin = 3 SA mana; 1 SA mana = 60 (small) gin.

Describes the second oldest known mathematical table, a table for the areas of small squares, expressed in the complicated Sumerian metrological system for areas. (From Adab, about 2400 B.C.)

153. Powell, Marvin A. "Sumerian Area Measures and the Alleged Decimal Substratum." *Zeitschrift für Assyriologie* 62 (1972), 165-221.

Contains, in particular, an interesting chapter on the "development and interrelation of surface units" and a table of third millennium area notations. Part of the discussion is based on Sumerian texts concerned with various forms of grain allotment for agricultural work. In a number of informative footnotes, the author criticizes the ideas of, in particular, Neugebauer 1927, item 142, with regard to the origin of the sexagesimal system, and he also rejects the incorrect evaluation in van der Meer 1936, item 146, of the units in the capacity system of the Jemdet Nasr period.

154. Limet, Henri. *Etude de documents de la période d'Agadé.* (Bibliothèque de la Faculté de Philosophie et Lettres de l'Université de Liège, 206.) Paris: Société d'Editions Les Belles Lettres, 1973, 67 pp., 10 pls.

Contains a presentation and preliminary discussion of four short but mathematically very interesting exercise texts of a very early date (about 2200 B.C.). In one of these texts appears what is probably the earliest documented deliberate use of "regular" sexagesimal numbers in order to get an exactly solvable mathematical exercise problem.

155. Powell, Marvin A. "The Antecedents of Old Babylonian Place Notation and the Early History of Babylonian Mathematics." *Historia Mathematica* 3 (1976), 417-439.

In this extremely important paper the author tries "to show that Babylonian place notation, far from being a creation of the Old Babylonian period (ca. 2000-1600 BC), actually has roots deep in the third millennium and was, in fact, invented before the end of the Third Dynasty of Ur (ca. 2100-2000 BC)." He also claims that "the origins of Babylonian mathematics can now be traced back to the middle of the third millennium BC." In addition to a rich documentation in support of these claims, the paper contains a number of historically interesting notes and remarks. See also item 1855.

156. Schmandt-Besserat, Denise. "The Earliest Precursor of Writing." *Scientific American* 238:6 (1978), 50-59.

One of several papers in which the author presents evidence of the existence in the Middle East, as early as from the ninth millennium and onwards, of a widespread system of counting by use of clay tokens. The author also argues that from the similarity of the tokens to certain proto-Sumerian number signs and signs for commodities there must have been a close link between the use of such tokens and the invention of writing.

157. Friberg, Jöran. "The Third Millennium Roots of Babylonian Mathematics, I: A Method for the Decipherment, through Mathematical and Metrological Analysis, of Proto-Sumerian and Proto-Elamite Semi-Pictographic Inscriptions." Department of Mathematics, CTH-GU (Göteborg, Sweden) 1978, 56 pp. Preprint.

Gives the first complete description of the numerational and metrological number systems in use in the still largely undeciphered account texts from the Jemdet Nasr period in Mesopotamia and from contemporary Iran. The texts considered show examples of quite complicated computations and yield important information about the origin of the sexagesimal and decimal number systems, and about the origin of some characteristic traits in Babylonian mathematics.

158. ———. "The Early Roots of Babylonian Mathematics, II: Metrological Relations in a Group of Semi-Pictographic Tablets of the Jemdet Nasr Type, Probably from Uruk-Warka." Department of Mathematics, CTH-GU (Göteborg, Sweden) 1979, 80 pp. Preprint.

Contains, in particular, a detailed discussion of the important proto-Sumerian "bread-and-beer-text" considered in Falkenstein 1937, item 148, a text which yields additional confirmation of the results in Friberg 1978, item 157, and which, because of its complexity, is of considerable interest in its own right.

Babylonian Mathematics

In 1851, H.C. Rawlinson published his *Memoir on the Babylonian and Assyrian Inscriptions*, and in 1857 Rawlinson, Talbot, Hincks, and Oppert independently of each other translated an inscription on a clay cylinder of the Assyrian king Tiglath Pileser I, an event which marks the beginning of Assyriology. Right from the start mathematical and metrological questions attracted considerable attention among the Assyriologists.

159. Rawlinson, Henry Creswicke. "Notes on the Early History of Babylonia." *Journal of the Royal Asiatic Society* 15 (1855), 217, n. 4.

Contains the following statement: "that the Babylonians did really make use both of the centesimal and sexagesimal notation, as stated by Berosus, is abundantly proved by the monuments; and from the same sources we can illustrate the respective uses of the *Sarus*, the *Nerus*, and *Sossus* in the calculations of the higher numbers." Goes on to verify this claim by presenting "the concluding portion of a table of squares, which extends from 1 to 60" (on one of the "tablets from Senkereh" excavated by Loftus a few years earlier, BM 92 680, see Neugebauer, item 174, I, 71, n. 2; II, 3). See also Wieleitner 1930, item 167.

160. Lepsius, Karl Richard. "Die Babylonisch-Assyrischen Längenmasse nach der Tafel von Senkereh." *Abhandlungen der Königlichen Akademie der Wissenschaften zu Berlin, Phil.-Hist. Klasse* (1877), 105-144, 2 pls.

Contains a discussion of a "metrological list" on a second tablet from Senkereh, as well as some important remarks concerning the absence of the sexagesimal point and any special signs for final or intermediate zeros in the Babylonian sexagesimal notation. The author also correctly identifies the special cuneiform signs for *šar* (60 x 60) and *ner* (10 x 60).

161. Oppert, Jules. "Sechshundert drei und fünfzig. Eine babylonische magische Quadrattafel." *Zeitschrift für Assyriologie* 17 (1903), 60-74.

Gives an entirely unfounded "mystic-cabbalistic" interpretation of an admittedly far from trivial mathematical text from Scheil's excavations at Sippar, published in 1894. See H.V. Hilprecht, item 140, p. 25.

162. Scheil, Vincent. "Les tables 1 igi x gal-bi, etc." *Revue d'Assyrologie* 12 (1915), 195-198.

Uses the example of a table of reciprocals "not later than the time of Hammurabi" to refute Hilprecht's hypothesis about the "Number of Plato" and to describe the real character of a table of reciprocals.

163. Weidner, Ernst F. "Die Berechnung rechtwinkliger Dreiecke bei den Akkadern um 2000 v. Chr." *Orientalistische Literatur-Zeitung* 19 (1916), 257-263.

Gives for the first time a correct mathematical interpretation of an Old Babylonian mathematical problem text (VAT 6598), thus opening the way for a better understanding of the special terminology used in the mathematical cuneiform texts. The text in question is concerned with two different methods for the approximative solution of the problem to find the length of the diagonal in a rectangle. It is not claimed here that the method presupposes the knowledge of the "Pythagorean theorem." (Cf., however, Neugebauer, "Zur Geschichte des Pythagoräischen Lehrsatzes." *Nachrichten von der Gesellschaft der Wissenschaften zu Göttingen, Math.-Phys. Klasse* [1927], 45-48.)

164. Lutz, Henry Frederick. "A Mathematical Cuneiform Tablet." *American Journal of Semitic Languages and Literature* 36 (1920), 249-257.

Presents and analyzes a combined multiplication table. In the transcription of the text the author still makes use of decimal notation and common fractions instead of a straightforward rendering of the sexagesimal notation of the original.

165. Neugebauer, Otto. "Zur Geschichte der babylonischen Mathematik." *Quellen und Studien zur Geschichte der Mathematik, Astronomie und Physik* B1:1 (1929), 67-80.

Discusses a number of unusually sophisticated geometric problems on six Old Babylonian tablets published in C. Frank, *Strassburger Keilschrifttexte*, 1928. Some of the problems lead to quadratic equations, but these equations are not solved in the text. However, the author already announces here the result which will be published later in Schuster 1931, item 169, namely, that the Old Babylonian mathematicians were in possession of a formula for the solution of quadratic equations.

166. ———, and Vasilij Vasil'evič Struve. "Über die Geometrie des Kreises in Babylonien." *Quellen und Studien* B1:1 (1929), 81-92.

Considers some of the problems on the important text BM 85 194 (published by the British Museum in 1900). Among the findings are Babylonian approximating formulas for the circumference and area of the circle ($\pi \simeq 3$), and for the volume of a truncated cone. In addition, a correct formula for the length of the chord of a circle implies that Babylonian mathematicians were familiar with the Pythagorean theorem.

167. Wieleitner, Heinrich. "Zur Geschichte der Entdeckung des babylonischen Sexagesimalsystems." *Historische Studien und Skizzen zu Natur- und Heilwissenschaft, Festgabe Sticker*. Berlin: Springer, 1930, 11-17.

Contains some historical notices about the respective roles of British, French, and German scholars in the earliest studies of the Babylonian sexagesimal system.

168. Neugebauer, Otto. "Sexagesimalsystem und babylonische Bruchrechnung." I: *Quellen und Studien* B1:2 (1930), 183-193; II: B1:4 (1931), 452-457; III: B1:4 (1931), 458-463; IV: B2:2 (1933), 199-210.

Parts I and II: Here it is made clear, with departure from the example of the combined multiplication table in Lutz 1920, item 164, how Babylonian tables of reciprocals and combined multiplication tables were constructed. It is shown that, with very few exceptions, the "head numbers" of multiplication tables are identical with the reciprocals of suitably restricted "regular" sexagesimal numbers, and that the numbers appearing in the first column of the reciprocal tables are by necessity just such regular numbers. Part III: Considers more unusual types of tables of reciprocals. Part IV: Contains a detailed, although not quite convincing, explanation of how the enormous table of reciprocals AO 6456 (originally published by Thureau-Dangin in *Tablettes d'Uruk*, Paris, 1922) may have been constructed.

169. Schuster, Hans-Siegfried. "Quadratische Gleichungen der Seleukidenzeit aus Uruk." *Quellen und Studien* B1 (1931), 194-200.

It is established here for the first time without doubt that the Babylonian mathematicians of the Seleucid period (the last three centuries B.C.) as well as of the Old Babylonian period were in possession of correct methods for finding the (positive) roots of general quadratic equations.

170. Vogel, Kurt. "Zur Berechnung der quadratischen Gleichungen bei den Babyloniern." *Unterrichtsblatt für Mathematik und Naturwissenschaften* 39 (1933), 76-81.

Introduces the important concept of a Babylonian "normal form" for quadratic equations, and demonstrates the irrelevance of the question of whether the Babylonian mathematicians were aware of the existence of the "second root" of a quadratic equation.

171. (*G*) Neugebauer, Otto. *Vorlesungen über Geschichte der antiken mathematischen Wissenschaften, I: Vorgriechische Mathematik.* Berlin: Springer, 1934, 240 pp.

Published in connection with the editing of volume I of Neugebauer, item 174. Contains, in addition to a chapter on Babylonian mathematics in general, three particularly interesting chapters on "Babylonian techniques of computation," "General history, language, and writing," and "Number systems."

172. Thureau-Dangin, François. "La mesure des volumes d'après une tablette inédite du British Museum." *Revue d'Assyrologie* 32 (1935), 1-28.

Presents and discusses the big combined problem text BM 85 196, which contains, apart from a number of stereometric problems, a certain "cane-against-a-wall-problem," known also from a text from the Seleucid period (BM 34 568, see Neugebauer, item 174, III), and which contains an explicit application of the Pythagorean theorem.

173. Neugebauer, Otto. "Der Verhältnisbegriff in der babylonischen Mathematik." *Analecta Orientalia* 12 (1935), 235-258.

Considers, in particular, a group of 34 equations on YBC 4712, the 13th tablet of a "series text," which gives important new insights into Babylonian mathematical terminology.

174. (*S*) ———. "Mathematische Keilschrift-Texte." *Quellen und Studien* A 3. Issued in separate volumes, I, II: Berlin: Springer, 1935; III: Berlin: Springer, 1937. Reprinted Berlin, Heidelberg, New York: Springer, 1973, 516 + 64 + 83 pp., 69 + 6 pls.

The classical work in the field. Presents and discusses all mathematical cuneiform texts available to the author at the time of publication, many of them published here for the first time. Complete with photographs and autographs of the more important new texts.

I: Chapter 1 gives a systematic survey of all known Babylonian mathematical table texts; the following chapters are devoted to mathematical problem texts or series texts from the Louvre, the British Museum, the museums of Berlin, the Yale Babylonian Collection, etc. II: Contains indispensable photographs and autographs, a text concordance, a bibliography, a date list, Sumerian and Akkadian vocabularies, etc., and a few new texts. III: Contains some complementary material and an interesting "Rückblick" (postscript).

175. Thureau-Dangin, François. "L'équation du deuxième degré dans la mathématique babylonienne d'après une tablette inédite du British Museum." *Revue d'Assyrologie* 33 (1936), 27-48.

Presents and discusses BM 13 901, an early Old Babylonian mathematical problem text, interesting because of its obvious close connection with the (somewhat younger) so-called "series texts" in which, however, no solutions are given to the systematically enumerated problems.

176. Gandz, Solomon. "The Origin and Development of the Quadratic Equations in Babylonian, Greek, and Early Arabic Algebra." *Osiris* 3 (1937), 405-557.

This important comparative study is based solely on Neugebauer, item 174, and Thureau-Dangin, item 175.

177. (*S*) Thureau-Dangin, François. *Textes mathématiques babyloniens.* (Ex Oriente Lux, I.) Leiden: J. Brill, 1938, 243 pp.

Duplicates more or less the presentation in Neugebauer, item 174, but contains many improvements of linguistic and other details.

178. (*S*) Neugebauer, Otto, and Abraham Joseph Sachs. *Mathematical Cuneiform Texts*. New Haven, Conn.: American Oriental Society and American Schools of Oriental Research, 1945, 177 pp., 49 pls.

An important continuation and complement to Neugebauer, item 174, mainly occupied with cuneiform mathematical texts from American collections. The first chapter contains interesting information about Babylonian metrology and examples of metrological computations; the second chapter presents some new types of mathematical table texts. Chapter 3, Problem texts, contains in particular a presentation of the important tablet "Plimpton 322," called "the oldest preserved document in ancient number theory." Chapter 4, by A. Goetze, "The Akkadian Dialects of the Old-Babylonian Texts," is devoted to a study of the provenance of mathematical cuneiform texts from clandestine excavations (the majority of all mathematical cuneiform texts). The volume ends with indexes, a vocabulary, etc.

179. Sachs, Abraham Joseph. "Babylonian Mathematical Texts," published in three parts: I: "Reciprocals of Regular Sexagesimal Numbers." *Journal of Cuneiform Studies* 1 (1947), 219-240; II: "Approximations of Reciprocals of Irregular Numbers in an Old-Babylonian Text"; III: "The Problem of Finding the Cube Root of a Number." *Journal of Cuneiform Studies* 6 (1952), 151-156.

I: A most interesting analysis of three important texts (two of them from Hilprecht's excavations in Nippur around the turn of the century) making it clear how the Babylonian mathematicians could compute the reciprocals of many-place regular numbers by use of an iterative algorithm. II: Presents a tablet with a unique Babylonian table of reciprocals of non-regular numbers. The table contains information about what may be the sign of the errors in the round-offs. III: Presents a text with a badly confused example of the applications of a factorization method for finding the cube root of a given number.

180. Gandz, Solomon. "Studies in Babylonian Mathematics, I: Indeterminate Analysis in Babylonian Mathematics." *Osiris* 8 (1948), 12-40.

Makes an interesting comparison of some parallels in Diophantus to a number of problems in Babylonian mathematics leading to indeterminate equations of the type $x^2 - y^2 = a$ or $x^2 - y^2 = y^2 - z^2$. The author concludes: "There seems to be more depth and significance in Babylonian mathematics than heretofore imagined" and "Babylonian geometry is algebraic and their algebra extends its roots deep into geometry and mensuration."

181. *(S)* Baqir, Taha. "An Important Mathematical Problem Text from Tell Harmal." *Sumer* 6 (1950), 39-54, 3 pls.; "Another Important Mathematical Text from Tell Harmal," 130-148, 5 pls.; "Some More Mathematical Texts," 7 (1951), 28-45, 13 pls.

Contains the presentation and preliminary discussion of eleven newly excavated mathematical texts from the Old Babylonian period, of considerable mathematical interest in themselves, and even more interesting because they are among the first Babylonian mathematical texts with a known provenance, and therefore possible to date with some precision. One of the tablets, IM 55357, is devoted to a problem involving a decreasing sequence of similar right triangles obtained by means of a recursive procedure.

182. (*G*) Neugebauer, Otto. *The Exact Sciences in Antiquity*. Item 91.

Treats Babylonian mathematics and astronomy, in particular investigating the possibility of Babylonian influence on Egyptian and Greek or Hellenistic scientific achievements. See also item 245.

183. Bruins, Evert M. "Revision of the Mathematical Texts from Tell Harmal." *Sumer* 9 (1953), 241-253.

Comments on the publications of T. Baqir, item 181.

184. ———. "Three Geometrical Problems." *Sumer* 9 (1953), 255-259; "Some Mathematical Texts," 10 (1954), 55-61.

Contains an interesting discussion of some unpublished mathematical tablets in the Iraq Museum in Baghdad.

185. Vaĭman, A.A. "Ermitažnaya klinopisnaya matematičeskaya tablička No 015189" (The cuneiform mathematical tablet No. 015189 from the Hermitage). *Epigrafika vostoka* 10 (1955), 71-83.

Presents a curious tablet from the collections in Leningrad, with a sequence of identical geometric drawings illustrating the use of similarity in setting up a series of geometrical problems from a single set of data. The basis problem is a "trapezoid partition problem" (cf. Neugebauer 1929, item 165, and Gandz 1948, item 180), and the solutions are what the author calls "Babylonian numbers" (rational solutions of a certain indeterminate equation, closely related to the more well-known Pythagorean equation).

186. Vygodskiĭ, M.Y. "Proishoždenie znaka nulya v vavilonskoĭ numeracii" (The origin of the zero sign in Babylonian numeration). *Istoriko-Matematicheskie Issledovaniia* 12 (1959), 393-420.

Contains an interesting and thorough discussion of documentations of the use of special signs for the intermediate zero in Babylonian mathematics.

187. Saggs, Henry William Frederick. "A Babylonian Geometrical Text." *Revue d'Assyrologie* 54 (1960), 131-145.

Publishes an important join to the unique and very curious text BM 15 285 in Neugebauer, item 174, with its series of drawings to geometrical problems concerned with the cutting up of squares into geometric figures of smaller area.

188. (*S*) Bruins, Evert M., and Marguerite Rutten. *Textes mathématiques de Suse*. (Mémoires de la Mission Archéologique en Iran, XXXIV.) Paris: P. Geuthner, 1961, 136 pp., 39 + 12 pls.

Presents and discusses 26 mathematical tablets from Susa (the Old Babylonian period), several of unusual sophistication. In addition to being of considerable mathematical interest in itself, this collection proves that the knowledge of "Babylonian mathematics" was not confined to Babylonia alone. Of particular interest are two catalogue texts, a text concerned with an "iterated trapezoid partition problem," and some texts where computations with regular polygons are involved. Complete with photographs, autographs, and a vocabulary of mathematical terms.

189, (*G*) Van der Waerden, Bartel Leendert. *Science Awakening, I*. 2nd ed. in English translation. Leyden: Noordhoff International Publishers, and New York: Oxford University Press, 1961, 37-45, 62-81.

A brief but well-written introduction to the subject of Babylonian mathematics, accompanied by plates with photographs of some of the more important Babylonian mathematical tablets.

190. (*G*) Bruins, Evert M. "Interpretation of Cuneiform Mathematics." *Physis* 4 (1962), 277-371.

Gives the author's view of Babylonian mathematics by summing up what he has written in a number of shorter articles on the subject, with special attention to the tablets from Susa in Bruins and Rutten 1961, item 188.

191. Baqir, Taha. "Tell Dhiba'i: New Mathematical Texts." *Sumer* 18 (1962), 11-14, 3 pls.

Presents a newly excavated text, with an explicit application of the Pythagorean theorem.

192. Aaboe, Asger. "Some Seleucid Mathematical Tables (Extended Reciprocals and Squares of Regular Numbers)." *Journal of Cuneiform Studies* 19 (1965), 79-86.

An interesting discussion of a class of tablets and fragments featuring tables of reciprocals and squares of many-place regular sexagesimal numbers related to the parade example AO 6456 (Neugebauer 1933, item 168). The fragments were originally published in A.J. Sachs, *Late Babylonian Astronomical and Related Texts, copied by T.G. Pinches and J.N. Strassmaier* (Providence, 1955). See also Vaĭman 1961, item 151.

193. (G) Goetsch, H. "Die Algebra der Babylonier." *Archive for History of Exact Sciences* 5 (1968-1969), 79-152.

A systematic survey of texts of algebraic content in Neugebauer 1935-1937, item 174, Thureau-Dangin 1938, item 177, and Neugebauer and Sachs 1945, item 178.

194. Friberg, Jöran. "Methods and Traditions of Babylonian Mathematics, I: Plimpton 322, Pythagorean Triples, and the Babylonian Triangle Parameter Equations." *Historia Mathematica* 8 (1981), 277-318.

Tries to explain the method of construction of the tables on Plimpton 322 by reference to the Babylonian factorization method which is based on the use of regular sexagesimal numbers. Also discusses the importance of the "cane-against-a-wall-problem" (see Thureau-Dangin 1935, item 172) as an exactly solvable problem, and the way in which the Pythagorean theorem may originally have been discovered.

195. ———. "Methods and Traditions of Babylonian Mathematics, II: An Old Babylonian Catalogue Text with Equations for Squares and Circles." *Journal of Cuneiform Studies* 33 (1981), 57-64.

Gives an analysis of a text originally published as a mathematical text "dealing with areas" (BM 80 209), but which is really a systematic list of linear and quadratic equations.

Personalia

196. Dhorme, Edouard. *Hommage à la mémoire de l'éminent assyriologue François Thureau-Dangin (1872-1944)*. (Ex Oriente Lux, 8.) Leiden: E.J. Brill, 1946, 35 pp.

Contains, in particular, information about the two books and about 50 articles of various lengths which were published by Thureau-Dangin between the years 1896 and 1940 and which deal with questions concerning Babylonian and Sumerian mathematics or metrology.

197. Schneider, Ivo. "Ein Leben für die Wissenschaftsgeschichte: Kurt Vogel." In *Beiträge zur Geschichte der Arithmetik*. Edited by K. Vogel. Munich: Minerva Publications, 1978, 7-18, 88-96, 3 pls.

In the bibliography to this biographical article are listed more than 80 books and articles in the history of ancient mathematics field, many of them about various aspects of Babylonian mathematics, written between the years 1907 and 1978. Published together with the biographical article is a short paper by Vogel, "Das Fortleben babylonischer Mathematik bei den Völkern des Altertums und Mittelalters" (pp. 19-34).

198. Sachs, Janet, and G.J. Toomer. "Otto Neugebauer, Bibliography, 1925-1979." *Centaurus* 22 (1979), 258-280, 1 pl.

Lists close to 300 books and articles written by Neugebauer between 1925 and 1979 concerned with the history of mathematics, many of them on Babylonian mathematics or astronomy. (Neugebauer's works on Babylonian astronomy have for lack of space not been mentioned in this annotated bibliography. See, for instance, O. Neugebauer, *A History of Ancient Mathematical Astronomy*, item 246.)

GREEK MATHEMATICS

This bibliography treats two aspects of the extensive literature that has been produced on the history of ancient Greek mathematics: roughly half of the entries record either the best available editions of the texts themselves or reliable translations of these into a modern language, while the other half lists histories and studies of various problems or methods of ancient Greek mathematics.

With regard to the editions and translations of the texts, available space precludes even the listing of one translation of each known Greek mathematical work; here the aim has been a comprehensive coverage of only the major treatises with a representative selection of the minor ones. In the case of secondary literature, emphasis has been given to works in English of which the researcher should be aware. From these, studies have been chosen that treat areas where considerable amounts of work still remain to be done, such as the history of numerical methods in Greek mathematics, ancient methods of dealing with arcs of circles on the sphere, the development of geometry that led up to the composition of Euclid's *Elements* (including the still-murky history of incommensurables and the theory of proportions), infinitesimal methods in ancient Greece, and the many problems still surrounding Archimedes' works.

Thus much of the work undertaken by older generations of scholars such as Bretschneider, Hankel, Tannery, and Zeuthen has been omitted, as well as a certain amount of modern European work. However, the reader who pursues the references given here will rapidly discover which of those works are important for his purposes. Finally, the reader will also want to search through the recent literature for further contributions in such periodicals as *Archive for History of Exact Sciences*, *Archives Internationales d'Histoire des Sciences*, *Centaurus*, and *Historia Mathematica*.

General Reference Works

199. Archibald, Raymond Clare. *Outline of the History of Mathematics*. Item 61.

Contains a 13-page summary of Greek mathematics and a particularly valuable eight-page bibliography on the subject.

200. Hammond, N.G., and H.H. Scullard, eds. *Oxford Classical Dictionary*. 2nd ed. New York: Oxford University Press, 1970.

Gives good coverage of Greek mathematics (especially biographical aspects) and incorporates results of recent research.

201. Mugler, C. *Dictionnaire historique de la terminologie géométrique des Grecs*. Paris: C. Klincksieck, 1958-1959. 2 vols.

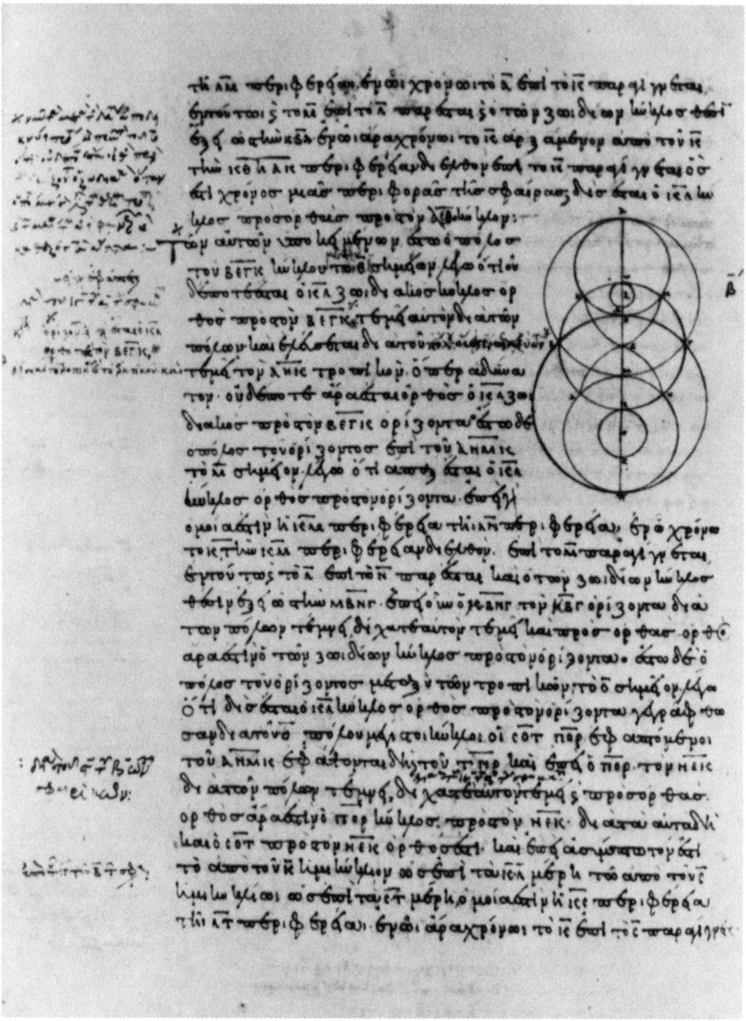

Figure 5. A page from Euclid's *Phenomena* in the manuscript Vat. gr. 204, fol. 61v. The figure with the text represents the celestial sphere according to the principle that one circle represents the horizon, anything inside it belongs to the visible hemisphere of the heavens, and anything outside it belongs to the invisible hemisphere. Some scholars believe that the mathematical expression of this principle led to stereographic projection. Further details on pp. 751–755 of item 246.

Records the changing meanings of Greek geometrical terms by supplying, along with definitions, examples of usage from such writers as Euclid, Archimedes, Apollonios, and Pappos, as well as Plato and Aristotle.

202. *Paulys Real-Encyclopädie der classischen Altertums-Wissenschaft*. Edited by G. Wissowa et al. Stuttgart: A. Druckenmüller, 1958. 34 vols. + 15 vols. of Supplement.

This monumental work (known as Pauly-Wissowa) was published from 1894 to 1978 and its various articles represent a scholarly introduction to what is known (and conjectured) about every aspect of classical antiquity. While its emphasis is on biography there are also short topical articles. The user should know the following: Volumes 1-24 were published from 1894 to 1963 and cover the letters A-Q. Beginning in 1914 a "Second Series" was published, completed in 1972, covering the letters R-Z. The 15-volume supplement, published from 1903 to 1978, incorporates more recent research relevant to the various articles, and this, with its index, must be consulted if one wants a reasonably recent view of many of the subjects.

203. *Der kleine Pauly*. Edited by K. Ziegler and W. Sontheimer. Stuttgart: A. Druckenmüller, 1964. 5 vols.

This is an abridged version of item 202.

204. Sezgin, Fuat. *Geschichte des arabischen Schrifttums*, Vols. 5 (Mathematics) and 6 (Astronomy). Leiden: E.J. Brill, 1974 (Vol. 5), and 1978 (Vol. 6).

These two volumes are an indispensable source for the study of Arabic manuscripts and printed editions of Greek mathematical and astronomical works (some of which are no longer extant in Greek versions), as well as for translations and studies of these in a variety of languages. See also items 27, 272.

See also items 11, 13, 14, and 17.

Source Materials

Anthologies

205. Bulmer-Thomas, Ivor. *Selections Illustrating the History of Greek Mathematics*. Cambridge, Mass.: Harvard University Press, 1941. 2 vols. Vol. I revised 1980.

A useful anthology of Greek mathematical texts with English translations in which one may find important passages not easily available elsewhere, and a great help to a person trying to learn to read mathematical Greek.

206. Cohen, M.R., and I.E. Drabkin. *A Source Book in Greek Science*. McGraw-Hill, 1948.

Practically all the branches of Greek mathematics are well represented by English translations of the relevant texts.

Individual Mathematicians

Apollonios

207. Heath, T.L. *Apollonius of Perga. Treatise on Conic Sections, Edited in Modern Notation with Introductions including an Essay on the Earlier History of the Subject*. Cambridge: The University Press, 1896. Reprinted New York: Dover and Cambridge: W. Heffer, 1961.

The only available English version of the seven surviving books of *The Conics*, but in no sense a translation. Related propositions have been grouped into one, the order of the propositions has been altered and "modern notation" (as of 1896) has been used throughout. Still a useful work.

208. Heiberg, J.L. *Apollonii Pergaei quae Graece exstant cum commentariis antiquis*. Leipzig: Teubner, 1891, 1893. 2 vols.

The principal edition of the Greek text, with Latin translation, of Books I-IV of Apollonios's *Conics* and the ancient commentaries. Since Eutocios's commentary on Apollonios has not been translated into a modern language the reader with no Greek must rely on Heiberg's Latin translation.

209. Nix, L.M. Ludwig, ed. and tr. *Das Fünfte Buch der Conica des Apollonius von Perga in der arabischen Übersetzung des Thabit ibn Corrah*. Leipzig: W. Drugulin, 1889.

Contains an introduction and the Greek definitions to the fifth book of Apollonios's *Conics*, together with an Arabic text of the book that breaks off in the middle of Proposition 6 and a German translation that ends in the middle of Proposition 7.

210. Ver Eecke, Paul. *Les coniques d'Apollonius du Perge ... avec une introduction et des notes*. Bruges: Desclée, De Brouwer et Cie., 1923.

Provides as literal a translation as possible of the seven extant books of *The Conics* as well as a scholarly introduction summarizing the work, citing the ancient testimonies on Apollonios and recounting the later history of the theory of conic sections.

Archimedes

211. Dijksterhuis, E.J. *Archimedes*. Copenhagen: Ejnar Munksgaard, 1956.

An edition and study of Archimedes' known works in which theorems are translated literally while a special notation, avoiding some of the implications of algebraic symbolism, is used to present paraphrases of the proofs. Subsidiary lemmas in the various treatises are grouped together in one chapter. Two chapters on the life of Archimedes and the tradition of the manuscripts open this important study.

212. Dold-Samplonius, Y., H. Hermelink, and M. Schramm. *Archimedes, Über einander berührende Kreise. Aus dem Arabischen übersetzt und mit Anmerkungen versehen.* Stuttgart: Teubner, 1975.

This is the fourth volume of the set *Archimedis Opera Mathematica* (see item 214). Dold-Samplonius has published a study of this treatise in *Sudhoffs Archiv* 57 (1973), Heft 1, 15-40, in which she concludes that it is a part of a larger Archimedean work *On Circles*.

213. Heath, T.L. *The Works of Archimedes with the Method of Archimedes.* Cambridge: The University Press, 1897 and 1912. Reprinted New York: Dover, n.d.

Presents the mathematical arguments in a symbolism that would have been familiar to a 19th-century mathematician, but still a standard source. The useful introductory chapters on "Arithmetic in Archimedes," "On the Problems Known as ΝΕΥΣΕΙΣ," and "The Terminology of Archimedes" must be supplemented by a study of more recent works on these topics.

214. Heiberg, J.L. *Archimedis opera omnia cum commentariis Eutocii.* 2nd ed. Leipzig: Teubner, 1910-1915. 3 vols.

This is the standard edition of the Greek text, with Latin translation, of Archimedes' works and is the one most frequently cited. There is, however, a recent reprint of the text, with some corrections by E. Stamatis, published by Teubner (3 volumes), Stuttgart, 1972, and a four-volume edition of the text, with a French translation by C. Mugler, published by Les Belles Lettres, Paris, 1970-1972.

215. Ver Eecke, Paul. *Les oeuvres complètes d'Archimède, suivies des commentaires d'Eutocius d'Ascalon.* 2nd ed. Liege: Vaillant-Carmanne, 1960. 2 vols.

The most faithful of the modern translations, this is the only one which contains the commentaries of Eutocios of Ascalon, an important source for any serious study of Archimedes' works.

Aristarchos

216. Heath, T.L. *Aristarchus of Samos, the Ancient Copernicus, a History of Greek Astronomy to Aristarchus Together with Aristarchus' Treatise on the Sizes and Distances of the Sun and Moon. A New Greek Text with Translation and Notes.* Oxford: Clarendon Press, 1913.

Though we should have liked Aristarchos's treatise on the heliocentric system to have been preserved, it is, instead, this earlier treatise which has survived. Its use of geometrical methods to handle problems for which trigonometry was later employed is especially interesting.

Aristotle

217. Heath, T.L. *Mathematics in Aristotle*. Oxford: Clarendon Press, 1949. Reprinted 1970.

A collection of passages from the writings of Aristotle bearing on mathematical topics. See also item 2250.

Autolykos

218. Czwalina, A., tr. *Autolykos Rotierende Kugel und Aufgang und Untergang der Gestirne. Theodosios von Tripolis Sphaerik.* (*Ostwald's Klassiker*, No. 232.) Leipzig: Akademie Verlagsgesellschaft, 1931.

The first two works, by Autolykos of Pitane, are examples of Greek spherics contemporary with Euclid, and the two books composing the second of these works have been shown by O. Schmidt ("Some Critical Remarks about Autolycus' On Risings and Settings," *Den. 11. skandinaviske matematikerkongress i Trondheim 22-25 August, 1949*, 202-209) to be simply two versions of the same work. For the third work by Theodosios, see item 240. There is also an English translation, whose renderings cannot always be trusted, by F. Bruin and A. Vondjidis, *The Books of Autolykos, On a Moving Sphere and On Risings and Settings* (Beirut: A.U.B. Press, 1971).

Diocles

219. Toomer, G.J. *Diocles on Burning Mirrors: The Arabic Translation of the Lost Greek Original*. Berlin: Springer-Verlag, 1976.

Provides an English translation (with historical introduction) of Diocles' work on burning mirrors. Here are also bibliographical references to the other major ancient source on burning mirrors, the treatise of Anthemios of Tralles. See also item 308.

Diophantos

220. Heath, T.L. *Diophantus of Alexandria; A Study in the History of Greek Algebra*. Cambridge: The University Press, 1885. Reprinted New York: Dover, 1964.

Really a historical study of the methods and influence of Diophantos's work, the book closes with an 80-page appendix abstracting the problems and solutions in six books of the *Arithmetica* and the treatise *On Polygonal Numbers*. The reprint is of the second edition (1910), which contains a supplement on the contributions of Fermat and Euler to Diophantine problems.

220a. Sesiano, J. *Books IV to VII of Diophantos' Arithmetica in the Arabic Translation of Qusṭā ibn Lūqā*. New York: Springer Verlag, 1982.

This is an English translation, together with an edition of the Arabic text, of what most scholars now regard as Books IV-VII of Diophantos's *Arithmetica*. One result is that we now have 10 of the 13 books of this work, and what Tannery thought were Books IV-VI are, in fact, from the last six books of the *Arithmetica*. Accompanying the text are commentaries of considerable historical, philosophical, and mathematical interest.

221. Tannery, Paul. *Diophanti Alexandrini opera omnia cum Graecis commentariis*. Leipzig: Teubner, 1893-1895. 2 vols.

This is the standard edition of the Greek text of what prior to very recent times seemed to be the first six books of Diaphantos's *Arithmetica*, by a distinguished French scholar. It also includes Diophantos's *On Polygonal Numbers*.

222. Ver Eecke, Paul. *Diophante d'Alexandrie*. Paris: Albert Blanchard, 1959. Reprinted from the original edition of 1926.

A careful French translation of Diophantos's extant works, both the *Arithmetic* and *On Polygonal Numbers*.

Euclid

223. Archibald, Raymond Clare. *Euclid's Book on Divisions of Figures: A Restoration Based on Woepcke's* [Arabic] *Text and on the Practica Geometriae of Leonardo Pisano*. Cambridge: The University Press, 1915.

The statements of the 36 propositions as well as the proofs of four of them were found by Fr. Woepcke in an Arabic treatise. The Greek text is lost. Archibald restored the remaining proofs on the basis of a medieval Latin work.

224. Chasles, Michel. *Les trois livres de Porismes d'Euclide*. Paris: Mallet-Bachelier, 1860.

A reconstruction of a lost Euclidean treatise, by a leading 19th-century French geometer, based on references by Pappos, Proclos, Ibn al-Haytham, and others.

225. Heath, T.L. *The Thirteen Books of Euclid's Elements*. 2nd ed. Cambridge: The University Press, 1926. Reprinted New York: Dover, 1956. 3 vols.

A good English translation of this basic source, including an account of the spurious Books XIV and XV. Heath's notes on *The Elements* take into account all the important ancient and modern commentaries, and his chapters on various aspects of historical studies of *The Elements* are good preliminaries to more modern works.

226. Heiberg, J.L., and H. Menge, eds. *Euclidis opera omnia*. Leipzig: Teubner, 1883-1916. 8 vols. + Supplement.

Besides the Greek text with Latin translation of genuine and spurious works it gives the Latin translation by Gerard of Cremona of the commentary on the first ten books of *The Elements* written by al-Nayrīzī. This series is being re-issued, minus the Latin translation, but with notes incorporating those of Heiberg and recent research by E.S. Stamatis. The first volume, *Euclidis Elementa* I, Libri I-IV, was issued by Teubner (Leipzig, 1969).

227. Nokk, A. *Euklid's Phaenomene* (an Incomplete Translation and Notes). Beilage zum Freiburger Lyceums-Programm von 1850. Freiburg, 1850, 58 pp. + 3 pls.

One of the surviving works by Euclid, closely connected with *On the Moving Sphere* by Autolykos.

228. Thaer, C. *Die Data von Euklid*. Berlin: Springer, 1962.

A German translation of the Greek text of an influential Euclidean work.

229. Ver Eecke, Paul. *Euclide: L'Optique et la Catoptrique*. Paris: A. Blanchard, 1959.

Only the first of the two works mentioned is by Euclid, and this French translation contains, as well, Theon's recension of *The Optics*. The introduction surveys the history of Greek optics and the history of the text.

Heron

230. Bruins, Evert M., ed. *Codex Constantinopolitanus, Palatii Veteris No. 1*, Parts 1-3. Leiden: E.J. Brill, 1964.

The three parts of this work contain photographs of the codex, an edition of the Greek text, and a translation of the same. The work itself is a collection of the metrical writings of Heron of Alexandria.

231. Heiberg, J.L., et al. *Heronis Alexandrini Opera quae supersunt omnia*. Leipzig: Teubner, 1899-1914. 5 vols.

The principal edition of the Greek or Arabic texts of the known works of Heron with German translations, containing also numerous selections from other classical writers in the Heronic tradition.

Menelaos

232. Krause, M., tr. and ed. "Die Sphärik von Menelaos aus Alexandrien in der Verbesserung von Abū Naṣr Manṣūr b. ᶜAlī b. ᶜIrāq." *Abhandlungen der Gesellschaft der Wissenschaften zu Göttingen, philol.-hist. Klasse*, 3. F., 17 (1936). Issued separately, Berlin: Weidmannsche Buchhandlung, 1936.

Since the Greek text of Menelaos's *Spherics* no longer exists, this edition of the Arabic text, accompanied by a German translation and a thorough study of the history of the text among the Arabs, is especially valuable. See also item 315.

Nicomachus

233. D'Ooge, M.L., F.E. Robbins, and L.C. Karpinski. *Nicomachus of Gerasa: Introduction to Arithmetic*. New York: The Macmillan Co., 1926.

Contains an English translation of this late Greek text on number theory as well as studies of Greek arithmetic by the two last-named authors.

Pappos

234. Hultsch, F., ed. *Pappi Alexandrini Collectionis quae supersunt*. Berlin: Weidmann, 1876-1878. Reprinted Amsterdam: A.H. Hakkert, 1965. 3 vols.

The edited Greek text, with Latin translation and commentary, of the eight existing books of this 4th-century A.D. guide to the classical literature of mathematics. For the eighth book the reader should also consult the translation by D.E.P. Jackson of the Arabic version of this book (which contains material not in the present Greek text) in his Cambridge dissertation, soon to be published by Springer. (For further details, see Jackson, "The Arabic Translation of a Greek Manual of Mechanics," *Islamic Quarterly* 16 [1972], 96-103.)

235. Thomson, W., and G. Junge. *The Commentary of Pappus of Alexandria on Book X of Euclid's Elements*. Cambridge, Mass.: Harvard University Press, 1930. Reprinted New York: Johnson Reprint Corporation, 1968.

Since the Greek text of this work is lost, the present translation is based on an Arabic version of the work. See also item 321.

236. Ver Eecke, Paul, tr. *Pappus d'Alexandrie, La Collection Mathematique*. Paris, Bruges: Desclée, De Brouwer et Cie., 1933.

This is the only translation of this work into a modern language and one that faithfully reflects the Greek text established in item 234.

Proclos

237. Manitius, K., tr. and ed. *Procli Diadochi Hypotyposis astronomicarum positonum*. Leipzig: B.G. Teubner, 1909, and Morrow, G.R., tr. *Proclos Diadochos: A Commentary on the First Book of Euclid's Elements*. Princeton: Princeton University Press, 1970.

These two works by the fifth-century scholar Proclos are valuable sources of historical information. The former work is a good introduction to the planetary models found in Ptolemy and the latter contains among much other valuable material a summary of the early history of Greek geometry composed by Aristotle's pupil Eudemos.

Ptolemy

238. Manitius, K., tr. *Ptolemäus: Handbuch der Astronomie*. Leipzig: B.G. Teubner, 1963. 2 vols.

This second edition of Manitius's excellent German translation of Ptolemy's *Almagest* is accompanied by a brief foreword and corrections by O. Neugebauer. The reader more comfortable with English will want to use G. Toomer's translation, *Ptolemy's Almagest* (London: Duckworth, 1984).

239. Nobbe, C.F.A., ed. *Claudii Ptolemaei Geographia*. Leipzig: C. Tauchnit, 1843-1845. Reprinted Leipzig: Holze, 1898, and Hildesheim: G. Olms, 1966. 3 vols.

A German translation of Book I can be found in H. von Mžik, *Des Klaudios Ptolemaios Einführung in die darstellende Erdkunde, Teil I*, Vienna (1938); this contains F. Hopfner's study of Ptolemy's map projections on pp. 93-105, and of Marinos of Tyre's projection on pp. 87-89.

Theodosios

240. Ver Eecke, Paul. *Les sphériques de Théodose de Tripoli*. Bruges: Desclée, De Brouwer et Cie., 1927.

A reliable French translation of a first century B.C. source for the study of great circles and parallel circles on a sphere, although Theodosios was from Bithynia, not Tripoli.

General Histories of Greek Mathematics

241. Aaboe, Asger. *Episodes from the Early History of Mathematics*. New York: Random House, 1964.

Lets Euclid, Archimedes, and Ptolemy speak for themselves (if only in paraphrase) to provide a good introduction to Greek mathematics for the reader who is unfamiliar with it.

242. Becker, Oscar. *Das mathematische Denken der Antike*. 2nd ed. Göttingen: Vandenhoeck & Ruprecht, 1966.

The bulk of this book is a presentation of a wide selection of pieces of ancient mathematics, chosen to illustrate important problems and methods from those of ancient Babylon and Egypt to Diophantos. The book contains a thorough discussion of Archytas's duplication of the cube and both the theory of homocentric spheres and the general proportion theory of Eudoxos. For a much longer discussion of Eudoxos's contributions, the reader should consult Becker's "Eudoxos-Studien, I-V," in *Quellen und Studien zur Geschichte der Mathematik, Astronomie und Physic*, Abt. B2 (1933), 311-333, 369-388; 3, 236-244, 370-388, and 389-410, but he should be aware that there is considerable debate about some of Becker's conclusions.

243. Cantor, Moritz B. *Vorlesungen über Geschichte der Mathematik*. Leipzig: Teubner, 1880-1908. Reprinted New York: Johnson Reprint Corporation, 1965. 4 vols.

The first volume of this work, concerned with the history of mathematics to the year 1200 A.D., is a useful first reference for the history of Greek mathematics although much of it is now outdated since, for example, Cantor wrote before Archimedes' treatise *The Method* was even known, and Heron is placed in the first century B.C. See also items 74, 327, 466, 622.

244. Heath, T.L. *A History of Greek Mathematics*. Oxford: Clarendon Press, 1921, 2 vols. Reprinted New York: Dover, 1981.

The only complete history of almost 2000 years of Greek mathematics available in English. Based on a careful study of primary and secondary sources it is especially useful as an introduction to a study of the mathematical texts themselves, but not for the Oriental background to Greek mathematics. Its use of the concept "geometrical algebra" has been recently attacked (see item 264), but it is still a major point of departure for all serious historical studies of Greek mathematics.

245. Neugebauer, Otto. *The Exact Sciences in Antiquity*. 1st ed. Copenhagen: E. Munksgaard and Princeton: Princeton University Press, 1951; also Princeton, 1952. 2nd ed. Providence, R.I.: Brown University Press, 1957. Reprinted New York: Harper, 1962.

Deals with the origin and transmission of Hellenistic exact sciences, i.e., mathematics and mathematical astronomy, although only the last chapter and two appendices are specifically concerned with Greek mathematics. An important work whose value is increased by the copious bibliographic notes at the end of each chapter. See also items 91, 182.

246. ———. *A History of Ancient Mathematical Astronomy*. 3 parts. Berlin, Heidelberg, New York: Springer-Verlag, 1975.

Gives an account of the "numerical, geometrical and graphical methods devised to control the mechanism of the planetary system" from about the fifth century B.C. to the seventh A.D. The results of a lifetime of scholarship, this work will be a standard tool for scholars of the ancient exact sciences for years to come. See also items 135, 1806.

247. Van der Waerden, Bartel Leendert. *Science Awakening, I*. 2nd ed. Leyden: Noordhoff International Publishers and New York: Oxford University Press, 1961.

Aims at showing how Babylonian mathematics was transformed by early Greek mathematicians and then brought by Theaetetos and Eudoxos to the state found in Euclid's *Elements*. Photographs and illustrations exhibit the interaction of ancient mathematics and culture. The considerable controversy surrounding many of the conclusions indicates the important place the work occupies among modern studies of Greek mathematics. See also item 189.

Studies

248. Beckmann, F. "Neue Gesichtspunkte zum 5. Buch Euklids." *Archive for History of Exact Sciences* 4 (1/2) (1967), 1-144.

Examines closely the logical structure of Euclid's Book 5 from the standpoint of modern mathematics and argues that it is not "ratio" but magnitudes, and their relation of "Having a ratio," which are fundamental in the theory of that book. Contains an extensive bibliography.

249. Berggren, J.L. "Spurious Theorems in Archimedes' *Equilibrium of Planes: Book I*." *Archive for History of Exact Sciences* 16 (1976), 87-103.

Argues, on the basis of an analysis of the logical structure of *Equilibrium of Planes: Book I*, that five, and possibly seven, theorems in this work are not due to Archimedes but to later editors.

249a. ———. "History of Greek Mathematics: A Survey of Recent Research." *Historia Mathematica* 11 (4) (1984).

Surveys the literature of the past three decades on the history of ancient Greek mathematics and contains an extensive bibliography.

250. Clagett, Marshall. *The Science of Mechanics in the Middle Ages*. Madison: University of Wisconsin Press, 1959.

Includes a study of ancient mechanics, based on a distinction between a dynamical and statical approach to the problems of equilibrium, as well as a translation of the Arabic treatise, ascribed to Euclid, "The Book on the Balance." See also item 496.

251. Hintikka, J., and U. Remes. *The Method of Analysis: Its Geometrical Origin and Its General Significance*. Dordrecht, Boston: Reidel, 1974.

The suthors present a view of the method of analysis (as explained and practiced by Pappos) which contradicts that of Mahoney, item 258, in several important points. They argue that analysis should be seen primarily as an analysis of geometrical figures involving a search for "concomitants" rather than "consequences." The book is a highly readable attempt to use the philosophy of mathematics to shed light on historical questions.

252. Klein, J. *Greek Mathematical Thought and the Origin of Algebra*. Cambridge, Mass.: MIT Press, 1968.

Traces the concept of "number," with strong emphasis on the philosophical aspects, from early Greek mathematics through the works of Diophantos, arguing that these represent the tradition of theoretical logistic. The second half of the book is concerned with post-Greek mathematics.

253. Knorr, Wilbur R. *The Evolution of the Euclidean Elements: A Study of the Theory of Incommensurable Magnitudes and Its Significance for Early Greek Geometry*. Dordrecht: Reidel, 1975.

Aims to give a new account of the pre-Euclidean development in the field of incommensurable magnitudes. Among other things the book contains a reconstruction of proofs by Theodoros which led to difficulties in proving the side of a square measuring 17 feet is incommensurable with its diagonal and suggests the nature of the new approach by Theaetetos that resolved the problem. Argues Books II, IV, X, and XIII of Euclid's *Elements* stem from Theodoros, Theaetetos, and Eudoxos while Books I, III, and VI are older, going back to Hippocrates of Chios.

254. ———. "Archimedes and *The Elements*: Proposal for a Revised Chronological Ordering of the Archimedean Corpus." *Archive for History of Exact Sciences* 19 (1978), 211-290.

The pivot of a series of studies of Archimedes' works by the same author, this paper takes issue with the usually accepted ordering and argues on the basis of language and techniques for a quite different sequence and a developmental view of the Archimedean corpus into "early" and "late" works. Uses re-ordering to conclude that mechanical works were less important in Archimedes' development than others have thought.

255. ———. "Archimedes and the Measurement of the Circle: A New Interpretation." *Archive for History of Exact Sciences* 15 (2) (1976), 115-140.

This is part of an extensive literature on the bounds for the ratio of the circumference to the diameter of a circle which Heron reports he found in a treatise of Archimedes. The author presents his version of these bounds and how Archimedes may have arrived at them.

256. Lasserre, F. *The Birth of Mathematics in the Age of Plato.* Larchmont, N.Y.: American Research Council, 1964.

Argues that abstract mathematics was born in "the quarter century which followed the introduction of mathematics into the program of studies at Plato's Academy." Supporting translations from Greek texts are present in abundance but the work has no bibliography and relies almost completely on the views of T.L. Heath.

257. Luckey, P. "Das Analemma von Ptolemäus." *Astronomische Nachrichten* 230, No. 5498 (1927), cols. 17-46.

A fundamental paper analyzing the nomographic procedures Ptolemy describes in *On the Analemma* for solving some problems of spherical astronomy.

258. Mahoney, Michael S. "Another Look at Greek Geometrical Analysis." *Archive for History of Exact Sciences* 5 (1968), 318-348.

Asserts that instead of being a single method Greek analysis was a growing corpus of techniques and theorems for the generation and solution of problems, such as, for example, the technique of "verging" constructions.

258a. Mueller, I. *Philosophy and Deductive Structure in Euclid's Elements.* Cambridge, Mass.: MIT Press, 1981.

Although the book is, in a sense, a survey of *The Elements*, Mueller is principally concerned to illustrate some aspects of Euclid's conception of mathematics by an examination of the logical structure of Euclid's chief work. Thus, a key feature of the study of a single proposition, a group of propositions, or even of a whole book is the study of how it is used elsewhere in *The Elements*.

259. Neugebauer, Otto. "The Equivalence of Eccentric and Epicyclic Motion According to Apollonius." *Scripta Mathematica* 24 (1959), 5-21.

Presents an account of two sections in *The Almagest* that give a method of transforming an epicyclic model into an eccentric model and vice versa. The purpose is to argue that both selections are based on the work of Apollonios, whom Neugebauer regards as the real originator of Greek mathematical astronomy.

260. Pedersen, Olaf. "Logistics and the Theory of Functions: An Essay in the History of Greek Mathematics." *Archives Internationales d'Histoire des Sciences* 24, No. 94 (1974), 29-50.

Contends that the numerical procedures taken for granted in *The Almagest* constitute a theory of functions of one or more variables which was known to astronomers however much Greek mathematicians may have ignored it.

261. ———. *A Survey of the Almagest.* Odense: Odense Universitetsforlag, 1974.

A fine introduction to the study of Ptolemy's greatest work, emphasizing the mathematical techniques employed by Ptolemy in building his geometrical models out of observational data. See, however, the review by G. Toomer in *Archives internationales d'histoire des sciences* 100 (1977), 137-150, for corrections to some points in Pederson's work.

262. Riddell, R.C. "Eudoxan Mathematics and the Eudoxan Spheres." *Archive for History of Exact Sciences* 20 (1) (1979), 1-19.

Demonstrates that three discoveries credited to Eudoxos may be extracted from the basic kinematic element in Eudoxos's model of homocentric spheres for the movement of the planets, the sun, and the moon.

263. Schneider, Ivo. *Archimedes: Ingenieur, Naturwissenschaftler und Mathematiker*. Darmstadt: Wissenschaftliche Buchgesellschaft, 1979.

Contributes an account of the past century of research on Archimedes' life and works, with an emphasis on recent studies, in which Schneider argues for the importance of Archimedes' early mechanical investigations in his intellectual development. The 19-page bibliography is virtually exhaustive.

264. Szabó, Árpád. *The Beginnings of Greek Mathematics*. Dordrecht, Boston: D. Reidel, 1978.

Studies the appearance and use of certain key words in context to investigate early Greek mathematics. Makes a strong case for the dominant role of music in the early history of proportion theory and of Eleatic dialectic in the beginnings of rigorous mathematics. An appendix argues that the commonly used term "geometric algebra" is quite misleading.

265. Toomer, G.J. "The Chord Table of Hipparchus and the Early History of Greek Trigonometry." *Centaurus* 18 (1973), 6-28.

Shows how Hipparchus could have constructed his chord table (based on steps of 7-1/2°) using only theorems known very early in Greek mathematics and argues against any important role for Archimedes in the early history of trigonometry. (See also N. Swerdlow in *Mathematical Reviews* 58 (6) [December 1979], #26714.)

MEDIEVAL MATHEMATICS

ISLAMIC MATHEMATICS

A clear, short introduction to the subject is provided by E.S. Kennedy, item 334. More detailed surveys may be found in the *Cambridge History of Iran*, items 266 and 335, and in the book by A.P. Youschkevitch, item 355. The introductions in Sezgin, item 272, and Sarton, item 348, are also useful. S.H. Nasr, item 343, is well illustrated. Since the mathematical sciences of the Islamic world relied heavily on Greek achievements, Greek mathematical texts--many of them edited by J.L. Heiberg (or F. Hultsch) and translated into English by T. Heath or into French by ver Eecke--are essential. O. Neugebauer's item 246, and individual articles in general encyclopedias like the *Dictionary of Scientific Biography*, item 10, and Pauly-Wissowa, item 202, are perhaps the best reference works for Greek mathematics and astronomy. References to

Byzantine mathematics are to be found in C. Krumbacher, *Geschichte der byzantinischen Litteratur von Justinian bis zum Ende des Oströmischen Reiches* (Munich, 1897), as well as in more modern works. Nor should Indian sources be forgotten (see especially the article by D. Pingree, item 404. On the general question of translations into Arabic, consult M. Steinschneider, item 284, and O'Leary, item 344. Major centers for the study of Arabic science also issue materials from time to time, especially the Institute for the History of Arabic Science (Aleppo, Syria) and the former Smithsonian Institution Project in Medieval Islamic Astronomy based at the American Research Center in Egypt (Cairo). A new Institute for the History of Islamic-Arabic Science has been founded at the J.W. Goethe University in Frankfurt, and a department of History of Turkish-Islamic Science has been established at the Technical University, Istanbul. Orientalist and scientific journals regularly publishing material relevant to the history of Arabic mathematics and astronomy include the following:

Abhandlungen zur Geschichte der mathematischen Wissenschaften
Al-Abhath
Al-Andalus
Al-Mashriq
Ankara Üniversitesi Dil ve Tarih-Coğrafya Fakültesi Dergisi
Archive for History of Exact Sciences
Archives internationales d'histoire des sciences
Bibliotheca Mathematica
Centaurus
Historia Mathematica
Isis
Istoriko-Matematicheskie Issledovaniia
Istoriko-Astronomicheskie Issledovaniia
Janus
Journal for the History of Arabic Science
Journal for the History of Astronomy
Journal of Near Eastern Studies
Journal of the American Oriental Society
The Mathematics Teacher
Osiris
Physis
Quellen und Studien zur Geschichte der Mathematik (Abteilung A: Quellen; Abteilung B: Studien)
Zeitschrift für Geschichte der arabisch-islamischen Wissenschaften

Handbooks and General Bibliographies

In addition to standard reference works and general histories of mathematics already described in Sections I, II, and III of this bibliography, the following works contain information on Arabic mathematics and science.

266. *The Cambridge History of Iran*. Cambridge, England: Cambridge University Press, 1968. 8 vols.

Generally construed as a cultural history of the Iranian region. Volume I: demography and geography; Volumes II-VII: chronological history; Volume VIII: bibliography, survey of research, indexes. See especially Kennedy, item 335.

267. *Encyclopaedia of Islam*. 2nd ed. Leiden: E.J. Brill, and London: Luzac and Company, 1960.

A general encyclopedia, but conceived on a broader plan than the first edition; alphabetically arranged, with cross-references and bibliographies compiled by an international group of editors and authors. The second edition has reached K/Ḳ.

267a. Kennedy, E.S., et al. *Studies in the Islamic Exact Sciences.* Edited by D.A. King and M.-H. Kennedy. Beirut: American University of Beirut Press, 1983.

Numerous articles by E.S. Kennedy, colleagues, and students are reprinted in this volume. Currently in preparation is another volume edited by D.A. King and G. Saliba, containing over thirty articles by the leading scholars in the history of Islamic astronomy and mathematics, which will appear as a *Festschrift* in honor of E.S. Kennedy.

267b. King, D.A. "The Exact Sciences in Medieval Islam: Some Remarks on the Present State of Research." *Bulletin of the Middle East Studies Association of North America*, 4 (1980), 10-26.

268. Lorch, R. "Arabic Mathematics and Astronomy." In *Middle East and Islam. A Bibliographical Introduction.* Edited by Diana Grimwood-Jones and D. Hopwood. Zug, Switzerland: Inter Documentation Co., 1972, 139-145, plus one page of "additions."

This bibliography was issued as volume 15, *Bibliotheca Asiatica* (1972), and runs to 368 pages.

268a. Maeyama, Y., and W.G. Saltzer, eds. *Prismata: Naturwissenschaftliche Studien: Festschrift für Willy Hartner.* Wiesbaden: F. Steiner, 1977.

269. Nasr, S.H. *An Annotated Bibliography of Islamic Science.* Vol. I. Tehran: Imperial Iranian Academy of Science, 1975.

The first of five projected volumes; two sections cover "General works" and "Biographical and bibliographical studies of Muslim men of science." There are some rather serious omissions, but this bibliography is a useful supplement to others.

270. Pearson, J.D., ed. *Index Islamicus.* 1906-1955, 1956-1960, 1961-1965, 1966-1970, 1971-1975; since 1977 the *Index* has appeared quarterly in *The Quarterly Index Islamicus.* (1976 is included in 1977.)

271. ———. *Oriental and Asian Bibliography. An Introduction with Some Reference to Africa.* London: Lockwood, 1966.

Oriental bibliography in general.

272. Sezgin, Fuat. *Geschichte des arabischen Schrifttums.* Leiden: E.J. Brill, 1967-.

Replaces Brockelmann, item 274, for the period up to about 430 H/1040 A.D., citing many additional manuscripts and giving introductory chapters. Volumes 5 (1974), 6 (1979), and 7 (1980) are respectively devoted to mathematics, astronomy, and astrology. See also items 27, 204. Essay review of parts of this work by D. King, "On the Sources for the Study of Early Islamic Mathematics: An Essay Review," *Journal of the American Oriental Society*, 99 (1979), 450-459; "Early Islamic Astronomy: An Essay Review," *Journal for the History of Astronomy*, 12 (1981), 55-59.

273. Suter, Heinrich. "Die Mathematiker und Astronomen der Araber und ihre Werke." *Abhandlungen zur Geschichte der mathematischen Wissenschaften* 10 (1900), ix + 278 pp. Reprinted New York: Johnson Reprint Corporation, 1971. Additions and corrections in *Ibid.* 14 (1902), 155-185, and by H.J.P. Renaud, *Isis* 18 (1932), 166-183. A similar work in Russian has been prepared by B.A. Rosenfeld and G.P. Matvievskaya: *Matematiki i astronomi musulmanskogo srednevekovya i ikh trudi (VII-XVII vv).* Moscow: Nauk, 1983. 3 vols.

Suter's bio-bibliographical reference work includes essays on the life and work of 528 Islamic scientists who lived between 750 and 1600. Complete bibliographic information is given on all works known at the time (i.e., up to 1900). It is still the most important guide for the study of the history of medieval mathematics in the East after the tenth century. See also Suter, item 28.

Manuscripts

In addition to F. Sezgin, item 272, and A.T. Grigorian et al., item 328, the following are useful for finding manuscripts.

274. Brockelmann, C. *Geschichte der arabischen Literatur.* Leiden, 1898-1942; 2nd ed. Leiden: E.J. Brill, 1943-1949. 2 vols. Supplement, 1937-1942. 3 vols.

Gives the locations of manuscripts of all known works in Arabic. This is a fundamental work on the history of Arabic literature, containing vast biographical and bibliographical information on the mathematicians of medieval Islam, and the extant manuscripts of their works. See Sezgin, item 272.

275. Huisman, A.J.W. *Les manuscrits arabes dans le monde. Une bibliographie des Catalogues.* Leiden: E.J. Brill, 1967.

Can now be replaced by the lists in Sezgin, item 272; see especially volume 6.

276. King, D.A. *A Catalogue of the Scientific Manuscripts in the Egyptian National Library.* Cairo: General Egyptian Book Organization, 1981 and 1984. 2 vols.

These volumes are in Arabic and constitute a handlist of the collection with extracts arranged chronologically by subject. This should be used in conjunction with D.A. King, *A Survey of the Scientific Manuscripts in the Egyptian National Library, Publications of the American Research Center in Egypt* (Malibu, Cal.: Undena Press, 1984). The *Survey* is in English and arranged chronologically by author as a supplement to items 272, 273, and 274.

277. Krause, M. "Stambuler Handschriften islamischer Mathematiker." *Quellen und Studien zur Geschichte der Mathematik, Astronomie und Physik*, B3 (1936), 437-532.

278. Steinschneider, M. *Die arabische Literatur der Juden; ein*

Beitrag zur Literaturgeschichte der Araber, grossenteils aus handschriftlichen Quellen. Frankfurt: J. Kauffmann, 1902. Reprinted Hildesheim: G. Olms, 1964.

279. Storey, C.A. *Persian Literature. A Bio-Bibliographical Survey*. London: Luzac and Co., 1927-1953.

Similar to C. Brockelmann, item 274, but for works in Persian. Volume II, Part I, deals with mathematics and astronomy.

Translations from Arabic

280. Carmody, F.J. *Arabic Astronomical and Astrological Sciences in Latin Translation; a Critical Bibliography*. Berkeley and Los Angeles: University of California Press, 1956.

Lists the Latin and some Arabic manuscripts, with other data, for each work mentioned. Not always reliable, but useful when used in conjunction with Sezgin, item 272, or Thorndike and Kibre's list of Latin incipits, item 465a.

281. Goldstein, Bernard R. "The Survival of Arabic Astronomy in Hebrew." *Journal for the History of Arabic Science* 3 (1979), 31-39.

See item 433.

282. Haskins, Charles Homer. *Studies in the History of Mediaeval Science*. Cambridge, Mass.: Harvard University Press, 1924. 2nd ed. 1927. Reprinted New York: Ungar Publishing Co., 1960.

See item 565.

283. Millás Vallicrosa, J. "Translations of Oriental Scientific Works (to the End of the Thirteenth Century)." *Journal of World History* 2 (1954-1955), 395-428. Reprinted in *The Evolution of Science*. Edited by G. Metraux and F. Crouzet. New York: Mentor, 1963.

284. Steinschneider, M. *Die arabischen Übersetzungen aus dem griechisen*. Leipzig: O. Harrassowitz, 1889-1893. Reprinted Graz: Akademische Druck- und Verlagsaustalt, 1960. 2 vols.

Papers collected and reprinted from various journals.

285. ———. *Die europäischen Übersetzungen aus dem arabischen bis Mitte des 17. Jahrhunderts*. Graz: Akademische Druck- und Verlagsaustalt, 1956.

Articles reprinted from Proceedings of the Royal Academy of Sciences, Vienna, from the years 1904 and 1905.

286. ———. *Die hebraeischen Uebersetzungen des Mittelalters und die Juden als Dolmetscher*. Graz: Akademdemische Druck- und Verlagsaustalt, 1956.

Originally published in 1893. This work often gives Arabic manuscripts as well as Hebrew translations. See also item 445.

Texts and Commentaries (Specific Authors)

Numerous Arabic texts are published at Hyderabad. Sometimes these represent the only published material available for many authors, including Abū Naṣr ibn ᶜIrāq, al-Ṭusī, Ibn al-Haytham, Thābit ibn Qurra, etc. However, these editions are often taken from one manuscript, and so are not always reliable. Several texts have been published in facsimile with Russian translation (sometimes with facsimiles of Arabic manuscripts), mostly in Moscow or Tashkent. For details, see A.P. Juškevič (Youschkevitch), item 331, and S. Kh. Sirazhdinov and G.P. Matvievskaya, item 349. What follows is a listing of the major edited texts and studies currently available.

Abū-l-Wafā'

287\. Saidan, A.S., ed. *Arabic Arithmetic. The Arithmetic of Abū al-Wafā' al-Būzajānī, 10th Century. Mss. Or. 103 Leiden and 42 M Cairo.* Edited with introduction, commentaries, and ample reference to the arithmetic of al-Karajī (11th century), Ms. 855 Istanbul. Amman: Jordanian University, 1971. In Arabic.

In addition to the original text of the arithmetical treatise of the great tenth-century mathematician Abū-l-Wafā' al-Būzajānī, this book provides a detailed essay on the history of Arabic arithmetic. See also items 1879, 1880.

287a. Sesiano, J. "Lés methodes d'analyse indeterminée chez Abū Kāmil." *Centaurus* 21 (1977), 89-105.

Abū Kāmil

287b. Levey, M. *The Algebra of Abū Kāmil, Kitāb fī'l-jabr wa'l-muqābala in a Commentary by Mordecai Finzi.* Hebrew Text, Translation and Commentary with Special Reference to the Arabic Text. Madison-Milwaukee, and London, 1966.

287c. Suter, H. "Die Abhandlung des Abū Kāmil Shogāᶜ b. Aslam über das Fünfeck und Zehneck." *Bibliotheca Mathematica*, 3. F., 10 (1909-1910), 15-42.

Abū Naṣr

288\. Samsó Moya, J. *Estudios sobre Abū Naṣr b. ᶜAlī b. ᶜIrāq.* Madrid, Barcelona: Asociación para la Historia de la Ciencia Española, 1969.

Abū Naṣr was one of the astronomers credited with the sine-theorem in spherical trigonometry. This book is a study of his trigonometry and includes a Spanish translation of several of his works.

See also Krause, item 315.

Al-Battānī

289\. Al-Battānī, Muḥammad b. Jābir. *Opus astronomicum.* Edited by C.A. Nallino. Milan: Osservatorio astronomico di Breva N. 40, 1899-1907. Reprinted Frankfurt: Minerva, 1969, and New York, 1976.

The Arabic text of Al-Battānī's *zīj* with modern Latin translation, introduction, and commentary (also in Latin).

Al-Bīrūnī

290. Al-Bīrūnī, Muḥammad b. Aḥmad, Abū Raihān. *Alberuni's India.* Translated by E. Sachau. London, 1910. Reprinted New Delhi: S. Chand, 1964. 2 vols.

This text was first edited by Sachau (London, 1888); another edition was published at Hyderabad (Osmania Oriental Publishing Bureau, 1958). There is also a Russian translation (Tashkent, 1963).

291. ———. *Al-Qānūnu'l-Masᶜūdī.* Hyderabad: Osmania Oriental Publishing Bureau, 1954-1956. 3 vols.

This edition is partly based on the preliminary work of Max Krause. An aid in using this enormous astronomical compendium is the chapter-by-chapter analysis by E.S. Kennedy, "Al-Bīrūnī's Masudic Canon," *Al-Abhath* 24 (December 1971), 59-81. See item 234.

292. ———. *The Determination of the Coordinates of Positions for the Correction of Distances between Cities.* A translation from the Arabic of al-Bīrūnī's *Kitāb Taḥdīd Nihāyāt al-Amākin Li-taṣḥīḥ Masāfāt al-Masākin* by Jamil Ali. Beirut: The American University of Beirut, 1967.

The most significant medieval work on mathematical geography. The text was edited by P. Bolgakoff (Cairo: The Cultural Department of the Arab League, 1962). There is also a very useful commentary by E.S. Kennedy, item 296.

293. ———. *The Exhaustive Treatise on Shadows.* Translation and commentary by E.S. Kennedy. Aleppo: University of Aleppo, 1976. 2 vols.

The English translation of al-Bīrūnī's text (Volume I) and extensive commentaries (Volume II) represent one of the most important recent studies of the history of mathematics and astronomy in the lands of Islam.

294. ———. *Izbrannie proizvedenija* (Selected works). Tashkent: Fan, 1957-1976. 7 vols.

Russian and Uzbek translations of selected works of al-Bīrūnī, including the *Al-Qānūnu 'l-Masᶜūdī.*

295. Debarnot, M.-T. "Les clefs de l'astronomie (Kitāb maqālīd ᶜilm al-hay'a) d'Abū al-Rayḥān Muḥammad b. Aḥmad al-Bīrūnī. La trigonométrie sphérique chez les Arabes de l'est à la fin du Xᵉ siècle." *Thèse de 3ème cycle.* Paris: Université de Paris, 1980.

Studies the question of priority in finding the sine-theorem and considers its relation to the *zīj* of Ḥabash al-Ḥāsib.

Principally a text and French translation of al-Bīrūnī's *Maqālīd*, plus commentary. This thesis (to be published by the French Oriental Institute of Damascus) is the starting point for all future study of Arabic spherical trigonometry and replaces von Braunmühl, item 363, for this period.

296. Kennedy, E.S. *A Commentary upon Bīrūnī's Kitāb taḥdid* [*nihāyāt*] *al-amākin, an 11th Century Treatise on Mathematical Geography*. Beirut: The American University of Beirut, 1973.

See item 292.

See also Rosenfeld, item 347, Auluck, item 1869.

Al-Biṭrūjī

297. Al-Biṭrūjī, Nūr al-Dīn. *On the Principles of Astronomy*. Edited by B.R. Goldstein. (Yale Studies in the History of Science and Medicine.) New Haven: Yale University Press, 1971.

Arabic text, Moses ibn Tibbon's Hebrew translation, English translation, analysis, and glossary. Michael Scot's Latin was published by F.J. Carmody in 1952. Together with Ibn Rushd's commentaries on the *De Caelo* and the *Metaphysics* (see the article by F.J. Carmody in *Osiris* 10 [1952], 556-586), the treatise gave rise to much discussion in the Latin West of the physical possibility of the Ptolemaic system.

Al-Karajī (or al-Karkhī)

298. Al-Karajī. *L'Algèbre al-Badī^{c} d'al-Karagī. Manuscrit de la Bibliothèque Vaticane Barberini Orientale 36, 1*. Edition, introduction et notes par Adel Anbouba. Beirut: Publications de l'Université libanaise, 1964.

298a. Sesiano, J. "Le traitement des équations indeterminées dans le *Badī^{c}fī'l-ḥisāb* d'Abū Bakr al-Karajī." *Archive for History of Exact Sciences* 17 (1977), 297-379.

298b. Woepcke, F. *Extrait du Fakhvi, traité d'algèbre par Abou Bekr précédé d'un mémoire sur l'algèbre indeterminée chez les Arabes*. Paris, 1853.

French translation plus a few extracts in Arabic.

See also the book by A.S. Saidan on Abū-l-Wafā', item 287.

Al-Kāshī

299. Al-Kāshī, Ghiyāth al-Dīn Jamshīd Mas^{c}ūd. *Miftāḥ al-Ḥisāb. Miftāḥ al-Ḥisāb d'Al-Kāshī*. Edition, Notes et Introduction par Nabulsi Nader. Damascus: Université de Damas, 1977.

Though late (ca. 1400 A.D.), this work represents the culmination of Islamic mathematics and is of great interest because of its wide-ranging content and great influence on subsequent textbooks. There is a Russian translation by B.A. Rosenfeld, item 300, which also includes a translation of al-Kāshī's *Al-Risāla al-Muḥīṭīya*. There is also a valuable analysis of the *Miftāḥ* by P. Luckey, item 301, and a translation of the *R. Muḥīṭīya* in item 302.

300. ———. *Klyuch aritmetiki. Traktat ob okruzhnosti* (The key of arithmetic. The treatise on the circumference). Translated by B.A. Rosenfeld. Edited by V.C. Segal and A.P. Juškevič. Commentaries by A.P. Juškevič and B.A. Rosenfeld. Moscow: Gos. izd-vo Tekhniko-teoret. lit-ry, 1956.

Arabic texts in facsimile and Russian translations of the two works by al-Kāshī.

301. Luckey, P. *Die Rechenkunst bei Ǧamšīd b. Mas^cūd al-Kāšī mit Rückblicken auf die ältere Geschichte des Rechnens*. (Abhandlungen für die Kunde des Morgenlandes, 31.) Wiesbaden: F. Steiner, 1951, 143 pp.

This work provides a thorough analysis of the arithmetical methods described in "The Key of Arithmentic" (*Miftāḥ al-Ḥisāb*) of al-Kāshī.

302. ———. *Der Lehrbrief über den Kreisumfang von Ǧamšīd b. Mas^cūd al-Kāšī*. Translated by P. Luckey. Edited by A. Siggel. Berlin: Akademie Verlag, 1953.

Arabic text and German translation of the "Treatise on the Circumference" of Jamshīd al-Kāshī with the commentaries. A determination of π to 16 significant decimal digits, this book is a landmark in the history of computational mathematics. See Neugebauer, item 245, 23.

Al-Khwārizmī

303. Al-Khwārizmī, Muḥammad b. Mūsā. "The Astronomical Tables of al-Khwārizmī." Edited by O. Neugebauer. *Kongelige Danske Videnskabernes Selskab, hist.-filos. Skrifter* (Copenhagen), 4 (2) (1962).

Uses the Latin of Adelard of Bath to give the English translation and commentary. Suter's edition of the Latin translation should also be noted.

304. ———. *El comentario de Ibn al-Mutannā' a las Tablas astronómicas de al-Jwārizmī*. Edited by E. Millás Vendrell. Madrid, Barcelona: Asociación para la Historia de la Ciencia Española, 1963.

Contains a critical edition of Hugo Sanctallensis's Latin translation and a substantial commentary.

305. ———. *Ibn al-Muthannā's Commentary on the Astronomical Tables of al-Khwārizmī*. Edited and translated by B.R. Goldstein. (Yale Studies in the History of Science and Medicine, 2.) New Haven: Yale University Press, 1967.

Two Hebrew versions, with commentary.

305a. *Al-kitāb al-mukhtaṣar fi ḥisāb al-jabr wa'l-muqābala*. Edited by Musharrafa and Muhammad Aḥmad. Cairo, 1939, 1968.

This text is perhaps most easily studied in Karpinski's edition of Robert of Chester's Latin translation (New York, 1915) or in G. Libri's edition of Gerard of Cremona's translation (Paris, 1938).

305b. Vogel, K. *Muhommed ibn Musa Alchwarizmi's Algorismus. Das frühste Lehrbuch zum Rechnen mit indischen Ziffern*. Aalen, 1963.

This arithmetic is lost in Arabic.

See also items 373c and 516.

Al-Uqlīdisī

306. Al-Uqlīdisī. *The Arithmetic of al-Uqlīdisī*. Translated and Annotated by A.S. Saidan. Dordrecht and Boston: Reidel, 1978.

Saidan also published the Arabic text (Amman: Jam^{c}iyyat ^{c}ummāl al Matabi^{c}, 1971). See also item 425.

Archimedes

307. Clagett, Marshall. *Archimedes in the Middle Ages*. Vol. 1: Madison: University of Wisconsin Press, 1964; Volumes 2-4 (in multiple parts): Philadelphia: American Philosophical Society, 1976, 1978, 1980.

With this book the Arabic tradition of Archimedes may be studied through Latin translations. Clagett includes *inter alia* an edition of the *Verba filiorum* (i.e., of the banū Mūsā) in Gerard's translation. See also item 493.

Azarquiel (Zargālī)

See Millás Vallicrosa, item 340.

Diocles

308. Diocles. *On Burning Mirrors*. The Arabic translation of the lost Greek original, edited, with English translation and commentary, by G.J. Toomer. (Sources in the History of Mathematics and Physical Sciences, 1.) Berlin: Springer-Verlag, 1976.

On conic sections. See also item 219.
Review: Rashed, R., *Archives Internationales d'Histoire des Sciences* 28 (1978), 329-333.

Diophantos

308a. *Books IV to VII of Diophantus' "Arithmetica" in the Arabic Translation Attributed to Qustā ibn Lūqā*. Edited by J. Sesiano. Berlin, Heidelberg, New York: Springer-Verlag, 1982.

Euclid

309. *Anaritii in decem libros priores Elementorum Euclidis commentarii. Ex interpretatione Gherardi Cremonensis in codice Cracoviensi 569 servata*. Edited by M. Curtze. Leipzig: Supplement to Euclid's *Opera Omnia*, 1899.

309a. *Codex Leidensis 399.1. Euclidis Elementa ex interpretatione al-Hadschdschadschii cum commentariis al-Narizi*. Edited by R.O. Besthorn, J.L. Heiberg, et al. Copenhagen, 1897-1932.

The text appears to be an unreliable witness to the Hajjāj translation or translations, but it is all we have in Arabic, apart from quotations, unless Books XI-XIII of the Isḥāq version are accepted as identical to the Ḥajjāj.

309b. *The First Latin Translation of Euclid's "Elements" Commonly Attributed to Adelard of Bath. Books I to VIII; X 36-XV 2*. Edited by H.L.L. Busard. Toronto: Pontifical Institute of Medieval Studies (Studies and Texts 64), 1983.

Apparently a translation and not a reworking. It appears to belong to the tradition of the Arabic of Ḥajjāj rather than that of Isḥāq.

309c. *The Latin Translation of the Arabic Version of Euclid's "Elements" Commonly Ascribed to Gerard of Cremona*. Edited by H.L.L. Busard. Leiden, 1983.

This translation appears to have been reworked, especially in the enunciations, and to have been based on a manuscript of the Isḥāq tradition, though with additions from other versions.

309d. De Young, G.R. "The Arithmetic Books of Euclid's *Elements* in the Arabic Translation: An Edition, Translation, and Commentary." Dissertation, Harvard University, May 1981.

309e. Engroff, J.W., Jr. "The Arabic Tradition of Euclid's *Elements*: Book V." Dissertation, Harvard University, May 1980.

309f. Klamroth, M. "Ueber den arabischen Euklid." *Zeitschrift der deutschen morgenländischen Gesellschaft* 35 (1881), 270-326, 788.

Still the basic work on the Arabic Euclid.

Ibn al-Bannā'

310. Ibn al-Bannā', Aḥmad b. Muḥammad. *Talkhis aᶜmāl al-ḥisāb*. Edited by M. Souissi. Tunis: Université de Tunis, 1969.

Text with French translation and commentary.

Ibn al-Haytham

311. Rashed, R. "Ibn al-Haytham et la mesure du paraboloïde." *Journal for the History of Arabic Science* 5 (1981), 262-291.

311a. Sabra, A.I. "An Eleventh-Century Refutation of Ptolemy's Planetary Theory." *Studia Copernicana* 16 (1978), 117-131.

A study of Ibn al-Haytham's critique of Ptolemy, with a partial translation of his *al-Shukūk ᶜalā Baṭlaymūs*, which was edited by Sabra and N. Shehady (Cairo, 1971).

Ibn al-Muthanna

See al-Khwārizmī, items 304 and 305.

Ibn al-Shāṭir

312. Ibn al-Shāṭir. *The Life and Work of Ibn al-Shāṭir, an Arab Astronomer of the Fourteenth Century*. Edited by E.S. Kennedy and I. Ghanem. Aleppo: University of Aleppo, 1976.

This volume contains biographical and bibliographical material on Ibn al-Shāṭir (14th century), as well as reprints of papers on the works of this prominent scientist, in Arabic and European languages.

Ibn Mu^{c}ādh

313. Villuendas, M.V. *La trigonometria en el siglo XI. Estudio de la obra de Ibn Mu^{c}ādh, El Kitāb mayhūlāt*. Barcelona: Instituto de Historia de la Ciencia de la Real Academia de Buenas Letras, 1979.

Gives Arabic text and Spanish translation, plus commentary. Ibn Mu^{c}ādh derives theorems for solving spherical triangles from Menelaos's theorem.

Ibn Turk

313a. *Abdülhamid Ibn Türk'ün katışık denklemlerde mantıckî zaruretler adlı yazısı ve zamanın(ın) cebri* (Logical necessities in mixed equations by ^{c}Abd al-Hamīd ibn Turk and the algebra of his time). Edited and translated by A. Sayili. Ankara, 1962.

Jaghmīnī

314. Jaghmīnī, Maḥmud b. Muḥammad b. ^{c}Umar. *Mulakhkhas fi'l-Hay'a*. G. Rudloff and A. Hochheim, trans. in "Die Astronomie des Maḥmud ibn Muḥammed ibn ^{c}Omar al-Ǵagmînî." *Zeitschrift der deutschen morgenländischen Gesellschaft* 47 (1893), 213-275.

Includes a German translation of Jaghmīnī's much-commented-upon compendium of astronomy of the 13th or 14th century.

Menelaos of Alexandria

315. Krause, M., ed. *Die Sphärik von Menelaos aus Alexandrien in der Verbesserung von Abu Naṣr Manṣūr b. ^{c}Alī b. ^{c}Iraq. Mit Untersuchungen zur Geschichte des Textes bei den islamischen Mathematikern. Abhandlungen der Geschichte der Wissenschaften zu Göttingen*, philol.-hist. Klasse, 3. F., 17 (1936).

Al-Ṭūsī's redaction of the *Spherics* is in volume 2 of the Hyderabad edition of his *Rasā'il* (1940-1941). Halley based his Latin translation of 1758 on Arabic as well as Hebrew versions (translated from Arabic by Jacob b. Makhir). See also A.A. Bjørnbo's study, including a German translation, in *Abhandlungen zur Geschichte der mathematischen Wissenschaften* 14 (1902), 1-154, item 543. See also item 232.

Nāṣīr al-Dīn al-Ṭūsī

316. Nāṣīr al-Dīn al-Ṭūsī. *Traité du quadrilatère*. Edited by A. Carathéodory. Constantinople, 1891.

Text and French translation of Nāṣīr al-Dīn's book on Menelaos's theorem. Contains many historical remarks. On the history of this theorem see Burger and Kohl, item 325.

317. Mamedbeyli, G.D. *Osnovatel Maraginskoy observatorii Nasireddin Tusi* (The founder of the observatory in Maragha Nasireddin Tusi). Baku: Izdat. Akademii Nauk Azerbaijan SSR, 1961.

This book contains a biography and analysis of the mathematical works of Nāṣīr al-Dīn al-Ṭūsī (1201-1274).

Nicomachus of Gerasa

318. Kutsch, W., ed. *Tābit ibn Qurra's arabische Übersetzung der* 'Αριθμητικὴ *Εἰσαγωγή des Nikomachus von Gerasa*. Beirut: Recherches de l'Institut de Lettres Orientales de Beyrouth, 1959.

Contains a full Arabic-Greek and Greek-Arabic glossary as well as the text.

ᶜUmar Khayyām

319. ᶜUmar Khayyām. *Traktaki*. Edited by B.A. Rosenfeld and A.P. Juškevič. Moscow-Leningrad: Izdat. vostochnoi Lit., 1961.

Russian translation of the mathematical works of ᶜUmar Khayyām. The Arabic text has been published by R. Rashed and A. Djebbar (*L'oeuvre algébrique d'al-Khayyām*. Aleppo: IHAS, 1981). For a biography and analysis of the mathematical works of ᶜUmar Khayyām, see Rosenfeld and Juškevič, item 345.

320. Woepcke, F. *L'algèbre d'Omar Alkhayyami*. Paris: B. Duprat, 1851.

The first European translation of ᶜUmar Khayyam's *Algebra*. The works of F. Woepcke (1826-1864) on the history of mathematics in the Islamic world are still useful (see bibliographies of his contributions listed in Sarton, item 348, Juškevič, item 332, Matvievskaya, item 338, etc. For an Arabic edition of the *Algebra*, see the publication by R. Rashed and A. Djebbar noted in item 319.

Pappos

321. Thomson, W., ed. *The Commentary of Pappus on Book X of Euclid's Elements*. Cambridge, Mass.: Harvard Semitic Series #8, 1930.

Arabic text and translation by W. Thomson; commentary by Thomson and G. Junge. The Arabic translation is by al-Dimishqī. See also item 235.

Ptolemy

322. *Der Almagest des Claudius Ptolemäus; die Syntaxis Mathematica des Claudius Ptolemäus in arabisch-lateinischer Überlieferung.* Edited by P. Kunitzsch. Wiesbaden: Harrassowitz, 1974.

The author's *Habilitationsschrift*, Munich. Particularly good on the fixed stars. A new English translation of the *Almagest* by G.J. Toomer (London: Duckworth, 1984) takes cognizance of the Arabic translations. Olaf Pedersen's *A Survey of the Almagest*, item 262, is a welcome help in understanding it.

323. Goldstein, Bernard R. "The Arabic Version of Ptolemy's Planetary Hypotheses." *Transactions of the American Philosophical Society* New Series, 57 (4) (1967), 1-55.

The text is a facsimile of the British Museum manuscript with variants from two other manuscripts. There is an English translation of the part not extant in Greek, as well as some commentary.

323a. Ibn al-Salāh. *Zur Kritik der Koordinatenüberlieferung im Sternkatalog des Almagest.* Edited and translated by Paul Kunitzsch. Göttingen: *Abhandlungen der Akademie der Wissenschaften in Göttingen*, philol.-hist. Klasse, 3. F., Nr. 94, 1975.

Thābit ibn Qurra

324. Thābit ibn Qurra. *The Astronomical Works of Thābit b. Qurra.* Edited by F.J. Carmody. Berkeley: University of California Press, 1960.

Several of Thābit's works in Latin translation with specimens of Arabic and an introduction. Commentary not always reliable (see the review by O. Neugebauer in *Speculum* 37 [1962], 99-103). An edition, by Régis Morelon, of Thābit's astronomical works in Arabic is to appear in Paris (Les Belles Lettres).

325. Burger, H., and K. Kohl. "Thabits Werk über den Transversalensatz." *Abhandlungen zur Geschichte der Naturwissenschaft und der Medizin* (Erlangen) 7 (1924).

Thābit's *De figura sectore* Bjφrnbo's edition of Gerard's Latin (Bjφrnbo leaves out most of the rehearsal of the modes) plus a discussion of the whole question of Menelaos's theorem in the hands of Arabic writers.

For Thābit's Arabic translation of Nicomachus, see Kutsch, item 318.

General Works and Biographies

326. Bulgakov, P.S. *Zhizn i trudi Beruni* (The life and works of Bīrūnī). Tashkent: Izdat. "Fan," Uzbek. SSR, 1972. In Russian.

The most detailed essay on the life and scientific activity of Abū'l-Rayhān al-Bīrūnī. An exhaustive bibliography is given.

327. Cantor, Moritz B. *Vorlesungen über Geschichte der Mathematik.* Vols. 1 and 2. Leipzig: Teubner, 1894; 2nd ed., 1907-1913.

This work contains valuable factual material on the development of mathematics in medieval Islam, but many of its assertions and conclusions are now out of date. See also item 74, 243, 466, 622.

327a. Djebbar, A. *Enseignement et recherche mathématiques dans le Maghreb des XIIIe-XIVe siècles*. Université de Paris-Sud: Département de Mathématique, 1980.

327b. ———. *L'analyse combinatoire dans l'enseignement d'Ibn Mun^{c}im (XIIe-XIIIe siècles)*. Université de Paris-Sud: Département de Mathématique, 1982.

An important contribution to a previously unexplored aspect of Islamic mathematics.

328. Grigorian, A.T., and A.P. Juškevič, eds. *Fiskio-mathematicheskie nauki v stranakh Vostoka. Sbornik statjey i publikaziy* (Physical and mathematical sciences in the East. Collected articles and publications). Moscow: Izdat. "Nauka," 1966. In Russian.

This volume contains Russian translations from a number of medieval Arabic mathematical treatises, as well as papers by T.N. Kary-Niyazov, B.A. Rosenfeld, and others. Arabic and Persian manuscripts in the libraries of the Soviet Union are listed (compiled by B.A. Rosenfeld).

329. Hartner, W. *Oriens, Occidens: ausgewählte Schriften zur Wissenschafts- und Kulturgeschichte*. Hildesheim: G. Olms, 1968.

This book reprints a collection of the author's papers on mathematics and astronomy in Islam. A second volume, edited by Y. Maeyama, is to be published in June 1984.

330. Ibn al-Nadīm. *Fihrist*. Translated by B. Dodge. New York: Columbia University Press, 1970.

Originally a bookseller's list and later extended, this is one of the standard Arabic bibliographical sources for the history of Arabic science. Dodge uses manuscripts as well as Flügel's edition of 1871.

331. Juškevič, Adolf P. "Soviet Investigation on the History of Mathematics for the Sixty Years (1917-1977)." *Istoriko-Matematicheskie Issledovaniia* 24 (1979), 66-76.

332. ———. *Geschichte der Mathematik im Mittelalter*. Leipzig: B.G. Teubner, 1964.

A German translation of the original Russian version of 1961. This book offers the most comprehensive account of achievements of the mathematicians of the Islamic world in the Middle Ages. An extensive bibliography is given. See also items 355 and 467.

333. Kary-Niyazov, T.N. *Astronomicheskaya shkola Ulugbeka* (The astronomical school of Ulugh Beg). Moscow, Leningrad: Izdat. Akad. Nauk SSSR, 1950. In Russian.

This book contains a detailed discussion of the accomplishments of mathematicians and astronomers of the scientific school at

Samarqand in the fifteenth century, including Ulugh Beg, al-Rūmī, al-Kāshī, and others.

334. Kennedy, Edward S. "The Arabic Heritage in the Exact Sciences." *Al-Abhath* 23 (1-4) (1970), 327-344.

A general survey of mathematics, astronomy, mechanics, and physcis in the medieval Islamic world.

335. ———. "The Exact Sciences in Abbasid Iran," "The Exact Sciences in Iran under the Saljuqs and Mongols," and "The Exact Sciences in Timurid Iran," all chapters in *The Cambridge History of Iran*, item 266.

335a. King, D.A. "On Medieval Islamic Multiplication Tables." *Historia Mathematica* 1 (1974), 317-323.

Supplemented with additional material by the same author in *Historia Mathematica* 6 (1979), 405-417.

336. Kubesov, A. *Matematicheskoye nasledie al-Farabi* (The mathematical heritage of al-Fārābī). Alma-Ata: Izdat. "Nauka" Kax. SSR, 1974. In Russian.

The mathematical works of the great 10th-century philosopher and scientist al-Fārābī are considered.

337. Matvievskaya, G.P. *K istorii matematiki Sredney Azii* (On the history of mathematics in Central Asia). Tashkent: Izdat. Akademii Nauk Uzbek. SSR, 1962. In Russian.

A brief survey of mathematics in Central Asia from the 9th to 15th centuries, with biographies of some of the prominent scientists of this period. There are also an historiographical essay and an annotated bibliography of works published in European languages on Islamic mathematics to 1962.

338. ———. *Uchenie o chisle na srednevekovom Bliznem i Srednem Vostoke* (Studies on number in the medieval Near and Middle East). Tashkent: Izdat. "Fan" Uzbek. SSR, 1967. In Russian.

The major doctrines on the concept of number in medieval Islamic mathematics are considered: theoretical and practical arithmetic, algebra (especially the theory of quadratic irrationalities in the Arabic commentaries to Book X of Euclid's *Elements*), and the theory of ratios. An extensive bibliography is given.

339. Mieli, A. *La science arabe et son rôle dans l'evolution scientifique mondial*. Leyden: Brill, 1938.

Survey of medieval Arabic science, including mathematics, with an estimation of its influence on the development of world science. Biographical and bibliographical information is given on the major Islamic mathematicians.

340. Millás Vallicrosa, J. *Estudios sobre Azarquiel*. Madrid: Escuelas de Estudios Arabes de Madrid y Grana, 1950.

This book is the standard work on Zarqālī.

341. ———. *Estudios sobre historia de la ciencia española.* Barcelona: Consejo Superior de Investigaciones Científicas, Instituto Luis Vives, Estudios No. 2, 1949.

342. ———. *Nuevos estudios sobre historia de la ciencia española.* Barcelona: Consejo Superior de Investigaciones Científicas, Instituto Luis Vives, No. 7, 1960.

This, like its predecessor, item 341, deals mainly with astronomical topics, but there are chapters on translations and mathematics in general.

343. Nasr, S.H. *Islamic Science, an Illustrated Survey.* London: World of Islam Festival Publishing Co., 1976.

General introduction to Arabic mathematics and astronomy. See the essay review of the chapters in Nasr's book on astronomy and mathematics by D. King, "Islamic Astronomy and Mathematics: An Essay Review," *Journal for the History of Astronomy* 9 (1978), 212-219; reprinted in *Bibliotheca Orientalis* 35 (1978), 339-343.

344. O'Leary, De Lacy. *How Greek Science Passed to the Arabs.* 4th printing London: Routledge and Kegan Paul, 1964.

345. Rosenfeld, B.A., and A.P. Juškevič. *Omar Khayam.* Moscow: Izdat. "Nauka," 1965.

Biography of the poet, philosopher, and scientist ᶜUmar Khayyām (1048-1131), with analysis of his mathematical works. See also item 319.

346. ———. "Matematika v srednie veka. Strani islama" (Mathematics in the Middle Ages. Islamic World). *Istoria matematiki s drevneyshikh vremyon do nachala XIX stoletiya* (The history of mathematics from ancient times to the beginning of the XIXth century). Vol. 1 edited by A.P. Juškevič. Moscow: Izdat. "Nauka," 1970, 205-244. In Russian.

Surveys achievements of Islamic mathematicians of the 9th to 16th centuries.

347. Rosenfeld, B.A., M.M. Rozhanskaya, and Z.K. Sokolovskaya. *Abu-r-Rayhan al-Biruni, 973-1048.* Moscow: Izdat. "Nauka," 1973.

Biography and analysis of the scientific works of al-Bīrūnī.

348. Sarton, George. *Introduction to the History of Science.* Baltimore: Williams and Wilkins, 1927-1948. 3 vols.

A systematic survey of the development of the history of science to the 14th century, with considerable attention paid to Islamic mathematicians; biographical essays and extensive bibliography. See also items 469, 2324.

349. Sirazhdinov, S. Kh., ed. *Iz istorii tochnikh nauk na srednevekovom Blizhnem i Srednem Vostoke* (On the history of the exact sciences in the medieval Near and Middle East). Tashkent: Izdat. "Fan," 1972. In Russian.

A collection of papers devoted to various questions on the history of the physical and mathematical sciences in the Middle Ages. A paper by G.P. Matvievskaya describes the mathematical and astronomical manuscripts in the Institute of Oriental Studies at the Academy of Sciences of the Uzbek SSR.

350. ———. *Matematika i astronomiya v trudakh uchenikh srednevekovogo Vostoka* (The mathematics and astronomy in the works of the scientists of the medieval East). Tashkent: Izdat. "Fan" Uzbek. SSR, 1977. In Russian.

A collection of papers devoted to the history of the exact sciences in the medieval Near and Middle East. It also contains translations of some treatises from Arabic into Russian.

351. ———. *Matematika na srednevekovom Vostoke* (The mathematics of the medieval East). Tashkent: Izdat. "Fan" Uzbek. SSR, 1978. In Russian.

A collection of papers similar to item 350.

352. ———, and G.P. Matvievskaya. "On the Study of the History of Mathematics in Central Asia." *Istoriko-Matematicheskie Issledovaniia* 21 (1976), 51-60.

Surveys of Soviet studies of Arabic mathematics in Central Asia.

353. Sobirov, G. *Tvorcheskoye sotrudnichestvo uchenikh Sredney Azii v samarqandskoy nauchnoy shkole Ulugbeka* (The creative collaboration of the scientists of Middle Asia in Ulugh Beg's scientific school in Samarqand). Dushanbe: Izdat. "Irfon," 1973. In Russian.

Devoted to the scientific activity of Ulugh Beg's school and to its influence on the further development of mathematics and astronomy in the East.

354. Wiedemann, E. *Aufsätze zur arabischen Wissenschaftsgeschichte*. Hildesheim: Olms, 1970.

This is a reprinting, in two volumes, of Wiedemann's articles on the physical and mathematical sciences in the Islamic world during the Middle Ages, all of which originally appeared in the *Sitzungsberichte der phys.-mediz. Sozietät* in Erlangen as "Beiträge zur Geschichte der Naturwissenschaften" (1902-1928). Many of Wiedemann's articles are still hidden in obscure journals, but this collection is invaluable. Many of the articles are based in large measure upon Arabic manuscripts.

355. Youschkevitch, Adolf P. *Les mathématiques arabes (XVIIe-XVe siècles)*. Paris: J. Vrin, 1976.

A French translation of the relevant sections of the German version of 1964, item 332. The original Russian was printed in Moscow in 1961.

Algebra

356. Rashed, Roshi. "L'arithmétisation de l'algèbre au XIIème siècle." *Actes du XIIIème Congrès d'Histoire des Sciences*, Sections 3 and 4 (Moscow), 3 (1974), 63-69.

357. ———. "Resolution des équations numériques et algèbre: Sharaf-al-Dīn-al-Ṭusī-Viète." *Archive for History of Exact Sciences* 12 (1974), 244-290.

See item 1181.

358. ———. *L'art de l'algèbre de Diophante*. Cairo: Edition de la Bibliothèque nationale, 1975.

359. ———. "Les recommencements de l'algèbre aux XIème et XIIème siècles." In *The Cultural Context of Medieval Learning*. Edited by J. Murdoch and E. Sylla. Boston: Reidel, 1975, 33-60.

360. ———. "L'extraction de la racine n^ième et l'invention des fractions décimales (XIe-XIIe siècles)." *Archive for History of Exact Sciences* 18 (1978), 191-243.

361. ———. "L'analyse diophantienne au Xème siècle: l'example d'al-Khāzin." *Revue d'Histoire des Sciences* 32 (1979), 193-222.

362. ———, and S. Ahmad. *Al-Bāhir en Algèbre d'al-Samaw'al*. Damascus: Université de Damas, 1972.

Trigonometry

363. Braunmühl, A. von. *Vorlesungen über die Geschichte der Trigonometrie*. Vol. 1. Leipzig: B.G. Teubner, 1900.

In this comprehensive study of the history of trigonometry, considerable attention is given to medieval mathematicians of the Islamic world. For medieval Islam, however, this work has recently been superseded by Debarnot, item 295.

364. Rozhanskaya, M.M. *Mekhanika na srednevekovom Vostoke* (Mechanics of the medieval East). Moscow: Izdat. "Nauka," 1976. In Russian.

Contains much information on other parts of mathematics, especially spherical trigonometry.

For other Arabic works or studies of Islamic trigonometry, see the items already cited by Burger and Kohn, item 325; Debarnot, item 295; King, item 373c; Krause, item 315; Samsó Moya, item 288; and Villuendas, item 313.

Geometry

365. Hogendijk, J.P. "On the Trisection of an Angle and the Construction of a Regular Nonagon by Means of Conic Sections in Medieval Islamic Geometry." Second International Symposium for the History of Arabic Science, Aleppo, April, 1979. Preprint.

366. Rosenfeld, B.A. *History of Non-Euclidean Geometry*. Moscow: "Nauka," 1976. In Russian.

     Chapters 1-4 contain studies of spherical trigonometry, the theory of parallel lines, geometrical transformations, geometrical algebra, and the philosophy of space in Islamic mathematics.

366a. ———, and A.P. Youschkevitsch. *Teoriya parallelnikh linii na srednevekovom vostoke IX-XIV vv*. Moscow: Nauka, 1983.

     On the theory of parallels in Islamic mathematics.

367. Schoy, K. "Die Gnomonik der Araber." Part of *Die Geschichte der Zeitmessung und der Uhren*. Edited by E. von Basserman-Jordan. Berlin: De Gruyter, 1923.

     A major study of Islamic gnomonics. See also King, items 373b, 373c, 380a, 380b.

For additional works previously listed which deal with Geometry, see Clagett, item 307; Luckey, item 301; Mamedbeyli, item 317; Matvievskaya, items 337 and 338; Rosenfeld, items 345 and 346; Millás Vallicrosa, items 341 and 342.

Indeterminate Equations

See Bashmakova, item 791; Rashed, items 358, 360, 361; Sesiano, items 287a, 298b; Woepcke, item 298a.

Astronomy

368. Delambre, J.B.J. *Histoire de l'astronomie du moyen âge*. Paris: V. Courcier, 1819. Reprinted New York: Johnson Reprint Corp., 1965.

     Still worth consulting.

369. Duhem, Pierre. *Le système du monde; histoire des doctrines cosmologiques de Platon à Copernic*. Paris: A. Hermann, 1913-1917. Reprinted Paris: A. Hermann, 1956-1963.

     Although not always reliable, Duhem's work is still worth consulting; volume 2 is devoted to Islamic astronomy.

370. Kennedy, Edward S. "A Survey of Islamic Astronomical Tables." *Transactions of the American Philosophical Society*, New Series 46 (ii) (1956), 121-177.

     Introduces the Arabic *zījes*, listing the known ones, and gives details of twelve of them. A basic reference work, in which the author bases his conclusions upon the study of unedited Arabic and Persian manuscripts; many important problems of the history of astronomy and mathematics in medieval Islam are resolved. See also King, item 373.

371. King, D.A. "Al-Khalīlī's *Qibla*-Table." *Journal of Near Eastern Studies* 34 (1975), 81-122.

Together with his article "Ḳibla" in the new edition of the *Encyclopaedia of Islam*, this article makes a good introduction to medieval methods for finding the *qibla* (or direction of Mecca) for prayer.

372. ———. "Ibn Yūnus' 'Very Useful Tables' for Reckoning Time." *Archive for History of Exact Sciences* 10 (1973), 342-394.

A detailed analysis of the corpus of tables for astronomical timekeeping which was used in medieval Cairo.

373. ———. "On the Astronomical Tables of the Middle Ages." *Colloquia Copernicana* 3 (1975), 37-56.

Classifies tables not listed in Kennedy, item 370.

373a. ———. "Al-Khalīlī's Auxiliary Tables for Solving Problems of Spherical Astronomy." *Journal for the History of Astronomy* 4 (1973), 99-110.

Shows the sophistication of trigonometrical techniques achieved by the fourteenth century.

373b. ———. "The Astronomy of the Mamluks." *Isis* 74 (1983), 531-555.

Surveys the achievements in mathematical astronomy in 13th- to 16th-century Egypt and Syria.

373c. ———. "Al-Khwārizmī and New Trends in Mathematical Astronomy in the Ninth Century." *New York University: Hagop Kevorkian Center for Near Eastern Studies: Occasional Papers on the Near East* 2 (1983).

Description and analysis of seven recently discovered minor works by al-Khwārizmī.

374. Kunitzsch, P. *Untersuchungen zur Sternnomenklatur der Araber*. Wiesbaden: O. Harrassowitz, 1961.

375. ———. *Arabische Sternnamen in Europa*. Wiesbaden: O. Harrassowitz, 1959.

Both this and the preceding work deal with names and nomenclature in Arabic for stars. See also Kunitzsch's work on Ptolemy, item 322.

376. Pingree, David. "The Greek Influence on Early Mathematical Astronomy." *The Journal of the American Oriental Society* 93 (1973), 32-44.

A survey article of great depth.

377. Sayili, A.M. *The Observatory in Islam and its Place in the General History of the Observatory*. Ankara: Turkish Historical Society Publications, series 7, #38, 1960.

The standard book on the subject. Good bibliography.

For additional material concerned with Arabic astronomy, see al-Battani, item 289; al-Bīrūnī, item 291; al-Bitruji, item 297; Carmody, items 280 and 324; Debarnot, item 295; Goldstein, items 281, 305, and 323; Hartner, item 329; Jaghmini, item 314; Kary-Niyazov, item 333; Kennedy, items 293, 296, 334, and 335; Millás Vendrell, item 304; Ptolemy, item 322; Sabra, item 311; Samsó Moya, item 288; Schoy, item 367; and Sirazhdinov, item 350.

Instruments

378. Brieux, A., and F.R. Maddison. *Répertoire des facteurs d'astrolabes et de leurs oeuvres*. Vol. 1, Islam (forthcoming).

    This work will largely supersede L.A. Mayer's *Islamic Astrolabists and Their Works* (Geneva: A. Kundig, 1956), which is now out of print, and Maddison's bibliography on instruments in item 381. The introduction to the *Répertoire* is a revision of the relevant part of the catalog (not printed) by Maddison and A.J. Turner of the exhibition "Science and Technology in Islam" at the Science Museum, London, for the World of Islam Festival in 1976.

379. Gibbs, S., J. Henderson, and D. Price. *A Computerized Checklist of Astrolabes*. New Haven: Yale University Department of the History of Science and Medicine, 1973.

380. Gunther, R.T. *The Astrolabes of the World*. Oxford: Oxford University Press, 1932. Reprinted London: The Holland Press, 1976.

    Largely, but not entirely, based on the Lewis Evans collection in Oxford. There are checklists of astrolabes such as those compiled by Gibbs, item 379, and by R. Webster, M. Webster, and D. Pingree, to appear. Gunther includes treatises on the astrolabe as well. For Pseudo-Messahala's treatise, see Gunther's *Early Science in Oxford*, Vol. V, and an article by P. Kunitzsch in *Archives internationales d'histoire des sciences* (1981).

380a. King, D. "A Fourteenth-Century Tunisian Sundial for Regulating the Times of Muslim Prayer." In *Prismata*. Edited by W. Saltzer and Y. Maeyama. Item 268a, 187-202.

380b. ———. "Three Sundials from Islamic Andalusia." *Journal for the History of Arabic Science* 2 (1978), 331-357.

    Both this and the preceding item present mathematical analyses of the markings on four medieval Islamic sundials and investigate their mode of construction.

381. Maddison, Francis R. "Early Astronomical and Mathematical Instruments: A Brief Survey of Sources and Modern Studies." *History of Science* 2 (1963), 17-50.

    See item 574 and Brieux, item 378.

382. Michel, Henri. *Traité de l'astrolabe*. Paris: Gauthier-Villars, 1947, viii + 202 pp., 24 pls.

    For a brief elementary explanation of the astrolabe in English, see J.D. North, "The Astrolabe," *Scientific American* 230 (1) (January 1974), 96-106. See also item 1804.

INDIAN MATHEMATICS

Mathematics in India can be traced back to the Śulva works (6th century B.C. and later) which are compilations of geometrical rules for the construction of Vedic altars. But the main sources for our knowledge about Indian mathematics are found in gaṇita or computational science which, in Sanskrit literature, is traditionally classified as one of the three branches of jyotiḥśāstra; the other two are horā or horoscopic astrology and saṃhitā or divination. Gaṇita includes mathematics (also called gaṇita) and mathematical astronomy; major treatises on the latter, called siddhānta, often include a few chapters on mathematics. The fundamental operations of arithmetic, such as addition, etc., and geometry are contained in pāṭī, a compilation of algorithms or, in other words, rules for completely mechanical solutions of different mathematical problems. Since about the 8th century, algebra, under the name of bījagaṇita, or seed mathematics, has been the source of pāṭī so far as it was used to generate algorithms for pāṭī. Trigonometry, in which there were noteworthy developments in India until the 18th century, was traditionally associated with astronomy. The present bibliography covers studies of these subjects and editions of texts. General reference works cited here will greatly facilitate further reference.

General Reference Works

383. Pingree, David. "Census of the Exact Sciences in Sanskrit." *Memoirs of the American Philosophical Society*, Series A, 81 (1970), 1-60; 86 (1971), 1-147; 111 (1976), 1-208; 146 (1981), 1-447.

An exhaustive work in progress to "provide all available bibliographical information concerning works in jyotiḥśāstra and related fields and biographical information concerning their authors." The first four volumes contain articles on the authors whose names begin with the vowel *a* to the labial *m* in the order of the Sanskrit alphabet. See also articles by the same author in the *Dictionary of Scientific Biography* (ed. C.C. Gillispie), which give ample biographical and bibliographical data for major Indian mathematicians and astronomers. See also item 19, and especially item 404.

384. Raja, K. Kunjunni. "Astronomy and Mathematics in Kerala: Account of the Literature." *Brahmavidyā* (Adyar Library Bulletin, Madras) 27 (1963), 118-167.

385. Sarma, K.V. *A History of the Kerala School of Hindu Astronomy: In Perspective*. Hoshiarpur: Vishveshvaranand Institute, 1972.

Most pages are devoted to bibliographical (manuscripts) and biographical information.

386. Sen, S.N., A.K. Bag, and S.R. Sarma. *A Bibliography of Sanskrit Works on Astronomy and Mathematics, Part 1: Manuscripts, Texts, Translations, and Studies*. New Delhi: National Institute of Sciences of India, 1966.

Source Materials

387. *Āpastambaśulvasūtra*. Edited by A. Bürk. "Das Āpastamba-Śulba-Sūtra." *Zeitschrift der deutschen morgenländischen Gesellschaft* 55 (1901), 543-591; and 56 (1902), 327-391.

With an introduction and German translation.

388. Āryabhaṭa I. *Āryabhaṭīya*. Edited by K.S. Shukla. New Delhi: Indian National Science Academy, 1976.

With the commentary of Bhāskara I and the editor's introduction. For a recent study and modern translation of the second chapter "Mathematics," see Kurt Elfering, *Die Mathematik des Āryabhaṭa I* (Munich: Wilhelm Fink Verlag, 1975).

389. Āryabhaṭa II. *Mahāsiddhānta*. Edited by S. Dvivedin. (Benares Sanskrit Series Nos. 148, 149, and 150.) Benares: Braj Bhushan Das & Co., 1910.

With editor's commentary in Sanskrit. No modern translation of the two chapters devoted to mathematics exists.

390. *Bakhṣālī Manuscript*. See G.R. Kaye, *The Bakhshālī Manuscript*. Item 420.

391. Bhāskara II. *Līlāvatī*. Edited by V.G. Āpṭe. (Ānandāśrama Sanskrit Series No. 107.) Poona: Ānandāśrama Press, 1937. Also edited by K.V. Sarma. Hoshiarpur: Vishveshvaranand Vedic Research Institute, 1975.

Āpṭe's edition is accompanied by the commentaries of Gaṇeśa and Mahīdhara; Sarma's by those of Śaṅkara and Nārāyaṇa. For an English translation, see H.T. Colebrooke, *Algebra with Arithmetic and Mensuration from the Sanskrit of Brahmegupta and Bhāscara*. London: John Murray, 1817.

392. ———. *Bījagaṇita*. Edited by V.G. Āpṭe. (Ānandāśrama Sanskrit Series No. 99.) Poona: Ānandāśrama Press, 1930.

With the commentary of Kṛṣṇa. For an English translation, see the work by Colebrooke cited in item 391.

393. Brahmagupta. *Brāhmasphuṭasiddhānta*. Edited by S. Dvivedin. "The Brāhmasphutasiddhānta and Dhyānagrahopadeśādhyāya." *The Paṇdits*, New Series 23 (1901), 309-324, 389-404, 453-468, 517-532, 581-596, and 645-660; New Series 24 (1902), 1-16, 65-80, 137-142, 209-240, 273-312, 321-360, 385-416, 465-496, 529-576, 593-624, 657-688, and 721-755. Reprinted in one volume, Benares: Medical Hall Press, 1902.

With editor's commentary in Sanskrit. For an English translation of two chapters devoted to mathematics see Colebrooke as cited in item 391.

394. Kamalākara. *Siddhāntatattvaviveka*. Edited by S. Dvivedin. (Benares Sanskrit Series Nos. 1, 2, 3, 6, and 14.) Benares: Braj Bhushan Das & Co., 1880-1885. Revised by Muralīdhara Jhā, Benares, 1924-1935.

No modern translation available. For a sporadic study, see below R.C. Gupta, "Addition and Subtraction Theorems," item 415.

395. Mahāvīra. *Gaṇitāsarasaṅgraha*. Edited by M. Raṅgācārya. Madras: Government Press, 1912.

With an English translation. For a discussion on Mahāvīra's contribution, see B.S. Jain, "On the Gaṇitasārasaṃgraha of Mahāvīra (*c.* A.D. 850)." *Indian Journal of History of Science* 12 (1977), 17-32.

396. Nārāyaṇa. *Gaṇitakaumudī*. Edited by P. Dvivedi. (The Princess of Wales Saraswati Bhavana Texts No. 57.) Benares: Government Sanskrit Library, 1936 (part 1) and 1942 (part 2).

No modern translation available. For a study of a chapter "Bhadragaṇita" (computation of magic squares), see S. Cammann, "Islamic and Indian Magic Squares," item 422.

397. Nīlakaṇṭha. *Tantrasaṅgraha*. Edited by K.V. Sarma. (Panjab University Indological Series No. 10.) Hoshiarpur: Panjab University, 1977.

With editor's introduction and two commentaries, the Yuktidīpikā and Laghuvivṛti. No modern translation available.

398. Śrīdhara. *Pāṭīgaṇita*. Edited by K.S. Shukla. Lucknow: Lucknow University, 1959.

With an anonymous commentary and English translation.

399. ———. *Triśatikā*. Edited by S. Dvivedin. Benares: Chandraprabha Press, 1899.

For its partial English translation, see Ramanujacharia and G.R. Kaye, "The Triśatikā of Śrīdharācārya," *Bibliotheca Mathematica* 3 (13) (1912-1913), 203-217.

400. Śrīpati. *Gaṇitatilaka*. Edited by H.R. Kāpadīā. (Gaekwad's Oriental Series No. 78.) Baroda: Oriental Institute, 1937.

With the commentary of Siṃhatilakasūri and editor's introduction. No modern translation available.

401. ———. *Siddhāntaśekhara*. Edited by B. Miśra. Calcutta: University of Calcutta, 1932 (part 1) and 1947 (part 2).

With the commentaries of Makkibhaṭṭa (in chapters 1-4) and of the editor (in chapters 5-20) and an introduction by N.K. Majumdar. No modern translation available.

General Histories

402. Bag, Amulya Kumar. *Mathematics in Ancient and Medieval India*. Varanasi: Chaukhambha Orientalia, 1979.

Important for its rare attempt to trace the development of mathematics in India.

403. Datta, Bibhutibhusan, and A.N. Singh. *History of Hindu Mathematics*. Lahore: Motilal Banarsidass, 1935 (part 1) and 1938 (part 2). Reprinted in 1 vol. Bombay: Asia Publishing House, 1962.

A source book arranged according to topics. Part 1 deals with numeral notation and arithmetic and part 2 with algebra. Part 3, which was to contain geometry, trigonometry, etc., did not see the light of day. The authors' posthumous manuscript entitled "Hindu Geometry" was revised and published by K.S. Shukla, *Indian Journal of History of Science* 15 (1980), 121-188.

404. Pingree, David. "History of Mathematical Astronomy in India." *Dictionary of Scientific Biography.* Vol. 15. Edited by C.C. Gillispie. New York: Charles Scribner's Sons, 1978, 533-633.

A monumental work that traces the roots, and throws light on the development, of the subject with special attention to its transmission from other civilizations into India.

405. ———. *Jyotiḥśāstra: Astral and Mathematical Literature.* (A History of Indian Literature, Vol. 6, fasc. 4.) Edited by J. Gonda. Wiesbaden: Otto Harrassowitz, 1981.

Chapters I, Śulbasūtras, II, Astronomy, III, Mathematics, and IX, Transmission of Jyotiḥśāstra, are especially important for students of the history of Hindu mathematics. Marvelous historical sketch of the literature with useful notes. Reveals, among other things, that the date of Śrīdhara is fixed between Brahmagupta (fl. A.D. 628) and Govindasvāmin (fl. A.D. 850).

Algebra

406. Ganguli, Saradakanta. "India's Contribution to the Theory of Indeterminate Equations of the First Degree." *Journal of the Indian Mathematical Society/Notes and Questions* 19 (1931-1932), 110-120, 129-142, and 153-168.

An analysis of the development of the kuṭṭaka theory in India up to Bhāskara II (12th century). Author's opinion about "India's contribution" may not be approved by every reader.

407. Selenius, Clas-Olof. "Rationale of the Chaklavāla Process of Jayadeva and Bhāskara II." *Historia Mathematica* 2 (1975), 167-184.

Throws new light on the mathematical meaning and validity of the cakravāla method from the viewpoint of the theory of the ideal continued fractions. See also C.-O. Selenius, "Kettenbruchtheoretische Erklärung der zÿklischen Methode zur Lösung der Bhāskara-Pell-Gleichung," *Acta academica Aboensis, mathematica et physica*, 23 (10) (1963), 1-44.

408. Shukla, Kripa Shankar. "Ācārya Jayadeva, the Mathematician." *Gaṇita* 5 (1954), 1-20.

A historically important report on quotations from Jayadeva's work (11th century or earlier), which deals with the varga-prakṛti, a theory of general Pellian equations, including the cakravāla method.

Geometry

409. Datta, Bibhutibhusan. *The Science of the Śulba.* Calcutta: University of Calcutta, 1932.

A basic study of the mathematical contents of the Śulva

literature. Not always reliable. See also G. Thibaut, "On the Śulva-sūtras," *Journal of the Asiatic Society of Bengal* 44 (1875), 227-275.

410. Michaels, Axel. *Beweisverfahren in der Vedischen Sakralgeometrie.* (Alt- und Neu-Indische Studien [Universität Hamburg] No. 20.) Wiesbaden: Franz Steiner Verlag GmbH, 1978.

The author declares that Vedic geometry embodied in the *Śulvasūtras* is logic-free and non-axiomatic, but provable, and that the proofs are given in the texts "by a demonstration of the appropriate patterns of action." Useful appendix: concordance of the editions of the *Śulvasūtras*.

411. Pottage, John. "The Mensuration of Quadrilaterals and the Generation of Pythagorean Triads: A Mathematical, Heuristical and Historical Study with Special Reference to Brahmagupta's Rules." *Archive for History of Exact Sciences* 12 (1974), 299-354.

Investigates possible derivations of Brahmagupta's mensuration rules from Hindu mathematics of those days.

412. Sarasvatī, T.A. *Geometry in Ancient and Medieval India.* New Delhi: Motilal Banarsidass, 1979.

Composed of ten chapters: introduction, Śulva geometry, Jaina geometry, trapezium, quadrilateral, triangle, circle, solids, geometrical algebra, and shadow problems and others. Mostly based on the author's own research in primary sources.

413. Seidenberg, A. "The Ritual Origin of Geometry." *Archive for History of Exact Sciences* 1 (1962), 488-527.

Seidenberg's main thesis is that the elements of geometry in ancient Greece, Babylonia, Egypt, India, and China are derived from a system of ritual practices as disclosed in the *Śulvasūtras*. See also A. Seidenberg, "The Origin of Mathematics," *Archive for History of Exact Sciences* 18 (1978), 301-342, as well as item 1868.

Trigonometry

414. Gupta, R.C. "Bhāskara I's Approximation to Sine." *Indian Journal of History of Science* 2 (1967), 121-136.

An analysis of a remarkable approximation to sine by a rational function.

415. ———. "Addition and Subtraction Theorems for the Sine and the Cosine in Medieval India." *Indian Journal of History of Science* 9 (1974), 164-177.

416. Sarasvatī, T.A. "The Development of Mathematical Series in India after Bhāskara II." *Bulletin of the National Institute of Sciences of India* 21 (1963), 320-343.

Main topics are the infinite series for π and the sine and cosine power series developed in India up to the 16th century. See also C.T. Rajagopal and M.S. Rañgāchāri, "On an Untapped Source of Medieval Keralese Mathematics," *Archive for History of Exact Sciences* 18 (1978), 89-102.

Pāṭī and Others

417. Ganguli, Saradakanta. "The Indian Origin of the Modern Place-Value Arithmetical Notation." *American Mathematical Monthly* 39 (1932), 251-256, 389-393; and 40 (1933), 25-31, 154-157.

A close examination of both mathematical and non-mathematical literatures as well as of inscriptions.

418. Gupta, R.C. "Some Important Indian Mathematical Methods as Conceived in Sanskrit Language." *Indological Studies* (University of Delhi) 3 (1974), 49-62.

Rule of three, method of iteration, and method of average.

419. Hayashi, Takao. "An Introduction to the Japanese Translation of the Līlāvatī." *Indo Tenmongaku-Sūgaku Shū*. Edited by M. Yano. (Kagaku no Meicho Series No. 1.) Tokyo: Asahi Press, 1980, 141-196. In Japanese.

Through an analysis of the *Līlāvatī* and *Bījagaṇita* of Bhāskara II, the author points out that pāṭī is characterized as consisting of algorithms for non-negative numbers, many of which algorithms are obtained by means of bījagaṇita, a theory of equations in the domain of positive and negative numbers, with the help of symbols for unknown numbers.

420. Kaye, G.R. *The Bakhshālī Manuscript: A Study in Medieval Mathematics*. (Archaeological Survey of India, New Imperial Series No. 43.) Parts 1 and 2, Calcutta: Government of India, Central Publication Branch, 1927; and part 3, New Delhi: Manager of Publication, 1933.

Consists of a study, facsimiles of the whole extant text, and transliteration. The study and transliteration are not always reliable. See also B. Datta, "The Bakhshālī Mathematics," *Bulletin of the Calcutta Mathematical Society* 21 (1929), 1-60; and A.A.K. Ayyangar, "The Bakshālī Manuscript," *The Mathematics Student* 7 (1939), 1-16.

421. Srinivasan, Saradha. *Mensuration in Ancient India*. New Delhi: Ajanta Publication, 1979.

A good study of weights and measures found in the literatures of different fields as well as in archaeological excavations.

Transmission

422. Cammann, Schuyler. "Islamic and Indian Magic Squares." *History of Religion* 8 (1968-1969), 181-209, 271-299.

One of a series in the author's comparative study of magic squares in China, India, and the Islamic world, with special attention to their transmission and religious background.

423. Chakrabarti, Gurugovinda. "Typical Problems of Hindu Mathematics." *Annals of the Bhandarkar Oriental Research Institute* 14 (1932-1933), 87-102.

Significant and hitherto rare approach to the comparative study of typical problems widely spread among ancient and medieval nations. See also H. Hermelink, "Arabic Recreational

Mathematics as a Mirror of Age-Old Relations between Eastern and Western Civilizations," *Proceedings of the First International Symposium for the History of Arabic Sciences*, part 2 (English section), Aleppo, 1976, 44-52.

424. Pingree, David. "The Indian and Pseudo-Indian Passages in Greek and Latin Astronomical and Astrological Texts." *Viator* (University of California Press) 7 (1976), 141-195.

An amazing work tracing the transmission of the subjects among Sanskrit, Arabic, Greek, and Latin texts.

425. Saidan, A.S. *The Arithmetic of al-Uqlīdisī*. Dordrecht and Boston: D. Reidel Publishing Company, 1978.

English translation of the earliest extant Arabic work of Hindu arithmetic, *Kitāb al-Fuṣūl fī al-Ḥisāb al-Hindī* of al-Uqlīdisī (10th century), with an introduction and commentary.

HEBREW MATHEMATICS

Parallel to the main streams of medieval learning which flowed in Arabic and Latin was a smaller but vigorous one in Hebrew. Because Jews were spread through both Islam and Christendom, intellectual contact between the two was sometimes mediated through Hebrew, and there are Arabic works that survive only in Hebrew translations. Moreover, a significant number of these Hebrew treatises were translated into Latin and were consulted by European scholars for several centuries. It should also be noted that many original contributions were made by these authors (especially Levi ben Gerson), not all of which have been adequately explored in the modern secondary literature. Because so much of the material remains in manuscript form, the bibliographies (see, for example, items 443, 444, and 445) contain much information not available elsewhere.

The bibliography that follows covers the Hebrew tradition in mathematics from antiquity until early modern times when Jewish scientists began to write in European languages. Because of its close identification with mathematics some literature in astronomy is also cited here.

In its earliest phases, Hebrew mathematics derived from Oriental and Babylonian origins. In the Middle Ages it was nourished from Arabic translations of Greek works and developed close links with developments in the Muslim world. Hebrew scientific works were written primarily in Spain and southern France, but a number of authors lived in other parts of the Mediterranean basin, e.g., Greece, Turkey, Italy, Egypt, and some in other places, e.g., Yemen. The period most strongly represented is the one extending from the 12th to the 15th centuries. Interest was centered on algebra, geometry, number theory, trigonometry, and mathematical astronomy (including mathematical tables for calculating planetary positions).

General Studies

426. Feldman, W.M. *Rabbinical Mathematics and Astronomy*. Reprinted New York: Hermon Press, 1965.

Although marred by omissions and errors, this book can serve as an introduction to the field. A more complete treatment of the calendrical work of Maimonides is given in item 435 below.

427. Gandz, Solomon. *Studies in Hebrew Astronomy and Mathematics.* Selected, with an Introduction, by S. Sternberg. New York: Ktav, 1970, 544 pp.

A reprint of 15 essays that appeared between 1927 and 1951. Gandz's approach to the Hebrew sources was to see them in the context of Babylonian, Greek, and Arabic traditions. Perhaps the most important of these essays is his edition and commentary of the *Mishnat ha-Middot*, an early treatise on geometry in Hebrew that Gandz dated ca. 150 A.D., but that Sarfatti (item 442 below, pp. 58-60) argues persuasively belongs to the early Islamic period. There are also essays on the Jewish calendar, the astrolabe, Hebrew numerals, and Saadya Gaon (10th century) as a mathematician.
Review: Goldstein, B.R., *Speculum* 47 (1972), 124-125.

428. ———. "The Mishnat ha-Middot, the First Hebrew Geometry about 150 C.E., and the *Geometry of al-Khowarizmi*, the First Arabic Geometry (ca. 820) Representing the Arabic Version of the Mishnat ha-Middot. Texts with English Translation, Introduction and Notes." *Quellen und Studien zur Geschichte der Mathematik und Physik*, A2. Berlin: Springer, 1932.

Gandz relates the *Mishnat ha-Middot* to Heron and his school, placing it in the Oriental tradition of devising mathematical tools for practical applications rather than exploring theory (pure science) for its own sake.
This must be read in conjunction with the articles by Gad B. Sarfatti, "The Mathematical Terminology of the *Mishnat ha-Middot*" (Hebrew), *Leshonenu* 23 (5719), 156-171, and 24 (5720), 88-94, which also contain an improved text. Sarfatti refutes Gandz's early dating and ascribes the *Mishnat ha-Middot* to the 9th century or even later. This work by Gandz is reprinted in item 427.

429. ———. "The Invention of the Decimal Fractions and the Application of the Exponential Calculus by Immanuel Bonfils of Tarascon (c. 1350)." *Isis* 25 (1936), 16-45.

A classic study of work relating to the introduction of decimal fractions by Immanuel Bonfils (14th century, southern France). Gandz's claims for Bonfils have been modified by subsequent work on Arabic mathematics, and Bonfils is no longer considered to have played the crucial role in this development (see P. Luckey, *Die Rechenkunst bei Ǧamšid b. Mas^{c}ūd al-Kāšī*, item 301, especially p. 120 ff, and R. Rashed, "L'extraction de la racine n^{ième} et l'invention des fractions décimales," item 360). In addition to the manuscripts cited by Gandz, see MS Paris, Bibliothèque Nationale, Hebrew, 903, fol. 138a.

430. Goldstein, Bernard R. "The Astronomical Tables of Levi ben Gerson." *Transactions of the Connecticut Academy of Arts and Sciences* 45 (1974), 1-285.

An edition of the astronomical tables of Levi ben Gerson (14th century, southern France) with an introduction to his other astronomical achievements, including his wide-ranging

discussion of possible astronomical models and a description of his profound modifications of Ptolemy's lunar model. Unlike most medieval table-makers, Levi did not simply copy tables from the works of his predecessors; rather he recomputed them anew often changing the underlying parameters based on his own observations. The mathematical structure of the tables is displayed in the editor's notes and, as a result of this study, it is clear that Levi was truly outstanding among those who wrote in Hebrew in the Middle Ages. Indeed his work compares favorably with contemporary achievements by those who wrote in Latin and Arabic.
Review: Moesgaard, K.P., *Centaurus* 21 (1977), 197-199.

431. ———. *Ibn al-Muthannā's Commentary on the Astronomical Tables of al-Khwārizmī*. New Haven: Yale University Press, 1967, 406 pp.

An edition with translation and commentary of the two surviving Hebrew versions of this text. The earliest phase of Islamic astronomy was deeply influenced by Indian sources, but it was largely displaced with the introduction of Greek methods in the 9th century. In reconstructing the early phase we are forced to depend on fragments preserved in later texts as well as commentaries on treatises that no longer survive. In this case, the Arabic text of Ibn al-Muthannā (10th century) is only preserved in Hebrew, but its information on the original form of al-Khwārizmī's tables is most valuable. The commentator attempted unsuccessfully to explain Hindu procedures in terms of the Greek models. In one of the Hebrew versions, Ibn Ezra (12th century, Spain) added an introduction which gives his understanding of the transmission of Indian science to the Islamic world. An edition of the Latin version of this text has also appeared: E. Millás Vendrell, *El Commentario de Ibn al-Mutannā a las Tablas Astronomicas de al-Jwārizmī ... en la version de Hugo Sanctallensis* (Madrid, Barcelona: Consejo Superior de Investigaciones Cientificas, 1963).
Review: Pingree, D., *Speculum* 43 (1968), 722-724.

432. ———. "The Role of Science in the Jewish Community in Fourteenth Century France." *Annals of the New York Academy of Sciences* 314 (1978), 39-49.

A survey of intellectual achievements of a highly productive Jewish community and the relationship with their Spanish predecessors. Both translations and original works are discussed, including those of Samuel ben Judah of Marseille, Kalonymos ben Kalonymos of Arles, Levi ben Gerson of Orange, Immanuel ben Jacob Bonfils of Tarascon, and Shelomo ben Davin of Rodez.

433. ———. "The Survival of Arabic Astronomy in Hebrew." *Journal for the History of Arabic Science* 3 (1979), 31-39.

Argues that Hebrew manuscripts are an important source for Arabic science in three categories: (1) Arabic texts written in Hebrew characters, (2) translations (where the original does not always survive), and (3) original Hebrew treatises based on Arabic models. A number of texts are identified here for the first time, including a 19th-century copy of Ibn al-Shātir's *al-zīj al-jadīd*, originally composed in the 14th century.

434. Juschkewitsch, Adolf P. *Geschichte der Mathematik im Mittelalter*. Leipzig: B.G. Teubner, 1964.

Contains some material on Jewish mathematicians, relating their work to overall developments. See also items 332, 467.

435. Maimonides. *The Code of Maimonides: Sancitification of the New Moon*. Translated by S. Gandz, introduction by J. Obermann, commentary by O. Neugebauer. New Haven: Yale University Press, 1956.

An edition with astronomical commentary of Maimonides' description of the rules for fixing the Jewish calendar which shows that Maimonides was in full control of the technical aspects of Ptolemaic astronomy. In his commentary Neugebauer points out the relationship of the parameters used in Part III of this work with the values in al-Battānī's *zīj* (set of astronomical tables with an explanatory introduction).

In formulating the rules of the Jewish calendar, Maimonides dealt with one of the most difficult problems of medieval astronomy, namely, to determine criteria for the visibility of the new crescent. This gives an insight into the most advanced attainments of medieval lunar theory. Neugebauer's commentary is an invaluable aid.

An important contribution towards clearing up a difficulty in Maimonides' table of co-efficients for the parallax--a difficulty left unresolved by Neugebauer--is given by Kalman Kalikstein in his "Mathematical and Astronomical Commentary" to Maimonides' work (Hebrew with English summary) (New York: Post Talmudic Research Institute, 5738 [1978]).

436. Millás Vallicrosa, J.M. *La obra Sefer Ḥeshbon mahlekot ha-kokabim de R. Abraham Bar Ḥiyya ha-Bargeloni*. Barcelona: Instituto Arias Montano, 1959, 270pp.

An edition with translation and notes of a 12th-century Hebrew introduction to a set of astronomical tables. This text played a significant role in the transmission of Arabic astronomy to the Jewish communities of northern Spain and southern France where Arabic was not widely known. Bar Ḥiyya depends very heavily on the work of al-Battānī (d. 929). A glossary of technical terms is included. See also items 283, 340, 341, 342, 447, and 448.

437. Rabinovitch, Nachum L. "Early Antecedents of Error Theory." *Archive for History of Exact Sciences* 13 (1974), 348-358.

Citations from the Talmud and medieval rabbinic and astronomical authors, especially Levi ben Gerson, show that ancient and medieval scientific observers as well as artisans and lawyers were concerned with experimental and observational error, and devised rudimentary statistical rules to take it into account. These include replication of measurements and averaging data.

438. ———. *Probability and Statistical Inference in Ancient and Medieval Jewish Literature*. Toronto: University of Toronto Press, 1973, 205 pp.

Discusses rabbinic logic, random mechanisms in the Bible, statistical inference in the Talmud, and the treatment of combinations and permutations by medieval Hebrew writers. See also item 2027.
Review: Zabell, S., *Journal of the American Statistical Association* 71 (1976), 996-998.

439. Renan, E. "Les rabbins français du commencement du XIVe siècle." *Histoire Littéraire de la France* 27 (1877), 431-764.

This is a bio-bibliographical study, arranged by author, of Jewish scholars in France, based on both printed and manuscript sources with detailed references to the primary and secondary literature. Since a significant proportion of the authors treated wrote on mathematical subjects, this work is extremely valuable.

440. ———. Les écrivains juifs français du XIVe siècle. *Histoire Littéraire de la France* 31 (1893), 351-789.

A continuation of item 439.

441. Romano, D. "La transmission des sciences arabes par les juifs en Languedoc." *Juifs et judaisme de Languedoc*. Edited by M.-H. Vicaire and B. Blumenkranz. Toulouse: E. Privat, 1977, 363-386.

A discussion of the role of the Jews, particularly of the Kimḥi and Ibn Tibbon families, in the transmission of Arabic science to southern France in the 12th to 14th centuries. The most prolific translator, Moses Ben Tibbon (13th century), receives extensive treatment.

442. Sarfatti, G.B. *Mathematical Terminology in Hebrew Scientific Literature of the Middle Ages*. Jerusalem: Magnes Press, 1968, 265 pp. In Hebrew with English summary.

A discussion of the development of Hebrew scientific terms in ancient and medieval times starting with the Bible and the Talmud. The authors whose works are most extensively treated are: Abraham Bar Ḥiyya, Abraham Ibn Ezra, Maimonides, Moses Ben Tibbon, Isaac Israeli, and Levi ben Gerson. There is also a chapter on Hebrew translations with examples from Euclid, Nicomachus of Gerasa, and Archimedes. Terms in arithmetic and geometry are dealt with exhaustively, whereas trigonometry and mathematical astronomy receive less attention. Distinctions are made among various methods for enlarging the stock of scientific terms, including the modification of the meaning of an earlier word, semantic borrowing, and introducing foreign words. An extensive index is provided.

443. Steinschneider, M. *Mathematik bei den Juden*. 2nd ed., with an index by A. Goldberg. Hildesheim: George Olms, 1964, 221 pp.

Originally published as a series of articles (1893-1901). In this study the author lists in chronological order Jewish mathematicians and astronomers who wrote primarily in Hebrew. Some biographical remarks and a description of their works, mostly unpublished, are based on Steinschneider's own investigations of the manuscript sources. This bibliographic work sets a very high standard for the history of medieval science.

444. ———. "Mathematik bei den Juden (1551-1840)." *Monatsschrift für Geschichte und Wissenschaft des Judentums* 49, Neue Folge 13 (1905), 78-95, 193-204, 300-314, 490-498, 581-605, 722-743.

A continuation of item 443. Due to the author's death, the discussion does not go beyond the end of the eighteenth century.

445. ———. *Die hebraeischen Übersetzungen des Mittelalters und die Juden als Dolmetscher*. Berlin: Kommissionsverlag des Bibliographischen Bureaus, 1893, 1112 pp.

The basic bibliographic source for translations of scientific and philosophical texts into Hebrew in the Middle Ages. It is arranged by subject and divided into sections on Christian, Muslim, and Jewish authors. In addition to extensive references to manuscripts, both published and unpublished, there are summaries of the contents of the texts. See also item 286.

446. Wolfson, H.A. "The Classification of Sciences in Medieval Jewish Philosophy." *Hebrew Union College Jubilee Volume*. Cincinnati: Hebrew Union College, 1925, 263-315. "Additional Notes." *Hebrew Union College Annual* 3 (1926), 371-375.

The classification of the sciences was traditionally attributed to Plato and Aristotle, and the schemes found in Jewish literature reflect the Arabic versions of these traditions. The classifiers whose schemes are treated include Isaac Israeli, al-Mukammas, Baḥya ibn Pakuda, Abraham Ibn Ezra, and, especially, Maimonides. It is worth noting that al-Ghazali, Abraham Ibn Ezra, and Afendopolo put astrology under physics, while Abraham Bar Ḥiyya, Falaquera, and Rieti made it coordinate to astronomy and put it under mathematics.

Works on Individual Mathematicians

Abraham bar-Ḥiyya (died ca. 1136)

The texts by Abraham bar-Ḥiyya have been edited and translated into Spanish by J.M. Millás Vallicrosa. See item 436.

447. Levey, Martin. "Abraham Savasorda and His Algorism: A Study in Early European Logistic." *Osiris* 11 (1954), 50-64.

448. ———. "The Encyclopedia of Abraham Savasorda: A Departure in Mathematical Methodology." *Isis* 43 (1952), 257-264.

Levey attempts to show that Abraham bar-Ḥiyya combined the practical and the theoretical aspects of mathematics. Quotations from Fibonacci indicate his dependence on bar-Ḥiyya, whose method Fibonacci adopted and elaborated.

Abraham Ibn Ezra (1089-1164)

449. Ginsburg, J. "Rabbi ben Ezra on Permutations and Combinations." *The Mathematics Teacher* 15 (1922), 347-356.

450. Silberberg, Moritz. *Sefer ha-Mispar. Das Buch der Zahl ein hebraisch-arithmetisches Werk des R. Abraham ibn Ezra.* Halle a. S.: C.A. Kaemmerer & Co., 1891.

Crescas (died 1412)

451. Rabinovitch, Nachum L. "Rabbi Hasdai Crescas (1340-1410) and Numerical Infinities." *Isis* 61 (1970), 222-230.

Dissatisfaction with Aristotle's proof that an actual infinite does not exist led Crescas to redefine "finite" and "infinite" and then to the recognition that an arithmetic of infinite quantities is possible.

Finzi, Mordecai (fl. ca. 1460)

452. Levey, Martin. *The Algebra of Abū Kāmil in a Commentary by Mordecai Finzi.* Hebrew text with English translation and a commentary, with foreword by Marshall Clagett. Madison: University of Wisconsin Press, 1966.

An example of the work of Hebrew translators in preserving and transmitting mathematical science from East to West.

Immanuel Tov-Elem (Bonfils) (14th Century)

453. Gandz, Solomon. "The Invention of the Decimal Fractions and the Application of the Exponential Calculus by Immanuel Bonfils of Tarascon (c. 1350)." *Isis* 25 (1936), 16-45.

See item 429 above.

However, the origin of decimal fractions is placed much earlier by the two following authors.

454. Rabinovitch, Nachum L. "An Archimedean Tract of Immanuel Tov-elem (14th Century)." *Historia Mathematica* 1 (1974), 13-27.

455. Rashed, Roshdi. "L'extraction de la racine n^ième^ et l'invention des fractions décimales (XIe-XIIe siècles)." *Archive for History of Exact Sciences* 18 (1978), 191-243.

Levi ben Gerson (1288-1344)

456. Carlebach, J. *Lewi ben Gerson als Mathematiker.* Berlin: L. Lamm, 1910.

Includes the Latin text of *De Numeri harmonicis.*

457. Curtze, M. "Die Abhandlung des Levi ben Gerson über Trigonometrie und den Jacobstab." *Bibliotheca Mathematica* (1898), 77-112.

458. ———. "Urkunden zur Geschichte der Trigonometrie im Christlichen Mittelalter." *Bibliotheca Mathematica* (1900), 321-416.

459. Espenshade, P.H. "A Text on Trigonometry by Levi ben Gershon." *The Mathematics Teacher* 60 (1967), 628-637.

Based on Curtze (item 458) as well as manuscripts in Latin and Hebrew in an English translation.

460. Goldstein, Bernard R. "The Astronomical Tables of Levi ben Gerson." *Transactions of the Connecticut Academy of Arts and Sciences* 45 (1974), 1-285.

See item 430 above.

461. Lange, G. *Sefer Maasei Choscheb. Die Praxis der Rechners, ein hebraisch-arithmetisches Werk des Levi ben Gerschom aus dem Jahre 1321*. Frankfurt am Main: n.p., 1909.

462. Rabinovitch, Nachum L. "Rabbi Levi ben Gershon and the Origins of Mathematical Induction." *Archives for History of Exact Sciences* 6 (1970), 237-248.

Contains a translation of some theorems in Levi's *Maasei Hoshev*, mainly on combinations and permutations, as well as earlier work on this subject by Abraham Donnolo (913-970).

Maimonides

See item 435.

Māshā'allah (Menasheh ben Athan) (754-813)

See item 528, in which two chapters on stereographic projections from *Māshā'allah's Treatise on the Astrolabe* are translated into English. Recent scholarship has argued, however, that the Arabic original of this Latin text on the astrolabe ascribed to Māshā'allah is probably due to al-Majriti (a Spanish Muslim, d. 1007) or his school: cf. P. Kunitzsch, *Typen der Sternverzeichnissen* ... (Wiesbaden: O. Harrassowitz, 1956), 7. The entire text also appears in R.T. Gunther, *Chaucer and Mesella on the Astrolabe* (Oxford: Oxford University Press, 1929).

Zacut, Abraham (1452-ca. 1530)

463. Cantera Burgos, F. *Abraham Zacut*. Madrid: M. Aguilar, 1935, 225pp.

A biography and description of Zacut's works as well as a bibliography and selections from his astronomical and astrological treatises. Zacut lived in Spain until 1492 and then traveled to Portugal, North Africa, and Palestine. His best known scientific work is a set of astronomical tables written in Hebrew (*Ha-ḥibbur ha-gadol*) and translated into Latin (*Almanac Perpetuum*, Leiria, 1496; reprinted Berne, 1915), Arabic, Spanish, and Ladino. He was consulted by Vasco da Gama before his voyage to India and he influenced a number of later figures including his pupils Vizinho and Ricius, and Abraham Gascon (16th century, Cairo). For the most recent study of his work, see B.R. Goldstein, "The Hebrew Astronomical Tradition: New Sources," *Isis* 72 (1981), 237-251.

LATIN WEST

The boundaries of medieval Latin mathematics are difficult to define. A large part of medieval mathematics consists of the use and adaptation of earlier mathematical works. Consequently, the following bibliography includes a number of items concerning the medieval Latin versions of Euclid or Archimedes or of various Islamic mathematicians and so forth. Thus medieval Latin mathematics cannot be strictly separated from Greek, Islamic, or Hebrew mathematics. Another problem concerns how widely the net should be cast among medieval Latin works. In medieval divisions of the sciences, mathematics is often taken to include not only arithmetic and geometry, but also astronomy and music, i.e., all of the so-called quadrivium. Since much of the more complicated medieval mathematics and much of the more sophisticated *use* of mathematics in the Middle Ages occurred as a part of astronomy, one cannot obtain a balanced view of medieval mathematics without considering mathematical astronomy. Medieval musical treatises cast important light on the medieval conception of ratio, which may otherwise be less easily understandable. If one includes mathematical astronomy and music as parts of mathematics, then there is a temptation to include also the so-called *scientia mediae* or sciences half-way between mathematics and physics, such as optics, statics or the science of weights, and, later, kinematics and dynamics. There are also medieval philosophical discussions of concepts that may be considered mathematical, such as the concepts of infinity and continuity or of indivisibles. The following listings are more complete for mathematics more narrowly defined than for these other areas, but a selection of works concerning the broader areas has also been included.

Catalogues and Reference Works

464. Jayawardene, S.A. "Western Scientific Manuscripts before 1600: A Checklist of Published Catalogues." *Annals of Science* 35 (1978), 143-172.

A bibliography of catalogues of scientific manuscripts.

465. Lutz, L., et al. *Lexikon des Mittelalters*. Munich, Zurich: Artemis-Verlag, 1977-.

This dictionary of the Middle Ages was started in 1977. The first five fascicles contain mathematics-related articles on Adelard of Bath, Albert of Saxony, Leon Battista Alberti, Albertus Magnus, and Alcuin.

465a. Thorndike, Lynn, and Pearl Kibre. *A Catalogue of Incipits of Mediaeval Scientific Writings in Latin*. Rev. ed. (Mediaeval Academy of America Publications 29.) London: Mediaeval Academy of America, 1963. 1938 columns.

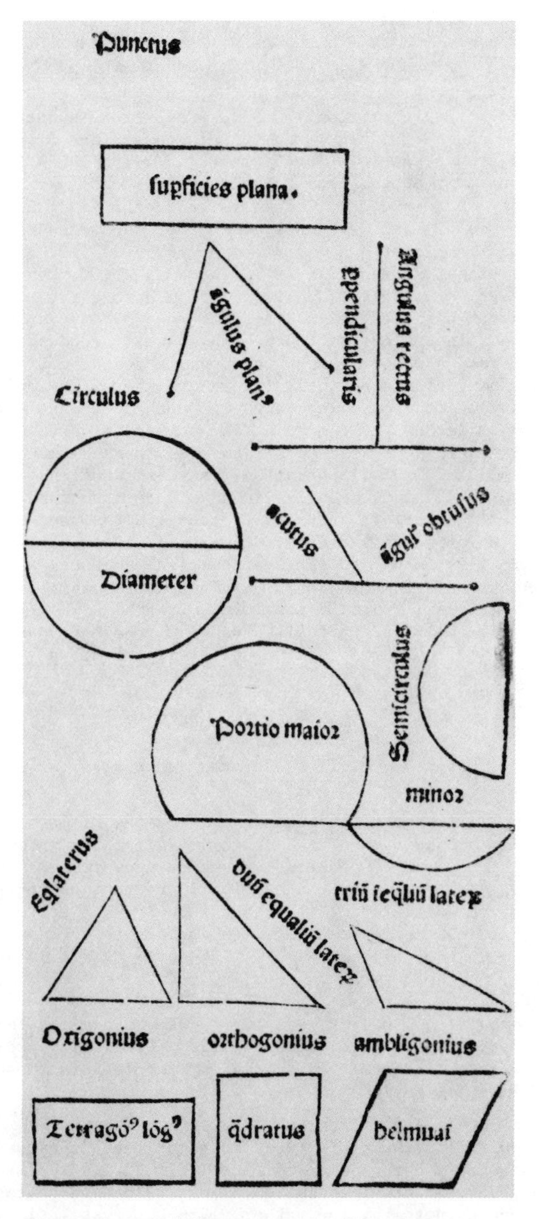

Figure 6. Detail from page 1 of *Preclarissimus Liber Elementorum Euclides . . . in Artem Geometriae* (1482), the first printed edition of Euclid, demonstrating the meaning of various Latin geometrical terms.

Fundamental reference work for the medievalist. Mathematical manuscripts may be located with the help of the name/subject index (cc. 1717-1938). Subject headings include: abacus, algorismus, arithmetic, astrolabe, calendar, computus, geometry, Euclid, measure, motion, number, proportion, quadrant, sphere.

The census of Latin mathematical manuscripts of the Middle Ages begun by Axel Anthon Bjørnbo (1874-1911) was interrupted by his untimely death. His index (some 1600 cards) is now at the Academy of Sciences in Stockholm. For details of his studies in codicology see Menso Folkerts, "Der Nachlass Axel Anthon Björnbos," *Historia Mathematica* 5 (1978), 333-339. Dorothea Waley Singer's "Handlist of Western Scientific Manuscripts in Great Britain and Ireland Dating from Before the Sixteenth Century," compiled in 1920 (card index in boxes arranged by subject, including: astronomy, calendar, computus, mathematics, measures and weights) has been deposited in the British Library, Department of Manuscripts. (There are microfilm copies at the Library of Congress, Washington, and the Warburg Institute, London.) Warren Van Egmond's "The Commercial Revolution and the Beginnings of Western Mathematics in Renaissance Florence, 1300-1500" (Ph.D. dissertation, University of Indiana, 1976), pp. 417-588, contains a catalogue of some two hundred arithmetical manuscripts in Rome, Florence, Siena, and New York City (Columbia University). Other projects to collect and catalogue information about medieval mathematical manuscripts are being conducted by Menso Folkerts of the Ludwig-Maximilians-Universität München (see *Historia Mathematica* 10 [1983], 98-99) and by Laura Toti Rigatelli and Raffaella Franci of the Centro Studi della Matematica Medioevale of the University of Siena.

General Textbooks

466. Cantor, Moritz B. *Vorlesungen über Geschichte der Mathematik*. Leipzig: Teubner, 1880-1908. 4 vols.

Cantor was one of the leading historians of mathematics in Germany at the turn of the century. He is best known for the once highly praised *Vorlesungen* which, despite many contemporary emendations, has not been equaled in content and extent. Although now dated, it is still informative on the subject of medieval mathematics, and gave a definite impetus to the development of the history of mathematics in general as a scholarly discipline. See also items 74, 243, 327, 622.

467. Juschkewitsch, Adolf P. *Geschichte der Mathematik im Mittelalter*. Leipzig: B.G. Teubner, 1964. Originally published in Russian as *Istoriia matematiki v srednie veka*. Moscow: Fizmatgiz, 1961.

The author's research, as well as other extensive studies of medieval mathematics, were systematized and generalized in this book (first Russian edition 1961; revised German translation 1964; French translation of part 3 on Arab mathematics, with added notes by the author, 1976). A systematic exposition of enormously rich material, this work is also a source of new ideas on medieval mathematics. It provides an up-to-date account of the history of medieval mathematics in China, India, Arabia, and Europe. See also item 332.

468. Lindberg, David, ed. *Science in the Middle Ages*. Chicago, London: University of Chicago Press, 1978.

Contains an excellent survey of medieval mathematics by Michael S. Mahoney as well as chapters on cognate fields such as astronomy, the science of weights, the science of motion, and optics. The Mahoney chapter is especially recommended for putting the mathematics into its medieval context. The chapter by Olaf Pedersen on astronomy should be used to supplement this bibliography with regard to mathematical astronomy.

469. Sarton, George. *Introduction to the History of Science*. Baltimore: Williams and Wilkins, 1927-1948. 3 vols.

Fundamental source for outline biographies and associated bibliographies. Volume I covers the beginnings through the eleventh century. Volume II in two parts covers the twelfth and thirteenth centuries. Volume III in two parts covers the fourteenth century. See also items 348, 2324.

470. Tropfke, Johannes. *Geschichte der Elementarmathematik*. 4th ed. Vol. 1. *Arithmetik und Algebra*. Vollständig neu bearbeitet von Kurt Vogel, Karin Reich, Helmuth Gericke. Berlin, New York: Walter de Gruyter, 1980, 724 pp.

This is the first of a new, completely revised, three-volume edition (the others will be devoted to geometry and analysis) of Tropfke's classic work on elementary mathematics. The first edition appeared in 1902, and the third between 1930 and 1940. This edition introduces a new structure to the text to improve the ease of use. Substantially more detail than in earlier editions is given in the section on the applied art of reckoning. Also, an index to people and subjects is provided, as well as an updated bibliography. Subjects include: number theory, numeration, algebra, and recreational mathematics. See also items 638, 644, 1185.

Editions of Texts

Included in this section are items consisting largely of texts, but many texts are accompanied by valuable introductions and commentaries.

The Centro Studi della Matematica Medioevale of the University of Siena has begun publishing a series of brief books, the *Quaderni del Centro Studi della Matematica Mediovale*, containing medieval mathematical texts. Through 1983, five of these quaderni had appeared:

471a. Tommaso della Gazzaia. *Praticha di geometria e tutte misure di terre dal ms. C.III.23 della Biblioteca Comunale di Siena*. Trascrizione di Cinzia Nanni, Introduzione di Gino Arrighi. 1982, 77 pp.

471b. M^{o} Benedetto da Firenze. *La reghola de algebra amuchabale dal Codice L.IV.21 della Biblioteca Comunale di Siena*. A cura e con introduzione di Lucia Salomone. 1982, xviii + 104 pp.

471c. Giovanni di Bartolo. *Certi chasi nella trascelta a cura di Maestro Benedetto secondo la lezione del Codice L.IV.21 (sec. XV) della Biblioteca degli Intronati di Siena.* A cura e con introduzione di Marisa Pancanti. 1982, ix + 107 pp.

471d. Filippo Calandri. *Una raccolta di ragioni dal Codice L.VI.45 della Biblioteca Comunale di Siena.* A cura e con introduzione di Daniela Santini. 1982, x + 50 pp.

471e. M^{o} Biagio. *Chasi exemplari alla regola dell'algibra nella trascelta a cura di M^{o} Benedetto dal Codice L.IV.21 della Biblioteca Comunale di Siena.* A cura e con introduzione di Licia Pieraccini. 1983, xv + 143 pp.

472a. Arrighi, G., ed. *M^{o} Guglielmo, vescovo di Lucca, De arithmetica compendiose tractata.* Pisa: Domus Galilaeana, 1964, 84 pp.

Arithmetic as found in the curriculum of the twelfth-century cathedral school of Lucca.

472b. ———. *Paolo Dell'Abaco. Trattato d'arithmetica.* Pisa: Domus Galilaeana, 1964, 164 pp.

Practical mathematics as taught by an independent practitioner who lived ca. 1281-1374.

473. Baron, R. "Hugonis de Sancto Victore Practica geometrie." *Osiris* 12 (1956), 176-224.

A critical edition of this important source from the first half of the 12th century.

474. Benjamin, Francis S., Jr., and G.J. Toomer. *Campanus of Novara and Medieval Planetary Theory. Theorica planetarum.* Madison: University of Wisconsin Press, 1971.

Edition of, translation of, and commentary on the first detailed account of the Ptolemaic astronomical system to appear in the West. Biography and bibliography of Campanus (for which see also the article in the *Dictionary of Scientific Biography*, item 10).

475. Blume, F., K. Lachmann, and A. Rudorff. *Die Schriften der römischen Feldmesser.* Berlin, 1848-1852. Reprinted Hildesheim: G. Olms, 1967. 2 vols.

Despite its age, still an essential edition of the writings of Roman surveyors, with commentary.

476. Boethius. *Anicii Manlii Torquati Severini Boetii de institutione arithmetica libri duo, de institutione musica libri quinque, accedit geometria quae fertur Boetii.* Edited by G. Friedlein. Leipzig: Teubner, 1867.

Although the text is not always reliable, because Friedlein used only a few manuscripts, there is no more recent edition for the *Arithmetic* and *Music*, which were widely circulated in the Middle Ages.

477. Boncompagni, B., ed. *Scritti di Leonardo Pisano.* Rome: Tipografia delle scienze matematiche e fisiche, 1857-1862. 2 vols.

A monumental edition of all of Leonardo's writings. The first volume contains the *Liber abaci* (459 pages, without notes); the second includes, among other works, the *Practica geometriae* (224 pages). See also item 683.

478. Bond, J.D. "Quadripartitum Ricardi Walynforde de Sinibus Demonstratis." *Isis* 5 (1923), 99-115.

Edition from a single manuscript, no notes.

479. ———. "Richard of Wallingford's Quadripartitum (English Translation)." *Isis* 5 (1923), 339-363.

This and the preceding item are now superseded by John North's edition listed below. Translation of part of the work appears in Edward Grant, *A Source Book in Medieval Science*, item 511, pp. 188-198.

480. Bubnov, N. *Gerberti postea Silvestri II papae opera mathematica (972-1003)*. Berlin: R. Friedländer, 1899. Reprinted Hildesheim: G. Olms, 1963.

This monumental work contains not only editions of all of Gerbert's writings related to the quadrivium, but edits numerous other mathematical texts (to the 12th century, and contains a large section on the transmission of knowledge related to surveying. The book is essential for works on medieval mathematics before Arabic-Latin translations.

481. Busard, H.L.L., ed. *The Translation of the Elements of Euclid from the Arabic into Latin by Hermann of Carinthia (?), Books VII-XII*. Leiden: E.M. Brill, 1968. Reprinted Mathematical Centre Tracts 84, Amsterdam: Mathematisch Centrum, 1977.

This publication completes the edition in *Janus* 54 (1967), 1-140, in which Books I-VI of the translation of Euclid's *Elements* ascribed to Hermann of Carinthia appeared. The single extant manuscript, Paris Bibl. Nat. Latin 16646, does not contain Books XIII-XV of the *Elements*. However, the author of the anonymous manuscript Vat. Reg. Lat. 1268, which contains Books V, VI and X-XI.4 of the *Elements* in folios 72r-113v, was acquainted with the original Hermann version. As the proofs in the Vatican manuscript which resemble Hermann's have been more fully elaborated, the Hermann version as we have it today, is in all likelihood a very succinct one. See also item 309.

482. ———. "L'algèbre au Moyen Âge: Le 'Liber mensurationum' d'Abū Bekr." *Journal des Savants* (April-June 1968), 65-124.

The Arabic gromatic tradition is principally represented by two works translated into Latin in the twelfth century. The first of these is the *Liber Embadorum* of Savasorda, the second is the *Liber mensurationum* of an unknown Abū Bakr, translated into Latin by Gerard of Cremona. Abū Bakr's formulas are almost exactly the same as those of Savasorda.

483. ———. "Die Vermessungstraktate Liber Saydi Abouthmi und Liber Aderamati." *Janus* 56 (1969), 161-174.

Belongs to the same category of works as the *Liber mensurationum* (see item 444). A brief introduction is followed by transcriptions of the two works.

484. ———. "Der Traktat *De isoperimetris*, der unmittelbar aus dem Griechischen ins Lateinische übersetzt worden ist." *Medieval Studies* 42 (1980), 61-88.

The Latin *De ysoperimetris* was translated in the early thirteenth century from an anonymous Greek text (cf. Pappus, *Collectio*, ed. of F. Hultsch, Vol. 3, pp. 1138-1165) and is extant in many manuscripts. It has been demonstrated that the circle is the most capacious of isoperimetric plane figures and the sphere of solid isoperimetric figures.

485. ———. "Ein mittelalterlicher Euklid-Kommentar, der Roger Bacon zugeschrieben werden kann." *Archives Internationales d'Histoire des Sciences* 24 (95) (1974), 199-218.

Presents the text of a Latin manuscript, Florence, Bibl. Naz. Conv. Soppr. J. IX. 26, of a Euclid commentary formerly attributed to Adelard of Bath. The text is in an appendix.

486. ———. "Het rekenen met breuken in de Middeleeuwen, in het bijzonder bij Johannes de Lineriis." *Mededelingen van de Koninklijke Vlaamse Academie voor Wetenschappen, Letteren en Schone Kunsten van België, Klasse der Wetenschappen* (Brussels), XXX, 7 (1968), 36 pp.

In the *Algorismus de minutiis* John of Lignères simultaneously treated physical and vulgar fractions. Its great success is attested to by the number of manuscripts in which it is preserved. The treatise was printed twice, together with, and after, the *Algorismus de integris* of Prosdocimus de Beldomandis. First edition Padua, 1483; second, Venice, 1540.

487. ———. "Die 'Arithmetica speculativa' des Johannes de Muris." *Scientiarum Historia* 13 (1971), 103-132.

About 1323 John of Murs wrote an arithmetic of the Boethian type. It was printed in Vienna (1515), and was a popular textbook for a considerable time (at least in the German countries).

488. ———. "Die Traktate *De proportionibus* von Jordanus Nemorarius und Campanus." *Centaurus* 15 (1971), 193-227.

The *Liber de proportionibus*, anonymous in most manuscripts, but attributed to Jordanus in one and to Thābit ibn Qurra in another, and the *Tractatus de proportione et proportionalitate*, attributed to Campanus in two manuscripts, to Alkindi in two others and to Ametus filius Iosephi in another one, are concerned with the eighteen cases of six quantities in proportion where the ratio of one quantity to another is composed of ratios of the four remaining quantities. Campanus's Propositions 2 and 3 are essentially the same as Propositions 2 and 3 of the Jordanian text and Proposition 1.13 of Witelo's *Perspectiva*.

489. ———. "The *Practica Geometriae* of Dominicus de Clavasio." *Archive for History of Exact Sciences* 2 (1965), 520-575.

The *Practica* was a popular work during the Middle Ages and has survived in numerous manuscript versions. It served as a model for a *Geometriae culmensis*, written in both Latin and German near the end of the 14th century. The *Practica* was composed in 1346 and is divided into an introduction and three books.

490. ———. *Nicole Oresme Quaestiones super Geometria Euclidis.* Leiden: E.J. Brill, 1961.

Oresme's *Quaestiones* concentrate their attention on four not especially Euclidean, but typically scholastic, subjects: (1) mathematics and the infinite, in particular, infinite convergent and divergent series, (2) the notions of commensurability and incommensurability, (3) the Oresmian speciality of the "graphic" representation of intensible and remissible forms or the geometry of qualities, and (4) the nature and continuity properties of angles. See also the review with corrections by John Murdoch in *Scripta mathematica* 27 (1964), 67-91.

491. ———. "Der *Tractatus proportionum* von Albert von Sachsen." *Österreichische Akademie der Wissenschaften, mathematische-naturwissenschaftliche Klasse, Denkschriften* (Vienna), Vol. 116, Part 2 (1971), 43-72.

The *Tractatus proportionum* is divided into two parts, the first of which deals with proportions and is dependent upon the *Tractatus de proportionibus* of Bradwardine. The second part deals with speeds and other mechanical problems. This book was quite popular and provides the skeleton for the much longer treatise *De proportionibus velocitatum in motibus* of Symon de Castello.

492. ———. "Der *Traktat De sinibus, chordis et arcubus* von Johannes von Gmunden." *Österreichische Akademie der Wissenschaften, mathematische-naturwissenschaftliche Klasse, Denkschriften* (Vienna), Vol. 116, Part 3 (1971), 74-113.

A treatise on the computation of sines and chords. His point of departure is an arc of 15° called a *kardaga*. Peurbach's *Tractatus super propositiones Ptolemaei de sinibus et chordis* is an abbreviated version of John's work.

493. Clagett, Marshall. *Archimedes in the Middle Ages.* Vol. 1, Madison: University of Wisconsin Press, 1964. Vols. 2 (2 parts), 3 (4 parts in 3), and 4 (2 parts). Philadelphia: American Philosophical Society, 1976, 1978, 1980.

Includes editions and translations of all the known Latin Archimedean texts stemming from the twelfth century through 1565. Essential and fundamental work. See also item 307.

494. ———. "The *Liber de motu* of Gerard of Brussels and the Origins of Kinematics in the West." *Osiris* 12 (1956), 73-175.

An edition and translation of what is in many ways the most impressive medieval work of mathematical physics, accomplishing the equivalent of an integration of velocities over the parts of rotating lines, planes, and bodies of various shapes in order to determine their average velocities. For a necessary correction to Clagett's interpretation see Wolfgang Breidert, *Das aristotelische Kontinuum in der Scholastik*, item 545, pp. 49-61, or Edith Sylla, *The Oxford Calculators and the Mathemathics of Motion, 1320-1350. Physics and Measurement by Latitudes* (unpublished Ph.D. dissertation, Harvard University, 1970), Appendix B.

495. ———, ed. *Nicole Oresme and the Medieval Geometry of*

Qualities and Motions. A Treatise on the Uniformity and Difformity of Intensities Known as Tractatus de configurationibus qualitatum. Madison: University of Wisconsin Press, 1968.

Definitive edition, translation, and commentary upon the work in which Oresme most fully developed his theory of the configurations of qualities, providing two-dimensional representations of linear qualities and analyzing these representations geometrically. See also item 1550.

496. ———. *The Science of Mechanics in the Middle Ages.* Madison: University of Wisconsin Press, 1959; 2nd printing 1961.

A collection of texts covering statics, kinematics, and dynamics, with English translations most prominent, but including also general introductions, Latin texts, and commentaries. Shows the extent and limitations of the use of mathematics in these areas during the Middle Ages. See also item 250.

497. Crosby, H. Lamar, Jr. *Thomas of Bradwardine. His Tractatus de Proportionibus. Its Significance for the Development of Mathematical Physics.* Madison: University of Wisconsin Press, 1955; 2nd printing 1961.

Edition and translation of a highly influential work applying the theory of ratios to dynamics.

498. Curtze, Maximilian. *Der Algorismus Proportionum des Nicolaus Oresme; zum ersten Male nach der Lesart der Handschrift R. 4°.2. der königlichen Gymnasialbibliothek zu Thorn.* Berlin: J. Draeger (C. Feicht), 1868.

Important for understanding the effect of notation and methods of manipulation on medieval conceptions of ratio. Part I of the work re-edited on the basis of thirteen manuscripts in Edward Grant, "The Mathematical Theory of Proportionality of Nicole Oresme," Ph.D. dissertation, University of Wisconsin, 1957, 331-339. Partial English translation by Edward Grant in *Isis* 56 (1965), 335-341, and in *A Source Book in Medieval Science*, item 511, pp. 150-157.

499. ———. "Commentar zu dem '*Tractatus de numeris datis*' des Jordanus Nemorarius." *Historisch-litterarische Abteilung der Zeitschrift für Mathematik und Physik* 36 (1891), 1-23, 41-63, 81-95, 121-138.

Partially translated into English in Edward Grant, *A Source Book in Medieval Science*, item 511, pp. 111-114. New edition in the dissertation of Barnabas Hughes, item 611, listed below.

500. ———. *Petri Philomeni de Dacia. Algorismum vulgarem Johannis de Sacrobosco Commentarius una cum Algorismo ipso edit.* Copenhagen: A.F. Høst, 1897.

Edition of the widely distributed textbook on arithmetic by Sacrobosco, and the commentary of Petrus de Dacia. Sacrobosco's *Algorismus vulgaris* is partially translated into English in Edward Grant, *A Source Book in Medieval Science*, item 511, pp. 94-101.

501. ———. *Anaritii in decem libros priores Elementorum Euclidis commentarii.* Leipzig: Teubner, 1899, 1-252.

A commentary on Books I-X of the *Elements* of Euclid (incomplete [?]) in the Latin translation of Gerard of Cremona. A large number of extracts of Geminus's and Heron's *Commentary on Euclid's Elements* are included.

502. ———. "Der *Liber Embadorum* des Abraham bar Chijja Savasorda in der Übersetzung des Plato von Tivoli." *Urkunden zur Geschichte der Mathematik im Mittelalter und der Renaissance*. Leipzig: Teubner, 1902; also *Abhandlungen zur Geschichte der mathematischen Wissenschaften* 12 (1902), 1-183.

Savasorda's most influential work by far is his Hebrew treatise on practical geometry, the *Ḥibbūr ha-meshīḥah we-ha-tish-boret*, translated into Latin as *Liber Embadorum* by Plato of Tivoli in 1145. The *Liber Embadorum* is the earliest exposition of Arab algebra written in Europe, and it contains the first complete solution in Europe of the quadratic equation. The *Hibbūr* was also among the earliest works to introduce Arab trigonometry and mensuration into Europe and influenced Leonardo Fibonacci who devoted an entire section of his *Practica geometriae* to division of figures.

503. ———. "Jordani Nemorarii Geometria vel de Triangulis Libri IV." *Mitteilungen des Copernicus-Vereins für Wissenschaft und Kunst zu Thorn* 6 (1887), xvi + 50 pp., 5 pp. of diagrams.

Clagett has distinguished two versions of the *De triangulis*, a shorter and presumably first edition, called the *Liber phyloteigni/phylotegni de triangulis*, and a longer version. The short version is completely different from the text of Curtze. The longer version, edited by Curtze, saw the re-arrangement and expansion of Book 2, as well as the addition of Propositions IV.12 to LV.28. Professor Clagett is editing and planning to publish both the longer and the shorter versions.

504. Elie, Hubert. *Le traité "De l'infini" de Jean Mair*. Paris: J. Vrin, 1938.

Edition and translation of a scholastic treatise asking whether there is an infinite in act and whether God can produce an infinite in act. Uses arguments concerning proportional parts, spiral lines, diagonals, angles, etc.

505. Folkerts, M. "Pseudo-Beda: *De arithmeticis propositionibus*, eine mathematische Schrift aus der Karolingerzeit." *Sudhoffs Archiv* 56 (1972), 22-43.

A critical edition (with commentary) of the manuscript (in existence in the 8th century), which among other things treats negative numbers.

506. ———. *Die älteste mathematische Aufgabensammlung in lateinischer Sprache: Die Alkuin zugeschriebenen Propositiones ad acuendos iuvenes*. Überlieferung, Inhalt, Kritische Edition. Vienna: Springer in Kommission, 1978.

A critical edition of the oldest collection of mathematical exercises, attibuted to Alcuin. Published as volume 116 (part 6) of *Österreichische Akademie der Wissenschaften, mathematisch-naturwissenschaftliche Klasse, Denkschriften*.

507. ———. *"Boethius" Geometrie II, ein mathematisches Lehrbuch des Mittelalters*. Wiesbaden: Steiner, 1970.

A critical edition of the geometry (in two books) ascribed to Boethius, with extensive notes to its contents and textual tradition. Euclidean quotations in other works taken from the translation by Boethius are also edited here.

507a. ———. "Die mathematischen Studien Regiomontans in seiner Wiener Zeit." *Regiomontanus-Studien*. Edited by G. Hamman. *Akademie der Wissenschaften, philosophischen-historische Klasse, Sitzungsberichte* (Vienna), Vol. 364 (1980), 175-209.

Unpublished and previously partially unknown texts are used to elucidate the early mathematical studies of Regiomontanus. Emphasized are arithmetic, algebra, and volumetrics (*ars visorie*).

507b. ———. "Regiomontans Euklidhandschriften." *Sudhoffs Archiv* 58 (1974), 149-164.

Regiomontanus's Euclidean manuscripts.

508. ———, and A.J.E.M. Smeur. "A Treatise on the Squaring of the Circle by Franco of Liège of about 1050." *Archives Internationales d'Histoire des Sciences* 26 (1976), 59-105, 225-253.

509. Grant, Edward. *Nicole Oresme and the Kinematics of Circular Motion. Tractatus de commensurabilitate vel incommensurabilitate motuum celi*. Madison: University of Wisconsin Press, 1971.

Edition and translation of a text in which Oresme attempts to undermine predictive astrology by showing that the motions of the heavens are probably incommensurable. His argument is based ultimately upon Bradwardine's theory of ratios of motions and on his own extension of that theory. Extensive introductory analysis.

510. ———. *Nicole Oresme: De proportionibus proportionum and Ad pauca respicientes*. Madison: University of Wisconsin Press, 1966.

The most important source for Oresme's theory of the ratios of ratios. An earlier version of some of the same material in "Nicole Oresme and His *De proportionibus proportionum*," *Isis* 51 (1960), 293-314.

511. ———, ed. *A Source Book in Medieval Science*. Cambridge, Mass.: Harvard University Press, 1974.

Contains English translations of eighteen short mathematical texts. Selections from Isidore of Seville, Boethius, Roger Bacon, Jordanus de Nemore, al-Khwārizmī, Leonardo of Pisa, Nicole Oresme, Campanus of Novara, Pseudo-Bradwardine, Albert of Saxony, the Banu Musa, Dominicus de Clavasio, and Richard Wallingford. Approximately 130 pages in all. Good place to start.

512. Halliwell, J.O., ed. *Rara mathematica, Vol. 1, Joannis de Sacro-Bosco: Tractatus de arte numerandi*. London, 1839, 1941. Reprinted Hildesheim: Olms, 1977.

Includes also "A Method Used in England in the Fifteenth Century for Taking the Altitude of a Steeple or Other Inaccessible Object," and other treatises on algorism.

513. Hughes, Barnabas. *Regiomontanus on Triangles. De triangulis omnimodis by Johann Müller, Otherwise Known as Regiomontanus*. Madison: University of Wisconsin Press, 1967.

Written in 1464 and first published in 1533, this work represents a "uniform foundation and a systematic ordering of trigonometric knowledge." Photomechanical reproduction of the original Latin with English translation and notes. See also item 751.

514. Hugonnard-Roche, Henri. *L'Oeuvre Astronomique de Themon Juif Maitre Parisien du XIVe Siècle*. Geneva: Droz, and Paris: Minard, 1973.

Discusses an interesting use of the mathematical techniques usually associated with medieval terrestrial mechanics in discussing the motion of the moon. Edits Themon's Commentary and Questions on the *Sphere* of Sacrobosco and his Question on the Motion of the Moon and Other Planets (Utrum necessarium sit, supposita veritate theoricarum, lunam vel aliquem planetarum inequalibus gradibus [motus], vel uniformiter difformiter vel difformiter difformiter, moveri).

515. Ito, Shuntaro. *The Medieval Latin Translation of the Data of Euclid*. Tokyo: University of Tokyo Press, and Boston: Birkhaüser, 1980.

An edition and translation of the text with a brief introduction.

516. Karpinski, Louis C. *Robert of Chester's Latin Translation of the Algebra of al-Khowarizmi*. (University of Michigan Studies, Humanistic Series, 11. Contributions to the History of Science, part 1.) First published New York, 1915. Ann Arbor: University of Michigan Press, 1930.

Latin text and English translation with introduction and short glossary.

517. Kaunzner, Wolfgang. "Über einige algebraische Abschnitte aus der Wiener Handschrift Nr. 5277." *Österreichische Akademie der Wissenschaften, mathematisch-naturwissenschaftliche Klasse, Denkschriften* (Vienna), Vol. 116, Part 4 (1972), 115-184 + iv.

Contains a description of the contents of Codex Vind. Pal. 5277 and the text of some algebraic parts.

518. Libri, G. "Liber Maumeti filii Moysi alchoarismi de algebra et almuchabala." *Histoire des sciences mathématiques en Italie*. Vol. 1. Paris: J. Renouard et Cie., 1838, 253-299.

The anonymous Latin version of the *Algebra* of al-Khwārizmī printed by Libri is probably that of Gerard of Cremona, but the problem is complicated by the existence of another Latin text which is a free adaptation of al-Khwārizmī's *Algebra*, whose translation is expressly ascribed to Gerard of Cremona.

519. Moody, Ernest, and Marshall Clagett. *The Medieval Science of Weights (Scientia de Ponderibus). Treatises Ascribed to Euclid, Archimedes, Thabit ibn Qurra, Jordanus de Nemore and Blasius of Parma*. Madison: University of Wisconsin Press, 1952, 1960.

Contains editions, translations, and discussions of the major texts of medieval statics.

520. Murdoch, John, and Edward A. Synan. "Two Questions of the Continuum: Walter Chatton (?), O.F.M. and Adam Wodeham, O.F.M." *Franciscan Studies* 25 (1966), 212-288.

Treatises reflecting the early fourteenth-century controversy over whether a continuum might be composed of indivisibles.

521. Nikolaus von Cues. *Die mathematischen Schriften, übersetzt von Joseph Hofmann*. Leipzig: F. Meiner, 1951; Hamburg, F. Meiner, 1952. 2nd ed., 1979.

All mathematical writings (including those not previously published) are translated. The extensive introduction by J.E. Hofmann evaluates the attempts by Cusanus to square the circle. This work has also been reprinted under the following title: *Nikolaus von Kues. Die mathematischen Schriften*.

522. North, John. *Richard of Wallingford. An Edition of His Writings with Introductions, English Translations and Commentary*. Oxford: Clarendon Press, 1976. 3 vols.

Superbly done, these volumes provide an extensive if not complete edition of Wallingford's astronomical and related mathematical works. For the more important works there are English translations. The second volume contains commentaries on the astronomical and mathematical texts, setting them into historical context. The third volume consists of numerous figures, tables, appendices, glossaries, bibliography, indices, and plates.

523. Picutti, Ettore. "Il Libro dei Quadrati di Leonardo Pisano e i Problemi di analisi indeterminata nel Codice Palatine 577 della Biblioteca Nazionale di Firenze." *Physis* 21 (1979), 195-339.

Contains a medieval Italian translation of Fibonacci's *Book of Square Numbers* with accompanying problems.

524. Steele, Robert, ed. *Communia Mathematica Fratris Rogeri*. (Opera hactenus inedita Rogeri Baconi, Fasc. 16.) Oxford: Clarendon Press, 1940.

Latin text with very brief editorial material. In Part I Bacon defends the importance and usefulness of mathematics and describes its various parts. In Part II he defines integers, fractions, ratios, proportionalities, and the types of mathematical demonstration.

525. ———. *The Earliest Arithmetics in English*. London: Oxford University Press for the Early English Text Society, 1922.

Short fifteenth-century works based on the *de algorismo* of Alexander de Villa Dei and on the *de arte numerandi* of Sacrobosco. Also a work on a type of abacus from Robert Record's *Arithmetic* of 1543 and Alexander de Villa Dei's *Carmen de Algorismo*.

526. Tannery, Paul. "Le traité du Quadrant de Maître Robert Anglès (Montpellier, XIIIe siècle). Texte latin et ancienne traduction grecque." *Notices et extraits des manuscrits de la Bibliothèque nationale* 35: 561-640. Reprinted in *Mémoires scientifiques* 5, Toulouse: Edouard Privat, and Paris: Gauthier-Villars, 1922, 118-197.

This work, also ascribed to Johannes Anglicus, re-edited in the dissertation of Nan Britt [Hahn] listed below, item 607.

527. ———, and M. Clerval. "Une correspondance d'ecolatres du XIe siècle." *Notices et extraits des manuscrits de la Bibliothèque nationale* 36 (2) (1901), 487-543. Reprinted *Mémoires scientifiques* 5 (1922), 229-303.

Treats the discussion between Lothringian and Rheinish scholars at the beginning of the 11th century on geometrical questions.

528. Thomson, Ron B. *Jordanus de Nemore and the Mathematics of Astrolabes: De plana spera*. Toronto: Pontifical Institute of Mediaeval Studies, 1978.

An edition, translation, and commentary on three versions of the named work with a brief history of stereographic projection and editions of related works. Contains translations of two chapters on stereographic projections from Masha'allah's (Menasheh ben Athan, 754-813) *Treatise on the Astrolabe*.

529. Unguru, Sabetai. *Witelonis Perspectivae Liber Primus, Book I of Witelo's Perspectiva. An English Translation with Introduction and Commentary and Latin Edition of the Mathematical Book of Witelo's Perspectiva*. (Studia Copernicana XV.) Warsaw: Ossolineum, The Polish Academy of Sciences Press, 1977.

Composed by Witelo to provide the mathematics necessary for optics, this work is a good indicator of the state of mathematical knowledge in late thirteenth-century Europe.

530. Ver Eecke, Paul. *Léonard de Pise: de Livre des nombres carrés*. Bruges: Desclée, De Brouwer, 1952.

French translation of Leonardo Pisano's *Liber quadratorum* with notes. English translation in part in Edward Grant, *A Source Book in Medieval Science*, item 511, pp. 114-129.

531. Victor, Stephen K. *Practical Geometry in the High Middle Ages. Artis Cuiuslibet Consummatio and the Pratike de Geometrie*. Philadelphia: American Philosophical Society, 1979.

Edition, translation, and study of the first named work and an edition of the second named work, which is a vernacular version of the first.

532. Vogel, Kurt. *Die Practica des Algorismus Ratisbonensis*. Munich: Beck, 1954.

Vogel has not only edited the collection of exercises from St. Emmeram (Regensburg), but also goes into the origins and history of individual problems.

533. ———. *Der Donauraum, die Wiege mathematischer Studien in Deutschland*. Munich: Fritsch, 1973.

The author gives an overview of the mathematical studies in Bavaria and Austria in the late Middle Ages, and edits a calculating book (on lines), a mathematical collection of exercises, and Peurbach's writing on the calculation of the height of the sun.

534. ———. *Ein byzantinisches Rechenbuch des frühen 14. Jahrhunderts*. Vienna: Böhlau in Kommission, 1968.

Volume 6 of *Wiener byzantinistische Studien*. Greek and German on opposite pages. Original manuscript in the Bibliothèque Nationale, Paris, Cod. Par. Suppl. Gr. 387. Bibliography, pp. 164-168.

535. Zoubov, V.P. "Jean Buridan et les concepts du point au quatorzième siècle." *Medieval and Renaissance Studies* 5 (1961), 43-95.

Edits Jean Buridan's *Questio de puncto*, concerning whether points are separate entities in addition to lines.

Studies

Readers are reminded that some of the most recent and best work on biography and bibliography is to be found in *Dictionary of Scientific Biography* (item 10), especially for Leonardo of Pisa, Jordanus, Campanus, and Swineshead. For example, various nineteenth-century studies of Leonardo Pisano by Boncompagni, Woepke, and others (not cited here for lack of space) can be readily accessed through the *Dictionary of Scientific Biography* article.

536. Beaujouan, Guy. "L'enseignement de l'arithmétique élémentaire à l'université de Paris aux XIIIe et XIVe siècles." In *Homenaje a Millas-Vallicrosa*. Vol. 1. Barcelona: Consejo Superior de Investigaciones Cientificas, 1954-1956, 93-124.

Extracted from his unpublished dissertation, a summary of which appears in "Recherches sur l'histoire de l'arithmétique au Moyen-Âge," *École Nationale des Chartes. Positions des thèses ... de 1947* (Paris, 1947), 17-22.

537. ———. "L'enseignement du 'quadrivium.'" *Settimane di studio ..., 19, La Scuola nell'Occidente latino dell'alto medioevo*. Spoleto, 1972, 639-667.

Points out that the mathematics of the early Middle Ages was relatively independent from classical texts and relied on pedagogical tools such as the monochord, the game rithmomachia, and the use of parts of the hand for computation. Argues that before the rediscovery of classical texts in the twelfth century there was little distinction between theory and practice in mathematics.

538. ———. "Calcul d'expert en 1391, sur le chantier du Dome de Milan." *Le Moyen Âge* 69 (1963), 555-563.

Addresses the problems of the relation of theory to practice.

539. ———. "Le symbolisme des nombres à l'époque romane." *Cahiers de Civilisation Médiévale, X-XIIe Siècles* 4 (1961), 159-169.

Discusses the influence of Boethius's *De institutione arithmetica* on ideas of the properties of numbers, numbers in the Bible, and religious number symbolism in the works of Eudes de Morimond, Guillaume d'Auberive, Geoffroy d'Auxerre, and Thibaud de Langres.

540. ———. "Science livresque et art nautique au XVe siècle." *Le navire et l'économie maritime travaux du cinquième colloque international d'histoire maritime tenu à Lisbonne les 14-16 Septembre 1960*. Paris: Bibliothèque générale de l'Ecole pratique des hautes études, 1966.

Further discussion of theory and practice, this time in connection with navigation.

541. ———. "Réflexions sur les rapports entre théorie et pratique au Moyen Âge." In *The Cultural Context of Medieval Learning*. Edited by John Murdoch and Edith Sylla. (Boston Studies in the Philosophy of Science, vol. 26.) Dordrecht: Reidel, 1975, 437-484, including discussion.

Contains an astute discussion of the relations of medieval mathematical theory and practice, including the use of instruments.

542. Benedict, S.R. *A Comparative Study of the Early Treatises Introducing into Europe the Hindu Art of Reckoning*. Concord, N.H.: n.p., 1914; The Rumford Press, 1916.

Based on a study of algorisms available in print.

543. Bjǿrnbo, A.A. "Studien über Menelaos' Sphärik." *Abhandlungen zur Geschichte der mathematischen Wissenschaften* 14 (1902), 1-154.

Treats not only the *Spherics* of Menelaos, but is a general treatise on trigonometry in antiquity and the Middle Ages.

544. Bond, J.D. "The Development of Trigonometric Methods down to the Close of the Fifteenth Century." *Isis* 4 (1921-1922), 295-323.

Introduction to his edition of Wallingford's *Quadripartitum*. Later in the same volume (pp. 459-465) he provides a brief biography of Wallingford, now superseded by John North's study of Wallingford, item 522.

545. Breidert, Wolfgang. *Das aristotelische Kontinuum in der Scholastik. Beitrage zur Geschichte der Philosophie und Theologie des Mittelalters. Texte und Untersuchungen*, Neue Folge, Bd. 1. Munster: Aschendorff, 1970.

Valuable survey of scholastic discussions, e.g., of the continuity of reflected motions (must there be a period of rest between them?), of measures of rotations (including a good analysis of the work of Gerard of Brussels), and so forth, and of their relation to Cavalieri's method of indivisibles.

546. Busard, H.L.L. "Die Quellen von Nichole Oresme." *Janus* 58 (1971), 161-193.

An expository article discussing the sources of the works of Oresme.

547. Cantor, Moritz B. *Die römischen Agrimensoren und ihre Stellung in der Geschichte der Feldmesskunst*. Leipzig: B.G. Teubner, 1875.

This work indicates the connections between the Roman surveyors and Greek authors (above all, Heron). Bubnov's book, item 480, must also be consulted for corrections and additional information.

548. Clagett, Marshall. "The Medieval Latin Translations from the Arabic of the Elements of Euclid, with special emphasis on the versions of Adelard of Bath." *Isis* 44 (1953), 17-42.

Important study now available also in Clagett's collected articles listed next.

549. ———. *Studies in Medieval Physics and Mathematics*. London: Variorum, 1979.

A useful reprinting of thirteen articles including "The Use of Points in Medieval Natural Philosophy," "A Medieval Treatment of Hero's Theorem on the Area of a Triangle in Terms of its Sides," the previous entry, and four articles related to the medieval Archimedean tradition.

550. Eneström, G. "Über die 'Demonstratio Jordani de Algorismo.'" *Bibliotheca Mathematica*, Series 3, 7 (1906-1907), 24-37.

Includes the Latin text of the definitions, but not the proofs. Edited from Berlin, MS. Lat. qu. 510 and Dresden, Ms. Db. 86.

551. ———. "Über eine dem Jordanus Nemorarius zugeschriebene kurze Algorismusschrift." *Bibliotheca Mathematica* 8 (3) (1907-1908), 135-153.

The introduction is printed in full along with the propositions, but only a few of the proofs are included. There is, however, a German analysis. Taken mainly from MS. Vatican, Ottob. lat. 309.

552. ———. "Das Bruchrechnen des Jordanus Nemorarius." *Bibliotheca Mathematica*, Series 3, 14 (1913-1914), 42-44, 48-53.

The introduction and the propositions are printed in full, but the proof only of Proposition 23 is given. Taken from Vatican, MS. Ottob. lat. 309.

553. ———. "Der 'Algorismus de integris' des Meisters Gernardus." *Bibliotheca Mathematica*, Series 3, 13 (1912-1913), 289-332.

Contains definitions, axioms, and 43 propositions. Latin text (pp. 291-327) taken from Vatican MS. Reg. lat. 1261.

554. ———. "Der 'Algorismus de minutiis' des Meisters Gernardus." *Bibliotheca Mathematica*, Series 3, 14 (1913-1914), 99-149.

Contains definitions and 42 propositions. Latin text (pp. 100-142) taken from Vatican, MS. Reg. lat. 1261.

555. Evans, Gillian R. "A Commentary on Boethius's Arithmetica of the Twelfth or Thirteenth Century." *Annals of Science* 35 (1978), 131-141.

Describes a commentary found in Munich MS. CLM 4643 that may possibly be associated with the successors of Hugh of St. Victor.

556. ———. "*Difficillima et ardua*: Theory and Practice in Treatises on the Abacus c. 950-c. 1150." *Journal of Mediaeval History* 3 (1977), 21-38.

Attempts by medieval authors to elevate the study of the abacus to the status enjoyed by other arts.

557. ———. "*Duc oculum*: Aids to Understanding in Some Mediaeval Treatises on the Abacus." *Centaurus* 19 (1976), 252-263.

Analyzes the pedagogical problems faced by authors of abacus treatises. See also item 1861.

558. ———. "From Abacus to Algorism: Theory and Practice in Medieval Arithmetic." *British Journal for the History of Science* 10 (1977), 114-131.

Describes the development from the use of the abacus to the use of algorism for computation during the twelfth and thirteenth centuries. The algorism reintroduces the use of a symbol for zero. At the same time astronomical fractions take over from Roman ones.

559. ———. "Introductions to Boethius's Arithmetica of the Tenth to the Fourteenth Century." *History of Science* 16 (1978), 22-41.

A description of a number of introductions which used the methods of the trivium to provide an approach to Boethius's works.

560. ———. "The Rithmomachia: A Mediaeval Mathematical Teaching Aid?" *Janus* 63 (1976), 257-273.

Argues that this medieval chess-like board game was probably used as a method of teaching mathematics in eleventh- and twelfth-century schools.

561. ———. "The 'Sub-Euclidean' Geometry of the Earlier Middle Ages, up to the Mid-Twelfth Century." *Archive for History of Exact Sciences* 16 (1976-1977): 105-118.

Traces the transmission of elementary geometrical ideas in works unconnected with Euclid's *Elements*, including works on practical geometry and surveying.

562. Folkerts, M. *Anonyme lateinische Euklidbearbeitungen aus dem 12. Jahrhundert*. Vienna, 1971. Reprinted *Österreichische Akademie der Wissenschaften, mathematisch-naturwissenschaftliche Klasse, Denkschriften* (Vienna), Vol. 116, Part 1 (1971), 1-42.

This work treats various Euclidean collections between the Greek-Latin translation of Boethius and the Arabic-Latin translation of Adelard of Bath.

563. ———. "Mathematische Aufgabensammlungen aus dem ausgehenden Mittelalter." *Sudhoffs Archiv* 55 (1971), 58-75.

Those problems are discussed in the dispersed collections of mathematical exercises that were collected in the monasteries of the late Middle Ages.

564. ———. "Die Entwicklung und Bedeutung der Visierkunst als Beispiel der praktischen Mathematik der frühen Neuzeit." *Humanismus und Technik* 18 (1974), 1-41.

This work gives an overview of the methods used in the late Middle Ages to determine the volumes of casks; it discusses the social position of the measurers and gives a list of the printed and unprinted texts of cask measurement.

564a. ———. "Regiomontanus als Mathematiker." *Centaurus* 21 (1977), 214-245.

This work gives a short overview of the contributions of Regiomontanus in the area of mathematics.

565. Haskins, Charles Homer. *Studies in the History of Mediaeval Science*. Cambridge, Mass.: Harvard University Press, 1924. 2nd ed. 1927. Reprinted New York: Ungar Publishing Co., 1960.

Contains fundamental studies on medieval translators from the Greek and Arabic into Latin. Also chapters on twelfth-century astronomy and on the abacus and the exchequer.

566. Hofmann, Joseph E. *Zum Winkelstreit der rheinischen Scholastiker in der 1. Hälfte des 11. Jahrhunderts*. Berlin: W. de Gruyter, 1942.

This work on the controversy over angles among Scholastics of the Rhineland also appeared as volume 8 in the series *Abhandlungen der Preussischen Akademie der Wissenschaften, mathematische-naturwissenschaftliche Klasse* (Berlin) 1942.

567. Hoskin, Michael, and A.G. Molland. "Swineshead on Falling Bodies: An Example of Fourteenth Century Physics." *British Journal for the History of Science* 3 (1966), 150-182.

Representative of the extent of Swineshead's use of mathematics.

568. Hughes, Barnabas. "Biographical information on Jordanus de Nemore up to Date." *Janus* 62 (1975), 151-156.

Recent review of an enigmatic subject.

569. Kaunzner, Wolfgang. "Über den Beginn des Rechnens mit Irrationalitäten in Deutschland: Ein Beitrag zur Geschichte der Rechenkunst im Ausgehenden Mittelalter." *Janus* 57 (1970), 241-260.

Discusses the period from about 1460 to the early 16th century.

570. ———. "Über einen frühen Nachweis zur symbolischen Algebra." *Österreichische Akademie der Wissenschaften, mathematisch-naturwissenschaftliche Klasse, Denkschriften* (Vienna), Vol. 116, Part 5 (1975), 3-12 + ii.

The author mentions the oldest (to date) known example of symbolic representations in algebra, which occurred before 1380 in Italy. See also item 1176.

571. ———. *Über die Handschrift Clm 26639 der Bayerischen Staatsbibliothek München*. Hildesheim: Gerstenberg Verlag, 1978.

The Munich manuscript is important, above all, because it is connected with the presentation of geometry in Widmann's textbook.

572. Kretzmann, Norman, ed. *Infinity and Continuity in Ancient and Medieval Thought*. Ithaca, N.Y.: Cornell University Press, 1982.

Contains papers covering ancient and medieval views, with contrasts between the Aristotelian tradition and anti- or quasi-Aristotelian positions.

573. Lindgren, U. *Gerbert von Aurillac und das Quadrivium. Untersuchungen zur Bildung im Zeitalter der Ottonen*. Wiesbaden: Steiner, 1976.

Treats Gerbert's education and teaching, his studies of the quadrivium and his relation to the political powers of his time.

574. Maddison, Francis R. "Early Astronomical and Mathematical Instruments: A Brief Survey of Sources and Modern Studies." *History of Science* 2 (1963), 17-50.

Describes medieval Europe, pp. 26-30. Long bibliography.

575. Maier, Anneliese. *Studien zur Naturphilosphie der Spätscholastik*. Rome: Edizioni di Storia e Letteratura, 1949-1958; later editions 1966-1968. 5 vols.

In volume I, *Die Vorläufer Galileis im 14. Jahrhundert*, Part II concerns "Mathematisch-physicalische Fragestellungen," including the concept of function, total versus instantaneous velocity, the continuum, minima, and actual infinites. Volume II, *Zwei Grundprobleme der Scholastischen Naturphilosophie*, includes a consideration of Nicole Oresme's doctrine of the configuration of forms. In Volume III, *An der Grenze von Scholastik und Naturwissenschaft*, Part III concerns "Die Mathematik der Formlatitudinen."

576. Millás Vallicrosa, J. *Assaig d'història de les idées físiques i matemàtiques a la Catalunya medieval*. Barcelona: Estudis Universitaris Catalans, sèrie monogràfica 1, 1931.

Important work on the mathematics and physics in medieval Catalan. It considers western manuscripts and Arabic sources.

577. Molland, A.G. "An Examination of Bradwardine's Geometry." *Archive for History of Exact Sciences* 19 (1978), 113-175.

Important article concerning Bradwardine's *Geometria speculativa*. Discusses its provenance, sources, and influence, provides a synopsis of the work, characterizes Bradwardine's views of the objects and methods of geometry, of measure and ratio, and deals with the special topics of star-polygons, isoperimetry, and filling space. See also item 1578.

578. ———. "Ancestors of Physics." *History of Science* 14 (1976), 54-75.

Review essay on Olaf Pedersen and Mogens Pihl, *Early Physics and Astronomy: A Historical Introduction* (London: Macdonald and Jane's, New York: American Elsevier, 1974). Discusses especially the problem of mathematical translation and what can be learned by paying attention to the form in addition to the content of past theories.

579. ———. "The Geometrical Background to the 'Merton School': An Exploration into the Application of Mathematics to Natural Philosophy in the Fourteenth Century." *British Journal for the History of Science* 4 (1968-1969), 108-125.

With Molland's other articles, one of the best introductions to medieval ratio theory as found in the work of Thomas Bradwardine.

580. ———. "Oresme Redivivus." *History of Science* 8 (1969), 106-118.

Essay review of Marshall Clagett, *Nicole Oresme and the Medieval Geometry of Qualities and Motions*, item 495, with valuable insights.

581. Murdoch, John. "Euclid: Transmission of the Elements." *Dictionary of Scientific Biography*. Vol. 4. Edited by C.C. Gillispie. New York: Scribner's, 1971, 437-459.

The most up-do-date summary, with extensive notes, of the state of knowledge concerning medieval Latin versions of Euclid's *Elements*.

582. ———. "Euclides Greco-Latinus: A Hitherto Unknown Medieval Latin Translation of the Elements Made Directly from the Greek." *Harvard Studies in Classical Philology* 71 (1966), 249-302.

Describes a word-for-word translation of the *Elements* from Greek to Latin most likely made in Sicily during the twelfth century by the same unknown translators who also translated Ptolemy's *Almagest* from the Greek.

583. ———. "*Mathesis in Philosophiam Scholasticam Introducta*. The Rise and Development of the Application of Mathematics in Fourteenth Century Philosophy and Theology." *Arts Libéraux et Philosophie au Moyen Âge*. Actes du Quatrième Congrès International de Philosophie Médiévale. Montreal: Institut d'Etudes Médiévales, and Paris: J. Vrin, 1969.

Describes discussions of the continuum and the infinite, of the quantification of motion, and of latitudes and the perfection of species, with an appendix on Swineshead's treatise *de loco elementi* as published in the article of Hoskin and Molland, item 567. The same volume also contains Pearl Kibre, "The *Quadrivium* in the Thirteenth Century Universities (with special reference to Paris)," pp. 175-191, and Jean Gagné, "Du Quadrivium aux *scientiae mediae*," pp. 975-986, as well as other studies of the quadrivium.

584. ———. "The Medieval Euclid: Salient Aspects of the Translations of the *Elements* by Adelard of Bath and Campanus of Novara." *Revue de Synthèse* 89 (1968), 67-94.

Argues that the additions and changes to the text appearing in these translations are generally of a didactic nature, attempting to clarify the text, to simplify comprehension, and to make the logical structure of the whole clear.

585. ———. "The Medieval Language of Proportions." In *Scientific Change*. Edited by A.C. Crombie. London: Heinemann, and New York: Basic Books, 1963, 237-271, and discussion, pp. 334-343.

Emphasizes the consequences of the medieval misunderstanding of the Eudoxan definition of the sameness of ratios.

586. ———. "Naissance et développement de l'atomisme au bas Moyen Âge Latin." *La science de la nature: théories et pratiques*. (Cahiers d'Etudes Médiévales, Vol. 2.) Edited by G.-H. Allard and J. Ménard. Montreal: Bellarmin, 1974, 11-32.

Survey of medieval mathematical indivisibilism, especially in the work of Henry of Harclay, Walter Chatton, Gerard of Odo, and Nicolas Bonettus.

587. ———. *"Rationes Mathematice": Un aspect du rapport des mathématiques et de la philosophie au Moyen Âge*. (Les Conférences du Palais de la Découverte.) Paris: Université de Paris, 1961.

Studies the application of mathematics in astrological prediction, in the question of the eternity of the world, and with regard to the composition of the continuum.

588. ———. "Superposition, Congruence and Continuity in the Middle Ages." In *Mélanges Alexandre Koyré*. Vol. 1. Paris: Hermann, 1964, 416-441.

Leads up to Thomas Bradwardine's use of superposition in his *Tractatus de continuo*.

589. North, John D. "Kinematics--More Ethereal than Elementary." In *Machaut's World: Science and Art in the Fourteenth Century*. Edited by Madeleine Pelner Cosman and Bruce Chandler. (*Annals of the New York Academy of Sciences*, Vol. 314.) New York: The New York Academy of Sciences, 1978, 89-102.

Compares the techniques used in medieval terrestrial and celestial kinematics.

590. Pedersen, Olaf. "The Corpus Astronomicum and the Traditions of Medieval Latin Astronomy." *Colloquia Copernicana*. Vol. 3. Warsaw: Zaklad Narodowy Imenia Ossolińskich, 1975, 57-96.

Detailed analysis of the development of the *corpus astronomicum*.

591. Rose, Paul Lawrence. *The Italian Renaissance of Mathematics: Studies on Humanists and Mathematicians from Petrarch to Galileo*. Geneva: Droz, 1975, 316 pp.

Traces the collaboration of scientists and humanists in bringing about a revival of Greek mathematics in fifteenth- and sixteenth-century Italy. Little internal analysis of the mathematics involved. See also item 628.

592. Shelby, Lon R. "The Geometrical Knowledge of Mediaeval Master Masons." *Speculum* 47 (1972), 395-421.

Well-documented discussion of the problem of the use of geometry by cathedral builders.

593. Skabelund, Donald, and Phillip Thomas. "Walter of Odington's Mathematical Treatment of Primary Qualities." *Isis* 60 (1969), 331-350.

Use of a relationship like Bradwardine's in an alchemical work.

594. Stamm, Edward. "Tractatus de continuo von Thomas Bradwardina." *Isis* 26 (1936), 13-32.

Surveys the treatise. Superseded by the edition in the dissertation by John Murdoch, item 613 listed below.

595. Sylla, Edith. "Medieval Concepts of the Latitude of Forms. The Oxford Calculators." *Archives d'Histoire Doctrinale et Littérarie du Moyen Âge* 40 (1973), 223-283.

Considers the use of latitudes to express functions (pp. 264-270).

596. ———. "William Heytesbury on the Sophism 'Infinita sunt finita.'" *Miscellanea Mediaevalia*, Band 13 (2). (*Sprache und Erkenntnis im Mittelalter*.) Berlin and New York: Walter de Gruyter, 1981, 628-636.

Describes the medieval application of logical analysis to problems of the infinite.

597. Tannery, Paul. *Mémoires scientifiques*. Vol. 5. *Sciences exactes au Moyen Âge*. Toulouse: Edouard Privat, 1922.

Contains numerous articles on medieval mathematics, some superseded, others not, for instance, "La géométrie au XIe siècle," pp. 79-102. Still the best source for the mathematical work of the Lotharingian schools.

598. Thomson, R.B. "Jordanus de Nemore: Opera." *Mediaeval Studies* 38 (1976), 97-144.

An up-to-date statement on the difficult subject of distinguishing the works genuinely belonging to Jordanus.

599. Ullman, B.L. "Geometry in the Mediaeval Quadrivium." In *Studi di bibliografia e di storia in onore di Tammaro de Marinis*. Band 4. Verona: Stamperia Valdonega, 1964, 263-285.

Argues that in the early medieval period works on surveying were used as a source for geometrical material to be used in teaching the quadrivium. There is sufficient information in early manuscripts and library catalogues to trace the history of gromatic collections. Corbie appears to have been a center for geometrical teaching.

600. Van Wijk, W.E. *Origine et développement de la computistique médiévale*. (Les Conférences du Palais de la Découverte, no. 29.) Paris: Université de Paris, 1954.

Describes medieval computus treatises, which taught the elementary astronomical calculations needed to determine the date of Easter.

601. Weisheipl, James A., ed. *Albertus Magnus and the Sciences. Commemorative Essays 1980*. Toronto: Pontifical Institute of Mediaeval Studies, 1980.

Includes A.G. Molland, "Mathematics in the Thought of Albertus Magnus," pp. 462-478, and Paul M.E.J. Tummers, "The Commentary of Albert on Euclid's Elements of Geometry," pp. 479-499.

602. Wieleitner, H. "Zur Geschichte der unendlichen Reihen im christlichen Mittelalter." *Bibliotheca Mathematica* 14 (1914), 150-168.

Infinite series in the work of Oresme, Alvarus Thomas, et al.

603. Wilson, Curtis. *William Heytesbury. Medieval Logic and the Rise of Mathematical Physics*. Madison: University of Wisconsin Press, 1956; 2nd printing 1960.

Description of Heytesbury's studies of temporal limits, maxima and minima, and measures of motions, with comparisons to related works. No texts. Although Heytesbury in effect sums infinite series and the like, his major techniques rely on logical distinctions more often than on mathematical calculations.

604. Zoubov, V.P. "Traktat Bradvardina O Kontinuume." *Istoriko-Matematicheskie Issledovaniia* 13 (1960), 385-440.

Contains the Latin of the major enunciations of the treatise; now superseded by the dissertation of John Murdoch, item 613 below.

605. ———. "Walter Catton, Gérard d'Odon, et Nicolas Bonet." *Physis* 1 (1959), 261-278.

Describes fourteenth-century disputes over the composition of continua.

Dissertations (Unpublished Unless Otherwise Noted)

606. Allard, A. "Les plus anciennes versions latines du 12e siècle issues de l'Arithmétique d'al-Khwārizmī." Université Catholique de Louvain, 1975.

607. Britt Hahn, F. Nan. "A Critical Edition of Tractatus Quadrantis." Emory University, 1972. Published as Hahn, Nan L., *Medieval Mensuration: Quadrans Vetus and Geometrie Due Sunt Partes Principales....* (Transactions of the American Philosophical Society, Vol. 72, part 8.) Philadelphia: American Philosophical Society, 1982.

608. Brown, Joseph E. "The *Scientia de Ponderibus* in the Later Middle Ages." University of Wisconsin, 1967.

609. Cunningham, T.J. "Book V of Euclid's Elements in the Twelfth Century: The Arabic-Latin Traditions." University of Wisconsin, 1972.

610. Goldat, G.D. "The Early Medieval Traditions of Euclid's Elements." University of Wisconsin, 1957.

611. Hughes, Barnabas B. "The De numeris datis of Jordanus de Nemore." Stanford University, 1970. Published as *Jordanus de Nemore. De numeris datis*. (Publications of the Center for Medieval and Renaissance Studies, Vol. 14.) Berkeley, Los Angeles, London: University of California Press, 1981.

612. Molland, A.G. "The Geometria speculativa of Thomas Bradwardine." University of Cambridge, 1967.

613. Murdoch, John. "Geometry and the Continuum in the Fourteenth Century: A Philosophical Analysis of Thomas Bradwardine's Tractatus de continuo." University of Wisconsin, 1957.

614. Schrader, Sister Walter Reginald, O.P. "The Epistola de proportione et proportionalitate of Ametus filius Josephi." University of Wisconsin, 1961.

615. Van Ryzin, Sister Mary St. Martin. "The Arabic-Latin Tradition of Euclid's Elements in the Twelfth Century." University of Wisconsin, 1960.

RENAISSANCE MATHEMATICS

The algebraic method, *algebra et almuchabala*, introduced to Europe through twelfth-century Latin translations of the works of al-Khowarizmi, was popularized by the *Liber abaci* (1202) of Leonardo Fibonacci of Pisa. Leonardo's work served as a model for generations of *maestri d'abbaco* who taught arithmetic to merchant apprentices in the commercial cities of central and northern Italy. It was in their schools and in their manuals that the rules of algebra were kept alive until they joined the mainstream of knowledge in the *Summa de arithmetica* (1494) of Luca Pacioli. The solution of the cubic and quartic equations (by Scipione dal Ferro, Tartaglia, and Ferrari) during the early sixteenth century prompted Cardano (1545) and Bombelli (1572) to write their respective treatises. The *Arithmetica* of Diophantos (partly revealed in Bombelli's *Algebra*) was published in a Latin translation by Xylander (Basle, 1575); a French version of Books 1-4 was included by Stevin in his *L'arithmêtique* (Leyden, 1585). Meanwhile German mathematicians had made considerable progress in algebraic notation. Algebra was now ready for the next stage of its development in the hands of Viète, who took the significant step of distinguishing between unknown quantities and parameters by providing separate symbols for them. What had begun as a collection of rules for solving arithmetical problems was now an independent discipline. The *arte della cosa* of the *maestri d'abbaco* had become the *ars analytica* of Viète and Harriot.

In the bibliography which follows, the number of primary sources has been kept to a minimum. Reliance has been placed on a variety of secondary sources to serve as entry points to the literature of the subject.

Bibliographies

616. May, Kenneth O. *Bibliography and Research Manual of the History of Mathematics*. Toronto: University of Toronto Press, 1973.

Renaissance mathematics is covered under various headings, including: Algebra (pp. 393-395), Arithmetic (pp. 402-406), Cubic equations (pp. 422-423), False position (p. 437), Quadratic equations (pp. 488-489), Quartic equations (pp. 489-490), XIIIth to XVIth centuries (pp. 640-643), Italy, (pp. 659-661), Bologna (p. 671). See also items 17 and 138.

617. Pancanti, M., and D. Santini. *Gino Arrighi storico della matematica medioevale: una bibliografia*. (Bibliografie e saggi, 1.) Siena: Centro Studi della Matematica Medioevale, 1983, 33 pp.

Bibliography of the writings of Gino Arrighi on late medieval mathematics in Italy, covering the period from Leonardo Fibonacci to Luca Pacioli. Includes editions of several *trattati d'abbaco*. Arrighi's work is now continued by the Center for the Study of

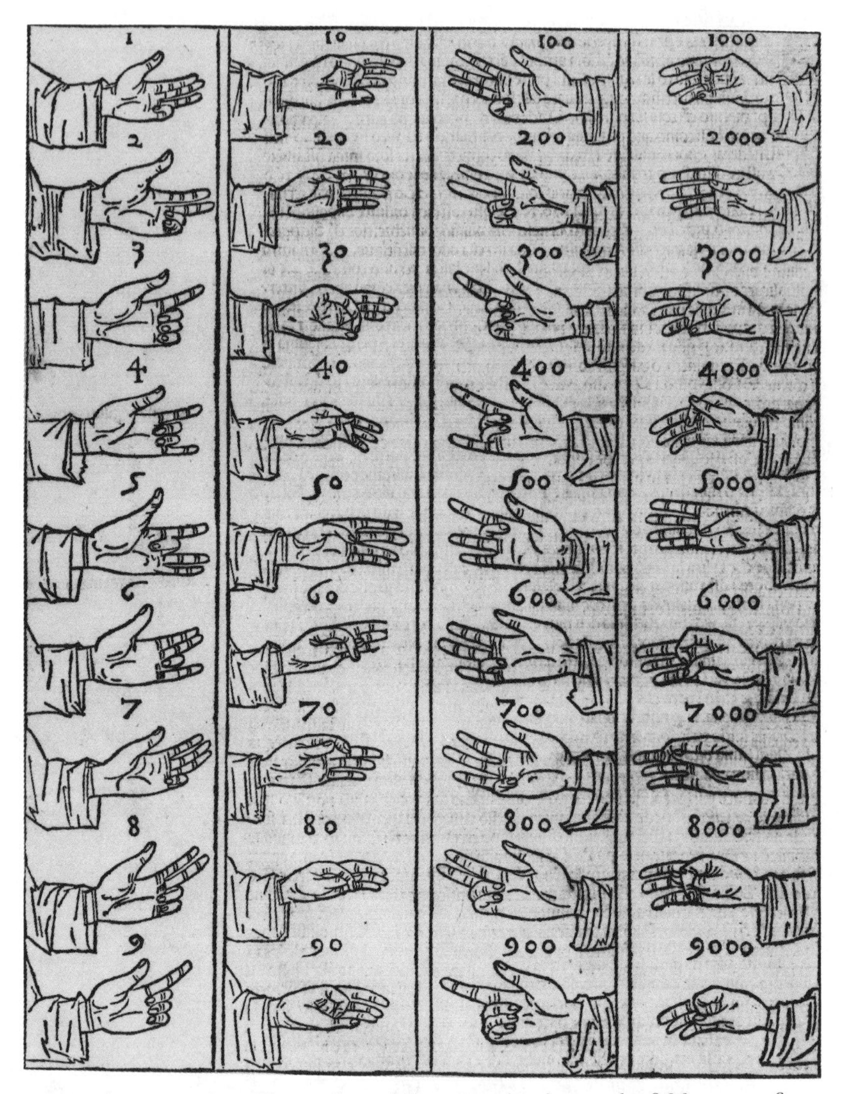

Figure 7. A woodcut illustration of finger reckoning on leaf 36 verso of Pacioli's *Suma de arithmetica, geometria, proportioni, et proportionalita* (item 737).

Medieval Mathematics at the University of Siena set up in 1981. It publishes studies of mathematical texts in *Quaderni*; see items 471a-e.

618. Rider, Robin E. *A Bibliogrpahy of Early Modern Algebra, 1500-1800.* (Berkeley Papers in History of Science, 7.) Berkeley, Calif.: Office for History of Science and Technology, University of California, 1982, 171 pp.

Chronologically arranged list of some 1450 published works (including articles in journals)--from Pacioli to Ruffini. This work is a sequel to the author's Ph.D. dissertation, "Mathematics in the Enlightenment: A Study of Algebra, 1685-1800" (University of California, Berkeley, 1980).

619. Rome, A. "Le R.P. Henri Bosmans, S.J. (1852-1928): notice biographique et index analytique de ses travaux historiques." *Isis* 12 (1929), 88-112.

Pages 94-100 contain an annotated bibliography of Bosmans's studies of the work of sixteenth-century mathematicians (including Gemma Frisius, Albert Girard, Guillaume Gosselin, Pedro Nuñez, Jacques Peletier, Adriaen van Roomen, Simon Stevin).

General Introductions and Histories

620. Boas, M. *The Scientific Renaissance, 1450-1630.* New York: Harper and Brothers, 1962.

A general, scholarly, and well-written account of science during the Renaissance, with two chapters of special interest for the history of mathematics. Chapter 12: "Mathematical Contributions in the Renaissance," pp. 231-249; and Chapter 13: "Arithmetic and Algebra in the Sixteenth and Seventeenth Centuries," pp. 250-284.

621. Bortolotti, E. *Storia della matematica nella Università di Bologna.* Bologna: N. Zanichelli, 1947.

Pages 35-80 offer a general account of the history of mathematics in Italy in the 16th century.

622. Cantor, Moritz B. *Vorlesungen über Geschichte der Mathematik.* Vol. 2: *Von 1200-1668.* 2nd ed. Leipzig: Teubner, 1900. Reprinted 1913.

Extensive studies of Renaissance algebraists on pp. 1-648. A series of additions and corrections, "Kleine Bemerkungen zur zweiten Auflage von Cantors *Vorlesungen* ..." by several scholars was published in *Bibliotheca Mathematica*, Series 3, 1 (1900) to 14 (1914). See also items 74, 243, 327, 466.

623. Crombie, A. *Augustine to Galileo.* London: Falcon Press, 1952. Reprinted Cambridge, Mass.: Harvard University Press, 1953.

Chapters 5 and 6 are devoted to topics related to the history of Renaissance mathematics in Europe.

624. Koyré, Alexandre. "Les mathématiques." *Histoire générale des sciences.* Vol. II. *La science moderne (de 1450 à 1800).* Sous la direction de René Taton. Paris: Presses Universitaires de France, 1958, 12-51.

Deals with the revival of mathematical studies and the development of algebra (from rhetorical to syncopated) during the Renaissance, in Germany, France, Italy, and the Low Countries. An English translation was published in 1964 by Thames and Hudson, London.

625. Libri, Guillaume. *Histoire des sciences mathématiques en Italie, depuis la renaissance des lettres jusqu'à la fin du dix-septième siècle*. Vols. 2 and 3. 2nd ed. Halle: Schmidt, 1865.

Libri was the first historian to stress the need to study the manuscripts of the *maestri d'abbaco*. Points out that they contain evidence of unsuccessful attempts at solving algebraic equations of the third and higher degrees. Appendices contain: Chapter XV of Leonardo's *Liber abaci* (algebra); extracts from two manuscript arithmetics (algebraic equations); Pacioli's *Summa*; Bombelli's *Algebra*.

626. Loria, Gino. *Storia delle matematiche dall'alba della civiltà al secolo XIX*. 2nd ed. Milan: Hoepli, 1950, 975 pp.

Chapter 12 (pp. 219-238), chapters 15-17 (pp. 287-352): General survey of arithmetic and algebra from Leonardo Fibonacci to Viète, with biographical sketches of the leading mathematicians. Short bibliography at the end of each chapter.

627. Mahoney, Michael Sean. *The Mathematical Career of Pierre de Fermat (1600-1665)*. Princeton, N.J.: Princeton University Press, 1973.

Chapter I (pp. 1-26) provides an excellent, concise overview of Renaissance mathematics. See also items 818, 1942.

628. Rose, Paul Lawrence. *The Italian Renaissance of Mathematics: Studies on Humanists and Mathematicians from Petrarch to Galileo*. Geneva: Droz, 1975, 316 pp.

An extensive survey of the work of generations of Renaissance scholars, rich in bibliographical information. Analyzes the mathematical sciences of the 15th and 16th centuries in Italy from three main aspects: internal developments; the "Renaissance sensibility" of mathematicians; and contacts between humanists and mathematicians. Contends that the leading mathematicians of the time were dominated by a desire to revitalize the mathematical sciences. Focuses attention on some 40 mathematicians, mostly of the humanistic tradition, and, unfortunately, ignores the work of the *maestri d'abbaco*. Exaggerates the role of the Italian algebraists in restoring Greek mathematics (pp. 143-158). See also item 591.

629. Smith, David Eugene. *History of Mathematics*. Vol. 2: *Special Topics of Elementary Mathematics*. Boston: Ginn, 1928.

Chapter VI (pp. 378-530) contains a well-documented topical survey of algebra, including that of the Renaissance. See also item 85.

630. Taylor, Eva G.R. *The Mathematical Practitioners of Tudor and Stuart England, 1485-1714*. Cambridge: Cambridge University Press, 1954, xi + 443 pp., 12 pls.

Chapters I and II (pp. 7-48) cover the period 1485-1600. Special attention is devoted to Thomas Digges, John Dee, Robert Norman, William Borough, Thomas Harriot, and Edward Wright. Methods, instruments, land surveying, and cartography, as well as vernacular textbooks are also discussed. See also items 801, 1809, 2216.

631. Wightman, W.P.S. *Science and the Renaissance. An Introduction to the Study of the Emergence of the Sciences in the Sixteenth Century*. Edinburgh: Olier and Boyd, 1962.

Included in this general history, special mention should be made of Chapter VI (pp. 87-99): "The Mathematical Disciplines," and Chapter VIII (pp. 129-247): "Mathematical Practitioners--Maps, Charts and Surveys."

632. ———. *Science in a Renaissance Society*. London: Hutchinson University Library, 1972.

Chapter 3 (pp. 42-56) treats mathematical themes, as does Chapter 10 (pp. 130-139): "Accent on Quantity." There is also an appendix on perspective.

633. Zeuthen, H.-G. *Geschichte der Mathematik im 16. und 17. Jahrhundert*. Leipzig: Teubner, 1903. Reprinted New York: Johnson Reprint Corporation, 1966, viii + 434 pp.

Though succeeded by more recent research, this is still a good place to begin for a general introduction, as its reprinting in 1966 indicates. See also item 774.

Special Studies

ALGEBRA

634. Cossali, Pietro. *Origine, trasporto in Italia, primi progressi in essa dell'algebra; storia critica di nuove disquisizioni analitiche e metafisiche arrichita*. Parma: Reale Tipografia, 1797-1799. 2 vols.

Studies of the work of Leonardo of Pisa, Pacioli, and other algebraists to Bombelli. Outdated, but still useful, especially for the observations on the treatment of the irreducible case by Cardano and Bombelli, and on Cardano's *De aliza regula*.

635. Hutton, Charles. "History of Algebra." *Tracts on Mathematical and Philosophical Subjects*. Vol. 2. London: F.C. and J. Rivington, 1812, 143-305.

Pages 195-274: Algebra from Leonardo of Pisa to Viète. Analyzes the important texts (Leonardo's *Liber abbaci* and the *Cartelli* of Ferrari and Tartaglia excepted). Discusses Cardano's and Viète's work at length. As a survey, still useful though outdated.

636. Russo, François. "La constitution de l'algèbre au XVIe siècle: étude de la structure d'une évolution." *Revue d'histoire des sciences* 12 (1959), 193-208.

Analyzes the following aspects of the development of algebra: notations; symbols for the unknown and its powers; known quantities and the use of symbols for them; and algebraic method.

637. Treutlein, P. "Die deutsche Coss." *Abhandlungen zur Geschichte der mathematischen Wissenschaften* 2 (1879), 1-124.

Surveys the German contribution to algebra (cossic terms and notation, the cossic algorithm, irrational quantities, solving equations, cossic texts). Needs updating with later studies of primary sources (see Tropfke, 4th ed., item 638).

638. Tropfke, Johannes. *Geschichte der Elementarmathematik.* 4th ed. Vol. 1. *Arithmetic und Algebra.* Vollständig neu bearbeitet von Kurt Vogel, Karin Reich, Helmuth Gericke. Berlin: Walter de Gruyter, 1980, 742 pp.

Completely revised edition. As in previous editions, treatment is topical (pp. 359-510 deal with algebra). No footnotes, but reference is made direct from the text to the bibliography on pp. 661-725. The bibliography, although not classified (arranged by author/editor/manuscript), is particularly useful as it contains editions of texts and studies published during the last 40 odd years. Name and subject indexes provided. See also items 470, 644, 1185.

ARITHMETIC

639. Glaisher, J.W.L. "On the Early History of the Signs + and - and on the Early German Arithmeticians." *Messenger of Mathematics* 51 (1921-1922), 1-148.

Includes a study of the German and Latin algebras in the Dresden Codex C.80, the cossic notation in Widman's *Rechnung*, the algebras of Riese, Rudolff, and Stifel.

640. Karpinski, Louis C. "An Italian Algebra of the Fifteenth Century." *Bibliotheca Mathematica*, Series 3, 11 (1910-1911), 209-219.

See item 1175.

641. Van Egmond, Warren. "The Commercial Revolution and the Beginnings of Western Mathematics in Renaissance Florence, 1300-1500." Dissertation, University of Indiana, 1976, 628 pp.

Study of the Florentine practical arithmetics, based on a preliminary examination of some 200 manuscripts. Although only a few pages are specially devoted to algebra (pp. 217-222, 256-270, 325-326), the appendices contain a catalogue of manuscripts (pp. 417-588) and bio-bibliographical sketches of the *maestri d'abbaco* (pp. 348-411), many of whom were responsible for popularizing the "rules of algebra."

TRIGONOMETRY

642. Zeller, M.C. "The Development of Trigonometry from Regiomontanus to Pitiscus." Dissertation, University of Michigan, 1944. Ann Arbor: Edwards Brothers, 1946. Lithoprint.

NOTATION AND TERMINOLOGY

643. Cajori, Florian. *A History of Mathematical Notations*. Chicago: Open Court, 1928-1929. 2 vols.

Volume 1, pp. 89-187, contains a brief outline of the symbols in arithmetic and algebra found in mathematical texts from the 13th century to the end of the 16th century. Pages 229-400 contain a topical survey of the use of notations; references to algebras of the Renaissance are found scattered throughout these pages. See also items 73, 986, 1821.

644. Tropfke, Johannes. *Geschichte der Elementar-Mathematik in systematischer Darstellung mit besonderer Berücksichtigung der Fachwörter*. Leipzig: Viet & Co., 1902-1903. 2 vols. 2nd ed. Berlin, Leipzig: Vereinigung Wissenschaftlicher Verleger, 1921-1924. 7 vols.; 3rd ed. rev. and enlarged. Vols. 2-3. Berlin: De Gruyter, 1930-1937, 266 pp. and 239 pp.

Volume 2, pp. 3-64: A. Die algebraische Ausdruckweise; pp. 64-70: B. Die Name Algebra; pp. 70-118: C. Die Entwicklung des Zahlbegriffes; pp. 118-204: D. Die algebraischen Operationen. Volume 3, pp. 3-22: A. Die Proportionen; pp. 22-198: B. Die Gleichungen. Appendices, pp. 199-202: Zeittafel zur Geschichte der modernen algebraischen Zeichenschrift; pp. 203-233: Zur Geschichte der mathematischen Schreibart. Zusammenstellung von Originalbeispielen; pp. 234-235: Zur Geschichte der kubischen Gleichungen. This work aims at overcoming the difficulties of consulting Cantor's *Vorlesungen*, items 74 and 622, when investigating the history of a particular topic. Although there is no index to this edition, concordances are provided with each volume to be used in conjunction with the name and subject indexes of the second edition (1924, vol. 7, 55-128). This work is still useful as the new (fourth) edition has been completely re-written. See also items 470, 638, 1185.

Individual Mathematicians and Primary Sources

645. Codex Dresdensis C 80, fol. 368-378^{v}. *Die erste deutsche Algebra aus dem Jahre 1481 nach einer Handschrift aus C 80 Dresdensis*. Herausgegeben und erläutert von Kurt Vogel. *Abhandlungen, Bayerische Akademie der Wissenschaften, math.-naturwiss. Klasse* (Munich), Neue Folge, Heft 160 (1981), 52 pp.

Annotated edition of the first algebraic text to be written in German (1481). The manuscript (471 folios), containing several texts of varying origin, is a rich source for the history of mathematics in Germany. Its contents are described by

W. Kaunzner in "Über Johannes Widmann von Eger: ein Beitrag zur Geschichte der Rechenkunst im ausgehenden Mittelalter," *Veröffentlichungen des Forschungsinstituts des Deutschen Museums für die Geschichte der Naturwissenschaften und der Technik* (Munich), Reihe C 7 (1968), 27-48.

Alberti, Leone Battista (1404-1472)

646. Michel, P.H. *La pensée de L.B. Alberti*. Paris: "Les Belles Lettres," 1930.

Contains an exhaustive bibliography.

Baldi, Bernardino (1553-1617)

647. Drake, S., and I.E. Drabkin. *Mechanics in 16th-Century Italy*. Madison, Wis.: University of Wisconsin Press, 1968.

Discusses Baldi's relation to central Italian mathematics.

648. Duhem, Pierre. *Etudes sur Léonardo de Vinci*. Paris: Hermann, 1906. Reprinted Paris: F. de Nobele, 1955.

Baldi's scientific work and influence are discussed in volume I, pages 89-156. Bibliography of Baldi's published works is included.

For additional material concerning Baldi, see Ugolino and Polidori, item 670, and Rose, item 628.

Benedetti, Giovanni Battista (1530-1590)

English translations of Benedetti's most important scientific writings are provided in Drake and Drabkin, item 647. For other other works treating Benedetti, see Duhem, item 648, Volume III, 214 ff, and A. Koyré, *Etudes galiléennes* (Paris: Hermann, 1939-1940, 3 vols.), I, 41-54.

649. Bordiga, G. "Giovanni Battista Benedetti, filosofo e matematico veneziano del secolo XVI." *Atti del Reale Istituto Veneto di Scienze, Lettere ed Arte* 85 (part 2) (1925-1926), 585-754.

650. Drabkin, I.E. "Two Versions of G.B. Benedetti's *Demonstratio*." *Isis* 54 (1963), 259-262.

651. Maccagni, C. "Contributi alla biobibliografia di G.B. Benedetti." *Physis* 9 (1967), 337-364.

652. ———. *Le speculazioni giovanili "de Motu" di G.B. Benedetti*. Pisa: Domus Galilaeana, 1967.

Bombelli, Rafael (1526-1572)

653. Bombelli, R. *L'algebra part maggiore dell'aritmetica divisa in tre libri*. Bologna, 1572, 650 pp. Reprinted Bologna: G. Rossi, 1579.

654. ———. *L'algebra, opera di Rafael Bombelli da Bologna. Libri IV e V comprendenti "La parte geometrica" inedita tratta dal manoscritto B. 1569,* [della] *Biblioteca dell'Archiginnasio di Bologna. Pubblicata a cura di Ettore Bortolotti.* Bologna: Zanichelli, 1929, 302 pp.

A so-called "edizione integrale"--a combined edition of items 653 and 654, edited by Umberto Forti, was published in 1966 by Feltrinelli in Milan.

Bombelli's *Algebra*, written some ten years before it was published, was one of the first attempts to formulate algebra as an independent discipline. Only Books I-III were published by the author. The promised "geometrical part" (Books IV and V) was not completed. It has since been found in a manuscript containing the whole work (B.1569 Biblioteca Comunale, Bologna) and published by Bortolotti with an introduction and an analysis of the 1572 edition. The unpublished Book III of the manuscript has been analyzed and compared with the 1572 text by Jayawardene, item 656. See also Hofmann, item 655.

655. Hofmann, Joseph E. "Bombellis Algebra--eine genialische Einzelleistung und ihre Einwirkung auf Leibniz." *Studia Leibnitiana* 4 (1972), 196-252.

Contains an analysis of items 653 and 654, followed by an account of Leibniz's study of Bombelli.

656. Jayawardene, S.A. "The Influence of Practical Arithmetics on the Algebra of Rafael Bombelli." *Isis* 64 (1973), 510-523.

Compares the unpublished manuscript of Book III of item 653 with the printed text, and offers both commentary and historical analysis.

657. ———. "Unpublished Documents Relating to Rafael Bombelli in the Archives of Bologna." *Isis* 54 (1965), 391-395.

658. ———. "Rafael Bombelli, Engineer-Architect: Some Unpublished Documents of the Apostolic Camera." *Isis* 56 (1965), 298-306.

Cardano, Girolamo (1501-1576)

659. Cardano, Girolamo. *Artis magnae, sive De regulis algebraicis liber unus.* Nuremberg: J. Petreius, 1545, 81 pp.

First systematic treatise on the solution of equations (up to and including the biquadratic). The second edition of this work, and its postscript, the *De aliza regula liber* ("algebraic logistics"), were included in Cardano's *Opus novum de proportionibus numerorum* ... (Basle: Officina Henricpetrina, 1570). The *Opera omnia* (Lyons, 1663), vol. 4, contains all the mathematical works. These include the hitherto unpublished *Sermo de plus et minus* (pp. 435-439), a commentary on Bombelli's *Algebra* (1572). The *Ars magna* has been translated into English: *The Great Art, or the Rules of Algebra*, translated by T. Richard Witmer (Cambridge, Mass.: MIT Press, 1968, 267 pp.).

660. Bellini, A. *Girolamo Cardano e il suo tempo*. Milan: U. Hoepli, 1947.

Includes extensive bibliographic references.

661. Bortolotti, E. "I contributi del Tartaglia, del Cardano, del Ferrari e della scuola matematica bolognese alla teoria algebrica delle equazioni cubiche." *Studi e memorie per la storia dell'Università di Bologna* 9 (1926), 55-108.

662. Kahn, D. *The Codebreakers*. New York: Macmillan, 1967.

Discusses Cardano's contributions to cryptology.

663. Ore, Øystein. *Cardano: The Gambling Scholar*. Princeton: Princeton University Press, 1953.

The most extensive study of Cardano to date; contains Gould's translation of *Liber de ludo aleae*. See also item 2026.

664. Tanner, R.C.H. "The Alien Realm of the Minus: Deviatory Mathematics in Cardano's Writings." *Annals of Science* 37 (1980), 159-178.

Examines the *De aliza regula* and the *Sermo de plus et minus* (see item 659) in light of the studies of the later algebraist, Thomas Harriot (1560-1621). For Harriot, see additional entries below.

665. Vacca, G. "L'opera matematica di Girolamo Cardano nel quarto centenario del suo insegnamento in Milano." *Rendiconti, Seminario matematico e fisico di Milano* 11 (1937), 22-40.

For additional sources related to Cardano, consult Cantor, items 74, 622, Volume II, 484-510 and 532-541; Duhem, item 648, 223-245.

Chuquet, Nicolas (fl. 1484)

666. Chuquet, N. "Le Triparty en la science des nombres par Maistre Nicolas Chuquet, parisien, d'après le manuscrit fonds français no. 1346 de la Bibliothèque Nationale de Paris." Edited by Aristide Marre. *Bullettino di bibliografia e storia delle scienze matematiche e fisiche* 13 (1880), 555-659, 693-814.

Introduction by Marre on pp. 555-592. The appendix to the *Triparty*, edited by E. Narducci, *Bullettino di bibliografia e storia* ... 14 (1881), 413-460, consists of a collection of problems of practical arithmetic and geometry (with applications of the rules of algebra). The manuscript of the *Triparty* was in the possession of Etienne de la Roche who used it in writing his *L'arismethique* (Lyons: Guillaume Huyon, 1520; Lyons: G. et J. Huguetan frères, 1538).

667. Lambo, C. "Une algèbre française de 1484." *Revue des questions scientifiques*, Series 3, 2 (1902), 442-472.

Provides an analysis of the entire work by Chuquet, item 666.

Commandino, Federico (1509-1575)

668. Riccardi, Pietro. *Biblioteca matematica italiana*. Milan: Goerlich, 1952.

Lists Commandino's publications. See also item 23.

669. Rosen, E. "John Dee and Commandino." *Scripta Mathematica* 28 (1970), 321-326.

670. Ugolino, F., and F.-L. Polidori, eds. *Versi e prose scelte di Bernardino Baldi*. Florence: Le Monnier, 1859.

Contains Baldi's biography of Commandino (written in 1587), pp. 513-537.

Copernicus, Nicholas (1473-1543)

671. Rossi, F.S. "Copernico matematico: La sua trigonometria piana." *Cultura Scuola* 12 (1973), 317-336.

Dee, John (1527-1608)

672. Cooper, T. "John Dee." *Dictionary of National Biography, V.* London: Oxford University Press, 1917. Reprinted 1921-1922, 721-729.

Contains an extensive list of Dee's original works.

673. Yates, F.A. *Theatre of the World*. Chicago: Chicago University Press, and London: Routledge and Kegan Paul, 1969.

Discusses Dee as a Renaissance philosopher, and his mathematical preface to Euclid.

For additional material concerning Dee and his mathematics, consult Rosen, item 669, and especially Taylor, item 630, which provides one of the best assessments of Dee's scientific work.

Dürer, Albrecht (1471-1528)

674. Panofsky, E. "Durer as a Mathematician." In *The World of Mathematics*. Edited by J.R. Newman. New York: Simon and Schuster, 1956, 603-621.

675. Steck, M. *Dürer's Gestaltlehre der Mathematik und der bildenden Künste*. Halle and Tübingen: M. Niemeyer, 1948.

Contains an extensive bibliography as well as a critical analysis of Dürer's mathematics.

676. ———. "Albrecht Dürer als Mathematiker und Kunst-theoretiker." *Nova Acta Leopoldina* 16 (1954), 425-434.

677. ———. "Theoretische Beiträge zu Dürer's Kupferstich 'Melancholia I' von 1514." *Forschungen und Fortschritte* 32 (1958), 246-251.

Digges, Thomas (1546?-1595)

678. Hall, A.R. *Ballistics in the 17th Century*. Cambridge, England: Cambridge University Press, 1952.

Includes some discussion of Digges and his mathematics.

679. Johnson, F.R., and S.V. Larkey. "Thomas Digges, the Copernican System, and the Idea of the Infinity of the Universe in 1576." *Huntington Library Bulletin* 5 (1934), 69-117.

Discusses Digges's works.

680. Koyré, Alexandre. *From the Closed World to the Infinite Universe*. New York: Harper, 1957. Reprinted Baltimore: Johns Hopkins Press, 1968.

Digges is discussed on pages 34-39.

Ferrari, Ludovico (1522-1565)

681. Ferrari, L., and Niccolò Tartaglia. *Cartelli di sfida matematica*. Riproduzione in facsimile delle edizioni originali 1547-1548. Edita con parti introduttorie da Arnaldo Masotti. Brescia: Ateneo di Brescia, 1974, cxciii + 202 pp.

This correspondence is the sequel to Tartaglia's indirect accusation (in his *Quesiti*) that Cardano had broken his promise of secrecy in regard to the formula for solving the cubic equation. Ferrari (Cardano's pupil) wrote in defense of his master and challenged Tartaglia to a disputation. Each party sent the other a series of problems--62 in all--of which 17 are related to arithmetic and algebra. Masotti's commentary--a labor of love--consists of a well-documented introduction, an index of names, a chronology, and a glossary of archaic terms.

682. Masotti, S.C. Arnaldo. "Sui cartelli di matematica disfida scambiati fra Ludovico Ferrari e Niccolò Tartaglia." *Rendiconti dell'Istituto lombardo di scienze e lettere* 94 (1960), 31-41.

Cites the most important secondary literature on Ferrari.

Fibonacci, Leonardo (Leonardo of Pisa), (ca. 1170-ca. 1240)

683. Boncompagni, B., ed. *Scritti di Leonardo Pisano* pubblicati da B. Boncompagni. I. *Il liber abaci*, secondo la lezione del Codice Magliabechiano, C.I.2616 Badia fiorentina, no. 73. II. *La practica geometriae*, secondo la lezione del Codice Urbinate no. 292 della Biblioteca Vaticana; *Opuscoli*, secondo la lezione di un codice della Biblioteca Ambrosiana di Milano contrassegnato E 75, parte superiore. Rome: Tipografia delle scienze matematiche e fisiche, 1857-1862. 2 vols.

This edition has neither commentary nor introduction. Boncompagni's extensive studies are listed by Kurt Vogel in his analysis of Leonardo's work, "Fibonacci, Leonardo," *Dictionary of Scientific Biography*, vol. 4 (1971), 604-613. See item 10 above. The *Liber abaci* (written in Latin, 1202) introduced to Italy the Hindu-Arabic numerals and the rules of algebra. Many copies of the manuscript seem to have been in circulation. The work was popularized by generations of *maestri d'abbaco*--teachers of practical arithmetic--who used it as a model for their *trattati* (written in Italian). See also item 477.

Francesca, Piero della

See *Piero della Francesca*.

Galilei, Galileo (1564-1642)

684. Boyer, Carl B. "Galileo's Place in the History of Mathematics." In *Galileo, Man of Science*. Edited by E. McMullin. New York: Basic Books, 1968.

685. Carruccio, E. "La filosofia della matematica nel pensiero di Galileo." In *Symposium Internazionale di Storia, Metodologia, Logica e Filosofia delle Scienze*. Florence: Gruppo Italiano di Storia delle Scienze, 1967.

686. Cassirer, E. "Mathematical Mysticism and Mathematical Science." In *Galileo, Man of Science*. Edited by E. McMullin. New York: Basic Books, 1968.

687. Clavelin, M. "Le problème du continu et les paradoxes de l'infini chez Galilée." *Thales* 10 (1959), 1-26.

688. Costabel, P. "The Wheel of Aristotle and French Consideration of Galileo's Arguments." *Mathematics Teacher* 61 (1968), 527-534.

This article compares the work and influences of Mersenne, Roberval, Fermat, and Descartes in response to Galileo's mathematics.

689. Drake, S. "Mathematics and Discovery in Galileo's Physics." *Historia Mathematica* 1 (1974), 129-150.

690. Eneström, G. "Ueber den Erfinder des Namens 'Zykloide.'" *Bibliotheca Mathematica*, Series 3, 13 (1912-1913), 272.

This note discusses the problem of whether or not there is any concrete evidence for attributing the naming of the cycloid to Galileo.

691. Frajese, A. "Concezioni infinitesimali nella matematica di Galileo." *Archimede* 16 (1964), 241-245.

This work was also presented at the International Symposium mentioned in item 685.

692. Itard, J. "Sur une pretendue erreur mathématique de Galilée." *Revue d'Histoire des sciences et de leurs applications* 1 (1947-1948), 355-356.

Discusses a supposed error of Galileo.

693. May, Kenneth O. "Galileo Sequences, a Good Dangling Problem." *American Mathematical Monthly* 79 (1972), 67-69.

May takes exception to the idea that Galileo's results on free fall were dependent on his application of arithmetic progressions.

694. Pogrebysskii, Iosif Bendiktovich. "Galilei i matematika." *Voprosȳ Istorii Estestvoznaniia i Tekhniki* 16 (1964), 34-37. In Russian.

Discusses Galileo's mathematics.

695. Procissi, A. "La matematica nell'opera galileiana." *Archimede* 16 (1964), 246-252.

696. ———. "La Teoria delle Proporzioni tra gradezze in Galileo e nella sua scuola." In *Symposium Internazionale di Storia, Metodologia, Logica e Filosofia delle Scienze*. Florence: Gruppo Italiano di Storia delle Scienze, 1967.

697. Quan, S. "Galileo and the Problem of Concentric Circles." *Annals of Science* 24 (1968), 313-338.

698. ———. "Galileo and the Theorem of Pythagoras." *Annals of Science* 31 (1974), 227-261.

699. Toyoda, T. "Galileo's Concepts of Infinity and Space." In *Symposium Internazionale di Storia, Metodologia, Logica e Filosofia delle Scienze*. Florence: Gruppo Italiano di Storia delle Scienze, 1967.

700. Viola, T. "Galileo e l'analisi matematica del '600." In *Symposium Internazionale di Storia, Metodologia, Logica e Filosofia delle Scienze*. Florence: Gruppo Italiano di Storia delle Scienze, 1967.

Gherardi, Paolo (fl. 1328)

701. Gherardi, P. *Paolo Gherardi. Opera matematica: Libro di ragioni e Liber habaci*: codici magliabechiani classe XI, numeri 87 e 88 (sec. XIV) della Biblioteca Nazionale di Firenze. A cura di Gino Arrighi, to appear.

The chapter on the rules of algebra in this work has been translated by Warren van Egmond: "The Earliest Vernacular Treatment of Algebra: the *Libro di Ragioni* of Paolo Gerardi (1328)," *Physis* 20 (1978), 155-189.

Guidobaldo del Monte (1545-1607)

702. Favaro, A. "Galileo e Guidobaldo del Monte." *Atti dell'Accademia di scienze, lettere ed arti di Padova* 30 (1914), 54-61.

703. Rose, Paul Lawrence. "Materials for a Scientific Biography of Guidobaldo del Monte." *Actes du XIIe Congrès International d'Histoire des Sciences* (Paris: A. Blanchard, 1968), 12 (1972), 69-72.

704. ———. "The Origins of the Proportional Compass." *Physis* 10 (1968), 54-69.

705. ———. "Renaissance Italian Methods of Drawing the Ellipse and Related Curves." *Physis* 12 (1970), 371-404.

Harriot, Thomas (1560-1621)

706. Harriot, Thomas. *Artis analyticae praxis ad aequationes algebraicas resolvendas*. Edited by W. Warner. London: Robert Barker, 1631, 181 pp.

Posthumously published from an imperfect manuscript. See Lohne, item 709.

707. Lohne, J.A. "Thomas Harriott." *Centaurus* 6 (1959), 113-121.

708. ———. "Thomas Harriott als Mathematiker." *Centaurus* 11 (1965), 19-45.

See item 821.

709. ———. "Dokumente zur Revalidierung von Thomas Harriot als Algebraiker." *Archive for History of Exact Sciences* 3 (1966), 185-205.

710. Pepper, J.V. "Harriott's Calculation of Meridional Parts as Logarithmic Tangents." *Archive for History of Exact Sciences* 4 (1968), 359-413.

711. Shirley, J.W. "Binary Numeration before Leibniz." *American Journal of Physics* 19 (1951), 452-454.

712. Stevens, H. *Thomas Harriot, the Mathematician, the Philosopher and the Scholar*. London: Chiswick Press, 1900.

713. Tanner, R.C.H. "Thomas Harriot as Mathematician." *Physis* 9 (1967), 235-247, 259-292.

714. ———. "The Ordered Regiment of the Minus Sign: Off-Beat Mathematics in Harriot's Manuscripts." *Annals of Science* 37 (1980), 127-158.

For additional work relating Harriot and Cardano, see Tanner, item 664 above, and consult item 822 as well.

Leonardo da Vinci (1452-1519)

715. Clagett, Marshall. "Leonardo da Vinci and the Medieval Archimedes." *Physis* 11 (1969), 100-151.

716. Marcolongo, R. "Le ricerche geometrico-meccaniche di Leonardo da Vinci." *Atti della Società italiana delle scienze, della dei XL* 23 (1929), 49-100.

717. ———. *Il trattato di Leonardo da Vinci sulle transformazioni dei solidi*. Naples: Stab. industrie editorial; meridionali, 1934.

718. Marinoni, A. "Le operazioni aritmetiche nei manoscritti vinciani." *Raccolta vinciana* (Milan), 19 (1962), 1-62.

719. ———. "La teoria dei numeri frazionari nei manoscritti vinciani. Leonardo e Luca Pacioli." *Raccolta vinciana* 20 (1964), 111-196.

720. ———. "L'aritmetica di Leonardo." *Periodico di matematiche* 46 (1968), 543-558.

721. ———. *L'essere del nulla*. Florence: L.S. Olschki, 1970.

Discusses the "principles of geometry" in Leonardo.

722. Panofsky, E. *The Codex Huygens and Leonardo da Vinci's Art Theory*. London: The Warburg Institute, 1940.

Studies Leonardo's mathematical interests related to theories of perspective and proportion.

723. Pedretti, C. "The Geometrical Studies." In *The Drawings of Leonardo da Vinci at Windsor Castle*. Vol. 1. Edited by K. Clark. London: Phaidon, 1968, xlix-liii.

724. ———. "Leonardo da Vinci. Manuscripts and Drawings of the French Period, 1517-1518." *Gazette des beaux-arts* (November 1970), 185-318.

Maurolico, Francesco (1494-1575)

725. Amodeo, F. "Il trattato sulle coniche di Francesco Maurolico." *Biblioteca Mathematica*, Series 3, 9 (1908-1909), 123-138.

726. Baron, Margaret E. *The Origins of the Infinitesimal Calculus*. Oxford: Pergamon, 1969, viii + 304 pp.

Maurolico's mathematical interests are discussed on pages 90-94. See also item 775.

727. Fontana, M. "Osservazioni storiche sopra l'aritmetica di Francesco Maurolico." *Memorie dell'Istituto nazionale italiana, fis.-mat. classe* (Bologna) 2 (1808), 275-296.

728. Rosen, E. "The Editions of Maurolico's Mathematical Works." *Scripta Mathematica* 24 (1959), 56-76.

729. Vacca, G. "Maurolycus, the First Discoverer of the Principle of Mathematical Induction." *Bulletin of the American Mathematical Society* 16 (1909-1910), 70-73.

For additional works treating various aspects of Maurolico's mathematics, consult Cajori, item 72, Vol. I, 349, 362, 402; Vol. II, 150; Cantor, item 74, Vol. II, 558-559, 575; Smith, item 85, Vol. I, 301-302; Vol. II, 622.

Mercator, Gerardus (Gerhard Kremer), (1512-1594)

730. *Gerhard Mercator--1512-1594: zum 450. Geburtstag*. Duisberg: W. Renckhoff, 1962.

A special publication dealing with Mercator's life and works, on the occasion of the 450th anniversary of his birth. Issued as Volume 6 of *Duisburger Forschungen* (1962). Articles in Dutch, English, French, and German, with summaries in languages other than the article's own. Bibliographic footnotes.

Napier, John (1550-1617)

731. Cajori, Florian. "History of the Exponential and Logarithmic Concepts." *American Mathematical Monthly* 20 (1913).

Includes discussion of Napier's contributions to the history of logarithms.

732. Hobson, E.W. *John Napier and the Invention of Logarithms*. Cambridge, England: Cambridge University Press, 1914.

733. Knott, C.G., ed. *Napier Tercentenary Memorial Volume*. London: Longmans, Green and Co., 1915, ix + 441 pp.

Much detail on the historical background to Napier's work. See also items 836, 837.

Nuñez Salaciense, Pedro (1502-1578)

734. da Silva, L.P. *As obras de Pedro Nuñez, sua chronologia bibliogrâfica*. Coimbra: Imprensa da Universidade, 1925.

735. ———. *Quarto centenârio da publicaçao de Tratado de sphera de Pedro Nuñez*. Lisbon: n.p., 1938.

736. Guimarães, R. *Sur la vie et l'oeuvre de Pedro Nuñes*. Coimbra: Imprimerie de l'Université, 1915.

Pacioli, Luca (1445-1517)

737. Pacioli, L. *Summa de arithmetica, geometria, proportioni, et proportionalita*. Venice: Paganinus de Paganinis, 1494, 308 pp. 2nd ed. Toscolano: Paganino de Paganini, 1523.

Mathematical encyclopedia, first printed work to contain the rules of algebra and their application to solving arithmetical problems. Pacioli taught mathematics in several Italian cities. An earlier version, written for his pupils in Perugia, is in the Vatican Library (Cod. Vat. lat. 3129).

738. Masotti Biggiogero, G. "Luca Pacioli e la sua 'Divina Proportione.'" *Rendiconti dell'Istituto lombardo di scienze e lettere*, Series A, 94 (1960), 3-30.

739. Taylor, R.E. *No Royal Road: Luca Pacioli and His Times*. Chapel Hill: University of North Carolina Press, 1942.

Not always reliable, but still an engaging biography as a general introduction to Pacioli.

Peuerbach, Georg von (1423-1461)

740. Ferrari d'Occhieppo, K. "Weitere Dokumente zu Peuerbachs Gutachten über den Kometen von 1456 nebst Bemerkungen über den Chronikbericht zum Sommerkometen 1457." *Sitzungsberichte der Österreichischen Akademie der Wissenschaften, math.-naturwiss.-Klasse*, 169 (1961), 149-169.

741. Lhotsky, A., and K. Ferrari d'Occhieppo. "Zwei Gutachten von Georg von Peuerbach über Kometen (1456 und 1457)." *Mitteilungen des Instituts für österreichische Geschichtsforschung* 68 (1960), 266-290.

742. Zinner, E. *Leben und Wirken des Johannes Müller von Königsberg gennant Regiomontanus*. 2nd ed. Osnabrück: O. Zeller, 1968.

Peuerbach is discussed on pages 26-49. Notes contain much information on manuscripts, and there is a thorough bibliography.

Piero della Francesca (c. 1410-1492)

743. Piero della Francesca. *Trattato d'abaco. Dal Codice Ashburnhamiano 280 (359*-291*) della Biblioteca Medicea Laurenziana di Firenze*. (Testimonianze di Storia della Scienza, 6.) A cura e con introduzione di Gino Arrighi. Pisa: Domus Galileiana, 1970. 270 pp.

Differs from the majority of *trattati d'abbaco* in that algebraic methods are introduced at a very early stage and the majority of the problems are solved by the rules of algebra. Contains unsuccessful attempts at solving equations of the third and higher degrees. Analyzed by S.A. Jayawardene, "The *Trattato d'abaco* of Piero della Francesca," *Cultural Aspects of the Italian Renaissance: Essays in Honour of Paul Oscar Kristeller*, edited by Cecil H. Clough (Manchester: Manchester University Press, 1976), 229-243.

744. Davis, M.D. *Piero della Francesca's Mathematical Treatises. The "Trattato d'abaco" and "Libellus de quinque corporibus regularibus."* Ravenna: Longo, 1977.

This study contains no texts. It considers the two works mentioned in the title by Piero, the first of which can be related to the education of architects, from the standpoint of art history. Shows that these works were used by Pacioli.

745. Longhi, R. *Piero della Francesca*. Translated by L. Penloch. London: F. Warne and Co., 1913. Reprinted 1930.

Detailed, analytical study of Piero's life and work.

Recorde, Robert (1510-1558)

746. Easton, J.B. "The Early Editions of Robert Recorde's *Ground of Artes*." *Isis* 58 (1967), 515-532.

747. ———. "A Tudor Euclid." *Scripta Mathematica* 27 (1966), 339-355.

748. Johnson, F.R., and S.V. Larkey. "Robert Recorde's Mathematical Teaching and the Anti-Aristotelian Movement." *Huntington Library Bulletin* 7 (1935), 58-87.

749. Kaplan, E. "Robert Recorde (c. 1510-1558): Studies in the Life and Works of a Tudor Scientist." Dissertation, New York University, 1960.

Available through University Microfilms, Ann Arbor, Michigan.

750. Karpinski, Louis C. "The Whetstone of Witte." *Bibliotheca Mathematica*, Series 3, 13 (1912-1913), 223-228.

Regiomontanus, Johannes (1436-1476)

751. Hughes, Barnabas. *Regiomontanus on Triangles. De triangulis omnimodis by Johann Müller, Otherwise Known as Regiomontanus.* Madison: University of Wisconsin Press, 1967.

Short introduction (pages 3-19) is followed by text and translation of *On Triangles* (1464) by Johannes Müller (Regiomontanus). It discusses especially the history of trigonometry and the contributions of Regiomontanus. See also item 513.

752. Rosen, E. "Regiomontanus's Breviarium." *Medievalia et Humanistica* 15 (1963), 95-96.

753. Schmeidler, F., ed. *Joannis Regiomontani opera collectanea.* Osnabrück: O. Zeller, 1949, 1972.

Includes a biography by the editor.

754. Zinner, E. *Leben und Wirken des Johannes Müller von Königsberg gennant Regiomontanus.* 2nd ed. Osnabrück: O. Zeller, 1968.

See item 742.

Stevin, Simon (1548-1620)

755. Depau, R. *Simon Stevin.* Brussels: J. Lebèque & Cie., 1942. In French.

Bibliography of articles on Stevin is also included.

756. Dijksterhaus, E.J. *Simon Stevin.* The Hague: M. Nijhoff, 1943. Translated into English and condensed by R. Hooykaas and M.G.J. Munaert, eds., as *Simon Stevin: Science in the Netherlands around 1600.* The Hague: M. Nijhoff, 1970.

Includes studies related to Stevin and his mathematics. See item 853.

757. Struik, Dirk J. *The Land of Stevin and Huygens; a Sketch of Science and Technology in the Dutch Republic during the Golden Century*. Dordrecht, Holland, and Boston, Mass.: D. Reidel, 1981, 162 pp.

First published in Dutch in 1958; several subsequent editions, revised. In addition to Stevin, considerable space is devoted to other influential figures of the period, including Descartes, Huygens, and Boerhaave. Chapter IV (pp. 32-51) is devoted to "The Dutch Teachers of Mathematics and Navigation." With 25 illustrations.

Stifel, Michael (1487-1567)

758. Stifel, Michael. *Arithmetica integra*. Nuremberg: J. Petreius, 1544, 319 pp.; later editions: 1545, 1546, and 1548.

Most important of the 16th-century German algebras. This work and Stifel's revised and enlarged edition of Christoff Rudolff's *Coss* (Königsberg: A. Behm von Luthomisl, 1553, 1571) served as models for the widely read *Algebra* (Rome: B. Zannettum, 1608) of Christoph Clavius. On Stifel, see Joseph E. Hofmann, *Michael Stifel 1487?-1567: Leben, Wirken und Bedeutung für die Mathematik seiner Zeit*, Sudhoffs Archiv, Beihefte 9, (Wiesbaden: Steiner, 1968), 1-42.

Tartaglia, Niccolò (1499-1557)

759. Tartaglia, Niccolò. *Quesiti et inventioni diverse*. Riproduzione in facsimile dell'edizione del 1554 edita con parti introduttorie da Arnaldo Masotti. (Pubblicazione celebrativa del quarto centenario della morte di Niccolò Tartaglia.) Brescia: Ateneo di Brescia, 1959, lxxxv + 128 pp.

Consists of problems on various subjects--from ballistics to algebra--set out in the form of dialogues and correspondence with various people. An important source of information on Tartaglia and his times. Divided into nine Books, Book 9 dealing with arithmetic, geometry, and algebra (in particular, the solution of the cubic equation). Tartaglia describes here his disputations with Antonio Maria Fior, and gives his version of the events which led to his quarrel with Cardano and Ferrari. Quesito 33 (f. 120^v) contains the well-known *terza rima* in which Tartaglia divulged to Cardano the formula for solving the cubic equation. Pages xvii-lxxxv contain an introduction, commentary, and notes by Masotti.

760. Masotti, S.C. Arnaldo. *Studi su N. Tartaglia*. Brescia: n.p., 1962.

Viète, François (1540-1603)

761. Viète, François. *Opera mathematica*. Recognita Francis à Schooten. Vorwort und Register von Joseph E. Hofmann. Hildesheim: Georg Olms Verlag, 1970, xxxix + 544 pp.

Facsimile reprint of the 1646 Leiden edition. Pages v-xxx contain an introduction by Hofmann.

762. ———. *Einführung in die neue Algebra*. Übersetzt und erlautert von Karin Reich und Helmuth Gericke. Munich: Werner Fritsch, 1973, 145 pp.

Translation in German of *In artem analyticem isagoge* and abridged translations of *Ad logisticem speciosam notae priores* and of *Zeteticorum liber quinque*. Each translation is preceded by a commentary. The introduction contains an historical outline of alphabetical notation in arithmetic and algebra from antiquity to Viète, a biographical notice, and an annotated bibliography of Viète's works.

763. Busard, H.L.L. "Über einige Papiere aus Viètes Nachlass in der Pariser Bibliothèque Nationale." *Centaurus* 9 (1964), 65-126.

764. Ferrier, Richard Delahide. "Two Exegetical Treatises of François Viète, Translated, Annotated, and Explained." Dissertation, Indiana University, 1980, 179 pp.

765. Grisard, J. "François Viète mathématicien de la fin du seizième siècle." Thèse de 3e cycle, Ecole pratique des hautes études, Paris, 1968.

766. Hofmann, Joseph E. "Über Viète's Beiträge zur Geometrie der Einschiebung." *Mathematische-physikalische Semesterberichte* (Göttingen) (1972), 191-214.

767. Reich, K. "Diophant, Cardano, Bombelli, Viète; ein Vergleich ihrer Aufgaben." *Rechenpfennige* (Munich) (1968), 131-150.

768. Ritter, F. "François Viète, inventeur d'algèbre moderne, 1540-1603." *Revue occidentale philosophique, sociale et politique*, Series 2, 10 (1895), 234-274, 354-415.

Viète's algebraic works, first published between 1591 and 1640, are analyzed here. Viète used symbols to denote both known and unknown (more than one) quantities. He introduced the "clear-cut distinction between the important concept of a parameter and the idea of an unknown quantity" (C.B. Boyer).

For additional works dealing with Viète, see Cantor, item 74, Vol. II, 582-591, 629-641; and Zeller, item 642, 73-85.

Wright, Edward (1561-1615)

769. Cajori, Florian. "On an Integration Ante-dating the Integral Calculus." *Bibliotheca Mathematica*, Series 3, 14 (1914), 312-319.

770. ———. "Algebra in Napier's Day and Alleged Prior Inventions of Logarithms." In *Napier Tercentenary Memorial Volume*. Item 733, 93-109.

See item 1172.

771. Parsons, E.J.S., and W.F. Morris. "Edward Wright and His Work." *Imago Mundi* 3 (1939), 61-71.

772. Waters, David W. *The Art of Navigation in England in Elizabethan and Early Stuart Times*. London: Hollis and Carter, 1958. London: Her Majesty's Stationery Office, 1978, xl + 696 pp., 87 pls. 2nd ed. in 3 vols.

See item 1811.

MATHEMATICS IN THE 17th CENTURY

This section has been divided into a number of categories. Much of the literature covered here represents a larger period than the 17th century, although in this section titles have only been annotated for their content about the 17th century. In the section devoted to individual mathematicians, only titles which supply a general impression of the more important mathematicians' significant works are included.

General

773. Whiteside, D.T. "Patterns of Mathematical Thought in the Later Seventeenth Century." *Archive for History of Exact Sciences* 1 (1960-1962), 179-388.

This very rich monograph approaches 17th-century mathematics with a primary interest in mathematical structures and methods of proof. It emphasizes the work of British mathematicians. The three largest sections concern the concept of function, geometry, and the calculus.

774. Zeuthen, H.-G. *Geschichte der Mathematik im 16. und 17. Jahrhundert*. Leipzig: Teubner, 1903. Reprinted New York: Johnson Reprint Corporation, 1966, viii + 434 pp.

More than three-quarters of this book is devoted to 17th-century mathematics. Though rather dated, the *Geschichte* is a useful general survey of the period, and it contains penetrating comments on a number of 17th-century mathematical texts. See also item 633.

Analysis and Calculus

775. Baron, Margaret E. *The Origins of the Infinitesimal Calculus*. Oxford: Pergamon, 1969, viii + 304 pp.

Traces the history of methods and techniques for the determination of tangents, maxima and minima, centers of gravity, areas, volumes and arc-lengths developed before Newton and Leibniz. The book gives many detailed examples from the 17th century. It ends with an epilogue about the early works of Newton and Leibniz.

776. Bos, Hendrik J.M. "Differentials, Higher-Order Differentials and the Derivative in the Leibnizian Calculus." *Archive for History of Exact Sciences* 14 (1974-1975), 1-90.

A study of the theory, the techniques, and the underlying concepts of the infinitesimal calculus as practiced by Leibniz and his early followers. It discusses in particular the higher-order differentials, the treatment of which revealed several conceptual difficulties about differentials (e.g., their indeterminate character). The difference between the Leibnizian and the modern calculus is stressed (pp. 34-35): the former was a calculus of variables, the latter is one of functions. See also item 926.

777. Boyer, Carl B. *The History of the Calculus and Its Conceptual Development*. New York: Dover, 1959; first published as *The Concepts of the Calculus, a Critical and Historical Discussion of the Derivative and the Integral*. New York: Columbia University Press, 1939, vi + 346 pp. Reprinted New York: Hafner, 1949.

This book is still the basic comprehensive history of the calculus. The comments on the foundations of earlier calculus methods are not the strongest side of the book, but it is valuable because of its extensive coverage of primary and (up to ca. 1940) secondary sources. Chapters IV and V (pp. 96-223) concern the 17th century. See also items 860, 2230.

778. Edwards, C.H., Jr. *The Historical Development of the Calculus*. New York, Heidelberg, Berlin: Springer Verlag, 1979, xii + 351 pp.

Special features of this history of calculus are its interest in techniques (evident, e.g., in a separate chapter on Napier's logarithms) and the inclusion of many exercises which, if worked out by the reader, convey parts of the story. Chapters 4-9 (pp. 98-267) concern 17th-century developments. See also item 988.

779. Goldstine, Herman H. *A History of the Calculus of Variations from the 17th through the 19th Century*. New York, Heidelberg, Berlin: Springer, 1980, xviii + 410 pp.

Chapter 1 (66 pp.) traces the origins from Fermat's principle to the work of the brothers Bernoulli. See also item 864.

780. Grattan-Guinness, Ivor, ed. *From the Calculus to Set Theory, 1630-1910. An Introductory History*. London: Duckworth, 1980, 306 pp.

Six chapters by different authors on the history of mathematical theories involving the infinite. Chapter I, "Techniques of the Calculus, 1630-1660" (pp. 10-48) by Kirsti Møller Pedersen, surveys calculus techniques before Newton and Leibniz, stressing the ideas underlying the various 17th-century methods of determining areas under curves and tangents to curves. Chapter II, "Newton, Leibniz and the Leibnizian Tradition" (pp. 49-93) by H.J.M. Bos, sketches Newton's and Leibniz's "discoveries" of the calculus and the development of the Leibnizian calculus till 1780, concentrating on the fundamental concepts. See also item 2112.

781. Hofmann, Joseph E. "Zur Entdeckungsgeschichte der höheren Analysis im 17. Jahrhundert." *Math.-phys. Semesterberichte* 1 (1950), 220-255.

Valuable survey, rich in factual information about persons and sources. But the characterizations of the achievements are often so compact (in particular because of the use of modern notation) that they can hardly be understood without further study of the sources.

782. ———. "Über Auftauchen und Behandlung von Differentialgleichungen im 17. Jahrhundert." *Humanismus und Technik* 15, (Part 3), (1972), 1-40.

The article traces the history of the so-called "problem of Debeaune" (in modern terms, the differential equation $ay' = x - y$) from Descartes to the Bernoullis, thus providing examples of various approaches to problems involving differential equations which were developed in the 17th century. See also item 1434.

783. Naux, C. *Histoire des logarithmes de Neper à Euler*. Paris: Blanchard, 1966-1971, 158 + 230 pp. 2 vols.

Extensive description of the invention of logarithms, the calculation of the first tables, and the study of the logarithmic relation within the development of analysis.

784. Reiff, R. *Geschichte der unendlichen Reihen*. Munich: Urban & Schwarzenberg, 1889. Photographically reprinted Wiesbaden: Martin Sändig, 1969, iv + 212 pp.

The book (of which Chapter I [pp. 4-63] treats of 17th-century developments) is useful as a collection of summaries of texts on series. It should be kept in mind that after 1889 our knowledge of the sources has grown considerably. See also item 935.

Geometry

785. Boyer, Carl B. *History of Analytic Geometry*. New York: Scripta Mathematica, 1956, ix + 291 pp.

A useful survey of the development of analytic geometry (Chapters IV-VII, pp. 54-191, deal with the 17th century), based on a wide range of primary and secondary sources (see the valuable "analytical bibliography"). The book traces primarily the history of those parts of analytic geometry that can be found in mid-20th-century college textbooks. This viewpoint occasionally leads to an anachronistic treatment and judgment of the material.

786. Coolidge, J.L. *A History of Geometrical Methods*. Oxford: Oxford University Press, 1940. Reprinted New York: Dover, 1963, xviii + 451 pp.

Concerning the 17th century the book contains short surveys of the works of Desargues and Pascal on projective geometry, and of the contributions of Fermat, Descartes, Wallis, Barrow, and Newton to algebraic geometry.

787. Fladt, Kuno. *Geschichte und Theorie der Kegelschnitte und der Flächen zweiten Grades*. Stuttgart: Klett, 1965, x + 374 pp.

This book comprises both a historical and a theoretical treatment of the conic sections and the surfaces of degree 2. Chapter 3 of the first part contains a sketch of the development from the 16th to the 18th century.

788. Mainzer, Klaus. *Geschichte der Geometrie*. Mannheim, Vienna, Zürich: Bibliographisches Institut, 1980, 232 pp.

Chapter 4 (pp. 92-133) deals with geometry in the 17th and 18th centuries.

788a. Loria, Gino. "Perspektive und darstellende Geometrie." In *Geschichte der Mathematik*. Edited by M. Cantor. Vol. 4. Leipzig: B.G. Teubner, 1908, 579-622. Revised and translated into Italian by Gino Loria. *Storia della geometria descrittiva*. Milan: Ulrico Hoepli, 1921, 1-96.

This chapter contains a description of 17th-century literature on the theory of perspective and perspective contructions.

Number Theory

No survey exists of the development of number theory in the 17th century; however, contributions of mathematicians from this period can be found in the following books:

789. Dickson, L.E. *History of the Theory of Numbers*. Washington, D.C.: The Carnegie Institution, 1919-1923. Reprinted New York: Stechert, 1934, and New York: Chelsea, 1952, 1971. 3 vols., xii + 486 + 803, iv + 313 pp.

See items 1048, 1899.

790. Ore, Øystein. *Number Theory and Its History*. New York, Toronto, London: McGraw-Hill, 1948, x + 370 pp.

See item 1917.

791. Bašmakova, I.G. *Diophant und diophanatische Gleichungen*. Basel, Stuttgart: Birkhäuser (without year), and East Berlin: VEB Deutscher Verlag der Wissenschaften, 1974, 97 pp.

In the second part of the book the author traces how the method of Diophantos was rediscovered and elaborated by Viète, Fermat, and later mathematicians.

792. Hofmann, Joseph E. "Über zahlentheoretische Methoden Fermats und Eulers, ihre Zusammenhänge und ihre Bedeutung." *Archive for History of Exact Sciences* 1 (1960-1962), 122-159.

Beginning with a summary of Fermat's essential results in the theory of numbers, the author discusses several selected problems (some of which he had studied in earlier papers in detail) before he describes Fermat's and Euler's general procedures. He finally gives a unified treatment by means of the Weierstrass $\wp$-function. See also item 1909.

Probability and Statistics

793. Hacking, Ian. *The Emergence of Probability*. Cambridge: Cambridge University Press, 1975, 209 pp.

Considers the philosophical aspects of early ideas of probability and statistical inference. Most of the book is devoted to 17th-century incidents.

794. Maistrov, L.E. *Probability Theory. A Historical Sketch*. Translated by S. Kotz from *Teoriià Veroiàtnosteĭ Istoricheskiĭ Ocherk*, Moscow, 1967. New York and London: Academic Press, 1974, xi + 281 pp.

The first two chapters contain about 30 pages treating the contribution of the 17th century to the development of probability theory.

795. Pearson, Karl. *The History of Statistics in the 17th and 18th Centuries*. Edited by E.S. Pearson. London: Charles Griffin, 1978, xix + 744 pp.

The book contains lectures given by Karl Pearson in the period 1921-1933. The first 124 pages are devoted to the 17th century and describe the founding of the English school of political arithmetic and the correspondence between Caspar Neumann and Edmund Halley.

796. Westergaard, Harald. *Contributions to the History of Statistics*. London: P.S. King & Son, 1932. Photomechanically reprinted 1969, vii + 280 pp.

Contains a chapter on political arithmetic in the seventeenth century.

Special Studies and Individual Mathematicians

797. Cajori, Florian. *A History of the Logarithmic Slide Rule and Allied Instruments*. New York: The Engineering News Publishing Company, 1909. Reprinted in *String Figures and Other Monographs*. Edited by W.W.R. Ball et al. New York: Chelsea Publishing Company, 1960, 1969, vi + 136 pp.

The first 24 pages of Cajori's booklet deal with the origin of the slide rule in the 17th century. In the reprinted Chelsea edition, Cajori's monograph follows three others; the volume is not consecutively paginated. Cajori's study is the last in the book, is well illustrated, and contains a catalogue of slide rules as well as a bibliography on the subject.

798. Goldstine, Herman H. *A History of Numerical Analysis from the 16th through the 19th Century*. New York, Heidelberg, Berlin: Springer-Verlag, 1977, xiv + 348 pp.

The first two chapters (pp. 1-118) provide a survey of 17th-century numerical techniques concerning logarithms, interpolation, finite differences, trigonometric tables, the "Newton-Raphson method," and methods of numerical integration. See also items 863, 1947.

799. Loria, Gino. *Spezielle algebraische und transscendente ebene Kurven. Theorie und Geschichte.* Leipzig: Teubner, 1902, xxi + 744 pp. 2 vols.

The book is very useful for the study of curves dealt with in the 17th century. It accounts for the history and the properties of the plane curves which have played an important role in the development of the theory of curves.

800. Schneider, Ivo. "Der Einfluss der Praxis auf die Entwicklung der Mathematik vom 17. bis zum 19. Jahrhundert." *Zentralblatt für Didaktik der Mathematik* 9 (1977), 195-205.

Schneider discusses the methodological difficulties in assessing the influence of practice on the development of mathematics, and he presents many examples in which this influence can be discerned. See also item 2213.

801. Taylor, Eva G.R. *The Mathematical Practitioners of Tudor and Stuart England, 1485-1714*. Cambridge: Cambridge University Press, 1954, xi + 443 pp., 12 pls.

Very important study of 17th-century mathematics teachers, textbook writers, and other practitioners of the mathematical arts. Offers a general survey of the development within this mathematical profession, as well as short biographies of 528 mathematical practitioners and an annotated bibliography listing 628 contemporary works. In 1966 a sequel to this book appeared, covering the 18th and early 19th centuries: E.G.R. Taylor, *The Mathematical Practitioners of Hanoverian England 1714-1840*, item 1810. See also items 630, 1809, 2216.

Barrow, Isaac (1630-1677)

802. Zeuthen, H.-G. "Notes sur l'histoire des mathématiques, VII. Barrow, le maître de Newton." *Oversigt over det Kgl. Danske Videnskabernes Selskabs Forhandlinger* (1897), 565-606.

Discusses the extent to which Barrow had realized the connection between quadratures and the determination of tangents.

Bernoulli, Jakob (1654-1705) and Johann (1667-1748)

803. Dietz, P. "Die Ursprünge der Variationsrechnung bei Jakob Bernoulli." *Verhandlungen der naturforschenden Gesellschaft in Basel* 70 (1959), 81-146.

This is an extensively annotated edition of those parts of Jakob Bernoulli's scientific diary which relate to variational problems.

804. Fleckenstein, J.O. *Johann und Jakob Bernoulli.* (Supplement No. 6 to the journal *Elemente der Mathematik.*) Basel: Birkhäuser, 1949, 24 pp.

This short brochure gives biographical information and provides a detailed discussion of a number of problems (especially in variational calculus) that were treated by the brothers Bernoulli.

805. Hofmann, Joseph E. *Ueber Jakob Bernoullis Beiträge zur Infinitesimalmathematik.* (*Monographies de l'enseignement mathématique*, No. 3.) Geneva: Institut de Mathématique, Université, 1956, 126 pp.

This compact monograph is not easily readable, but with its 389 notes (pp. 57-96) and its indices of names, sources, and subjects it is a very rich source of factual information about mathematics in the period around 1700.

Cavalieri, Bonaventura (1598-1647)

806. Cavalieri, Bonaventura. *Geometria degli indivisibili.* Translated and edited by Lucio Lombardo-Radice. Turin: Unione Tipografico-Editrice Torinese, 1966, 870 pp.

The *introduzione* and *nota bibliografica* (pp. 9-37) are useful for information about Cavalieri's mathematical work.

806a. Andersen, Kirsti. "Cavalieri's Methods of Indivisibles." Preprint, Department of History of Science, Aarhus University, 1983, 124 pp., forthcoming in *Archive for History of Exact Sciences.*

A detailed presentation of the fundamental ideas, concepts, and techniques involved in Cavalieri's method. The author further describes how Cavalieri's mathematical concept of "all the lines" has been misunderstood from the 17th through the 20th centuries.

806b. Giusti, Enrico. *Bonaventura Cavalieri and the Theory of Indivisibles.* Bologna: Edizioni Cremonese, 1980, 95 pp.

This is a special issue of the introduction to the 1980 reprint of Cavalieri's *Exercitationes geometricae sex* giving an excellent survey of Cavalieri's ideas and the difficulties inherent in his concepts.

807. Masotti, S.C. Arnaldo. "Commemorazione di Bonaventura Cavalieri." *Rendiconti dell'Istituto Lombardo di scienze e lettere, parte generale e atti ufficiali* 81 (1948), 43-86. Also as a separate print, Milano: Ulrico Hoepli, 1949.

The appendix I, *Scritti di Bonaventura Cavalieri* gives an exact list of Cavalieri's works, and the second appendix carefully surveys the secondary literature about Cavalieri up to 1948.

808. Piola, Gabrio. *Elogio di Bonaventura Cavalieri.* Milan: Giuseppe Bernadoni di Giovanni, 1844, xxxi + 155 pp.

This is the source for most later biographies on Cavalieri.

Desargues, Girard (1591-1661)

809. Taton, René. *L'oeuvre mathématique de G. Desargues.* Paris: Presses Universitaires de France, 1951, 232 pp.

The book contains a biography of Desargues and provides the text of his work on projective geometry, *Brouillon project d'une atteinte aux evenèmens des rencontres du cone avec un plan.* See also item 1588.

Descartes, René (1591-1650)

810. Bos, Hendrik J.M. "On the Representation of Curves in Descartes' Géométrie." *Archive for History of Exact Sciences* 24 (1981), 295-338.

This is a study of the role of curves and constructions in the *Géométrie*. It is argued that a fundamental contradiction underlies the *Géométrie* and its program, namely, the contradiction between the usefulness of algebra as a tool in geometry and the need for truly geometrical criteria for adequacy of constructions. See also item 1546.

811. Hofmann, Joseph E. "Descartes und die Mathematik." In *Descartes. Drei Vorträge*. Edited by H. Scholz et al. Münster: Aschendorff, 1951, 48-73.

Short survey of Descartes's early mathematical studies and of the content of the *Géométrie*. Hofmann stresses the role of Descartes's interest in method.

812. Milhaud, G. *Descartes savant*. Paris: Alcan, 1921, 249 pp.

In this collection of essays on Descartes's scientific work much attention is given to his mathematics. Particularly useful are Chapters I and III (pp. 25-46, 69-88) on Descartes's early mathematical studies, and Chapters VI and VII (pp. 124-175) on analytical geometry and infinitesimal methods.

813. Scott, J.F. *The Scientific Work of René Descartes (1596-1650)*. London: Taylor & Francis, [1952], vi + 211 pp. Reprinted 1976.

Chapters VI-IX, pp. 84-157, concern mathematics, in particular Descartes's *Géométrie* which is presented through paraphrase and occasional comments. See also item 1583.

814. Vuillemin, J. *Mathématiques et métaphysique chez Descartes*. Paris: Presses Universitaires de France, 1960, 188 pp.

Thought-provoking essays on themes from Descartes's mathematics (with considerable attention to transcendental relations and curves) and their metaphysical aspects. The arguments are often very speculative and the factual information cannot always be trusted.

Fermat, Pierre (1601-1665)

815. Hofmann, Joseph E. "Pierre Fermat. Eine wissenschaftsgeschichtliche Skizze." *Scientiarum Historia* 13 (1971), 198-238.

A survey of Fermat's life and work, without technical details.

816. ———. "Pierre de Fermat--ein Pionier der neuen Mathematik († 12.1.1665)." *Praxis der Mathematik* 7 (1965), 113-119, 171-180, and 197-203. Also as a separate print.

A short biography of Fermat and comments on some of his studies in analytical geometry, number theory, and infinitesimal calculus.

817. Itard, Jean. "Pierre Fermat." (Supplement No. 10 of the journal *Elemente der Mathematik*.) Basel: Birkhäuser 1950, 24 pp. In French.

A short biography of Fermat and a survey of his main contributions to mathematics.

818. Mahoney, Michael Sean. *The Mathematical Career of Pierre de Fermat (1601-1665)*. Princeton, N.J.: Princeton University Press, 1973, xviii + 419 pp.

The book gives a good picture of Fermat's contributions to mathematics. Fermat's role in the shift to use of coordinates in treating geometric problems and his role in the development of calculus is well described; however, because of the technical details, Mahoney's account is sometimes difficult to follow. See also items 627, 1942.

Gregory, James (1638-1675)

819. Scriba, C.J. *James Gregorys frühe Schriften zur Infinitesimalrechnung*. (Mitteilungen aus dem Mathematischen Seminar Giessen, Nr. 55.) Giessen: Selbstverlag des Mathematischen Seminars, 1957, 80 pp.

This study deals with Gregory's (often underestimated) merits in the development of the differential and integral calculus. It contains valuable surveys of the contents of Gregory's principal works and of some manuscripts and letters.

820. Turnbull, Herbert Westren, ed. *James Gregory Tercentenary Memorial Volume. Containing His Correspondence with John Collins and His Hitherto Unpublished Mathematical Manuscripts, Together with Addresses and Essays Communicated to the Royal Society of Edinburgh, July 4, 1938*. London: G. Bell & Sons, 1939, xii + 524 pp.

This volume consists mainly of an edition of Gregory's correspondence and his mathematical papers. A biographical sketch and short articles by several scholars on Gregory's published works are also included.

Harriot, Thomas (ca. 1560-1621)

821. Lohne, J.A. "Thomas Harriot als Mathematiker." *Centaurus* 11 (1966), 19-45.

Deals with some of Harriot's studies of the loxodrome, areas of spherical triangles, interpolations, the law of refraction, and coordinates. See also item 1950.

822. ———. "Essays on Thomas Harriot." *Archive for History of Exact Sciences* 20 (1979), 189-312.

The article consists of three parts: I. Billiard Balls and Laws of Collision. II. Ballistic Parabolas. III. A Survey of Harriot's Scientific Writings.

Hudde, Johann (1628-1704)

823. Haas, Karlheinz. "Die mathematischen Arbeiten von Johann Hudde (1628-1704), Bürgermeister von Amsterdam." *Centaurus* 4 (1955/1956), 235-284.

Contains a biography and a survey of Hudde's contributions to the theory of equations and the infinitesimal calculus.

Huygens, Christian (1629-1695)

824. Bell, A.E. *Christian Huygens and the Development of Science in the Seventeenth Century*. London: Arnold, 1947. Reprinted 1950, 220 pp.

Scientific biography of Huygens; there is no separate section on Huygens's mathematics.

825. Bos, Hendrik J.M., et al., eds. *Studies on Christiaan Huygens; Invited Papers from the Symposium on the Life and Work of Christiaan Huygens, Amsterdam, 22-25 August 1979*. Lisse: Swets & Zeitlinger, 1980, 321 pp.

The invited papers combine to form an overall scientific biography of Huygens. There is a paper on his mathematics (by H.J.M. Bos, pp. 126-146); papers on mechanics (by Alan Gabbey, pp. 166-199) and on the measurement of time and longitude at sea (by Michael S. Mahoney, pp. 234-270) are also informative about Huygens's use of mathematical methods.

826. Dijksterhuis, E.J. "Christiaan Huygens; An Address Delivered at the Annual Meeting at the Holland Society of Sciences at Haarlem, May 13th, 1950, on the Occasion of the Completion of Huygens's Collected Works." *Centaurus* 2 (1951/1953), 265-282.

Short sketch of Huygens's life and work.

Kepler, Johannes (1571-1630)

827. Kepler, Johannes. *Gesammelte Werke. Band IX. Mathematische Schriften*. Edited by Franz Hammer. Munich: C.H. Beck, 1960.

The *Nachbericht* (pp. 427-483) is a detailed account of the content and origin of the *Stereometria*, the *Messekunst*, and Kepler's work on logarithms. See also item 1399.

828. Hofmann, Joseph E. "Ueber einige fachliche Beiträge Keplers zur Mathematik." *Internationales Kepler Symposium, Weil der Stadt, 1971*. Edited by F. Krafft et al. Hildesheim: Dr. H.A. Gerstenberg, 1973, 261-284.

Discusses Kepler's mathematical background, his study of polyhedra, and his solution of some quadrature problems.

829. ———. "Johannes Kepler als Mathematiker." *Praxis der Mathematik* 13 (1971), 287-293 and 318-324.

Treats the same topics as the article in item 828 above and adds a short biography of Kepler.

Leibniz, Gottfried Wilhelm (1646-1716)

830. Child, J.M. *The Early Mathematical Manuscripts of Leibniz Translated from the Latin Texts Published by Carl Immanuel Gerhardt with Critical and Historical Notes*. Chicago, London: Open Court Publishing Company, 1920, iv + 238 pp.

Useful in providing texts that are otherwise not easily accessible. But the commentaries to the texts, written from an interest in settling the priority and plagiarism dispute, are superseded by later research and should not be trusted.

831. Hall, A. Rupert. *Philosophers at War. The Quarrel between Newton and Leibniz*. Cambridge, London, New York, New Rochelle, Melbourne, Sydney: Cambridge University Press, 1980, xiii + 338 pp.

Hall accounts for the ingredients of the famous controversy about the priority of the creation of the calculus; and in non-technical language he describes the difference between Newton's and Leibniz's approaches to the calculus. Further, Hall discusses earlier writings about the quarrel and reaches very sensible conclusions.

832. Hofmann, Joseph E. *Leibniz in Paris, 1672-1676, His Growth to Mathematical Maturity*. Cambridge: Cambridge University Press, 1974, xi + 372 pp.

This revised translation of Hofmann's *Die Entwicklungsgeschichte der Leibnizschen Mathematik während des Aufenthaltes in Paris (1672-1676)*, Munich, 1949, minutely traces Leibniz's studies leading up to his "invention of the calculus" in 1675. It aims at finally setting right many conflicting statements of historians and others about Leibniz's achievements, a very sensitive matter because of the priority and plagiarism dispute between Leibniz and Newton. Hofmann concludes that, as to the methods of the infinitesimal calculus, Leibniz "wholly uninfluenced by others, gained his crucial insights unaided" (p. 306).

833. Knobloch, E., ed. *Die mathematischen Studien von G.W. Leibniz zur Kombinatorik*. (Studia Leibnitiana Supplementa, Band XI.) Wiesbaden: Franz Steiner Verlag, 1973, xvi + 277 pp.

834. ———. *Die mathematischen Studien von G.W. Leibniz zur Kombinatorik. Textband*. (Studia Leibnitiana Supplementa, Band XVI.) Wiesbaden: Franz Steiner Verlag, 1976, xii + 339 pp.

The former volume is a detailed monograph based mostly on manuscript material. Its main topics are Leibniz's "Dissertatio de arte combinatoria," symmetric functions, and the theory of partitions. The second volume contains an edition of 60 manuscripts--the most important ones of those that are analyzed in the former volume. See also item 1887.

835. Zacher, Hans J. *Die Hauptschriften zur Dyadik von G.W. Leibniz. Ein Beitrag zur Geschichte des binären Zahlensystems*. Frankfurt am Main: Vittorio Klostermann, 1973, viii + 384 pp.

This monograph traces in detail the development of Leibniz's ideas about the binary system and the importance he assigned to it in his philosophical views. In an appendix (pp. 216-356) 28 important letters or manuscripts on the subject are published in critical edition.

Napier, John (1550-1617)

836. Gridgeman, N.T. "John Napier and the History of Logarithms." *Scripta Mathematica* XXIX (1973), 49-65.

An easily read survey of Napier's life and his logarithms.

837. Knott, C.G., ed. *Napier Tercentenary Memorial Volume*. London, New York, Bombay, Calcutta and Madras: Longmans, Green and Company, 1915, ix + 441 pp.

The first seven articles (pp. 1-137) deal with Napier and the development of logarithms and the law of exponents.

See also items 731, 732.

Newton, Sir Isaac (1642-1727)

838. Hofmann, Joseph E. "Der junge Newton als Mathematiker (1665-1675)." *Mathematisch-physikalische Semesterberichte* 2 (1951/1952), 45-70.

D.T. Whiteside's valuable work on Newton, items 841-844, has provided us with much more information than Hofmann had when writing this article; still, it contains useful comments on Newton's mathematics.

839. Scriba, Christoph J. "The Inverse Method of Tangents: A Dialogue Between Leibniz and Newton (1675-1677)." *Archive for History of Exact Sciences* 2 (1962-1966), 113-137.

See item 1352.

840. Westfall, Richard S. *Never at Rest. A Biography of Isaac Newton*. Cambridge: Cambridge University Press, 1980, xviii + 908 pp.

A comprehensive scientific biography including descriptions of Newton's mathematical discoveries. See item 1368.

841. Whiteside, D.T. "Isaac Newton: Birth of a Mathematician." *Notes and Records of the Royal Society of London* 19 (1964), 53-62.

Describes Newton's mathematical training up to 1664.

842. ———. "Newton's Marvellous Year: 1666 and All That." *Notes and Records of the Royal Society of London* 21 (1966), 32-41.

A non-technical description of Newton's mathematical achievements in the period 1664-1666.

843. ———, ed. *The Mathematical Works of Isaac Newton*. New York and London: Johnson Reprint Corporation, 1964-1967. 2 vols.

The introductions (vol. I, pp. vii-xix, and vol. II, pp. ix-xxvii) provide a concise summary of Newton's mathematical development. Unlike Horsley's edition of the *Opera*, this reprint contains those works that have been published in English.

844. ———, ed. *The Mathematical Papers of Isaac Newton*. Cambridge: Cambridge University Press, 1967-1981. 8 vols.

The introductions in the eight volumes give a survey of Newton's mathematical research. Exceptionally valuable are the notes provided by Whiteside throughout the eight volumes, offering a penetrating analysis of Newton's historical development as a mathematician.

Pascal, Blaise (1621-1662)

845. Bosmans, H. "Sur l'oeuvre mathématique de Blaise Pascal." *Mathesis* 38 (1924), Supplement. Also Extrait de la *Revue des Questions scientifiques*, (January and April 1924), 1-59, separately paginated.

Leisurely written survey of Pascal's mathematical works, composed in order to assess Pascal's merits as compared with other 17th-century mathematicians.

846. Schobinger, Jean-Pierre. *Blaise Pascals Reflexionen über die Geometrie im allgemeinen: "De l'esprit géométrique" und "De l'art de persuader."* Mit deutscher Übersetzung und Kommentar. Basel, Stuttgart: Schwabe & Co., 1974, 522 pp.

This scholarly edition contains a brief curriculum vitae of Pascal, 70 pages of French text with German translation of the two papers, approximately 350 pages of commentary and 50 pages of critical apparatus.

847. Taton, René, ed. *L'oeuvre scientifique de Pascal*. Paris: Presses Universitaires, 1964, ix + 311 pp.

Most of the fourteen chapters of the book are devoted to Pascal's mathematical work, particularly on projective geometry, mathematical induction, infinitesimal calculus, and the calculating machine.

Roberval, Gilles Personne de (1602-1675)

848. Auger, Léon. *Un savant méconnu: Gilles Personne de Roberval (1602-1675)*. Paris: Blanchard, 1962, 215 pp.

The book contains a short biography of Roberval, deals with his contributions to the infinitesimal calculus, and further discusses his work in mechanics and philosophy.

849. Pedersen, Kirsti Møller. "Roberval's Method of Tangents." *Centaurus* 13 (1968), 151-182.

Covers Roberval's kinematic method of tangents and its place in the history of kinematic arguments in the development of the infinitesimal calculus.

850. Walker, Evelyn. *A Study of the Traité des indivisibles of Gilles Personne de Roberval*. New York: Teachers College, Columbia University, 1932, 273 pp.

297

Des minimes de Paris

LA

GEOMETRIE.

LIVRE PREMIER.

Des problesmes qu'on peut construire sans y employer que des cercles & des lignes droites.

Ous les Problesmes de Geometrie ſe peuuent facilement reduire a tels termes, qu'il n'eſt beſoin par aprés que de connoiſtre la longeur de quelques lignes droites, pour les conſtruire.

Et comme toute l'Arithmetique n'eſt composée, que de quatre ou cinq operations, qui ſont l'Addition, la Souſtraction, la Multiplication, la Diuiſion, & l'Extraction des racines, qu'on peut prendre pour vne eſpece de Diuiſion : Ainſi n'a-t'on autre choſe a faire en Geometrie touchant les lignes qu'on cherche, pour les preparer a eſtre connuës, que leur en adiouſter d'autres, ou en oſter, Oubien en ayant vne, que ie nommeray l'vnité pour la rapporter d'autant mieux aux nombres, & qui peut ordinairement eſtre priſe a diſcretion, puis en ayant encore deux autres, en trouuer vne quatrieſme, qui ſoit à l'vne de ces deux, comme l'autre eſt a l'vnité, ce qui eſt le meſme que la Multiplication; oubien en trouuer vne quatrieſme, qui ſoit a l'vne de ces deux, comme l'vnité eſt

Commẽt le calcul d'Arithmetique ſe rapporte aux operations de Geometrie.

Pp

Figure 8. Descartes's *Géométrie* was first published in 1637 as an appendix to the *Discours de la Méthode.* Although its unified treatment of geometry and algebra soon opened the way to modern analytic geometry, in many respects the text was still closely wedded to the classical Greek tradition. Descartes saw the work as an application of his general philosophical method, which he conceived as a royal road to knowledge and wisdom.

Roberval's work on indivisibles is related to his other activities in mathematics and to the contributions of his contemporaries.

Van Schooten, Frans (ca. 1615-1660)

851. Hofmann, Joseph E. *Frans van Schooten der Jüngere*. Wiesbaden: Steiner, 1962, ii + 54 pp.

This densely written and extensively documented monograph deals with van Schooten's life and his scientific activities, and presents a selection from his mathematical works.

Sluse, René-François Walter de (1622-1685)

852. Le Paige, C. "Correspondance de René-François de Sluse publiée pour la première fois et précédée d'une introduction par M.C. le Paige." *Bullettino di bibliografia e di storia delle scienze matematiche e fisiche* 17 (1884). Reprinted New York & London: Johnson Reprint Corporation, 1964, 427-554, 603-726.

The first 75 pages of this edition contain a biography of of Sluse and a survey of his mathematical work.

Stevin, Simon (1548-1620)

853. Dijksterhuis, E.J. *Simon Stevin. Science in the Netherlands around 1600*. The Hague: Nijhoff, 1970, ix + 145 pp.

This is a very condensed English edition of Dijksterhuis's very thorough and detailed Dutch monograph *Simon Stevin* (The Hague: Nijhoff, 1943, xii + 379 pp.). Dijksterhuis himself started the English edition; after his death R. Hooykaas and M.G.J. Minnaert completed and edited it. There is a separate chapter on Stevin's mathematics, pp. 14-47.

853a. Struik, D.J., ed. *The Principal Works of Simon Stevin*. Vol. II A and B. Amsterdam: Swets & Zeitlinger, 1958, 976 pp.

In the introductions to Stevin's various mathematical works, Dirk Struik very usefully sets these in a historical perspective.

Torricelli, Evangelista (1608-1647)

854. *Opere di Evangelista Torricelli*. Edited by G. Loria and G. Vassura. Vol. I-III. Faenza: Montanari, 1919; Vol. IV, Faenza: Lega, 1944.

The *Introduzione*, Vol. I, pp. iii-xxx, gives biographic and bibliographic information about Torricelli.

855. Weis, F. "Evangelista Torricelli." *Archiv für Geschichte der Mathematik, der Naturwissenschaften und der Tecynik*, Neue Folge I, 10 (1927/1928), 250-281.

Contains a biography of Torricelli and an account of the history of his posthumous papers.

Wallis, John (1616-1703)

856. Kramar, F.D. "Integrationsmethoden von John Wallis." *Istoriko-Matematicheskie Issledovaniia* 14 (1961), 11-100. In Russian.

This detailed study concentrates on Wallis's pre-calculus methods for integration, as presented first in *Arithmetica Infinitorum* of 1656/1657 with its famous result, the infinite product for $\pi/4$.

857. Prag, A. "John Wallis, 1616-1703. Zur Ideengeschichte der Mathematik im 17. Jahrhundert." *Quellen und Studien zur Geschichte der Mathematik*, Abt. B: Studien 1 (1929), 381-412.

This study of Wallis's work, especially of the *Arithmetica Infinitorum* and the *Algebra*, is a valuable assessment of Wallis's achievements as compared with those of other 17th-century mathematicians.

858. Scott, J.F. *The Mathematical Work of John Wallis, D.D., F.R.S. (1616-1703)*. London: Taylor & Francis, 1938, xi + 240 pp.

This book consists mainly of summaries of Wallis's principal mathematical works. There are also two short biographical chapters. As in his book on Descartes, Scott presents the contents of Wallis's mathematical publications mostly through paraphrase, enriched by occasional comments and set within the framework of Wallis's academic life.

859. Scriba, C.J. *Studien zur Mathematik des John Wallis (1616-1703); Winkelteilungen, Kombinationslehre und zahlentheoretische Probleme. Im Anhang die Bücher und Handschriften von Wallis.* Wiesbaden: Steiner, 1966, xi + 144 pp.

In this study of a number of themes in Wallis's mathematical work, Scriba has made use of many hitherto unpublished manuscripts and letters of Wallis.

MATHEMATICS IN THE 18th CENTURY

Many of the problems and interests of mathematicians in the 18th and 19th centuries overlap, so that it is impossible to draw any definite boundary at 1800. In fact, this section of the bibliography, as well as the two following for the 19th and 20th centuries, draw their boundaries only in a general and pragmatic way. Certainly the fluidity and transition of mathematics from its scattered practitioners in the 18th century to its professionalization in the 19th century make it necessary to consult the sections dealing with later periods for a more comprehensive sense of relevant research materials, even if one's primary interests lie in the 18th century. More often than not, titles in the following sections of the bibliography range over issues that originate in the 18th century and subsequently experience major developments or find their resolution in the 19th century.

General Works

860. Boyer, Carl B. *The History of the Calculus and Its Conceptual Development*. New York: Dover, 1959, 346 pp. This is an unaltered reprint of *The Concepts of the Calculus, a Critical and Historical Discussion of the Derivative and the Integral*.

New York: Columbia University Press, 1939. Reprinted New York: Hafner, 1949.

Views the development of the calculus as an effort lasting 2500 years to explain a vague intuitive feeling for continuity. Argues that a hindrance to this development was the fact that concepts were not defined as concisely and formally as possible, and that the final answer was provided by the rigorous definitions of function and limit in the 19th century. See also items 777, 2230.

861. Cajori, Florian. "Frederick the Great on Mathematics and Mathematicians." *American Mathematical Monthly* 34 (1927), 122-130.

862. Gillispie, Charles Coulston. *The Edge of Objectivity: An Essay in the History of Scientific Ideas*. Princeton: Princeton University Press, 1960, 562 pp.

A good, general introduction to the 18th century in a chapter devoted to "Science and the Enlightenment." See also item 1473.

863. Goldstine, Herman H. *A History of Numerical Analysis from the 16th through the 19th Century*. (Studies in the History of Mathematics and Physical Sciences, Vol. 2.) New York, Heidelberg, Berlin: Springer-Verlag, 1977, xiv + 348 pp., especially pp. 119-260.

Provides a survey of 18th-century developments and contends that the work of Euler and Lagrange must be considered the foundation of numerical analysis. Gives an account of the contributions of Euler, Lagrange, Laplace, Legendre, and Gauss. See also items 798, 1947.

864. ———. *A History of the Calculus of Variations from the 17th through the 19th Century*. (Studies in the History of Mathematics and Physical Sciences, Vol. 5.) New York, Heidelberg, Berlin: Springer, 1980, xviii + 410 pp., especially pp. 50-150.

Is the only recent account of the history of the calculus of variations; devotes about 100 pages to 18th-century developments. See also item 779.

865. Grattan-Guinness, Ivor. *The Development of the Foundations of Mathematical Analysis from Euler to Riemann*. Cambridge, Mass.: MIT Press, 1970, 186 pp.

See item 1072.

866. Hofmann, Joseph E. *Classical Mathematics. A Concise History of the Classical Era in Mathematics*. Translated by H.O. Midonick. New York: Philosophical Library, 1959.

Chapter III is devoted to "Age of Enlightenment," pp. 115-154. Especially noteworthy is Chapter II, "The Late Baroque Period" (ca. 1665-1730), including discussion of infinitesimal mathematics in Japan, from 1650 to 1770 (pp. 101-114).

867. Kline, Morris. *Mathematical Thought from Ancient to Modern Times.* New York: Oxford University Press, 1972.

For the 18th century, chapters devoted to specific subjects include 18, "Mathematics as of 1700," pp. 391-399; 19, "Calculus in the Eighteenth Century," pp. 400-435; 20, "Infinite Series," pp. 436-467; 21, "Ordinary Differential Equations in the Eighteenth Century," pp. 468-501; 22, "Partial Differential Equations in the 18th Century," pp. 502-543; 23, "Analytic and Differential Geometry in the Eighteenth Century," pp. 544-572; 24, "The Calculus of Variations in the Eighteenth Century," pp. 573-591; 25, "Algebra in the 18th Century," pp. 592-613; and 26, "Mathematics as of 1800" begins on page 614. See also items 82, 990, 1106, 1317, 1571, 2145.

868. Montucla, J.F. *Histoire des mathématiques*. Nouv. ed. Paris: Henri Agasse, 1799-1802, vii + 739 + 718 + viii + 832 + 688 pp. 4 vols. Volumes 3 and 4 were edited by J. de la Lande. Photomechanically reprinted, with an introduction by Ch. Naux, Paris: Albert Blanchard, 1968.

Volumes 3 and 4 of this monumental treatise are devoted almost exclusively to the history of 18th-century mathematics. "Mathematics" is taken in its wide 18th-century denomination and includes calculus, optics, mechanics, machine building, astronomy, navigation, etc. The book provides a very valuable description of the state of the art in these subjects as seen through an educated contemporary eye.

869. Muir, Thomas. *The Theory of Determinants in the Historical Order of Development*. London: Macmillan, 1890; 2nd ed. of vol. 1, 1906; vols. 2-4, 1906-1923. Reprinted New York: Dover, 1960, 4 vols. in 2.

See item 1237.

870. Scott, J.F. "Mathematics Through the Eighteenth Century." *Philosophical Magazine* (1948), 67-90.

Emphasis is placed on mathematics in England.

Studies of Individual Mathematicians

Bayes, Thomas (1702-1761)

871. Bayes, T. *Facsimile of Two Papers, with Commentaries by E.C. Molina and W.E. Deming*. Washington, D.C.: The Graduate School, Department of Agriculture, 1940.

872. Barnard, G.A. "Thomas Bayes, A Biographical Note" as a preface to T. Bayes, "An Essay Towards Solving a Problem in the Doctrine of Chances." *Biometrika* 45 (1958), 296-315.

873. Savage, L.J. *The Foundations of Statistics*. New York: Wiley, 1954. 2nd rev. ed., New York: Dover, 1972.

Includes a very brief, explicitly historical "background" as part of the book's introduction on pages 1-4. Stresses foundational issues, aspects of the philosophical connections, and applications of statistics.

Bernoulli family

For general studies of the family, easily the most remarkable in the history of mathematics for the number and significance of its members from one generation to another, see the following.

874. Fleckenstein, J.O. *L'école mathématique baloise des Bernoulli à l'aube du XVIIIe siècle*. Paris: Libraire du Palais de la Découverte, 1959.

875. Spiess, O. "Die Mathematikerfamilie Bernoulli." In *Grosse Schweizer*. Edited by Huerlimann. Zürich: Atlantis Verlag, 1942, 112-119.

876. ———. "Bernoulli, Basler Gelehrtenfamilie." *Neue deutsche Biographie*. Berlin: Duncker & Humblot, 1955, 128-131.

Bernoulli, Daniel (1700-1782)

877. Wolf, R. "Daniel Bernoulli von Basel, 1700-1782." In *Biographien zur Kulturgeschichte der Schweiz*. Zürich, 1860, 151-202.

Bernoulli, Jakob (Jacques) I (1654-1705)

878. Fleckenstein, J.O. *Johann und Jakob Bernoulli*. *Elemente der Mathematik*. Basel: Birkhäuser, 1949, 24 pp.

See item 804.

879. Hofmann, Joseph E. *Über Jakob Bernoullis Beiträge zur Infinitesimalmathematik*. *(Monographies de l'enseignement mathématique*, No. 3.) Geneva: Institut de Mathématique, Université, 1956, 126 pp.

See item 805.

Bernoulli, Johann (Jean) I (1667-1748)

880. Hofmann, Joseph E. "Johann Bernoulli, Propagator der Infinitesimalmethoden." *Praxis der Mathematik* 9 (1967/1968), 209-212.

Carnot, Lazare-Nicolas-Marguerite (1753-1823)

881. Boyer, Carl B. "The Great Carnot." *The Mathematics Teacher* 49 (1956), 7-14.

882. Watson, S.J. *Carnot*. London: Bodley Head, 1954.

883. Youschkevitch, Adolf P. "Lazare Carnot and the Competition of the Berlin Academy in 1786 on the Mathematical Theory of the Infinite." In *Lazare Carnot Savant*. Edited by C.C. Gillispie. Princeton: Princeton University Press, 1971, 149-168.

Includes as an appendix a photoreproduction of Carnot's "Dissertation sur la théorie de l'infini mathématique" with notes by A.P. Youschkevitch.

Clairaut, Alexis-Claude (1713-1765)

884. Brunet, P. *La vie et l'oeuvre de Clairaut*. Paris: Presses Universitaires de France, 1952.

Condorcet, Marie-Jean-Antoine, Marquis de (1743-1794)

885. Granger, G.-G. *La mathématique sociale du Marquis de Condorcet*. Paris: Presses Universitaires de France, 1956.

886. Taton, René. "Condorcet et Sylvestre-François Lacroix." *Revue d'histoire des sciences* 12 (1959), 127-158, 243-262.

D'Alembert, Jean le Rond (1717-1783)

887. Grimsley, R. *Jean d'Alembert*. Oxford: Clarendon Press, 1963.

888. Hankins, Thomas. *Jean d'Alembert. Science and the Enlightenment*. Oxford: Clarendon Press, 1970.

See item 1298.

889. Vollgraff, J.A. "Christian Huygens et Jean le Rond d'Alembert." *Janus* 20 (1915), 269-331.

Discusses the historical development of d'Alembert's principle.

De Moivre, Abraham (1667-1754)

890. Schneider, Ivo. "Der Mathematiker Abraham de Moivre (1667-1754)." *Archive for History of Exact Sciences* 5 (1968-1969), 177-317.

An exhaustive account of the life and mathematical work of de Moivre, providing full bibliographic data (including unpublished materials). De Moivre was active in three fields: the theory of equations, to which he contributed trigonometric expressions for the roots of unity; series and recurrent expressions; and the theory of probability, containing de Moivre's derivation of the limit theorem for binomial distributions. See also item 2040.

891. Walker, H.M. "Abraham de Moivre." *Scripta Mathematica* 2 (1934), 316-333.

Diderot, Denis (1713-1784)

892. Krakeur, L.G., and R.L. Krueger. "The Mathematical Writings of Diderot." *Isis* 33 (1941), 219-232.

Euler, Leonhard (1707-1783)

893. Carathéodory, C. "Introduction to Euler." *Opera Omnia*, 24. Zurich: Orell Füssli, 1952.

Introductions to various individual volumes of Euler's *Collected Works* provide useful historical and critical surveys. Other useful commentaries include G. Faber on infinite series,

16 (1) (1935); G. Faber and A. Krazer on integrals, 19 (1) (1932); J.O. Fleckenstein on mechanics, 5 (2) (1957); A. Speiser on geometry, 26-29 (1) (1953-1956); and C. Truesdell, "The Rational Mechanics of Flexible or Elastic Bodies, 1630-1780," 11 (2) (1960).

894. Du Pasquier, L.G. *Leonhard Euler et ses amis*. Paris: J. Hermann, 1927.

895. Eneström, G. "Verzeichnis der Schriften Leonhard Eulers." *Jahresbericht der Deutschen Mathematiker-Vereinigung* 4 (1910-1913).

In three parts; Part I (organized by date of publication) is reprinted in Euler's *Opera omnia*, 35, Part I, 352-386. Parts II and III are arranged in order of dates of composition and by subject, respectively.

896. Spiess, O. *Leonhard Euler*. Frauenfeld, Leipzig: Huber & Co., 1929.

Among special collections of papers and commemorative issues, the most useful include the following.

897. *Leonhard Euler 1701-1783: Beiträge zu Leben und Werk: Gedenkband des Kantons Basel-Stadt*. Edited by E.A. Fellmann. Basel, Boston, Stuttgart: Birkhäuser Verlag, 1983.

A handsomely produced commemorative volume giving a comprehensive picture of current Euler scholarship. Articles in English, French, but mostly German.

898. *Festschrift zur Feier 200. Geburtstages Leonhard Eulers*. Leipzig, Berlin: B.G. Teubner, 1907.

Contains numerous articles on Euler published by the Berliner Mathematische Gesellschaft.

899. Deborin, A.M., ed. *Leonard Eyler, 1707-1783*. Moscow, Leningrad: n.p., 1935. In Russian.

900. Lavrentiev, M.A., A.P. Youschkevitch, and A.T. Grigorian, eds. *Leonard Eyler. Sbornik statey*. Moscow: Akademii͡a Nauk SSSR, 1958. In Russian.

901. Schröder, K., ed. *Sammelband des zu Ehren des 250. Geburtstages Leonhard Eulers ... vorgelegten Abhandlungen*. Berlin: Akademie-Verlag, 1959.

902. Winter, E., et al., eds. *Die deutsch-russische Begegnung und Leonhard Euler*. Berlin: Akademie-Verlag, 1958.

Lagrange, Joseph Louis (1736-1813)

903. Grabiner, J. *The Origins of Cauchy's Rigorous Calculus*. Cambridge, Mass.: MIT Press, 1981.

Offers an excellent introduction to the history of the calculus in the 18th century, with special attention given to detailed study of Lagrange and his importance for the later work of Cauchy.

904. Sarton, George. "Lagrange's Personality (1736-1813)." *Proceedings of the American Philosophical Society* 88 (1944), 457-496.

905. ———, R. Taton, and G. Beaujouan. "Documents nouveaux concernant Lagrange." *Revue d'histoire des sciences* 3 (1950), 110-132.

906. *J.L. Lagrange, Sbornik statey k 200-letiyu so dnya rozhdenia.* Moscow: Soviet Academy of Sciences, 1937. In Russian.

Articles celebrating the second centenary of Lagrange's birth. Contents listed in *Isis* 28 (1938), 199.

Lambert, Johann Heinrich

907. Peters, W.S. "Lamberts Konzeption einer Geometrie auf einer imaginären Kugel." *Kantstudien* 53 (1961-1962), 51-67.

Based on the author's dissertation, University of Bonn, 1961, 87 pp.

908. Sheynin, O.B. "J.H. Lambert's Work on Probability." *Archive for History of Exact Sciences* 7 (1971), 244-256.

909. Steck, M. *Bibliographia Lambertiana*. Berlin: G. Lüttke, 1943. Rev. ed., Hildesheim: H.A. Gerstenberg, 1970.

Contains Lichtenberg's biography, as well as Lambert's bibliography in chronological order and a partial bibliography of secondary literature on Lambert.

910. Speiser, A. [Preface on Lambert.] In J.H. Lambert: *Opera Mathematica*. Edited by A. Speiser. Berlin, 1946; Zurich: Orell Füssli Verlag, 1948. 2 vols.

With a frontispiece engraving of Lambert. Both volumes contain the "Vorrede des Herausgebers," pp. ix-xxxi. Volume I also contains the "Éloge de M. Lambert," pp. 1-15, by J.H.S. Formey, reprinted from the *Nouveaux mémoires de l'Académie des sciences de Berlin (année 1778)*, Histoire (1780), pp. 72-91.

Laplace, Pierre-Simon, Marquis de (1749-1827)

911. Andoyer, H. *L'oeuvre scientifique du Laplace*. Paris: Payot & Cie., 1922.

A short (192 pp.) but general introduction.

912. Dantzig, D. van. "Laplace probabiliste et statisticien et ses precurseurs." *Archives internationales d'histoire des sciences* 8 (1955), 27-37.

913. Petrova, S.S. "K istorii metoda kaskadov Laplasa." *Istoriko-Matematicheskie Issledovaniia* 19 (1974), 125-131. In Russian.

The article deals with Laplace's method of cascades.

914. ———. "Rannyaya istoria preobrazovania Laplasa." *Istoriko-Matematicheskie Issledovaniia* 20 (1975), 246-256. In Russian.

On the early history of the Laplace Transform.

915. Sheynin, O.B. "O poyavlenii delta-funktsii Diraka v trudakh P.S. Laplasa." *Istoriko-Matematicheskie Issledovaniia* 20 (1975), 303-308.

On the appearance of Dirac's delta function in Laplace's work.

916. ———. "P.S. Laplace's Work on Probability." *Archive for History of Exact Sciences* 16 (1976), 137-187.

An analysis, primarily, of Laplace's *Théorie analytique des probabilités*. See also item 2047.

917. ———. "Laplace's Theory of Errors." *Archive for History of Exact Sciences* 17 (1977), 1-61.

918. Stigler, S. "Napoleonic Statistics: The Work of Laplace." *Biometrika* 62 (1975), 503-517.

Maclaurin, Colin (1698-1746)

919. Turnbull, H.W. "Colin Maclaurin." *American Mathematical Monthly* 54 (1947), 318-322.

920. ———. *Bicentenary of the Death of Colin Maclaurin*. Aberdeen: University Press, 1951.

Maupertuis, Pierre Louis Moreau de (1698-1759)

921. Brunet, P. *Maupertuis*. Paris: Albert Blanchard, 1929. 2 vols.

Volume I: Etude Biographique.

Monge, Gaspard (1746-1818)

922. Aubry, P.-V. *Monge, le savant ami de Napoléon: 1746-1818*. Paris: Gauthier-Villars, 1954.

923. Taton, René. *L'oeuvre scientifique de Monge*. Paris: Presses Universitaires de France, 1951, 441 pp.

Discusses all aspects of Monge's work: descriptive, analytic, and synthetic geometry, and the geometric theory of partial differential equations. By embedding Monge's work in its historical context this book provides a fine survey of the development of geometry in the 18th century. See also items 1018, 1587.

Stirling, James (1692-1770)

924. Tweedie, C. *James Stirling. Sketch of His Life and Works*. Oxford: Clarendon Press, 1922.

MENSIS OCTOBRIS A. MDCLXXXIV. 467

NOVA METHODUS PRO MAXIMIS ET MInimis, itemque tangentibus, quæ nec fractas, nec irrationales quantitates moratur, & ſingulare pro illis calculi genus, per G.G. L.

SIt axis AX, & curvæ plures, ut VV, WW, YY, ZZ, quarum ordinatæ, ad axem normales, VX, WX, YX, ZX, quæ vocentur reſpective, *v*, vv, y, z; & ipſa AX abſciſſa ab axe, vocetur x. Tangentes ſint VB, WC, YD, ZE axi occurrentes reſpective in punctis B, C, D, E. Jam recta aliqua pro arbitrio aſſumta vocetur dx, & recta quæ ſit ad dx, ut *v* (vel vv, vel y, vel z) eſt ad VB (vel WC, vel YD, vel ZE) vocetur d*v* (vel d vv, vel dy vel dz) ſive differentia ipſarum *v* (vel ipſarum vv, aut y, aut z) His poſitis calculi regulæ erunt tales: TAB. XII.

Sit a quantitas data conſtans, erit da æqualis 0, & $d\,\overline{ax}$ erit æqu. a dx: ſi ſit y æqu. *v* (ſeu ordinata quævis curvæ YY, æqualis cuivis ordinatæ reſpondenti curvæ VV) erit dy æqu. d*v*. Jam *Additio & Subtractio*: ſi ſit z - y + vv + x æqu. *v*, erit $d\,\overline{z - y + vv + x}$ ſeu d*v*, æqu. dz - dy + dvv + dx. *Multiplicatio*, $d\,\overline{xv}$ æqu. x d*v* + *v* dx, ſeu poſito y æqu. x*v*, fiet d y æqu. x d*v* + *v* dx. In arbitrio enim eſt vel formulam, ut x*v*, vel compendio pro ea literam, ut y, adhibere. Notandum & x & dx eodem modo in hoc calculo tractari, ut y & dy, vel aliam literam indeterminatam cum ſua differentiali. Notandum etiam non dari ſemper regreſſum a differentiali Æquatione, niſi cum quadam cautione, de quo alibi. Porro *Diviſio*, $d\frac{v}{y}$ vel (poſito z æqu. $\frac{v}{y}$) dz æqu. $\frac{\pm v\,dy + y\,dv}{yy}$

Quoad *Signa* hoc probe notandum, cum in calculo pro litera ſubſtituitur ſimpliciter ejus differentialis, ſervari quidem eadem ſigna, & pro + z ſcribi + dz, pro - z ſcribi - dz, ut ex additione & ſubtractione paulo ante poſita apparet; ſed quando ad exegeſin valorum venitur, ſeu cum conſideratur ipſius z relatio ad x, tunc apparere, an valor ipſius dz ſit quantitas affirmativa, an nihilo minor ſeu negativa: quod poſterius cum ſit, tunc tangens ZE ducitur a puncto Z non verſus A, ſed in partes contrarias ſeu infra X, id eſt tunc cum ipſæ ordinatæ

N n n 3 z decre-

Figure 9. The incipit of Leibniz's first article on the differential calculus of October 1684. Marred with misprints and obscured by the notion of a finite differential, the article failed to attract any recognition at the time. At the end of the 1680s, Johann and Jakob Bernoulli were the first to find their way through to the basic ideas of Leibniz's "Nova Methodus. . . ."

Special Studies

ALGEBRA

925. Hamburg, R.R. "The Theory of Equations in the 18th Century: The Work of Joseph Lagrange." *Archive for History of Exact Sciences* 16 (1976/1977), 17-36.

Covers both the development of Lagrange's ideas on algebraic solvability of polynomial equations in connection with symmetric functions and his analytic work on numerical solutions. Argues that the analytic work did not lead to a unifed theory whereas the algebraic work was momentous for the later development of group theory.

CALCULUS

926. Bos, Hendrik J.M. "Differentials, Higher-Order Differentials and the Derivative in the Leibnizian Calculus." *Archive for History of Exact Sciences* 14 (1974/1975), 1-90.

Valuable discussion of the concept of "differential" as employed in early Leibnizian calculus. Focuses on the techniques underlying the daily use of the differential rather than its logical foundations; elaborates the crucial role of variables as opposed to functions in the conceptual framework of the differential calculus. Argues that in Euler's attempts to eliminate the indeterminacy of higher order differentials the derivative emerged as the fundamental concept of the calculus. See also item 776.

927. Cajori, Florian. *A History of the Conceptions of Limits and Fluxions in Great Britain from Newton to Woodhouse*. London; Chicago: Open Court, 1919, vii + 299 pp.

Argues that Berkeley's critique of the foundations of the fluxional calculus led to an early geometrical concept of limit (Robins). Thus the British treatment of the calculus in the 18th century gained clarity and logical rigor.

928. Youshkevitch, Adolf P. "Euler und Lagrange über die Grundlagen der Analysis." *Sammelband der zu Ehren des 250. Geburtstages Leonhard Eulers der Deutschen Akademie der Wissenschaften zu Berlin vorgelegten Abhandlungen*. Edited by K. Schröder. Berlin: Akademie Verlag, 1959, 224-244.

Argues that Euler's contributions to the foundations of the calculus have been effective and influential, although he had not taken up the basic view of differentials as actual zeros. Suggests that Lagrange's attempt to found the calculus on power series derives from Euler's results.

ANALYSIS

928a. Engelsman, S.B. *Families of Curves and the Origins of Partial Differentiation.* Dissertation, University of Utrecht, 1982. 234 pp. Rev. ed. (Mathematics Studies 93.) Amsterdam, New York, Oxford: North-Holland, 1984, ix + 238 pp.

Analyzes in detail how the concept of partial differentiation and the theorems on interchangeability of differentiation and integration and on the equality of mixed second order partial derivatives emerged in the work of Leibniz and the Bernoullis on trajectories in families of curves. Shows that Euler's work of the 1730s, related to partial differential equations, can be understood by considering it as part of the same tradition. Contains an edition of Euler's early manuscript "De differentiatione functionum duas pluresve variabiles quantitates involventium" (ca. 1730).

929. Greenberg, J.L. "Alexis Fontaine's 'Fluxio-differential Method' and the Origins of the Calculus of Several Variables." *Annals of Science* 38 (1981), 251-290.

Contends that the origins of Fontaine's partial differential calculus must be sought in his work on tautochrones in resistant media of the early 1730s.

930. Lützen, Jesper. "Funktionsbegrebets udvikling fra Euler til Dirichlet." *Nordisk Matematisk Tidskrift* 25/26 (1978), 4-32.

Argues that with Euler the concept of function acquired its central position in mathematics. Although a general definition of function as a correspondence had been published as early as 1755, in practice functions were always taken to be analytic expressions. This apparent contradiction is explained by the 18th-century belief in the generality of the calculus.

931. Ravetz, Jerome R. "Vibrating Strings and Arbitrary Functions." *The Logic of Personal Knowledge: Essays Presented to Michael Polanyi on his Seventieth Birthday, 11th March 1961.* London: Routledge and Paul, 1961, 71-88.

Deals with the famous controversy concerning the vibrating string, and describes different standpoints in terms of the relation which the abstract mathematical theory was considered to have with the experimental phenomena: should it explain all (D. Bernoulli), some (L. Euler, J.L. Lagrange), or none (J. d'Alembert) of the phenomena. Pursues the issue of arbitrary functions and trigonometric series through the work of Fourier of 1807.

932. Youshkevitch, Adolf P. "The Concept of Function up to the Middle of the 19th Century." *Archive for History of Exact Sciences* 16 (1976), 37-85.

This analysis of the development of the concept of function up to the general definitions set forth by Hankel and Dirichlet provides a prominent place to 18th-century extensions and revisions of the concept. It shows that Euler's definition of

1755--based merely on the idea of a correspondence--almost coincides with the final 19th-century definitions.

933. Woodhouse, R. *A History of the Calculus of Variations in the Eighteenth Century*. Bronx: Chelsea Publishing Company, 1964.

This book is an unaltered reprint of *A Treatise of Isoperimetrical Problems and the Calculus of Variations* (Cambridge, 1810). Originally intended as a summary of 18th-century Continental developments in the calculus of variations in order to update the knowledge of British mathematicians, this book still serves as a very readable and uniform account of this field.

INFINITE SERIES

934. Hofmann, Joseph E. "Um Eulers erste Reihenstudien." *Sammelband der zu Ehren des 250. Geburtstages Leonhard Eulers der Deutschen Akademie der Wissenschaften zu Berlin vorgelegten Abhandlungen*. Edited by K. Schröder. Berlin: Akademie Verlag, 1959, 139-208.

Provides a detailed chronology of the work on summation of series by the Bernoullis, Goldbach, Euler, and Stirling up to 1738. Contends that Euler's later contributions to the theory of series are a consolidation of the results arrived at in the period 1728-1738.

935. Reiff, R. *Geschichte der unendlichen Reihen*. Munich: Urban & Schwarzenberg, 1889. Photomechanically reprinted Wiesbaden: Martin Sändig, 1969, iv + 212 pp., especially pp. 64-159.

A very useful survey of the history of summation of series from the 17th century through 1850. The 18th century is characterized as the period of formal treatment of series. See also item 784.

COMPLEX ANALYSIS

936. Markushewitsch, A.I. *Skizzen zur Geschichte der analytischen Funktionen*. Berlin: VEB Deutscher Verlag der Wissenschaften, 1955, viii + 139 pp., especially pp. 1-45.

Eighteenth-century achievements are quite extensively discussed. They are viewed as ingredients and preliminaries of a more systematic development of the theory of analytic functions in the 19th century.

DIFFERENTIAL EQUATIONS

937. Demidov, Sergei S. "Differentsialnye uravneniya s chastnymi proizvodnymi v rabotakh zh. dalambera." *Istoriko-Matematicheskie Issledovaniia* 19 (1974), 94-124.

Provides a comprehensive and detailed survey of d'Alembert's contributions to the theory of partial differential equations, consisting of the coefficient method, separation of variables, and expansion in trigonometric series. Argues that d'Alembert revived Euler's interest in the subject.

938. ———. "Vozniknovenie teorii differentsialnykh uravnenii s chastnymi proizvodnymi." *Istoriko-Matematicheskie Issledovaniia* 20 (1975), 204-220.

Identifies five early occurrences of partial differential equations in the work of Euler (1734, 1760s) and d'Alembert (1743, 1747, 1749), which together constitute the early history of the theory of these equations. Provides a thorough analysis of the partial differential equations which occur in Euler's geometric work of 1734. See also item 1682.

939. Engelsman, Steven B. "Lagrange's Early Contributions to the Theory of First Order Partial Differential Equations." *Historia Mathematica* 7 (1980), 7-23.

Argues that both Lagrange's theory of singular solutions to ordinary differential equations and his revision of the concept of solution to partial differential equations in the 1770s originated in the idea that the entire set of solutions of a differential equation can be found by means of variation of constants from a solution containing an adequate number of constants. See also item 1091.

940. Simonov, N.I. "Sur les recherches d'Euler dans le domaine des équations différentielles." *Revue d'Histoire de Sciences* 21 (1968), 131-156.

Provides a brief survey of Euler's achievements in the theory of ordinary and partial differential equations. Deals with these matters from a strictly mathematical point of view.

GEOMETRY

941. Bonola, R. *Non-Euclidean Geometry. A Critical and Historical Study of Its Developments*. New York: Dover, 1955.

See item 1063.

942. Chasles, M. *Aperçu historique sur l'origine et le développement des méthodes en géométrie, particulièrement de celles qui se ràpportent à la géométrie moderne*. Brussels: Hayez, 1837. 2nd ed. Paris: Gauthier-Villars, 1875.

Valuable because it is quite extensive, and has the merit of nearness to the work described. The second edition of 1875 is an unaltered version of the first edition.

943. Kötter, E. "Die Entwickelung der synthetischen Geometrie von Monge bis auf Staudt." *Jahresbericht der Deutschen Mathematiker-Vereinigung* 5 (1896), 1-486.

Also issued separately as a book by Teubner in 1901.

944. Nielsen, N. *Géomètres français du XVIIIème siècle*. Copenhagen, Paris: Levin & Munksgaard, 1935, 437 pp.

Published posthumously and arranged alphabetically.

MATHEMATICAL PHYSICS/APPLIED MATHEMATICS

945. Bikerman, J.J. "Theories of Capillary Action." *Centaurus* 19 (1975), 182-206.

See item 1678.

946. Bos, Hendrik J.M. "Mathematics and Rational Mechanics." In *The Ferment of Knowledge*. Edited by G.S. Rousseau and R. Porter. Cambridge: University Press, 1980, 327-355.

Discusses changes in historiography of 18th-century mathematics and rational mechanics from Montucla's contemporary account through today in terms of changing pictures of that century. Recent developments are characterized as "taking concepts and context seriously."

947. Burkhardt, Heinrich. "Entwicklungen nach oscillirenden Functionen und Integration der Differentialgleichungen der mathematischen Physik." *Jahresbericht der Deutschen Mathematiker-Vereinigung* 10 (1901-1908), viii + 1804 pp. Also issued separately in 2 vols. Leipzig: B.G. Teubner, 1908, xii + xii + 1800 pp. Reprinted New York: Johnson Reprint Corporation, 1960.

For details, see Burkhardt, items 1418, 1679.

948. Dugas, René. *Histoire de la mécanique*. Paris: Griffon, 1950. Translated by J.R. Maddox: *A History of Mechanics*. Neuchatel: Editions du Griffon; New York: Central Book Co., 1955.

See item 1683.

949. Fox, R. "The Rise and Fall of Laplacian Physics." *Historical Studies in the Physical Sciences* 4 (1974), 89-136.

950. Frisinger, H. Howard. "Mathematics in the History of Meteorology: The Pressure-Height Problem from Pascal to Laplace." *Historia Mathematica* 1 (1974), 263-286.

See item 1734.

951. Gillmor, C. Stewart. *Coulomb and the Evolution of Physics and Engineering in 18th-Century France*. Princeton, N.J.: Princeton University Press, 1971, xviii + 328 pp.

See items 1474, 1689.

952. Grattan-Guinness, Ivor. "Mathematical Physics in France, 1800-1835." In *Epistemological and Social Problems of the Sciences in the Early Nineteenth Century*. Edited by M. Otte and H.N. Jahnke. Dordrecht: Reidel, 1981, 349-370.

953. ———. "Mathematical Physics in France, 1800-1840: Knowledge, Activity and Historiography." In *Mathematical Perspectives: Essays on Mathematics and Its Historical Development*. Edited by J. Dauben. New York: Academic Press, 1981, 95-138.

See item 1690.

954. Hankins, Thomas L. "Eighteenth-Century Attempts to Resolve the *Vis-Viva* Controversy." *Isis* 56 (1965), 281-297.

955. Hiebert, E.N. *Historical Roots of the Principle of Conservation of Energy*. Madison, Wis.: State Historical Society of Wisconsin, 1962.

956. Körner, T. "Der Begriff des materiellen Punktes in der Mechanik des achtzehnten Jahrhunderts." *Bibliotheca Mathematica* 5 (1904), 15-62.

957. Ravetz, Jerome R. "The Representation of Physical Quantities in 18th-Century Mathematical Physics." *Isis* 52 (1961), 7-20.

958. Szabó, Istvan. *Geschichte der mechanischen Prinzipien und ihrer wichtigsten Anwendungen*. Basel, Stuttgart: Birkhäuser, 1977, xvi + 491 pp.

Provides a lucid and valuable introduction to the fields that motivated large parts of 18th-century analysis: mechanics of rigid bodies, hydrodynamics, ballistics, elasticity, etc. Although the book is not limited to the 18th century, this period with its philosophical issues and mechanical problems and methods forms the core of the book. See also item 1794.

959. Todhunter, Isaac. *A History of the Mathematical Theories of Attraction and the Figure of the Earth, from the Time of Newton to that of Laplace*. London: Macmillan and Co., 1873, xxxvi + 476 + 508 pp. 2 vols. Reprinted in 1 vol.: New York: Dover, 1962.

This work is a chronicle of the research concerning gravitation and the figure of the earth from Newton's *Principia* through Laplace's *Mécanique céleste*. It is organized by scientist. See also item 2012.

960. Truesdell, Clifford A. *The Rational Mechanics of Flexible or Elastic Bodies 1638-1788: Introduction to Leonhardi Euleri Opera Omnia, series secunda, Volumina X et XI. Leonhardi Euleri Opera Omnia, series secunda, Volumen XI, sectio secunda*. Zurich: Orell Füssli, 1960, 435 pp.

Together with the introductions to volumes 12 and 13 of Euler's *Opera Omnia*, second series, this book is by far the most detailed and informative account of mathematical physics in Euler's time. See also item 1722.

961. ———. *Essays in the History of Mechanics*. Berlin: Springer Verlag, 1968, x + 384 pp.

See item 1703.

MATHEMATICS IN THE 19th CENTURY

The 19th century was one of immense growth in mathematics, and this bibliography of available literature, while partial, reveals astonishing gaps as well as some thorough treatments. The best way into the period may be biographical, in which case the *Dictionary of Scientific Biography*, item 10, should be consulted, or, if the inquiry is topic-based, through the German *Encyklopädie*, item 5, also cited below in item 967, or, for an overview, perhaps through the general histories of Klein, item 989, and Kline, items 82 and 990. The literature on the connections between mathematics and science in this period seems exceptionally scant.

Publications annotated here have been divided into several broad categories, including general reference works, journals especially useful for the history of mathematics in the 19th century, general histories, biographies of individual mathematicians, and specialized subject areas within mathematics. Within each group there has been a general attempt at classification, but the reader is advised to use pragmatic judgment in looking for items of interest. Thus, works on number theory have been listed under that heading, but readers should consult the other sections also. References to the collected works of individual mathematicians have not been given. Consult instead the listings on pages 15-20 of this bibliography.

Bibliographies

962. Karpinski, Louis C. *Bibliography of Mathematical Works Printed in America through 1850*. Ann Arbor: University of Michigan Press; London: Oxford University Press, 1940.

Covers books, pamphlets, broadsides, encyclopedias, reference works, journals, and newspapers with mathematical articles. Entries are arranged chronologically according to first editions. Annotations provide information on multiple editions of works, where appropriate, and list libraries holding copies of cited materials. The works are indexed according to authors and subjects. Additional indexes of printers and publishers and of non-English and Canadian works are included. Over 600 facsimiles of title or other pages of cited materials complement the text. See also item 16.

963. Poggendorff, Johann Christian, ed. *Biographisch-literarisches Handwörterbuch zur Geschichte der exacten Wissenschaften*. Leipzig: J.A. Barth, 1863-1940. Reprinted Ann Arbor, Mich.: Edwards Bros., 1944. 6 vols.

Brief bibliobiographies of mathematicians and scientists. Volumes 1 and 2 cover through 1858. Under different editors, later volumes include materials available through 1931. Entries are of uneven lengths; emphasis is on bibliography rather than biography. This handy reference tool is especially strong on continental mathematicians. See also items 20, 21.

964. Royal Society of London. *Catalogue of Scientific Papers*, 1800-1900. London: C.J. Clay, 1867-1902; volumes 13-19 have the imprint Cambridge: Cambridge University Press, 1914-1925. Reprinted New York: Johnson Reprint Corp., 1965. 19 vols.

Handy guide to periodical scientific literature of the 19th century. This valuable reference tool was inspired by Joseph Henry's suggestion that the British Association for the Advancement of Science sponsor a limited catalogue of contemporary scientific articles. It appeared in four series: first series (volumes 1-6), covering scientific literature published in article form during the period 1800-1863; second series (volumes 7-8), covering 1864-1873; third series (volumes 9-11), 1874-1883; and fourth series (volumes 13-19), 1884-1900. The first series contained references to all scientific articles appearing in the over 1,000 journals enumerated at the front of volume 1; with the second series the *Catalogue* was broadened to include inaugural addresses, biographies, and articles on the history of science. Volume 12 was a supplement which listed significant papers omitted from the preceding volumes because of publication in journals not previously indexed. All entries are arranged alphabetically according to authors; multiple entries under a particular author are listed chronologically. Prefaces to the various series supply historical and technical information. There is an incomplete *Subject Index* to the *Catalogue* (Cambridge: Cambridge University Press, 1908-1914. 4 vols.), the first volume of which covers "Pure Mathematics." See also item 25.

965. Wolffing, Ernst. *Mathematischer Bücherschatz*. I. Teil: *Reine Mathematik. Abhandlungen zur Geschichte der mathematischen Wissenschaften mit Einschluss ihrer Anwendungen*, no. 16. Edited by Moritz Cantor. Leipzig: B.G. Teubner, 1903.

Designed as a catalogue of the most important books on pure mathematics published during the 19th century. Entries are listed alphabetically under 313 headings, including history of mathematics and philosophy of mathematics. Although criticized as incomplete, this work is an excellent tool for exploration of the non-periodical literature of the 19th century, and as such nicely complements the Royal Society's *Catalogue of Scientific Papers* which covers the periodical literature of the same period. A second part (on applied mathematics) was promised but never published.

Source Books and Surveys

966. Birkhoff, Garrett, ed., with the assistance of Uta Merzbach. *A Source Book in Classical Analysis*. (Source Books in the History of the Sciences, edited by Edward H. Madden.) Cambridge, Mass.: Harvard University Press, 1973.

"A panoramic view" of the 19th-century development of systematic general theories of functions of real and complex variables. English translations of major papers on classical analysis, with general introductory remarks. Includes hitherto-untranslated material, and regularly preserves important original terms and words alongside translations. Also offers informative footnotes, and a short, partially annotated bibliography. Second half explores interactions between analysis and mathematical physics. See also item 52.

967. *Encyklopädie der mathematischen Wissenschaften mit Einschluss ihrer Anwendungen*. Leipzig: B.G. Teubner, 1898-1935. 6 vols. in 23 parts.

A major survey of turn-of-the-century mathematics and its applications, featuring review articles written by eminent (predominantly German) mathematicians, including Hilbert, Hölder, Pringsheim, and Study. The articles are divided into six major categories: 1. arithmetic and algebra, 2. analysis, 3. geometry, 4. mechanics, 5. physics, and 6. geodesy, geophysics, and astronomy; they are rich in footnotes and most contain bibliographies. The volumes on mechanics were edited by Klein and C.H. Müller. French mathematicians planned a revised (more historical) version of this encyclopedia, the *Encyclopédie des sciences mathématiques pures et appliquées*, edited by Jules Molk, only 32 installments of which appeared between 1904 and 1916. See also items 5, 1424.

968. Smith, David Eugene, ed. *A Source Book in Mathematics*. (Source Books in the History of the Sciences, edited by Gregory D. Walcott.) New York: McGraw-Hill, 1929.

Selections from major mathematical publications from roughly 1450 through 1900. Divided into five sections: Number; Algebra; Geometry; Probability; and Calculus, Functions, and Quaternions. Each selection appears in English and is preceded by brief remarks introducing it and its author. Rich in 19th-century materials, the collection includes excerpts from works by Abel, Bolyai, Chebyshev, Galois, Grassmann, Kummer, Riemann, and Wessel. At its publication this book was somewhat of a landmark since it featured many original English translations of important parts of mathematical classics done by distinguished historians of mathematics. It is still useful to beginning English-reading students of the history of mathematics as a handy introduction to the mathematical classics and to the flavor of mathematics through 1900. See also items 56, 1925.

Journals

969. *Acta Mathematica*.

Multilingual journal originally edited by Mittag-Leffler. Appearing in 1882, volume 1 carried articles by such eminent mathematicians as L. Fuchs, Goursat, Hermite, Picard, and Poincaré. Subsequent 19th-century contributors included: Beltrami, G. Cantor, Darboux, Hill, S. Kowalevski, Minkowski, M. Noether, Stieltjes, Chebyshev, and Weierstrass. Volume 10 contains a comprehensive name index. There is also an index to volumes 1 through 35, *Acta Mathematica, 1882-1912: Table générale des tomes 1-35*, edited by Marcel Riesz (Uppsala: Almqvist & Wiksells boktr., 1913). This journal is still published.

970. *Cambridge and Dublin Mathematical Journal.*

Successor to the *Cambridge Mathematical Journal*. Appeared in nine volumes between 1846 and 1854, under the editorship of W. Thomson (Lord Kelvin). Carried articles by such major British mathematicians as De Morgan and W.R. Hamilton. Was replaced in 1857 by the *Quarterly Journal of Pure and Applied Mathematics*.

971. *Cambridge Mathematical Journal.*

First major British journal devoted exclusively to mathematics. Founded in 1837 by D.F. Gregory and R. Ellis. Four volumes appeared between 1839 and 1845, carrying articles by Boole, De Morgan, and others. This journal is one of two major sources on Gregory, most of whose articles and whose main obituary notice were published in it. The journal continued from 1846 through 1854 as the *Cambridge and Dublin Mathematical Journal*, which was followed in 1857 by the *Quarterly Journal of Pure and Applied Mathematics*. Volume 1 of the *CDMJ* contained an index to the *CMJ*.

972. *The Gentleman's Magazine.*

English monthly periodical circulated between 1731 and 1907. Title varied. Originated the term "magazine" in publishing. Carried articles on wide-ranging topics, poetry, drawings, and numerous other features. Most items appeared anonymously. Of use for occasional reviews of mathematical books and obituary notices of leading 19th-century British mathematicians.

973. *Jahrbuch über die Fortschritte der Mathematik.*

Critical year-by-year review of major mathematical books, pamphlets, and articles. Appeared in 68 volumes from 1871 through 1944, covering non-German as well as German publications from 1868 through 1942. (Volume 2 [1872] reviewed the literature from 1869 and 1870.) Entries are drawn from pure and applied mathematics and from the history and philosophy of mathematics, and are arranged topically. A name index appears at the end of each volume. Although criticized for uneven quality, this periodical remains a handy guide to mathematical publications through the early 1940s and to contemporary opinions thereof. See also item 41.

974. *Journal de mathématiques pures et appliquées.*

One of the leading mathematical journals of the 19th century. Also known as the *Journal de Liouville* after Joseph Liouville, its founder and editor from 1836 through 1874. Authors of articles appearing in this journal during the 19th century included: Cauchy, Cayley, Duhem, Eisenstein, Hermite, Jacobi, Jordan, Painlevé, Plücker, Sturm, and Chebyshev. In 1846 (volume 11) the *Journal* carried posthumously Galois's major work, carefully edited by Liouville. This journal is still published today. Earlier volumes have been reprinted (Kraus: Nendeln/Liechtenstein, 1976).

975. *Journal für die reine und angewandte Mathematik.*

A major mathematical journal of the 19th century. Also known as *Crelle's Journal* after August Leopold Crelle who founded it in 1826 and edited volumes 1-52 (1826-1856). Carried articles by Abel, Cayley, Dirichlet, Eisenstein, Jacobi, Kronecker, Kummer, Möbius, Plücker, Steiner, and Weierstrass. Still published today.

976. *Mathematische Annalen.*

Multilingual journal founded in 1868 by Clebsch and Neumann. Nineteenth-century issues carried articles by such mathematicians as Beltrami, Bessel, Cayley, Hankel, Jordan, Klein, Lie, and M. Noether. An index appeared at the end of the century: A. Sommerfeld, *Generalregister zu den Bänden 1-50* (Leipzig: B.G. Teubner, 1898). This journal is still published.

977. *Monthly Notices of the Royal Astronomical Society.*

A major publication of the Royal Astronomical Society. Volume 1 appeared in 1827. Contains fairly detailed summaries of papers read before the Society and noteworthy obituary notices of 19th-century British mathematicians. Since in its early decades the RAS included eminent mathematicians and mathematical practitioners, this is an important source for 19th-century British mathematics. Still published.

978. *Quarterly Journal of Pure and Applied Mathematics.*

Journal which appeared in 1857 as replacement for the defunct *Cambridge and Dublin Mathematical Journal.* Nineteenth-century editors and assistants included: Sylvester, Stokes, Cayley, and Hermite. Articles by editors, Clifford, Cockle, De Morgan, Salmon, Spottiswoode, Tait, and W. Thomson (Lord Kelvin). Volume 15 contained an index to volumes 1-15; volume 30, an index to volumes 16-30. The last volume (33) was published in 1927. Reprinted Amsterdam: Swets & Zeitlinger, 1966.

979. *Transactions of the Cambridge Philosophical Society.*

An essential source for British mathematics. Sponsored by the Cambridge Philosophical Society which was founded in 1819 at the University of Cambridge for "the advancement of Natural Philosophy." Twenty-three volumes were published at irregular intervals from 1822 through 1928. Contributors included: Babbage, Cayley, Green, J.F.W. Herschel, Maxwell, Stokes, and Todhunter. Some British mathematicians, such as De Morgan, published substantial portions of their original research in this journal.

980. *Transactions of the Royal Irish Academy.*

Major journal of the early Royal Irish Academy, first published in 1787 and then irregularly through 1907. Important source of articles by 19th-century Irish mathematicians and scientists, including W.R. Hamilton, Lardner, and Lloyd.

General Histories

981. Ball, Walter W. Rouse. *A History of the Study of Mathematics at Cambridge*. Cambridge: Cambridge University Press, 1889, xvii + 264 pp.

A useful guide to mathematicians and mathematical trends associated with the University from the Middle Ages through the mid-third of the 19th century. Also provides an introduction to the evolution of mathematical education at Cambridge, including a chapter on the mathematical tripos. Valuable footnote references to primary sources. See also item 2189.

982. Bell, Eric Temple. *The Development of Mathematics*. New York: McGraw-Hill, 1940. 2nd ed. 1945.

"*Not* a history of the traditional kind, but a narrative of the decisive epochs in the development of mathematics." Entertaining and popular. Noted for its breadth, clarity, and (sometimes premature) synthesis, but marred by inaccuracies and scanty footnotes. Offers extensive coverage of the 19th century; emphasizes connections between mathematics and the physical sciences. See also item 66.

983. ———. *Men of Mathematics*. New York: Simon and Schuster; London: Gollancz, 1937.

Lively, entertaining collection of biographies designed for a general audience. Emphasis on personalities and personal lives, with some attention to general historical settings. Especially recommended by Sarton for Bell's penetrating mathematical remarks. Many chapters devoted to leading 19th-century mathematicians. Offers almost no documentation; some of its major points have been revised by later scholarship. See, e.g., Joseph Warren Dauben, *Georg Cantor: His Mathematics and Philosophy of the Infinite*, item 1097, especially pp. 1-3. See also item 65.

984. Bourbaki, Nicolas. *Eléments d'histoire des mathématiques*. Paris: Hermann, 1960; revised 2nd ed., 1969; new ed., rev. and augmented, 1974.

The view from Nancago. It mostly describes the history of those branches of mathematics which the author has discussed elsewhere in his *Eléments*, so it is one of the few accounts of nineteenth- and twentieth-century topics. Dieudonné's *Abrégé d'Histoire* is, not surprisingly, somewhat similar. See also items 69, 1103.

985. Boyer, Carl B. *A History of Mathematics*. New York: John Wiley & Sons, 1968.

Exceptional survey from prehistory through the 20th century, based on primary and secondary sources. Distinguished by careful interweaving of historical, biographical, and mathematical materials, and many valuable references, in footnotes and bibliographies. Whole chapters devoted to Gauss and Cauchy, 19th-century geometry, arithmetization of analysis, and the rise of abstract algebra. Written as a textbook for upper-class undergraduates; therefore also includes exercises and chronological table. See also item 70.

986. Cajori, Florian. *A History of Mathematical Notations*. Chicago: Open Court, 1928-1929. 2 vols.

Most important work of its kind. Covers the initial appearances, origins, reception, and spread of enduring and obsolete mathematical symbols. Volume 1 deals with the notations of elementary mathematics (1. arithmetic and algebra and 2. geometry); volume 2, with higher mathematics (1. arithmetic and algebra, 2. modern analysis, and 3. geometry). The latter volume ends with general reflections on mathematical symbolism. Both volumes are well documented with numerous footnotes; volume 1 is generously illustrated with about 100 facsimiles. See also items 73, 643, 1821.

987. ———. *The Teaching and History of Mathematics in the United States*. Washington, D.C.: Government Printing Office, 1890, 400 pp.

Far-reaching survey of American mathematics through the late 19th century. Covers English and later French influences, textbooks, journals, leading mathematicians, attitudes towards fluxional and differential calculus, and all levels of mathematical education. Especially interesting are statistical results of a survey on mathematical education conducted by the U.S. Bureau of Education in the late 19th century. Abundant references in text, footnotes, and an appended bibliography on American calculus enhance the work's usefulness as a beginning research tool. No index. See also item 2154.

988. Edwards, C.H., Jr. *The Historical Development of the Calculus*. New York, Heidelberg, Berlin: Springer Verlag, 1979, xii + 351 pp.

Ten chapters cover material from Babylonian calculation to the Lebesgue integral. A lucid explanation of individual works by Archimedes, Wallis, Newton, Leibniz, Euler, and others, but short on historical analysis. Should be compared with *From the Calculus to Set Theory, 1630-1910*, edited by I. Grattan-Guinness. See also item 778.

989. Klein, F. *Vorlesungen über die Entwicklung der Mathematik im 19. Jahrhundert*. Berlin: Springer Verlag, 1926-1927. Reprinted in 1 vol. New York: Chelsea, 1967. Also in English, translated by M. Ackerman, *Development of Mathematics in the 19th Century*. Brookline, Mass.: Math Sci Press, 1979.

An unsurpassed account, by one of the century's leading figures, enlivened by personal comments, and always lucid in its account of the mathematics. It considers Gauss's work in pure and applied mathematics, the French school around the Ecole Polytechnique, Crelle's *Journal* and German mathematics, algebraic geometry, the British school of mathematical physics, the function theory of Riemann and Weierstrass, algebraic curves, group theory and function theory (particularly Poincaré's automorphic functions) and, in an incomplete second part, linear invariant theory, special relativity, and Riemannian manifolds. The English translation contains as an appendix "Kleinian Mathematics from an Advanced Standpoint" by R. Hermann, pp. 363-630. See also items 1316, 1570, 2086.

990. Kline, Morris. *Mathematical Thought from Ancient to Modern Times*. New York: Oxford University Press, 1972.

Most comprehensive English-language history of mathematics and its applications. More than half deals with post-1800 mathematics, with the bulk devoted to the 19th century. This highly recommended book stresses concepts rather than biography, applied rather than pure mathematics. Mathematical explanations often accompany the history, permitting the author to describe the work as an "historical introduction to mathematics." In addition, there are useful bibliographies at the ends of chapters and some footnotes. Reviewers have lavishly praised this standard history, one singling out its treatment of 19th-century analysis as exemplary of the author's historical and expository talents. See also items 82, 867, 1106, 1317, 1571, 2145.

991. Merz, John Theodore. *A History of European Scientific Thought in the Nineteenth Century*. Edinburgh and London: William Blackwood & Sons, 1904-1912. Reprinted Gloucester, Mass.: Peter Smith, 1976. 4 vols.

Written as a contribution to the "unification of thought." Explores some of the interrelations among the mathematics, science, and philosophy of the 19th century. Volumes 1 and 2 focus on science; volumes 3 and 4 (entitled *A History of European Thought in the Nineteenth Century*), on philosophy. Mathematical developments are treated in Volume 2, Chapter 12: "On the Statistical View of Nature" and Chapter 13: "On the Development of Mathematical Thought during the Nineteenth Century." This work is recommended primarily as a "treasure-trove of bibliographical information." Abundant discursive footnotes cite valuable primary and secondary materials.

992. Prasad, Ganesh. *Some Great Mathematicians of the Nineteenth Century: Their Lives and Their Works in Three Volumes*. Benares, India: Benares Mathematical Society, 1933-1934. 2 vols.

Includes biographies, partial bibliographies, and generous selections from major works. More descriptive than interpretative; some brief analysis in the form of quotations from other secondary sources. Volume 1 covers Gauss, Cauchy, Abel, Jacobi, Weierstrass, and Riemann; Volume 2 covers Cayley, Hermite, Brioschi, Kronecker, Cremona, Darboux, Cantor, Mittag-Leffler, Klein, and Poincaré. Volume 3 was never published. No index.

993. Smith, David Eugene, and Jekuthiel Ginsburg. *A History of Mathematics in America before 1900*. (The Carus Mathematical Monographs, no. 5.) Chicago: The Mathematical Association of America and Open Court Publishing Company, 1934, x + 209 pp.

Survey of American mathematics. Chapters 3 and 4 deal with the 19th century, providing basic information on major American figures, journals, societies, and universities, as well as foreign influences. Chapter 4 stresses the "little less than revolutionary" change which American mathematics underwent during the final quarter of the 19th century, attibuting it to the educational reforms introduced by Daniel Coit Gilman and Charles W. Eliot, the founding of the American Mathematical Society, and German influence. Although written almost a half-century ago, this brief work is still occasionally cited as a starting-point for research, partly because it is replete with valuable materials such as lists of late 19th-century American doctoral dissertations in mathematics and subject-bibliographies of articles published by American mathematicians. See also item 2181.

994. Struik, Dirk J. *A Concise History of Mathematics*. New York: Dover, 1948; 3rd rev. ed., 1967, 195 pp.

Covers an immense amount with a lightness and sureness of touch that has scarcely been matched, shrewd in its selection, and unfailingly apt in its comments both on the technical mathematics and the broader historical issues. Covers Oriental, Greek, and Arabic mathematics, and mathematics in the West from its beginnings to the end of the nineteenth century. Third edition is attractively illustrated with pictures of many mathematicians. Also available in translations in numerous foreign languages. See also item 86.

995. Taylor, Eva G.R. *The Mathematical Practitioners of Hanoverian England, 1714-1840*. Cambridge: Cambridge University Press, 1966, 503 pp.

A sequel to the author's *Mathematical Practitioners of Tudor and Stuart England* (items 630, 801, 1809, 2216), this book is a pioneer reference work treating 2,282 great and not-so-great mostly British artisans, navigators, surveyors, mechanics, instructors in the mathematical arts, instrument makers, and the like. It is divided into two parts, the first of which presents a chronological overview of mathematical practice through 1840. By the latter date, the author implies, the professionalization of science and its division into pure and applied destroyed the common link among mathematical practitioners. Of major interest is part 2 which contains bibliobiographies of the practitioners, arranged alphabetically in successive decades. The work includes a brief statement on sources consulted and two indexes, but has been criticized as difficult to use since some bibliobiographies are not indexed and the biographies are usually undocumented. See also items 1810, 2217.

Biographical Studies

This section of the bibliography is highly selective; a thorough compilation of biographical literature on 19th-century mathematicians would be prohibitively long. Readers are again reminded to use standard guides like Gillispie's *Dictionary of Scientific Biography*, item 10, and Poggendorff, items 20 and 21, as preliminary introductions to specific figures.

STUDIES ON GAUSS, CARL FRIEDRICH (1777-1855)

996. Brendel, M., F. Klein, and L. Schlesinger. "Materialen für eine wissenschaftliche Biographie von Gauss." In Gauss's *Werke* X (2), (1922-1923).

Particularly valuable essays on Gauss's work in number theory, analysis, and astronomy, written by expert authors with an intimate knowledge of Gauss's *Nachlass*, which they edited.

997. Dieudonné, Jean. *L'oeuvre mathématique de C.F. Gauss*. Paris: Palais de la Découverte, 1961.

Stresses Gauss's abstract, unifying approach to mathematics. Re-worked in the author's "Carl Friedrich Gauss: A Bi-Centenary," *Bulletin Mathématique* 2 (1978), 61-70, which is the best recent introduction to Gauss's mathematics.

998. Dunnington, G.W. *Carl Friedrich Gauss, Titan of Science*. New York: Exposition Press, 1955.

The most thorough biography so far, useful on his career if somewhat superficial on his scientific achievements.

Other biographical studies of Gauss include:

999. Hall, T. *Carl Friedrich Gauss: a Biography*. Translated by A. Froderberg. Cambridge, Mass.: MIT Press, 1970, 176 pp.

1000. Wussing, H. *C.F. Gauss*. Leipzig: Teubner, 1974.

Works dealing with various aspects of Gauss's mathematics:

1001. Coxeter, H.S.M. "Gauss as a Geometer." *Historia Mathematica* 4 (1977), 379-396.

1002. Reichardt, H. *Gauss und die nicht-euklidische Geometrie*. Leipzig: Teubner, 1976.

1003. Waterhouse, W.C. "Gauss on Infinity." *Historia Mathematica* 6 (1979), 430-436.

Argues that Gauss's much-quoted objection to the use of completed infinities in mathematics was only an objection to treating infinite quantities like finite ones (and not as limits), not a statement of "finitist" views about mathematics.

BIOGRAPHICAL STUDIES ARRANGED ALPHABETICALLY BY AUTHOR

1004. Arago, Dominique François. *Oeuvres complètes*. Vols. 1-3. Paris: Gide et J. Baudry, 1854-1855.

Principally collections of biographical notices written by the author as Secrétaire perpétuel de l'Académie des Sciences. Covers Ampère, Carnot, Fourier, Fresnel, Monge, Poisson, and other European mathematicians and scientists. English translations of ten of these biographies, accompanied by new notes, appear in *Biographies of Distinguished Scientific Men*, translated by W.H. Smyth, Baden Powell, and Robert Grant (London: Longman, Brown, Green, 1857).

1005. Archibald, Raymond Clare, ed. *Benjamin Peirce, 1809-1880*. Oberlin, Ohio: Mathematical Association of America, 1925.

Thirty-page collection of valuable biographical and bibliographical material on Peirce. Includes reminiscences by his former Harvard students: C.W. Eliot, A.L. Lowell, W.E. Byerly, and A.B. Chace; a biographical sketch by the editor; a list of surviving portraits of Peirce; and a partially annotated bibliography. Recommended for the bibliography, anecdotes, and glimpses of Peirce as mathematics professor. No index. Also published in *American Mathematical Monthly* 32 (1925), 1-30.

1006. Cayley, Arthur. *The Collected Mathematical Papers of Arthur Cayley*. Edited by Arthur Cayley and A.R. Forsyth. Cambridge: Cambridge University Press, 1889-1897. 13 vols. and a supplementary vol.

Cayley's published papers, listed chronologically and numbered consecutively throughout the 13 volumes (from 1 through 967). As editor of the early volumes, Cayley listed but did not reprint papers which were "controversial" (contained serious errors) or were merely translations of ones already included in the collection. Cayley also added "Notes and References" to the papers in volumes 1 through 8. Volume 8 begins with a biographical notice of Cayley and a list of courses he taught. The collection contains an occasional hitherto-unpublished paper. The supplementary volume provides a list of all the papers in the collection, arranged according to their assigned numbers, and an index of subjects and names. This work has been reprinted (New York: Johnson Reprint, 1963).

1007. De Morgan, Sophia Elizabeth. *Memoir of Augustus De Morgan, with Selections from His Letters*. London: Longmans, Green, 1882.

Best available biography of De Morgan. Written by his wife. Not a scientific biography; but offers valuable information on De Morgan's personal life and professional career. Also includes substantial selections from his correspondence. Very useful bibliography.

1008. Dubbey, John M. *The Mathematical Work of Charles Babbage*. Cambridge: Cambridge University Press, 1978, 235 pp.

Pioneer study describing Babbage's work on the calculus of functions, symbolical algebra, mathematical notation, and computers, and, to a lesser extent, analysis, probability, and geometry. Offers an interesting perspective on the early 19th-century Cambridge mathematical community (including Peacock and J.F.W. Herschel) which spawned the Analytical Society. Shows, for example, that in 1821 Babbage produced a manuscript which contained the seeds of symbolical algebra later developed by Peacock. Also argues that Babbage formulated "the major ideas upon which all modern computers are constructed." Because of the rich original source material which Dubbey has discovered and described, this book is a prerequisite to research on British mathematics of Babbage's period. A reviewer has criticized its failure to retain all of Babbage's original notation. See also item 1197.

1009. Eisele, Carolyn. *Studies in the Scientific and Mathematical Philosophy of Charles S. Peirce: Essays by Carolyn Eisele.* Edited by R.M. Martin. (Studies in Philosophy, no. 29.) The Hague and Paris: Mouton, 1979.

Collection of 30 published and unpublished essays written from 1951 through 1976. Covers Peirce's mathematics, science, logic, philosophy, history of science, and mathematics textbook writing. Provides a good introduction to his varied mathematical interests which included non-Euclidean geometry, linear algebra, the four-color problem, the infinite, n-valued logic, and infinitesimals. Based on a careful analysis of primary sources, the collection argues that Peirce was "primarily a *mathematician*, logician, and philosopher" and that his mathematics influenced his philosophy. Scattered throughout the essays is some useful information on the American mathematical community of the late 19th and early 20th centuries. See also item 1017.

1010. Graves, Robert Perceval. *Life of Sir William Rowan Hamilton.* Dublin: Hodges, Figgis, 1882-1889. 3 vols.

Massive Victorian "life and letters," written by a friend. Offers a wealth of biographical detail, supplemented by selections from Hamilton's poems, correspondence, and assorted manuscripts. The author's sense of propriety, however, led him to restrict coverage of very personal details and remarks on persons still living; his mathematical limitations precluded any substantial analysis of Hamilton's major work. Yet this biography should be a starting point for Hamiltonian scholarship. Volume 3 features a complete bibliography of Hamilton's publications, as well as extensive extracts from the Hamilton-De Morgan correspondence. Each volume includes an index.

1011. Gregory, Duncan Farquharson. *The Mathematical Writings of Duncan Farquharson Gregory, M.A.* Edited by William Walton. Cambridge: Deighton, Bell, 1865.

Handy collection of Gregory's published papers on such topics as symbolical algebra, differential equations, conic sections, calculus of residues, and calculus of finite differences. Includes a biographical memoir by R. Ellis.

1012. Macfarlane, Alexander. *Lectures on Ten British Mathematicians of the Nineteenth Century.* (Mathematical Monographs, edited by Mansfield Merriman and Robert S. Woodward, no. 17.) New York: John Wiley & Sons, 1916.

Introductions to some of the leading British mathematicians of the 19th century: Boole, Cayley, Clifford, De Morgan, W.R. Hamilton, Kirkman, Peacock, Smith, Sylvester, and Todhunter.

While criticized as unfocused, Macfarlane's biographical sketches (originally delivered as lectures) contain some insightful comments on the personalities and major contributions of these mathematicians. Occasional juicy quotations, but no bibliography or footnotes.

1013. ———. *Lectures on Ten British Physicists of the Nineteenth Century*. (Mathematical Monographs, edited by Mansfield Merriman and Robert S. Woodward, no. 20.) New York: John Wiley & Sons, 1919.

A supplement to the author's *Lectures on Ten British Mathematicians of the Nineteenth Century*, item 1012, and characterized by the same strengths and weaknesses. Covers "mathematicians whose main work was in physics, astronomy, and engineering": Adams, Airy, Babbage, Herschel, Maxwell, Rankine, Stokes, Tait, W. Thomson (Lord Kelvin), and Whewell.

1014. Marie, Maximilien. *Histoire des sciences mathématiques et physiques*. Paris: Gauthier-Villars, 1883-1888. 12 vols.

Offers brief biographies with bibliographical notices of major and lesser mathematicians from ancient times through the 19th century. Volumes 8 through 12 cover the latter century. There is an index at the end of each volume and a general index of names at the back of the last volume. Especially useful for references to minor French mathematicians; weak on British mathematics.

1015. Mathews, Jerold. "William Rowan Hamilton's Paper of 1837 on the Arithmetization of Analysis." *Archive for History of Exact Sciences* 19 (1978), 177-200.

A discussion of the content and reception of the ideas on arithmetization of analysis contained in Hamilton's famous essay. Argues that the essay introduced concepts (including one related to the cut in the rationals) found in the later work of such mathematicians as Peano and Dedekind. Concludes, however, that Hamilton's essay exercised no direct influence on the arithmetization of analysis in the late 19th century. Attributes this lack of influence to the relative obscurity of the journal in which Hamilton published, his metaphysical style, and his mathematical isolation. Good bibliography. See also item 1330.

1016. Ore, Øystein. *Niels Henrik Abel*. Minneapolis: University of Minnesota Press, 1957, 277 pp.

An extremely valuable biography, but thin on the mathematics.

1017. Peirce, Charles S. *The New Elements of Mathematics by Charles S. Peirce*. Edited by Carolyn Eisele. The Hague and Paris: Mouton; Atlantic Highlands, New Jersey: Humanities Press, 1976. 4 vols. in 5 books.

Well-organized, well-indexed collection of Peirce's mathematical writings, most of which were hitherto unpublished. Each volume bears a separate title, descriptive of its contents.

Thus, Volume 1: *Arithmetic* contains material for an unpublished arithmetic textbook. Volume 2: *Algebra and Geometry* reproduces "Elements of Mathematics," another textbook manuscript, and Peirce's notes towards an updated version of an elementary geometry textbook originally published by his father, Benjamin. Described by one reviewer as "of greatest mathematical interest," Volume 3: *Mathematical Miscellanea* appears in two books, with Book 1 offering assorted manuscripts and Book 2 featuring Peirce's mathematical correspondence and items from *The Nation*. Displaying Peirce's mathematical breadth, this volume deals with trigonometry, probability, finite differences, Boolean algebra, the four-color problem, map projections, linear algebra, matrices, non-Euclidean geometry, n-valued logic, the mathematical infinite, and additional topics. Volume 4: *Mathematical Philosophy* emphasizes logic and the philosophy of science and mathematics. All volumes open with useful introductions by the editor and include indexes of names and subjects. See also item 1009.

1018. Taton, René. *L'oeuvre scientifique de Monge*. Paris: Presses Universitaires de France, 1951, 441 pp.

Reviews Monge's life, including his important work in setting up the École Polytechnique, and then his work in descriptive, analytic, and infinitesimal geometry and other subjects, including partial differential equations. See also items 923, 1587.

Special Studies

ALGEBRA

1019. Bottazzini, Umberto. "Algebraische Untersuchungen in Italien, 1850-1863." *Historia Mathematica* 7 (1980), 24-37.

Brioschi's work in invariant theory and Betti's on Galois theory are briefly described and shown to have been influenced by Cayley and Sylvester before 1858, when they made their celebrated journey with Casorati. After that time Betti turned to analysis, under Riemann's influence. See also item 1226.

1020. Crowe, Michael J. *A History of Vector Analysis: The Evolution of the Idea of a Vectorial System*. Notre Dame, Ind.: University of Notre Dame Press, 1967, 270 pp.

Pioneering history of vector algebra. See also item 1195.

1021. Dieudonné, Jean. "Le développement historique de la notion de groupe." *Bulletin de la Société Mathematique de Belgique* 28 (1976), 267-296.

1021a. ———. "The Difficult Birth of Mathematical Structures (1840-1940)." In *Scientific Culture in the Contemporary World*. Edited by V. Mathieu and P. Rosi. Milan: Scientia, in cooperation with UNESCO, 1970.

These articles trace the slow evolution of the idea of calculating on abstract objects, typically equivalence classes of better known objects (such as quadratic forms, classes modulo an ideal, or loops) from eighteenth-century origins in the theory of numbers and the theory of equations to modern mathematics. Dieudonné stresses the unity of mathematics which has thereby been achieved during a period of very rapid growth.

1022. Fearnley-Sander, D. "Hermann Grassmann and the Creation of Linear Algebra." *American Mathematical Monthly* 86 (1979), 809-817.

Discusses Grassmann's pioneering achievements in his *Ausdehnungslehre* (1844, revised 1862) under the headings of linear algebra, linear products, exterior and inner products, linear transformations, and applications to geometry.

1023. Hawkins, Thomas. "The Theory of Matrices in the 19th Century." *Proceedings of the International Congress of Mathematicians*. Vol. 2. Edited by R.D. James. Vancouver: Canadian Mathematical Congress, 1975, 561-570.

1024. ———. "Weierstrass and the Theory of Matrices." *Archive for History of Exact Sciences* 17 (1977), 119-163.

1025. ———. "Another Look at Cayley and the Theory of Matrices." *Archives Internationales d'Histoire des Sciences* 26:100 (1977), 82-112.

1026. ———. "Cauchy and the Spectral Theory of Matrices." *Historia Mathematica* 2 (1975), 1-29.

In the four items above, Hawkins argues for the central importance in the development of the spectral theory of matrices of the work of Weierstrass. The general theory of simultaneous diagonalization of two quadratic forms in n variables, with particular attention to special cases, is shown to be characteristic of the Berlin school (Weierstrass, Kronecker, Frobenius). Origins of the theory of matrices in the work of Gauss, Eisenstein, Cauchy, Hermite, and Cayley are discussed, with examples taken from number theory, differential equations, and quadratic forms.

1027. ———. "The Origins of the Theory of Group Characters." *Archive for History of Exact Sciences* 7 (1970/1971), 142-170.

1028. ———. "Hypercomplex Numbers, Lie Groups, and the Creation of Group Representation Theory." *Archive for History of Exact Sciences* 8 (1971/1972), 243-287.

See item 1205.

1029. ———. "New Light on Frobenius' Creation of the Theory of Group Characters." *Archive for History of Exact Sciences* 12 (1974), 217-243.

1030. ———. "The Creation of the Theory of Group Characters." In *History of Analysis*. Edited by R.J. Stanton and R.O. Wells.

(Rice University Studies 64, Nos. 2, 3.) Houston, Tex.: William Marsh Rice University, 1978, 57-71.

These papers (items 1027-1030) trace in detail Frobenius's route to his discovery of the theory of characters and group representations, drawing on the extensive Dedekind-Frobenius correspondence. Item 1028 compares Frobenius's creation of group representation theory with the work of Burnside and Molien and contemporary hypercomplex systems and Lie groups.

1031. Kiernan, B.M. "The Development of Galois' Theory from Lagrange to Artin." *Archive for History of Exact Sciences* 8 (1971/1972), 40-154.

Kiernan considers the study of the solvability of polynomial equations by Lagrange, Ruffini, Gauss, Cauchy, Abel, and Galois; traces the development of Galois's ideas in France (Serret and Jordan) as abstract group theory, and in Germany (Kronecker, Dedekind, H. Weber) as field theory; his discussion ends with Emil Artin.

1032. Mehrtens, Herbert. "Das Skelett der modernen Algebra. Zur Bildung mathematischer Begriffe bei Richard Dedekind." In *Disciplinae Novae. Zur Entstehung neuer Denk- und Arbeitsrichtungen in der Naturwissenschaft. Festschrift zum 90. Geburtstag von Hans Schimank*. Edited by C.J. Scriba. Göttingen: Vardenhoeck & Ruprecht, 1979, 25-43.

This work is more informative on Dedekind than is Kiernan, item 1031.

1033. Meyer, Franz M. "Bericht über den gegenwartigen Stand der Invariantentheorie." *Jahresbericht der Deutschen Mathematiker-Vereinigung* 1 (1890-1891), 79-288.

A valuable report on the state of the art (c. 1890) and thorough on recent, mostly German, developments, but makes almost no attempt to look at the British origins (Cayley and Boole). See also item 1235.

1034. Nový, Luboš. *The Origins of Modern Algebra*. Prague: Academia Publishing House of the Czechoslovak Academy of Sciences, 1973, 252 pp.

See item 1215.

1035. Purkert, Walter. "Zur Genesis des abstrakten Körperbegriffs." *NTM-Schriftenreihe für die Geschichte der Naturwissenschaften, Technik und Medizin*. Teil 1, 10 (1) (1973), 23-37; Teil 2, 10 (2) (1973), 8-20.

See item 1217.

1036. ———. "Ein Manuskript Dedekinds über Galois-Theorie." *NTM-Schriftenreihe für die Geschichte der Naturwissenschaften, Technik und Medizin*, Series 2, 13 (1976), 1-16.

Purkert has more information on Dedekind than does Kiernan, item 1031.

1037. Van der Waerden, Bartel Leendert. "Die Galois-Theorie von Heinrich Weber bis Emil Artin." *Archive for History of Exact Sciences* 9 (1972/1973), 240-248.

This paper discusses the work of Hilbert and his school, which is omitted by Kiernan, item 1031.

1038. Waterhouse, W.C. "The Early Proofs of Sylow's Theorem." *Archive for History of Exact Sciences* 21 (1980), 279-290.

Surveys Sylow's and Frobenius's proofs and their origins in Cauchy's work in order to highlight the difference between permutation-theoretic and abstract proofs as they arose in nineteenth-century group theory.

1039. Wussing, Hans. *Die Genesis des abstrakten Gruppenbegriffes. Ein Beitrag zur Entstehungsgeschichte der abstrakten Gruppentheorie.* Berlin: VEB Deutscher Verlag der Wissenschaften, 1969, 258 pp.

The standard work on the emergence of groups as an explicit mathematical structure, particularly thorough in its survey of implicit group-theoretical ideas in geometry and number theory. Permutation groups and Galois theory are considered in detail, with particular emphasis on Jordan's work, as is the emergence of transformation groups in Klein's Erlanger Programm and Lie's work. A final section considers the axiomatization of abstract group theory. See also item 1224.

BRITISH ALGEBRA

1040. Hankins, Thomas L. "Algebra as Pure Time: William Rowan Hamilton and the Foundations of Algebra." In *Motion and Time, Space and Matter: Interrelations in the History of Philosophy and Science.* Edited by P.K. Machamer and R.G. Turnbull. Columbus: Ohio State University Press, 1976, 327-359.

See items 1201, 2249.

1041. ———. "Triplets and Triads: Sir William Rowan Hamilton on the Metaphysics of Mathematics." *Isis* 68 (1977), 175-193.

Hamilton's discovery of quaternions in 1843 is connected with his Kantian metaphysical interest via his interest in Coleridge's philosophical formulations of various triads. Also considers algebra "as a science of pure time." Draws on unpublished Hamilton material in Dublin.

1042. Koppelman, Elaine. "The Calculus of Operations and the Rise of Abstract Algebra." *Archive for History of Exact Sciences* 8 (1971/1972), 155-242.

Discusses the origins of the calculus of operations in France ca. 1800 and its greater success in Britain among the algebraic formalists like Peacock, as well as De Morgan, Hamilton, and Boole. See also items 1207, 1319.

1043. Nagel, Ernest. "'Impossible Numbers': A Chapter in the History of Modern Logic." *Studies in the History of Ideas* 3 (1935), 429-474. Reprinted in Ernest Nagel, *Teleology Revisited and Other Essays in the Philosophy and History of Science*. (The John Dewey Essays in Philosophy, edited by the Department of Philosophy, Columbia University, no. 3.) New York: Columbia University Press, 1979.

An important essay which interweaves philosophical and historical insights. Presents Boole's mathematical formulation of logic as an extension of the algebraic work of De Morgan, D.F. Gregory, W.R. Hamilton, and Peacock. Suggests a connection between the problem of the negative and complex numbers and abandonment of the definition of mathematics as the science of quantity. Also relates the ideas of Playfair, Wallis, Warren, and Woodhouse to the development of symbolical algebra. Especially useful for footnote references and provocative original quotations. See also items 1214, 2260.

1044. Olson, Richard. "Scottish Philosophy and Mathematics 1750-1830." *Journal of the History of Ideas* 32 (1971), 29-44.

Argues that "the epistemological doctrines associated with the Common Sense philosophy of Thomas Reid and Dugald Stewart not only reinforced an appreciation of geometrical reasoning, but also provided a significant obstacle to the acceptance of analytical methods by Scottish mathematicians." Discusses Reid's ideas on the geometry of visibles.

1045. ———. *Scottish Philosophy and British Physics, 1750-1880: A Study in the Foundations of the Victorian Scientific Style*. Princeton: Princeton University Press, 1975, 349 pp.

Argues that Scottish Common Sense philosophy exerted a major influence on late 18th- and 19th-century British physics. Cites the physicists' predilection towards geometrical mathematics, drive for simplicity, and concern with hypotheses and analogies as evidence for the thesis. Part 1 deals with the ideas of Reid, Stewart, Brown, and W. Hamilton; part 2, with Common Sense elements in the physics of Robison, Playfair, Leslie, Brougham, Forbes, Waterston, J.F.W. Herschel, Rankine, and Maxwell. Especially interesting is Chapter 3: "Common Sense Concerns with the Nature of Mathematics" which explores the extent of the Common Sense philosophers' acceptance of mathematics as a paradigm for philosophy and the physical sciences. This chapter covers W. Hamilton's attack on the mathematics as an integral part of a liberal education and mentions Reid's geometry of visibles. The book contains many lengthy quotations, some with but brief commentary. A review by G.N. Cantor has attacked its major premise, claiming that it is "impossible to speak, in any unified sense, of Common Sense philosophy or of its influence" (*British Journal for the History of Science* 10 [1977], 81-83).

1046. Richards, Joan L. "The Reception of a Mathematical Theory: Non-Euclidean Geometry in England 1865-1883," to appear.

Clifford's empiricist interpretation of Riemann's theory of geometry met with opposition from Jevons, Henrici, Cayley, and others, who adhered to a view of mathematics as *a priori* and therefore exemplifying the possibility of certainty in knowledge which they wished to extend to religion and morals. Cayley's projective geometry is contrasted with Clifford's radical criticisms in mathematics and philosophy.

1047. ———. "The Art and the Science of British Algebra: A Study in the Perception of Mathematical Truth." *Historia Mathematica* 7 (1980), 343-365.

Argues that what distinguished British algebra from modern algebra, and accounts for the different development of British algebra, is a fundamental difference between nineteenth- and twentieth-century views of truth.

NUMBER THEORY

1048. Dickson, L.E. *History of the Theory of Numbers*. Washington, D.C.: The Carnegie Institution, 1919-1923. Reprinted New York: Stechert, 1934, and New York: Chelsea, 1952, 1971. 3 vols.

See items 789, 1899.

1049. Ellison, W. and F. "Théorie des nombres." In *Abrégé d'histoire des mathématiques*. Vol. I. Edited by J. Dieudonné. Paris: Herman, 1978, 165-334.

1050. Hilbert, David. "Die Theorie der algebraischen Zahlkörper." *Jahresbericht der Deutschen Mathematiker-Vereinigung* 4 (1897), 175-546. Also in D. Hilbert, *Gesammelte Abhandlungen*. Vol. 1. Berlin: Springer, 1932-1935, 63-363. 3 vols.

1051. Smith, H.J.S. *Report on the Theory of Numbers*, originally issued in six parts as a Report of the British Association. Reprinted in *The Collected Mathematical Papers of H.J.S. Smith* (1895). Also reprinted separately and with the *Mathematical Papers*. New York: Chelsea, 1952 and 1965.

Regarding items 1048 to 1051, Smith and Dickson are classics, thorough, and full of insight into the mathematics. They are also interesting as primary sources, and reflect the considerable learning and skill of their authors (first published in 1865 and 1919). The Ellisons trace the route from Gauss to Hilbert and modern class-field theory, and also look at recent work on Diophantine equations and sieves. Hilbert's is not strictly a history, of course, but a distillation and reformulation of the nineteenth-century work of decisive importance for the twentieth.

1052. Edwards, H.M. "The Background of Kummer's Proof of Fermat's Last Theorem for Regular Primes." *Archive for History of Exact Sciences* 14 (1975), 219-236.

1053. ———. "Postscript to 'The Background of Kummer's Proof.'" *Archive for History of Exact Sciences* 17 (1977), 381-394.

1054. ———. *Fermat's Last Theorem. A Genetic Introduction to Number Theory*. New York: Springer-Verlag, 1977.

These papers argue convincingly that Kummer was not led to formulate his proof of Fermat's Last Theorem for regular primes by a fallacious earlier proof, but had already (i.e., by 1847) discovered the failure of unique factorization for cyclotonic integers in his search for higher reciprocity laws. Nonetheless, Kummer's theory did rest on false hopes in its early years. Edwards also considers French attempts (c. 1850) on Fermat's Last Theorem. Edward's *Postscript* considers the paper submitted and withdrawn in 1844, which contains a different but related error, detected by Jacobi, and also the paper itself (not published in Kummer's *Collected Works*). These matters are treated "genetically" in Edward's book, item 1054, which also explains related work of Euler, Gauss, and Dirichlet. See also items 1937, 1938.

GEOMETRY AND TOPOLOGY

1055. Coxeter, H.S.M. "The Space-Time Continuum." *Historia Mathematica* 2 (1975), 289-298.

Describes the projective interpretation due to E. Study (1907) of Minkowski space, and also Schläfli's description (1858) of spaces of constant curvature.

1056. Dauben, Joseph Warren. "The Invariance of Dimension: Problems in the Early Development of Set Theory and Topology." *Historia Mathematica* 2 (1975), 273-288.

1057. Johnson, Dale M. "Prelude to Dimension Theory: The Geometrical Investigations of Bolzano." *Archive for History of Exact Sciences* 17 (1977), 261-295.

1058. ———. "The Problem of the Invariance of Dimension in the Growth of Modern Topology, Part I." *Archive for History of Exact Sciences* 20 (1979), 97-188; Part II, 25 (1981), 85-267.

The paper on Bolzano considers his philosophical and mathematical motivation for considering the problem, and gives his two solutions (1817, 1840s). Johnson's thorough-going second paper considers the views of Gauss and Riemann on geometry and manifolds as multiply extended manifolds, before treating Cantor and Dedekind's discussions in detail. Attempts to resolve Cantor's paradox (there is a one-to-one correspondence between the points of a square and an interval) by Lüroth, Thomae, Jürgens, and Netto are considered, and shown to lead (in the hands of Peano and Schoenflies) to the growth of point-set topology. Part II takes the invariance of dimension problem up to its solution by Brouwer, with emphasis given to the Brouwer-Lebesgue dispute. Documents and letters to Baire from Brouwer are included, along with "Glimpses of the Development of Dimension Theory after Brouwer." Errata to Part I are given in Part II, p. 267. See also item 2144.

1059. Manheim, Jerome H. *The Genesis of Point Set Topology*. Oxford: Pergamon Press, and New York: Macmillan, 1964.

Problems in Fourier series, the arithmetization of analysis, and Cantorian set theory are discussed. See also items 2120, 2146.

1060. Phillips, Esther R. "Karl M. Peterson: The Earliest Derivation of the Mainardi-Codazzi Equations and the Fundamental Theorem of Surface Theory." *Historia Mathematica* 6 (1979), 137-163.

Peterson's priority was established in Reich, item 1062, but this paper gives a much more detailed description of his work, notably his 1853 dissertation in which the theorem first appeared. Peterson was later an important influence on the Moscow school of mathematics. See also item 1715.

1061. Pont, Jean Claude. *La topologie algébrique, des origines à Poincaré*. Paris: Presses Universitaires de France, 1974.

A careful account of the origins, with particular emphasis on the contributions of Euler, Gauss, Listing, and Moebius. The development of algebraic topology is less well served but is still informative. The homological ideas of Riemann (to do with integration), the emergence of the concept of manifold, and Poincaré's own work are scarcely discussed (some of these are discussed in Scholz, items 1582, 1596). Reviewed by H. Freudenthal, *Historia Mathematica* 3 (1976), 350-352. See also item 2147.

1062. Reich, Karin. "Die Geschichte der Differentialgeometrie von Gauss bis Riemann (1828-1868)." *Archive for History of Exact Sciences* 11 (1973), 273-382.

Thorough survey with careful summaries and well-chosen extracts of the major papers. The subject is divided into the theory of curves; curvature of surfaces; applicability of surfaces; geodesics; minimal surfaces; orthogonal systems; and other special surfaces. Large bibliography and a useful index. See also item 1717.

NON-EUCLIDEAN GEOMETRY

1063. Bonola, R. *Non-Euclidean Geometry. A Critical and Historical Study of Its Development*. Translated by H.S. Carslaw. Chicago: Open Court, 1912, 1938; New York: Dover, 1955.

Originally published in German (Berlin: Teubner, 1908; 2nd ed. 1919, 1921). Includes translations by G.B. Halsted of J. Bolyai's *The Science of Absolute Space* (1832), and N.I. Lobachevskii's *Geometrical Researches on the Theory of Parallels* (1840).

1064. Engel, F., and P. Stäckel. *Die Theorie der Parallellinien von Euklid bis auf Gauss*. Leipzig: Teubner, 1895. 2 vols.

See item 1557.

1065. Gray, Jeremy. "Non-Euclidean Geometry--A Re-Interpretation." *Historia Mathematica* 6 (1979), 236-258.

1066. ———. *Ideas of Space*. Oxford: Clarendon Press; New York: Oxford University Press, 1979.

1067. Miller, Arthur I. "The Myth of Gauss' Experiment on the Euclidean Nature of Physical Space." *Isis* 63 (1972), 345-348.

Bonola's account (first published 1912) is still the classic history. It surveys Greek, Moslem, and Western attempts to analyze Euclid's parallel postulate, gives a very thorough account of the work of Schweikart, Taurinus, and Gauss, as well as the successful discoveries of Bolyai and Lobachevskii; it is based on the works of Engel and Stäckel. Gray argues more strongly than Bonola for the crucial role of mathematical methods (trigonometry, differential geometry) in making the discovery possible, and less for questions of logical consistency. Kline, item 990, especially Chapters 36 and 38, stresses the importance of the physical implications, emphasized by Bolyai and Lobachevskii. Miller scotches the myth that certain of Gauss's surveys of Hannover were used to investigate the physical nature of space.

1068. Sommerville, D.M.Y. *Bibliography of Non-Euclidean Geometry*. London: Harrison & Sons, 1911. Reprinted New York: Chelsea, 1970.

Interprets its brief generously, lists articles chronologically, with an indication of the subject matter.

ANALYSIS, CALCULUS, FUNCTIONS

1069. Bernkopf, Michael. "The Development of Function Spaces with Particular Reference to Their Origins in Integral Equation Theory." *Archive for History of Exact Sciences* 3 (1966/1967), 1-96.

See item 1137.

1070. ———. "A History of Infinite Matrices." *Archive for History of Exact Sciences* 4 (1967/1968), 308-358.

These papers trace the theory of infinite matrices from Hill and Poincaré to Hilbert's theory of infinite dimensional linear operators, the subsequent elaboration of Hilbert space (by Hilbert's students), and thence to more general function spaces. It is established that, as was the case in finite dimensions, the theory of infinite dimensional operators preceded the theory of the appropriate infinite dimensional spaces. See also item 1138.

1071. Grabiner, Judith V. "The Origins of Cauchy's Theory of the Derivative." *Historia Mathematica* 5 (1978), 379-409.

Cauchy's definition, a rigorous one in $\varepsilon - \delta$ terms, is traced back via Ampère to Lagrange, and the significance of this for the form of Cauchy's foundations of analysis is discussed.

1072. Grattan-Guinness, Ivor. *The Development of the Foundations of Mathematical Analysis from Euler to Riemann*. Cambridge, Mass.: MIT Press, 1970, 186 pp.

Discusses topics in the theory of limits and convergence from the early discussions of the problem of the vibrating string by Euler, d'Alembert, and Lagrange to the rigorous methods of the successors of Cauchy and Dirichlet. Some of its arguments, notably that Cauchy might have plagiarized Bolzano, and the distinction between limit avoidance and limit achievement, have not met with general acceptance, but the book is a vivid and stimulating account of an important subject. It also contains a thorough bibliography.

1073. ———, in collaboration with J.R. Ravetz. *Joseph Fourier, 1768-1830*. Cambridge, Mass.: MIT Press, 1972, xii + 516 pp.

A thorough discussion, complete with biography and bibliography, centered on Fourier's hitherto unpublished monograph of 1807 on heat diffusion. See also item 1761.

1074. Hawkins, Thomas. *Lebesgue's Theory of Integration, Its Origins and Development*. 2nd ed. Madison: University of Wisconsin Press, 1970. Reprinted New York: Chelsea, 1975.

Considers the theory of integration from Cauchy to Riemann and finally Lebesgue, paying particular attention to problems arising from Fourier and trigonometric series, Cantorian set theory, rectification, and, above all, theoretical reformulations of the fundamental theorem of the calculus. Aimed at placing Lebesgue's work in an historical context, it succeeds by animating a half-century of vigorous debate. See also items 1147, 2113.

1075. Lützen, Jesper. "Heaviside's Operational Calculus and the Attempts to Rigorise It." *Archive for History of Exact Sciences* 21 (1979/1980), 161-200.

Surveys Heaviside's achievement, and looks at attempts to formalize it either in terms of the theory of integral transforms or as an abstract algebraic theory. The connections with Schwartz's theory of distributions are also discussed, as are the origins of that theory in J. Lützen, "The Prehistory of the Theory of Distributions," Dissertation, Aarhus University, Denmark, 1979. See also items 1435, 1498.

1076. Monna, Antonie Frans. *Dirichlet's Principle. A Mathematical Comedy of Errors and Its Influence on the Development of Analysis*. Utrecht: Oosthoek, Scheltema en Holkema, 1975, vii + 138 pp.

Traces the uses and criticisms of Dirichlet's principle by Gauss, W. Thomson (Lord Kelvin), Dirichlet, Riemann, Weierstrass, and Hilbert, with a consideration of the modern functional analytic view. Generous quotations from original sources. See also item 1994.

1077. ———. "The Concept of Function in the 19th and 20th Centuries." *Archive for History of Exact Sciences* 9 (1972/1973), 57-84.

Chiefly devoted to the discussions of Baire, Borel, and Lebesgue on the definition of a function, and on when a function is known. The roots of this debate are traced from the vibrating string problem via Dirichlet and Riemann. Compare with the A.P. Youschkevitsch article in the same journal, "The Concept of Function up to the Middle of the 19th Century," item 932.

1078. Phillips, Esther R. "Nicolai Nicolaevich Luzin and the Moscow School of the Theory of Functions." *Historia Mathematica* 5 (1978), 275-305.

Luzin's role in the emergence of the Moscow school straddled the Russian Revolution; his interests in set theory and analysis are shown to have been decisive for the careers of Alexandrov, Khinchin, Souslin, and Kolmogorov, among others. See also item 1169.

COMPLEX ANALYSIS

1079. Brill, A., and M. Noether. "Die Entwicklung der Theorie der algebraischen Funktionen in älterer und neuerer Zeit." *Jahresbericht der Deutschen Mathematiker-Vereinigung* 3 (1894), 107-566.

The only history of this immense topic, it covers early ideas (Descartes to Bézout), the French school (Lagrange, Cauchy, and Puiseux), Gauss, the theory of elliptic functions (Abel to Weierstrass), Riemann and his school, other schools of algebraic geometry (Clebsch and Gordan, Brill and Noether), the theory of singular points, and various special topics.

1080. Dugac, Pierre. "Eléments d'analyse de Karl Weierstrass." *Archive for History of Exact Sciences* 10 (1973), 41-176.

Weierstrass lectured more than he published, and his ideas were disseminated via student notes. Dugac's article is of the greatest importance in bringing together what Weierstrass taught at different times, chiefly on the theory of functions and the theory of irrational numbers. Over half the article consists of generous extracts from primary sources never published before.

1081. ———. "Problèmes d'histoire de l'analyse mathématique au XIXème siècle. Cas de Karl Weierstrass et de Richard Dedekind." *Historia Mathematica* 3 (1976), 5-19.

1082. Geppert, H. "Bestimmung der Anziehung eines elliptischen Ringes, Nachlass zur Theorie der arithmetischen-geometrischen Mittels und der Modulfunktionen von C.F. Gauss." *Ostwald's Klassiker der exakten Wissenschaften*, No. 225. Leipzig: Akademie Verlagsgesellschaft, 1927.

A thorough and fascinating reconstruction of Gauss's theory of elliptic functions reconstructed from the *Nachlass*.

1083. Houzel, C. "Fonctions elliptiques et intégrales abéliennes." In *Abrégé d'histoire des mathématiques, 1700-1900*. Edited by J. Dieudonné. Paris: Hermann, 1978, vol. II, 2-113.

Thorough survey, divided for convenience into 24 subdivisions reflecting different aspects of the mathematics, densely written but full of information on the problems as seen and solved during the nineteenth century.

1084. Krazer, A. "Zur Geschichte des Umkehrproblems." *Jahresbericht der Deutschen Mathematiker-Vereinigung* 18 (1909), 44-75.

An accurate account of the work of Euler, Gauss, Abel, and Jacobi on the inversion of elliptic integrals.

1085. Manning, K.R. "The Emergence of the Weierstrassian Approach to Complex Analysis." *Archive for History of Exact Sciences* 14 (1975), 297-383.

Concentrated on Gudermann's approach to function theory, which in turn is connected to Lambert's work on hyperbolic functions. Gudermann was Weierstrass's only significant mathematics teacher and strongly emphasized the use of power series in the study of elliptic functions; it is argued that this influenced Weierstrass's approach to complex analysis. The paper also contains a useful annotated bibliography.

1086. Bottazzini, Umberto. "Riemann's Einfluss auf L. Betti und F. Casorati." *Archive for History of Exact Sciences* 18 (1977/1978), 27-37.

1087. ———. "Le funzioni a periodi multipli nella correspondenza tra Hermite e Casorati." *Archive for History of Exact Sciences* 18 (1977/1978), 39-88.

1088. Neuenschwander, E. "The Casorati-Weierstrass Theorem." *Historia Mathematica* 5 (1978), 139-166.

1089. ———. "Der Nachlass von Casorati (1835-1890) in Pavia." *Archive for History of Exact Sciences* 19 (1978), 1-80.

1090. ———. "Casoratis Gespräche mit Kronecker und Weierstrass in Berlin im Jahre 1864." University of Aarhus: History of Science Department, 1977. Preprint.

Concerning items 1086 to 1090, the Casorati-Weierstrass theorem clarifies the nature of an essential singular point of a complex function (of a single variable), which was not at first understood. Casorati reached it in 1868 as part of a deliberate attempt to give a rigorous treatment of function theory; Weierstrass reached it independently somewhat earlier. The Russian mathematician Sokhotskii also found the result independently at the same time.

The other papers comb the Casorati *Nachlass* for unpublished material on, e.g., multiply periodic functions, essential singularities, and natural boundaries, and reveal a wealth of correspondence between Casorati and many German mathematicians on mathematical topics.

DIFFERENTIAL EQUATIONS

1091. Engelsman, Steven B. "Lagrange's Early Contributions to the Theory of First Order Differential Equations." *Historia Mathematica* 7 (1980), 7-23.

Considers Lagrange's reformulation, in 1776, of Euler's definition of a complete solution of a first order differential equation, derived from an earlier approach using the method of "variation of constants." See also item 939.

1092. Gilain, Christian. "La théorie géométrique des équations différentielles de Poincaré et l'histoire de l'analyse." Dissertation, Université de Paris I, 1977, 145 pp.

Poincaré's geometric theory of real, ordinary differential equations is described, connected to his other attempts to study the solutions to differential equations globally (in complex function theory), and compared with the local theory due initially to Cauchy. The paper throws an interesting light on mathematicians' attitudes to real and complex analysis in the 19th century. See also item 1433.

1093. McHugh, J.A.M. "An Historical Survey of Ordinary Linear Differential Equations with a Large Parameter and Turning Points." *Archive for History of Exact Sciences* 7 (1970/1971), 277-324.

See item 1437.

1094. Schlesinger, L. *Handbuch der Theorie der Linearen Differentialgleichungen*. Leipzig: B.G. Teubner, 1895-1898. 2 vols. in 3.

1095. ———. "Bericht über die Entwickelung der Theorie der Linearen Differentialgleichungen seit 1865." *Jahresbericht der Deutschen Mathematiker-Vereinigung* 18 (1909), 133-266.

Schlesinger's article and book are strictly reports, not histories, but contain a remarkable number of references to original papers; item 1095 has a bibliography of 1,742 items, listed chronologically, analyzed under 10 headings, indexed by author, and preceded by a pertinent discussion of the main lines of mathematical development. Item 1094 covers the same material in more mathematical detail, particularly on the work of Gauss, Fuchs, and Poincaré.

1096. Schlissel, Arthur. "The Initial Development of the WKB Solutions of Linear Second Order Ordinary Differential Equations and Their Use in the Connection Problem." *Historia Mathematica* 4 (1977), 183-204.

Approximate solutions and asymptotic solutions to equations with non-analytic coefficients, with particular reference to the connection problem, as presented by Gans, Jeffreys, Kramer, and others in the 1920s. See also items 1155, 1441, 1698.

Foundations

1097. Dauben, Joseph Warren. *Georg Cantor: His Mathematics and Philosophy of the Infinite*. Cambridge, Mass.: Harvard University Press, 1979, 361 pp.

Traces the evolution of Cantorian set theory from its origins in problems to do with trigonometric series to its emergence as a fully fledged mathematical theory with paradoxes and problems of its own. Also interesting for its sensitive discussion of Cantor's personality. See also items 2102, 2240.

1098. Dugac, Pierre. *Richard Dedekind et les fondements des mathématiques*. Paris: Vrin, 1976.

This surveys Dedekind's work on the foundations of analysis, set theory, and algebraic structures. Dugac also provides a thorough biography, and 140 pages of letters by Dedekind and others on mathematics and mathematicians, mostly unpublished. Dauben's review (*Isis* 69 [1978], 141-143) corrects some mistakes. See also item 2104.

Institutions

1099. Biermann, Kurt-R. *Die Mathematik und ihre Dozenten an der Berliner Universität, 1810-1920. Stationen auf dem Wege eines mathematischen Zentrums von Weltgeltung*. Berlin: Akademie-Verlag, 1973, viii + 265 pp.

Invaluable source of information on the Berlin school of mathematics, the careers of Weierstrass, Kummer, Kronecker, Schwarz, Frobenius, and many others who attended the leading nineteenth-century mathematical center. See also item 2190.

1100. Ecole Normale Supérieure. *Le centenaire de l'Ecole Normale, 1795-1895*. Paris: Hachette, 1895.

Useful history of the Ecole Normale Supérieure. Contains section on the teaching of mathematics at the school and lists of instructors and students. (Faculty included: Darboux, Duhamel, Goursat, Hermite, Lagrange, Laplace, and Tannery.) Also features portraits of mathematicians and an article on the "Influence of Galois on the Development of Mathematics" by Sophus Lie. A table of contents appears at the back of the book.

1101. Ecole Polytechnique. *Livre du centenaire, 1794-1894*. Paris: Gauthier-Villars et Fils, 1895-1897. 3 vols.

Major source on the first century of the Ecole Polytechnique. Volume 1 deals with mathematics and science; it includes an institutional history and biographical sketches of such polytechnicians as Cauchy, Chasles, Duhamel, Liouville, Poinsot,

Poisson, and Serret. Especially striking are the portraits accompanying select biographies. A table of contents appears in each volume and a comprehensive name index is at the end of volume 3.

MATHEMATICS IN THE 20th CENTURY

Since 1900, mathematics has exhibited both points of continuity and points of discontinuity with its past. It has continued to grow exponentially and to develop at ever higher levels of abstraction and generalization the areas explored in the previous century. On the other hand, many new fields have arisen which would be barely comprehensible to a nineteenth-century mathematician. Writing the history of twentieth-century mathematics is impeded by the highly technical nature of the material and by the difficulty of gaining an adequate perspective on recent developments. Nevertheless, such recent history is often intriguing, thanks to the abundance of source material and to the possibility of oral history.

General Works
(Items that Discuss Broad Areas of Twentieth-Century Mathematics)

1102. Birkhoff, Garrett, ed. "Proceedings of the American Academy Workshop on the Evolution of Modern Mathematics...." *Historia Mathematica* 2 (1975), 425-615.

The 21 articles in this symposium volume are divided into five parts: historiography, history of foundations, contemporary foundations of mathematics, algebra, and analysis. The section on historiography includes papers by J.V. Grabiner (the historian vs. the mathematician), K.O. May (the nature of good history), E. Koppelman (progress in mathematics), H. Tropp (oral history), and M.J. Crowe (laws of history). The second section consists of two papers by I. Grattan-Guinness (the Axiom of Choice; Russell) and one by H. Freudenthal (Brouwer's topology). Contemporary foundations include papers by E. Bishop (constructivism), G. Sacks (against foundations), and H. Putnam (mathematical truth). The section on algebra consists of lectures by J. Dieudonné (fusion), A. Baker (Catalan's conjecture), G. Mackey (harmonic analysis), T. Hawkins (progress without fusion), and S. Abhyankar (algebraic geometry). Finally, the essays on analysis are by F.E. Browder (functional analysis), A. Zygmund (Fourier series), J.P. Kahane (Brownian motion), and J.B. Diaz (progress).

1103. Bourbaki, Nicolas. *Eléments d'histoire des mathématiques*. 2nd ed. Paris: Hermann, 1969.

Written to illuminate the predecessors of the polycephalic Bourbaki, this volume includes material on the twentieth century in almost every one of its 23 sections. However, the sections in which this century predominates are those on non-commutative algebra, topological spaces, uniform spaces, metric spaces, function spaces, topological vector spaces, integration, and Haar measure. See also items 69, 984.

1104. Browder, Felix E., ed. *Mathematical Developments Arising from Hilbert's Problems.* (Proceedings of Symposia in Pure Mathematics, Vol. 28.) Providence: American Mathematical Society, 1976.

Includes a translation of Hilbert's original lecture on mathematical problems, delivered in Paris in 1900. There follows a short essay, with 27 subsections, on important current problems in mathematics--from foundations to mathematical physics. The bulk of the volume is devoted to 23 essays on mathematical research and problems, especially recent ones, arising from those originally posed by Hilbert.

1105. Dieudonné, Jean. "Present Trends in Pure Mathematics." *Advances in Mathematics* 27 (1978), 235-255.

An attempt to discern some patterns in contemporary trends in mathematics, such as the use of structures, category theory, and sheaves. The particular subdisciplines treated (briefly and densely) are foundations, algebraic topology, differential geometry, differential equations, Banach spaces, harmonic analysis, Lie groups, algebraic geometry, and number theory. See also Dieudonné's *Panorama des mathématiques pures: Le choix bourbachique* (Paris: Gauthier-Villars, 1977).

1106. Kline, Morris. *Mathematical Thought from Ancient to Modern Times.* New York: Oxford University Press, 1972.

Chapters 43-51 primarily treat the twentieth century. These concern mathematics circa 1900 (the study of arbitrary structures), the theory of functions of real variables (measure and integration), integral equations (Hilbert's contribution), functional analysis (Volterra, Fréchet, and von Neumann), divergent series (summability), differential geometry (tensors), abstract algebra (groups, rings, fields), topology (point-set and combinatorial), and the foundations of mathematics (set theory, mathematical logic, philosophy of mathematics). See also items 82, 867, 990, 1317, 1571, 2145.

1107. May, Kenneth O. "Growth and Quality of the Mathematical Literature." *Isis* 59 (1968), 363-371.

Analyzes the exponential growth of the mathematical literature from 1800 to 1950. Studies one subfield, the history of determinants, in more depth. His results suggest the relative proportion of six categories: new ideas (theorems, proofs, methods), applications, systematization, textbooks, duplications, and trivia. See also May's "Quantitative Growth of the Mathematical Literature," *Science* 154 (1966), 1672-1673.

1108. ———. *Index of the American Mathematical Monthly.* Washington, D.C.: Mathematical Association of America, 1977, vi + 269 pp.

Includes annual, author, and subject indexes for the first 80 volumes (1894-1973) of the leading expository journal in

American mathematics. Many historical and quasi-historical articles are indexed. An expository preface (i-vi) explains the indexes, how to use them, plus two very informative sections, "How it was done," and "How it should have been done."

Studies of Individual Mathematicians

1109. Choquet, Gustave. "Notice sur les travaux scientifiques." *Historia Mathematica* 2 (1975), 153-160.

A summary of the influences on his research and of the research itself--on topology, measure theory, potential theory, and functional analysis. (Such a "Notice" exists for many French mathematicians and can often be found in their collected works.)

1110. Dick, Auguste. "Emmy Noether, 1882-1935." Beihefte zur Zeitschrift *Elemente der Mathematik*, No. 13. Basel: Birkhäuser, 1970.

A 45-page biography, including a bibliography of her publications and a list of those students who wrote dissertations under her direction. See also item 2364.

1111. Dugac, Pierre. "Notes et documents sur la vie et l'oeuvre de René Baire." *Archive for History of Exact Sciences* 15 (1976), 297-383.

A biography of Baire, together with many previously unpublished documents on his life and work. Especially important is his correspondence with V. Volterra.

1112. Fienberg, S.E., and D.V. Hinkley. *R.A. Fisher: An Appreciation*. (Lecture Notes in Statistics, No. 1.) New York: Springer, 1980.

A series of 18 essays discussing various aspects of Fisher's researches in statistics. Among these aspects are the design of experiments, statistical estimation, conditional inference, and the analysis of variance. Includes a brief biography by J. Box, one of Fisher's daughters. See also her full-length biography, *R.A. Fisher: The Life of a Scientist* (New York: Wiley, 1978), item 2032.

1113. Frewer, Magdalene. *Das wissenschaftliche Werk Felix Bernsteins*. Göttingen: Institut für Mathematische Statistik, 1977.

A full-length study of Bernstein's contributions to set theory, number theory, geometry, statistics, and genetics.

1114. Gentzen, Gerhard. *The Collected Papers of Gerhard Gentzen*. Edited by M.E. Szabo. (Studies in Logic.) Amsterdam: North-Holland, 1969.

Contains English translations of ten of Gentzen's papers on logic and includes a brief biographical sketch. A 23-page introduction gives a historical analysis of Gentzen's researches in Hilbert's proof theory.

1115. Hardy, G.H. *A Mathematician's Apology*. Cambridge: Cambridge University Press, 1967.

This *apologia pro vita sua* contains a 50-page preface by C.P. Snow, in effect an affectionate reminiscence and biography. See also the biography by E.C. Titchmarsh in Hardy's *Collected Papers*.

1116. Hilbert, David. *Gesammelte Abhandlungen*. Berlin: Springer, 1932-1935. 3 vols.

Besides Hilbert's articles, these volumes contain essays analyzing his work on algebraic number theory (by H. Hasse, vol. I, pp. 528-535), algebra (by B.L. van der Waerden, II, 401-403), the foundations of geometry (by A. Schmidt, II, 404-414), integral equations (by E. Hellinger, III, 94-145), and the foundations of arithmetic (by P. Bernays, III, 196-216). Finally, O. Blumenthal wrote a biography (III, 388-429), which is followed by a list of Hilbert's lectures, his doctoral students, and his writings not included in these volumes.

1117. Kennedy, Hubert C. *Peano: Life and Works of Giuseppe Peano*. Dordrecht: Reidel, 1980.

Based on extensive use of published and unpublished materials, this book provides a detailed personal and intellectual biography of Peano, especially his research in mathematical logic and his efforts on behalf of an international language. Includes appendices on Peano's professors, the mathematicians in his school of logic, the papers of others that he submitted to the Academy of Sciences at Turin, and a complete bibliography of his published writings. See also Kennedy's *Selected Works of Giuseppe Peano* (London: Allen and Unwin, and Toronto: University of Toronto Press, 1973).

1118. Kolata, G.B. "Hua Lo-Keng Shapes Chinese Math." *Science* 210 (1980), 413-414.

A semi-popular biography of the most influential mathematician in contemporary China. See also the article by A. Feferman, "Professor Lo-Keng Hua on Tour," *San Francisco Chronicle*, March 15, 1981, pp. 19-21.

1119. ———. "Gian-Carlo Rota and Combinatorial Math." *Science* 204 (1979), 44-45.

A brief biography, based on an interview with Rota. Discusses the rise of combinatorics in recent years, as well as its relationship to linguistics and statistical mechanics.

1120. ———. "Isadore Singer and Differential Geometry." *Science* 204 (1979), 933-934.

A semi-popular discussion, based on an interview with Singer, of his research in global differential geometry and its role in the physics of elementary particles.

1121. Lefschetz, Solomon. "Reminiscences of a Mathematical Immigrant in the United States." *American Mathematical Monthly* 77 (1970), 344-350.

A brief but informative synopsis of his education and of his institutional affiliations as a professor of mathematics. Little discussion of his research.

1122. Montgomery, Deane. "Oswald Veblen." *Bulletin of the American Mathematical Society* 69 (1963), 26-36.

An obituary which emphasizes Veblen's education and institutional affiliations, with a brief discussion of his contributions to projective and differential geometry.

1123. Mordell, L.J. "Reminiscences of an Octogenarian Mathematician." *American Mathematical Monthly* 78 (1971), 952-961.

Anecdotal reminiscences of the education and career of an eminent number-theorist.

1124. Painlevé, Paul. *Oeuvres*. Paris: Centre National de la Recherche Scientifique, 1972. 2 vols.

Includes a preface by R. Garnier discussing Painlevé's research, a chronology of his life (both mathematical and political), a bibliography of his speeches and writings (including many prefaces to the works of others), and finally a list of the articles written about him.

1125. Ramanujan, Srinivasa. *Collected Papers*. Cambridge: Cambridge University Press, 1927.

Contains articles, mostly on number theory, by Ramanujan as well as those that he wrote jointly with G.H. Hardy. Includes two brief biographies, one by Hardy and the other by S. Aiyar and R.R. Rao. An appendix contains excerpts from Ramanujan's letters to Hardy. See also the biography *Ramanujan: The Man and the Mathematician* (1967) by S.R. Ranganathan.

1126. Reid, Constance. *Hilbert*. New York: Springer, 1970.

A detailed but anecdotal biography, based on extensive interviews with Hilbert's colleagues and students. Also includes a reprint of H. Weyl's lengthy article analyzing Hilbert's many contributions to mathematics.

1127. ———. *Courant in Göttingen and New York: The Story of an Improbable Mathematician*. New York: Springer, 1976.

A very personal history, based on numerous interviews, of Courant and his institutional associations: the Mathematical Institute in Göttingen and the Courant Institute of Mathematical Sciences in New York City.

1128. Robinson, Abraham. *Selected Papers*. Edited by H.J. Keisler, S. Körner, W.A.J. Luxemburg, and A.D. Young. New Haven: Yale University Press, 1979. 3 vols.

A 20-page biography by G.B. Seligman appears at the beginning of each of the three volumes. The first volume (on model theory and algebra) contains an introduction by H.J. Keisler, the second (on non-standard analysis and philosophy), by W.A.J. Luxemburg, and the third (on aeronautics), by A.D. Young. See also item 1615.

1129. Sierpiński, Waclaw. *Oeuvres choisies*. Warsaw: Editions Scientifiques de Pologne, 1974-1976. 3 vols.

These volumes contain 11 articles by Sierpiński on number theory, 10 on analysis, and 261 on set theory or topology. In addition to a six-page biography and a complete list of his 895 publications, there are historical essays on his work in number theory (A. Schinzel), analysis (S. Hartman), abstract set theory (S. Hartman and E. Marczewski), general topology (E. Marczewski), and measure, category, and real functions (S. Hartman). Slightly abridged English translations of the biography by K. Kuratowski and the essay by Schinzel can be found in *Acta Arithmetica* 21 (1972), 1-13.

1130. Steinhaus, Hugo. "Stefan Banach, 1892-1945." *Scripta Mathematica* 26 (1961), 93-100.

Reminiscences together with a discussion of Banach's work in real and functional analysis.

1131. Traylor, D.R., et al. *Creative Teaching: Heritage of R.L. Moore*. Houston: University of Houston, 1972.

An extremely thorough, if somewhat partisan, study of Moore's life and work. Particular attention is paid to the innovative "Moore method" of teaching and to the controversy surrounding Moore's dismissal at the age of 87. Includes a lengthy list of Moore's academic descendants and their numerous publications.

1132. Troelstra, A.S. "The Scientific Work of A. Heyting." *Logic and Foundations of Mathematics*. Groningen: Wolters-Noordhoff, 1968, 3-12.

Traces Heyting's contributions to intuitionism, both as mathematics and as philosophy of mathematics. Includes a complete bibliography of his articles and books.

1133. Ulam, S.M. *Adventures of a Mathematician*. New York: Scribner's, 1976.

This semi-popular autobiography contains reminiscences of his early years at Lwow (as a prominent member of the Polish school), his years at Princeton and Harvard, his work on the atomic bomb at Los Alamos, and his return to academe at Boulder, Colorado. Photographs from throughout this period.

1134. ———, et al. "John von Neumann, 1903-1957." *Bulletin of the American Mathematical Society* 64 (May 1958), 1-129 (supplement).

This supplement was devoted exclusively to von Neumann's life and work. Ulam supplied a brief biography and discussed aspects of von Neumann's wide-ranging researches. The remainder of the supplement consists of essays on his contributions to lattice theory (G. Birkhoff), operator theory (F.J. Murray and R.V. Kadison), measure and ergodic theory (P.R. Halmos), quantum theory (L. Van Hove), game theory and mathematical economics (H.W. Kuhn and A.W. Tucker), and finally to automata theory (C.E. Shannon). See also Halmos's anecdotal biography in the *American Mathematical Monthly* 80 (1973), 382-394.

1135. Wiener, Norbert. *I Am a Mathematician*. New York: Doubleday, 1956.

An autobiography of his mathematical career, especially concerning quantum mechanics, the atomic bomb, and cybernetics. Includes reminiscences about many European and American mathematicians. For Wiener's youth and adolescence, see his earlier autobiographical volume, *Ex-Prodigy: My Childhood and Youth* (New York: Simon and Schuster, 1953).

Concepts and Disciplines
(Studies of Ideas and Subject Areas in Mathematics)

1136. Barone, Jack, and Albert Novikoff. "A History of the Axiomatic Formulation of Probability from Borel to Kolmogorov: Part I." *Archive for History of Exact Sciences* 18 (1978), 123-190.

Argues that E. Borel's important paper of 1909 on "denumberable probability"--containing his Zero-One Law, his Strong Law of Large Numbers, and his Continued Fraction Theorem--did not clearly grasp the relationship between probability and measure theory. Also discusses the polemic of 1912 between Borel and F. Bernstein, as well as the contribution of Hausdorff in 1914. See also Novikoff and Barone, "The Borel Law of Normal Numbers, the Borel Zero-One Law, and the Work of Van Vleck," *Historia Mathematica* 4 (1977), 43-65. See also item 2030.

1137. Bernkopf, Michael. "The Development of Function Spaces with Particular Reference to Their Origins in Integral Equation Theory." *Archive for History of Exact Sciences* 3 (1966/1967), 1-96.

A detailed treatment of some early work in functional analysis. Emphasizes D. Hilbert's researches on integral equations, M. Fréchet's theory of abstract spaces, the researches on Hilbert spaces by E. Schmidt and F. Riesz, and S. Banach's creation of Banach spaces. See also item 1069.

1138. ———. "A History of Infinite Matrices: A Study of Denumerably Infinite Linear Systems as the First Step in the History of Operators Defined on Function Spaces." *Archive for History of Exact Sciences* 4 (1967/1968), 308-358.

Emphasizes the origins of infinite matrices in the work of H. Poincaré and its continuation by H. von Koch. Culminates in J. von Neumann's study of Hermitian operators on Hilbert space. See also item 1070.

1139. Biggs, Norman L., E. Keith Lloyd, and Robin J. Wilson, eds. *Graph Theory 1736-1936*. Oxford: Clarendon, 1976.

Thirty-seven extracts from publications on graph theory, together with brief historical commentary. Twelve of these are from the twentieth century, including works by H. Prüfer, G. Polya, H. Tietze, O. Frink, O. Veblen, K. Kuratowski, H. Whitney, G.D. Birkhoff, and P. Franklin.

1140. Campbell, Paul J. "The Origin of 'Zorn's Lemma.'" *Historia Mathematica* 5 (1978), 77-89.

Discusses the development of maximal principles from F. Hausdorff (1907) to H. Kneser (1950). Includes excerpts from interviews with M. Zorn.

1141. Chern, S.-S. "Differential Geometry: Its Past and Its Future." *Actes du Congrès international des mathematiciens, Paris, 1970*. Paris: Gauthier-Villars, 1971, Vol. 1, 41-53.

Surveys some of the principal notions of differential geometry, including Lie groups, fiber bundles, and the use of variational methods.

1142. De Long, Howard. *A Profile of Mathematical Logic*. London: Addison-Wesley, 1970.

Contains a 25-page annotated bibliography on mathematical logic and the philosophy of mathematics in the twentieth century.

1143. Dieudonné, Jean. "The Historical Development of Algebraic Geometry." *American Mathematical Monthly* 79 (1972), 827-866.

A summary of the main themes in the history of algebraic geometry (particularly the notions of classification and transformation), described in terms of seven periods. The bulk of the article treats the fifth period ("Development and Chaos: 1866-1920"), which gave rise to the Riemann-Roch Theorem and to the Brill-Noether theory of linear systems; the sixth period ("New Structures: 1920-1950"), which saw generalized Stokes's theorems and abstract algebraic geometry; and the seventh period ("Sheaves and Schemes: 1950-"), which extended the Riemann-Roch Theorem to higher-dimensional varieties and utilized vector bundles. See also item 1228.

1144. Grattan-Guinness, Ivor. "Wiener on the Logics of Russell and Schröder...." *Annals of Science* 32 (1975), 103-132.

Describes Wiener's doctoral dissertation (done at Harvard), comparing Russell's logical system in *Principia Mathematica* with Schröder's in *Algebra der Logik*. Discusses Russell's later comments on the thesis, together with Wiener's rejoinders.

1145. ———. "Georg Cantor's Influence on Bertand Russell." *History and Philosophy of Logic* 1 (1980), 61-93.

Discusses the influence of Cantor's set theory on Russell's mathematical logic, especially insofar as that influence concerns irrational numbers, infinitesimals, cardinal numbers, ordinal numbers, the Axiom of Infinity, the paradoxes of logic, and the Axiom of Choice. See also item 2111.

1146. Harary, Frank. "On the History of the Theory of Graphs." *New Directions in the Theory of Graphs*. Edited by F. Harary. New York: Academic Press, 1973, 1-17.

A brief discussion of 12 major theorems concerning graphs, eight of them in the twentieth century. See also his "The Explosive Growth of Graph Theory," *Annals of the New York Academy of Sciences* 328 (1979), 5-11, which takes a quantitative approach.

1147. Hawkins, Thomas. *Lebesgue's Theory of Integration: Its Origins and Development*. 2nd ed. Madison: University of Wisconsin Press, 1970. Reprinted New York: Chelsea, 1975.

Chapters 4-6 and the Epilogue concern the development of measure and integration from 1900 to 1915. Beginning with Borel's theory of measure, these chapters culminate in the Lebesgue integral, its applications (especially the Riesz-Fischer Theorem), and Radon's extension of measure to more general spaces. See also items 1074, 2113.

1148. Mackey, George W. "Harmonic Analysis as the Exploitation of Symmetry--A Historical Survey." *Rice University Studies* 64 (1978), 73-228.

A grand tour through the mathematical developments that gave rise to harmonic analysis, by the patriarch of group representation theory. Sections 14-23 concern the twentieth century and range over a dense set of topics. These sections begin with the use of the Lebesgue integral in Fourier analysis (which H. Weyl then connected to group representations), the role of quantum mechanics, the rise of unitary group representations, N. Wiener's generalized harmonic analysis, Artin's and Hecke's application of group representation theory to number theory, and finally the Pontryagin-van Kampen duality theory.

1149. Moore, Gregory H. "Beyond First-Order Logic: The Historical Interplay between Mathematical Logic and Axiomatic Set Theory." *History and Philosophy of Logic* 1 (1980), 95-137.

This essay analyzes the historical relationship between mathematical logic and set theory during the period 1870-1930. It argues that set theory influenced Schröder, Löwenheim, and Hilbert to employ logics with infinitely long expressions. The question whether a logic stronger than first-order was needed for set theory culminated around 1930 in a debate between Zermelo and Gödel over the nature of proof. An appendix contains unpublished writings by Zermelo on the foundations of mathematics.

1150. ———. *Zermelo's Axiom of Choice: Its Origins, Development, and Influence*. (Studies in the History of Mathematics and the Physical Sciences 8.) New York: Springer, 1982.

A full-length historical study of the Axiom of Choice and its role in twentieth-century mathematics. The first chapter treats the prehistory of the axiom, particularly the use of arbitrary choices in analysis and the pivotal role of Cantor. The second chapter analyzes the controversy provoked by Zermelo's proof of Cantor's conjecture that every set can be well-ordered. In response to this controversy, Zermelo axiomatized set theory and embedded his proof of the Well-Ordering Theorem within it. The third chapter deals with this axiomatization, together with applications of the axiom in algebra and analysis. The fourth and final chapter details the growing use of the axiom and of its equivalents (such as Zorn's Lemma) in diverse fields of mathematics, as well as the results of K. Gödel on the consistency of the axiom. A brief epilogue discusses developments since 1940.

1151. Mostowski, Andrzej. *Thirty Years of Foundational Studies.* New York: Barnes and Noble, 1966, 180 pp.

These 16 lectures on the development of mathematical logic and the foundations of mathematics from 1930 to 1964 are expository but technical. They concern intuitionistic logic, Gödel's Incompleteness Theorem, Tarski's researches on semantics, Herbrand's and Gentzen's results in proof theory, the theory of models in first-order and higher-order languages, and the foundations of set theory (especially constructible sets and forcing). An excellent bibliography is appended. See also item 1663.

1152. Pier, J.-P. "Historique de la notion de compacité." *Historia Mathematica* 7 (1980), 425-443.

Analyzes the development of the notion of compact set, emphasizing particularly the work of Fréchet (1906) and Hausdorff (1914). Culminates in a discussion of compactness in a topological space by means of H. Cartan's filters.

1153. Polak, Elijah. "An Historical Survey of Computational Methods in Optimal Control." *SIAM* (Society for Industrial and Applied Mathematics) *Review* 15 (1973), 553-584.

Discusses the development of optimal control algorithms from 1948 to 1973, with brief excursions into earlier work.

1154. Saaty, Thomas L., and Paul C. Kainen. *The Four-Color Problem: Assaults and Conquest.* New York: McGraw-Hill, 1977.

A detailed technical study of variants of the four-color problem from Francis Guthrie, who formulated the problem and conjectured the sufficiency of four colors in 1852, to the problem's solution by K. Appel and W. Haken in 1976. Limited historical analysis.

1155. Schlissel, Arthur. "The Initial Development of the WKB Solutions of Linear Second Order Ordinary Differential Equations and Their Use in the Connection Problem." *Historia Mathematica* 4 (1977), 183-204.

Discusses the development up to 1930 of certain approximate solutions, called the WKB solutions, to a class of ordinary second order differential equations. See also items 1096, 1441, 1698.

1156. Steen, Lynn A. "Highlights in the History of Spectral Theory." *American Mathematical Monthly* 80 (1973), 359-381.

An overview of the spectral theorem in operator theory, as well as its roots in linear algebra and integral equations. Emphasizes the work of Hilbert and the central role of quantum mechanics. Concludes with the Gelfand-Naimark theorem.

1157. Taylor, August E. "The Differential: Nineteenth and Twentieth Century Developments." *Archive for History of Exact Sciences* 12 (1974), 355-383.

Treats the definitive formulation of the differential by O. Stolz in 1893, and the development of a differential suitable for functional analysis by M. Fréchet in 1911-1912.

1158. Van der Waerden, Bartel Leendert. "On the Sources of My Book *Moderne Algebra*." *Historia Mathematica* 2 (1975), 31-40.

A detailed reminiscence about those mathematicians (especially E. Noether) who influenced the material in the most influential textbook in abstract algebra. See also his articles on Galois theory and algebraic geometry in *Archive for History of Exact Sciences* 7 (1971), 171-180; 9 (1972), 240-256.

1159. Wilder, Raymond L. "Evolution of the Topological Concept of 'Connected.'" *American Mathematical Monthly* 85 (1978), 720-726.

Analyzes nineteenth- and early twentieth-century researches on connectedness, especially those of B. Bolzano, G. Cantor, C. Jordan, W.H. and G.C. Young, N. Lennes, F. Riesz, and F. Hausdorff.

1160. Willard, Stephen. *General Topology*. London: Addison-Wesley, 1970.

Although not a history, this textbook includes 25 pages of historical notes on general topology, together with an excellent bibliography.

Communities
(Studies of Institutions and of National Traditions in Mathematics)

1161. Archibald, Raymond Clare. "History of the American Mathematical Society, 1888-1938." *Bulletin of the American Mathematical Society* 45 (1939), 31-46.

Discusses the presidents of the American Mathematical Society and related institutional matters.

1162. Dauben, Joseph Warren. "Mathematicians and World War I: The International Diplomacy of G.H. Hardy and Gösta Mittag-

Leffler as Reflected in their Personal Correspondence." *Historia Mathematica* 7 (1980), 261-288.

Discusses the epistolary efforts of Hardy and Mittag-Leffler to achieve a rapprochement between Allied and German mathematicians shortly after the First World War. Includes letters from that correspondence.

1163. Dieudonné, Jean. "The Work of Nicolas Bourbaki." *American Mathematical Monthly* 77 (1970), 134-145.

One of the founders of the polycephalic Bourbaki explains its origins, structure, and rules. Dieudonné describes how the loss of young French mathematicians during the First World War created a gap for those, such as himself, who came into mathematics soon afterward. J. Hadamard's seminar on analysis, led after 1934 by G. Julia, helped to fill this gap. Van der Waerden's *Moderne Algebra* stimulated the Bourbaki group to write their synopsis of mathematics.

1164. Fitzgerald, Anne, and Saunders MacLane, eds. *Pure and Applied Mathematics in the People's Republic of China: A Trip Report of the American Pure and Applied Mathematics Delegation*. Washington, D.C.: National Academy of Sciences, 1977.

Discusses sympathetically the role that mathematics has played in contemporary China, particularly since the Cultural Revolution. In particular, it considers the political goals of Chinese mathematicians, as well as the place of mathematics in research institutes and universities. Includes abstracts of several lectures given by Chinese mathematicians to the American delegation.

1165. Kuratowski, Kazimierz. *A Half Century of Polish Mathematics: Remembrances and Reflections*. New York: Pergamon, 1980.

Not a detailed historical analysis but a series of reminiscences, this volume treats the development of Polish mathematics from 1920 to 1970. Both the Warsaw and Lwow schools are discussed, and brief biographies of nine of the leading mathematicians in those schools are given.

1166. Kuzawa, Mary Grace. *Modern Mathematics: The Genesis of a School in Poland*. New Haven: College and University Press, 1968.

Analyzes in detail (but somewhat uncritically) the forces that led to the flowering of the Polish school of mathematics under W. Sierpiński between the two world wars. Contains a translation from Polish of Z. Janiszewski's seminal article "The Needs of Mathematics in Poland," which helped to stimulate the founding of the school and of the journal *Fundamenta Mathematicae*.

1167. May, Kenneth O., ed. *The Mathematical Association of America: Its First Fifty Years*. Washington, D.C.: The Mathematical Association of America, 1972, vii + 172 pp.

This institutional history contains six essays: the background to the founding of the M.A.A. (P.S. Jones), its first quarter-century (C.B. Boyer), its role during the Second World War (E.P. Starke), and during the next two decades (H.F. Montague), as well as its financial history (H.M. Gehman). The final third of the volume consists of ten appendices, with data and brief histories on such topics as the *Mathematics Magazine* and the Putnam Competition, by G.H. Moore. See also items 2177, 2195.

1168. Mikolás, M. "Some Historical Aspects of the Development of Mathematical Analysis in Hungary." *Historia Mathematica* 2 (1975), 304-308.

Discusses the factors that led to the flowering of Hungarian mathematics in the twentieth century.

1169. Phillips, Esther R. "Nicolai Nicolaevich Luzin and the Moscow School of the Theory of Functions." *Historia Mathematica* 5 (1978), 275-305.

An analysis of the Moscow school, founded by Luzin, and its work on the theory of functions of a real variable and on descriptive set theory. See also item 1078.

1170. Tarwater, Dalton, ed. *The Bicentennial Tribute to American Mathematics: 1776-1976*. Washington, D.C.: The Mathematical Association of America, 1977.

Five of the essays in this volume concern twentieth-century mathematics: G. Birkhoff describes leaders in American mathematics from 1891 to 1940. J. Ewing and five co-authors depict the changing status of ten leading problems from as many mathematical subdisciplines. M.S. Rees writes of her years in the Office of Naval Research. R.W. Hamming supplies a history of computing in the United States, while P.D. Lax deals with external influences on American mathematics since 1940.

1171. Tropp, Henry S. "The Origins and History of the Fields Medal." *Historia Mathematica* 3 (1976), 167-181.

A study of J.C. Fields and the committee that established the Fields Medal in 1936. Lists the recipients of the medal through 1974. For more recent recipients, see the brief articles in *Science* 202 (1978), 297-298, 505-506, 612-613, 737-739.

V. THE HISTORY OF MATHEMATICS: SUB-DISCIPLINES

The following section of this annotated bibliography divides the history of mathematics into a number of specific categories. Not all are exclusive, and there is from time to time unavoidable overlapping with earlier sections of the bibliography, although cross-referencing has been used to avoid needless repetition. Where items appear that have been annotated previously, comments are limited to remarks appropriate to the specific category in question. For the most part, general reference works, periodicals, and other non-specialized materials have not been listed here, except for items of very specific relevance.

ALGEBRA

Theory of Equations (Classical Algebra)

1172. Cajori, Florian. "Algebra in Napier's Days and Alleged Prior Inventions of Logarithms." *Napier Tercentenary Memorial Volume*. Edited by C.G. Knott. Edinburgh, 1915, 93-109.

The paper discusses the situation of algebra in Napier's days with regard to the invention of logarithms.

1172a. Day, M.S. *Scheubel as an Algebraist. Being a Study of Algebra in the Middle of the Sixteenth Century, Together with a Translation of and a Commentary upon an Unpublished Manuscript of Scheubel's Now in the Library of Columbia University*. New York: Teachers' College, Columbia University, 1926.

The first part gives a brief survey of the general status of algebra before the sixteenth century, the contributions of the sixteenth century before Scheubel's time, as well as Scheubel's life, works, and influence.

1173. Hoe, John. "Les systèmes d'équations polynômes dans le Sìyúan Yùjiàn (1303)." *Mémoirs de l'Institut des Hautés Etudes Chinoises*. Paris: Collège de France, Institut des Hautes Etudes Chinoises, 1977, 341 pp.

Hoe's aim is to appreciate the contributions of China to mathematics. The eighth chapter, "Jade Mirror of the Four Unknowns: Sìyúan yùjiàn," is consecrated to the solution of systems of linear equations by way of reducing the coefficient matrix to a triangular form. See also item 2315.

1174. Huber, Engelbert. "Historische Entwicklung von Näherungsverfahren zur Lösung algebraischer Gleichungen." Dissertation der Ludwig-

Maximilians-Universität, Munich, 1978, 121 pp.

The author collects all methods which have appeared in the history of mathematics and classifies them according to underlying principles. He analyzes the development of the theory up to the beginning of the nineteenth century. Extensive bibliography of primary sources.

1175. Karpinski, Louis C. "An Italian Algebra of the Fifteenth Century." *Bibliotheca Mathematica* 11 (1910/1911), 209-219.

Guglielmo de Lunis's algebraical achievements are investigated by means of discussing a manuscript of 1464.

1176. Kaunzner, Wolfgang. "Über einen frühen Nachweis zur symbolischen Algebra." *Österreichische Akademie der Wissenschaften, mathematisch-naturwissenschaftliche Klasse, Denkschriften* (Vienna), Vol. 116, Part 5 (1975), 3-12 + ii.

The author shows that the medieval manuscripts *Vat. lat. 4606* and *Lyell 52* (Oxford), depending on an unknown source, contain early proofs of symbolical algebra. See also item 570.

1177. ———. "Über die Entwicklung der algebraischen Symbolik vor Kepler im deutschen Sprachgebiet." *Kepler-Festschrift 1971* Munich: Mittelbayerische Druckerei- und Verlags-Gesellschaft, 1972, 175-185.

The author proves that modern algebraic notation depends with regard to several points on German mathematicians who lived in the fifteenth and sixteenth centuries in southern Germany.

1178. ———. "Über das Zusammenwirken von Systematik und Problematik in der frühen deutschen Algebra." *Sudhoffs Archiv* 54 (1970), 299-315.

The paper appreciates some mathematical achievements of the late middle ages (1460-1550). The methods of solving algebraic equations arose from a rather low level to a level never attained before. This progress culminated in an almost complete symbolization of equations of degree one, two, and reducible higher equations.

1179. Knobloch, Eberhard. "Unbekannte Studien von Leibniz zur Eliminations- und Explikationstheorie." *Archive for History of Exact Sciences* 12 (1974), 142-173.

A discussion of Leibniz's most important manuscripts which are concerned with elimination theory. Most of these manuscripts are published in Eberhard Knobloch, *Der Beginn der Determinantentheorie, Leibnizens nachgelassene Studien zum Determinantenkalkül* (Textband. Arbor scientiarum Reihe B, Bd. 2, Hildesheim: Gerstenberg, 1980), xi + 332 pp.

1180. Nesselmann, Georg Heinrich Ferdinand. *Versuch einer kritischen Geschichte der Algebra I. Die Algebra der Griechen*. Berlin: G. Reimer, 1842. Reprinted Frankfurt: Minerva, 1969.

This partly antiquated book gives a survey of Greek algebraic achievements.

1181. Rashed, Roshdi. "Résolution des équations numériques et algèbre: Šaraf-al-Dīn al-Ṭūsī, Viète." *Archive for History of Exact Sciences* 12 (1974), 244-290.

The paper's aim is to prove two theses: the work of al-Ṭūsī, concerned with numerical equations and decimal fractions, is the result of a renovation begun by the algebraists of the eleventh and twelfth centuries. Al-Ṭūsī possessed a method which Viète's method essentially followed.

1182. Sanford, Vera. *The History and Significance of Certain Standard Problems in Algebra*. New York: Teachers College, Columbia University, 1927.

The book contains a collection and historical analysis of the mathematical literature, especially of the Middle Ages and Renaissance, as far as certain problems of daily life and recreation literature are concerned. Also published by Teachers College, Columbia University as *Contributions to Education*, No. 251 (1927).

1183. Scriba, Christoph J. "Zur Entwicklung und Verbreitung der Algebra im 17. Jahrhundert." *Mededelingen uit het Seminarie voor geschiedenis van de Wiskunde en de Natuurwetenschappen aan de katholieke Universiteit de Leuven* 4 (1971), 13-22.

The article gives an outline of the development and propagation of algebra in the 16th and 17th centuries (Italian school, Stifel, Wallis).

1184. Spiesser, M. "La résolution numérique des équations dans l'histoire des mathématiques chinoises jusqu'au XIVème siècle." *Séminaire d'Histoire des Mathématiques de Toulouse* (Cahiers No. 4, 1982), 77-98.

1185. Tropfke, Johannes. *Geschichte der Elementarmathematik*. Bd. 1. Arithmetik und Algebra. Bearbeitet von Kurt Vogel, Karin Reich, Helmuth Gericke. Kapitel 3: Algebra. Berlin; New York: Walter de Gruyter, 1980, 742 pp.

The third chapter (pp. 359-511) deals with the development of algebraic thinking and procedures (the art of finding unknown quantities by means of known quantities) from the Egyptians up to the proof of the fundamental theorem by Gauss. Very useful bibliography. See also items 470, 638, 644.

1186. Viète, François. *Einführung in die neue Algebra*. Übersetzt und erläutert von Karin Reich, Helmuth Gericke. Munich: Werner Fritsch, 1973, 145 pp.

The book contains an extensive introduction to the prehistory of reckoning by means of letters and to Viète's algebraic works. See also item 762.

1187. Vogel, Kurt. "Zur Geschichte der linearen Gleichungen mit mehreren Unbekannten." *Deutsche Mathematik* 5 (1940), 217-240.

The article deals with the fifth part of the twelfth chapter of Leonardo's *Liber abaci* where many problems are solved algebraically.

1188. ———. "Die Algebra der Ägypter des mittleren Reiches." *Archeion* 12 (1930), 126-162.

The article discusses the algebraical achievements of the Egyptians of the middle empire.

1189. Wappler, Hermann Emil. "Zur Geschichte der deutschen Algebra." *Abhandlungen zur Geschichte der Mathematik* 9 (1899), 537-554.

The article analyzes a German algebra of 1481 (*codex Dresdensis* C 80) together with additional remarks of Johann Widmann of Eger.

1190. ———. "Zur Geschichte der deutschen Algebra im 15. Jahrhundert." *Programm des Gymnasiums Zwickau 1886/7*. Zwickau, 1887.

This work investigates some major contributions to algebra by German mathematicians of the 15th century.

Algebraic Structures
(Modern, Abstract Algebra)

1191. Bašmakova, Isabella G. "Sur l'histoire de l'algèbre commutative." *Revue de Synthèse* 89 (1968), 185-202.

The paper investigates the two sources of commutative algebra: the theory of divisibility (from antiquity to the 19th century) and the reciprocal influence of the theory of algebraic numbers and algebraic functions.

1192. Birkhoff, Garrett. "Current Trends in Algebra." *The American Mathematical Monthly* 80 (1973), 760-782.

Survey of the development of modern algebra up to World War I, and especially the reign of modern algebra between 1930 and 1970.

1193. Burkhardt, Heinrich. "Die Anfänge der Gruppentheorie und Paolo Ruffini." *Abhandlungen zur Geschichte der Mathematik*. Heft 6. Leipzig: B.G. Teubner, 1892, 119-159.

The article investigates the development of the ideas underlying Ruffini's papers which are concerned with group theory (theory of substitutions). It stresses the importance of Ruffini in spite of Cauchy's achievements.

1194. Clock, Daniel Arwin. "A New British Concept of Algebra: 1825-1850." Dissertation, University of Wisconsin, 1964.

Explores the development of an "appreciation of the nature of abstract algebra" in the publications of Peacock, Gregory, De Morgan, and Boole.

1195. Crowe, Michael J. *A History of Vector Analysis: The Evolution of the Idea of a Vectorial System*. Notre Dame, Ind.: University of Notre Dame Press, 1967, 270 pp.

Noteworthy history of vector algebra. Concentrates on 3-dimensional vectorial systems. Traces development from the geometrical representation of complex numbers to W.R. Hamilton's discovery of the quaternions to the physical application of quaternions by Tait and (to a limited extent) Maxwell to the construction of a modern system of vectors by Gibbs and Heaviside. Argues that the latter two "forged modern vector analysis from quaternion (not Grassmannian) elements." Features insightful chapters on Hamilton and Grassmann, and statistical analysis of comparative reception of their work. This valuable book is based on extensive research into primary and secondary sources. Absence of a bibliography is compensated for by especially rich footnotes. Footnote format has been criticized as awkward and inconvenient. See also item 1020.

1196. Dubbey, John M. "Babbage, Peacock and Modern Algebra." *Historia Mathematica* 4 (1977), 295-302.

The paper discusses the ideas of Babbage and Peacock insofar as they initiated modern algebra in England.

1197. ———. *The Mathematical Work of Charles Babbage*. Cambridge: Cambridge University Press, 1978, 235 pp.

Pioneer study. Chapter 4 covers Babbage's invention of the calculus of functions. Chapter 5 explores and tries to account for the similarities between Babbage's unpublished "Philosophy of Analysis" of the early 1820s and Peacock's *Treatise on Algebra* of 1830. Considerable duplication between Chapter 5 and Dubbey, item 1196; but the former offers a more elaborate description of "The Philosophy of Analysis." See also item 1008.

1198. Dubreil, Paul. "La naissance de deux jumelles: La logique mathématique et l'algèbre ordonnée." *Revue de Synthèse* 89 (1968), 203-209.

The paper gives an outline of the history of the development of mathematical logic and ordered algebra from Leibniz to Boole (middle of the 19th century).

1199. Freudenthal, Hans. "L'algèbre topologique en particulier les groupes topologiques et de Lie." *Revue de Synthèse* 89 (1968), 223-243.

The paper gives an outline of the development of topological algebra from 1888 (Cayley) to 1957 (Freudenthal). It is especially concerned with Lie's achievements in group theory and the divorce and later reunion of algebra and topology.

1200. Hamilton, William Rowan. *The Mathematical Papers of Sir William Rowan Hamilton*. Vol. 3: Algebra. Edited by H. Halberstam and R.E. Ingram. Cambridge: Cambridge University Press, 1967.

Judiciously compiled sampler of Hamilton's algebraic writings. Divided into four parts. Part 1 features the famous essay of 1837 on complex numbers as number-couples and on algebra as the science of pure time. The selections of part 2 cover the discovery, development, and application (to geometry and mechanics) of the quaternions. Part 3 deals with algebraic equations and includes Hamilton's account (with corrections) of Abel's proof of the insolubility of the general quintic. The only section offering substantial hitherto-unpublished material, part 4 introduces Hamilton's reflections on the icosian calculus. The primary material is supplemented by the editors' Introduction, footnotes, and appendices clarifying mathematical and historical points.

1201. Hankins, Thomas L. "Algebra as Pure Time: William Rowan Hamilton and the Foundations of Algebra." Items 1040, 2249.

1202. ———. "Triplets and Triads: Sir William Rowan Hamilton on the Metaphysics of Mathematics." Item 1041.

1203. Harkin, Duncan. "The Development of Modern Algebra." *Norsk matematisk Tidsskrift* 33 (1951), 17-26.

The paper contains some general ideas with regard to the development of modern algebra and discusses mainly Hamilton's couples and Benjamin Peirce's *Linear Associative Algebra*.

1204. Hasse, Helmut. "Geschichte der Klassenkörpertheorie." *Jahresbericht der Deutschen Mathematiker-Vereinigung* 68 (1966), 166-181.

Discussion of the history of class field theory from 1853 (Kronecker) up to 1940 (Hilbert). Bibliography of the primary literature.

1205. Hawkins, Thomas. "Hypercomplex Numbers, Lie Groups, and the Creation of Group Representation Theory." *Archive for History of Exact Sciences* 8 (1971/1972), 243-287.

The paper investigates the history of hypercomplex numbers, matrices, and Lie groups, the structure theorems of Molien and Cartan, group algebra and group representation, and Maschke's discovery of complete reducibility. Extensive bibliography.

1206. Itard, Jean. "La théorie des nombres et les origines de l'algèbre moderne." *Revue de Synthèse* 89 (1968), 165-184.

The article investigates the theory of numbers as one of the many origins of modern algebra, beginning with Euclid, Books VII, VIII, IX, and ending with the middle of the nineteenth century (Kummer's ideal numbers).

1207. Koppelman, Elaine. "The Calculus of Operations and the Rise of Abstract Algebra." *Archive for History of Exact Sciences* 8 (1971/1972), 155-242.

The paper tries to explain why following the introduction of the differential notation, the first important English contributions to mathematics were made in algebra rather than in analysis. Its thesis is that the work in algebra was a direct response of the English to a specific aspect of the work of continental analysis. This subject came to be called, by the English, the calculus of operations. See also items 1042, 1319.

1208. MacLane, Saunders. "Origins of the Cohomology of Groups." *L'Enseignement Mathématique*, Série 2, 24 (1978), 1-29.

Essay on the development of a part of contemporary mathematics: origin of the theory of the cohomology of groups, essential steps in its development, effects of this development in related fields of mathematics (for example, spectral sequences). Useful bibliography.

1209. May, Kenneth O. "The Impossibility of a Division Algebra of Vectors in Three Dimensional Space." *American Mathematical Monthly* 73 (1966), 289-291.

Oft-cited outline of work on the number of possible division algebras over the reals. Covers the period from Gauss through Bott and Milnor. Ends with a simple demonstration of the impossibility of a 3-dimensional division algebra over the reals. Useful basic bibliography.

1210. Mehrtens, Herbert. *Die Entstehung der Verbandstheorie*. (Arbor scientiarum, Reihe A, Bd. 6.) Hildesheim: Gerstenberg, 1979, 363 pp.

The book is concerned with the history of lattice theory from the middle of the 19th century up to 1940. The author understands the development of a (mathematical) theory as a social process. Extensive bibliography. Some important results with regard to Dedekind's methodological attitude are repeated in "Das Skelett der modernen Algebra. Zur Bildung mathematischer Begriffe bei Richard Dedekind," in *Disciplinae Novae. Zur Enstehung neuer Denk- und Arbeitsrichtungen in der Naturwissenschaft. Festschrift zum 90. Geburtstag von Hans Schimank*, edited by C.J. Scriba (Göttingen: Vandenhoeck & Ruprecht, 1979, 25-43). See also item 1032.

1211. Merzbach, Uta. "Development of Modern Algebraic Structures from Leibniz to Dedekind." Dissertation, Harvard University, 1964, 183 pp.

The dissertation discusses the origins of some of the concepts which characterize twentieth-century algebra and the close relationship between today's abstract algebra and the classical algebra which is concerned with real polynomials. It deals especially with Peacock's symbolical algebra, Boole's work on the algebra of logic, the axiomatic approach of Dedekind in developing his theory of ideals, and the historic basis of Dedekind's work. Useful bibliography of primary and secondary literature.

1212. Miller, George Abram. "History of the Theory of Groups to 1900." In *The Collected Works*. Urbana: University of Illinois, 1935-1938. 2 vols., here vol. 1, 427-467.

The article gives an outline of the main developments in group theory up to 1900, beginning with developments antedating the beginning of the nineteenth century.

1213. ———. "Note on the History of Group Theory during the Period Covered by This Volume." *The Collected Works*. Urbana: University of Illinois, 1935-1938. 2 vols., here vol. 2, 1-18.

Description of the development of group theory during the first decade of the twentieth century.

1214. Nagel, Ernest. "'Impossible Numbers': A Chapter in the History of Modern Logic." *Studies in the History of Ideas* 3 (1935), 429-474.

See items 1043, 2260.

1215. Nový, Luboš. *Origins of Modern Algebra*. Prague: Academia Publishing House of the Czechoslovak Academy of Sciences, 1973, 252 pp.

The book is concerned with the evolution of algebra between 1770 and 1870. This period is considered as one of the important stages in the development of modern algebra. Its aim is not to present an exhaustive history of algebra in the period considered but to discover, to illustrate, and to discuss trends in the evolution of algebra. Very extensive bibliography of the primary and secondary literature.

1216. ———. "L'école algébrique anglaise." *Revue de Synthèse* 89 (1968), 211-222.

The efforts of the members of the Cambridge analytical society in the beginning of the nineteenth century to create a new conception of algebra are discussed.

1217. Purkert, Walter. "Zur Genesis des abstrakten Körperbegriffs." *NTM-Schriftenreihe für die Geschichte der Naturwissenschaften, Technik und Medizin*, Teil 1, 10 (1) (1973), 23-37; Teil 2, 10 (2) (1973), 8-20.

The paper investigates the two roots of the field concept (theory of algebraic equations and algebraic number theory). It ends with an article of Steinitz, published in 1930, when the development of the theory of commutative fields came to a relative end.

1218. Pycior, Helena M. "Benjamin Peirce's *Linear Associative Algebra*." *Isis* 70 (1979), 537-551.

Study of the context, content, and significance of Peirce's *Linear Associative Algebra* of 1870. Explores the relationship between this work and that of such early 19th-century British algebraists as Peacock, De Morgan, and W.R. Hamilton. Argues that Peirce's acceptance and extension of the symbolical approach to algebra were facilitated by his theological belief in a correspondence between human ideas and the physical universe.

1219. ———. "The Role of Sir William Rowan Hamilton in the Development of British Modern Algebra." Dissertation, Cornell University, 1976.

Partial mathematical biography of Hamilton, emphasizing 1828 through 1851. Discusses dominant trends in British algebra of the late 18th and early 19th centuries. Investigates Hamilton's relationship with the symbolical algebraists, principally Peacock and De Morgan, and the connections among Hamilton's mathematics, philosophy, and poetry. Uses some of Hamilton's unpublished manuscripts. Noteworthy bibliography.

1220. Richards, Joan L. "The Art and the Science of British Algebra: A Study in the Perception of Mathematical Truth." Item 1047.

1221. Van der Waerden, Bartel Leendert. "Die Algebra seit Galois." *Jahresbericht der Deutschen Mathematiker-Vereinigung* 68 (1966), 155-165.

Discussion of the three sources of modern or abstract algebra: ideal theory, group theory, field theory founded by Galois (1830) and Dedekind (1871). The analysis ends in 1934.

1222. Wussing, Hans. "Zur Entstehungsgeschichte der abstrakten Gruppentheorie." *NTM-Schriftenreihe für die Geschichte der Naturwissenschaften, Technik und Medizin* 2 (1965), 1-16.

The paper investigates the origin of abstract group theory as being the first emergence of an abstract algebra structure and of fundamental importance for the emergence of modern algebra.

1223. ———. "Über den Einfluss der Zahlentheorie auf die Herausbildung der abstrakten Gruppentheorie." *Beiheft zur NTM-Schriftenreihe für die Geschichte der Naturwissenschaften, Technik und Medizin* (Leipzig, 1964), 71-88.

The article investigates the influence of number theory on the creation of abstract group theory.

1224. ———. *Die Genesis des abstrakten Gruppenbegriffes. Ein Beitrag zur Entstehungsgeschichte der abstrakten Gruppentheorie*. Berlin: VEB Deutscher Verlag der Wissenschaften, 1969, 258 pp.

The book is an exhaustive study of the development of the group concept. It begins with the implicit group theoretical thinking in the theory of algebraic equations, number theory, and geometry and ends with the abstract group concept. Very extensive bibliography of primary and secondary literature (747 items). See also item 1039.

Special National Developments

1225. Bell, Eric Temple. "Fifty Years of Algebra in America, 1888-1938." *American Mathematical Society Semicentennial Publications*. Vol. 2: Semicentennial Addresses of the American Mathematical Society. New York, 1938, 1-34. 2 vols.

The influences that appear to have been mainly responsible for the evolution of abstract algebra in America are kept in view. Bell discusses the main contributions of the most important American algebraists of the period considered (especially Miller, Dickson).

1226. Bottazzini, Umberto. "Algebraische Untersuchungen in Italien, 1850-1863." *Historia Mathematica* 7 (1980), 24-37.

Algebra was particularly important in the advance of mathematics in mid-nineteenth-century Italy (Betti, Brioschi). See also item 1019.

1227. Mal'cev, Anatolii I. "On the History of Algebra in the USSR during the First Twenty-Five Years." *Algebra and Logic* 10 (1971), 68-75.

Survey of the Russian contributions to algebra from 1900 up to the time after the Second World War. It proves that in early years algebraic researches in Russia had been slower than in a number of other branches of mathematics.

Determinant Theory, Algebraic Geometry, Invariant Theory, and Logic (With Regard to Algebra)

1228. Dieudonné, Jean. "The Historical Development of Algebraic Geometry." *American Mathematical Monthly* 79 (1972), 827-866.

The paper gives a survey of the development of the seven main ideas of algebraic geometry (classification, transformation, infinitely near points, extending the scalar, extending the space, analysis and topology in algebraic geometry, commutative algebra and algebraic geometry). They correspond to seven periods from 400 B.C. up to the present time. See also item 1143.

1229. Günther, Siegmund. *Lehrbuch der Determinanten-Theorie für Studirende*. 2nd ed. Erlangen, 1877, 1-31.

The first 31 pages give an outline of the development of determinant theory from Leibniz to Hesse. The author especially pays attention to the development of an index notation.

1230. Hayashi, Tsuruichi. "The 'Fukudai' and Determinants in Japanese Mathematics." *Tokyo Sugaku-Buturigakkwai Kizi* (Proceedings of the Tokyo Mathematico-Physical Society), Series 2, 5 (1910), 254-271. Italian version: *Giornale di matematiche di Battaglini*, Serie 3, 50 (1912), 193-211.

The paper investigates the *Fukudai* problems and the determinant theory required in the solution of these problems, and analyzes the content of Seki Kowa's Fukudai-wo-kaisuru-ho, and the method of expanding a determinant in Japanese mathematics as well.

1231. Knobloch, Eberhard. "Zur Vorgeschichte der Determinantentheorie." *Akten des III. Internationalen Leibnizkongresses, Hannover 12.-17. November 1977*. Bd. 4. Naturwissenschaften, Technik, Medizin, Mathematik. *Studia Leibnitiana Supplementa* XXII. Wiesbaden, 1982, 96-118.

The paper analyzes the development of Leibniz's index notation with regard to determinant theory, its propagation and influence.

1232. ———. "Die entscheidende Abhandlung von Leibniz zur Theorie linearer Gleichungssysteme." *Studia Leibnitiana* 4 (1972), 163-180.

The paper analyzes and edits Leibniz's most important manuscript worked out in 1684, which is concerned with the solution of systems of linear equations.

1233. Laita, Luis Maria. "A Study of the Genesis of Boolean Logic." Dissertation, University of Notre Dame, 1976, 261 pp.

The origins of Boole's discovery of algebraic logic are discussed. The final chapter gives a list of the influences which Laita sees as leading to the construction of Boole's logic. Useful bibliography.

1234. Mellberg, Edvard Julius. *Teorin för Determinant-kalkylen*. Helsinki: J.C. Frenckell & Son, 1876.

The first part (pages 1-50) entitled "Historisk" gives an outline of the development of determinant theory from Leibniz up to 1870, after the first textbooks on this theory had been published.

1235. Meyer, Franz M. "Bericht über den gegenwärtigen Stand der Invariantentheorie." *Jahresbericht der Deutschen Mathematiker-Vereinigung* 1 (1890/1891), 79-288.

A concise history of invariant theory from 1841 to 1867 (older period) and from 1868 to 1890. Special emphasis is given to equivalence and relationship of forms. See also item 1033.

1236. Mikami, Yoshio. "On the Japanese Theory of Determinants." *Isis* 2 (1914-1919), 9-36.

The paper is concerned with the Japanese contributions to determinant theory which are independent of the progress made in the West (17th and early 18th centuries).

1237. Muir, Thomas. *The Theory of Determinants in the Historical Order of Development*. London: Macmillan, 1890. 2nd ed. of vol. 1, 1906; vols. 2-4, 1906-1923. Reprinted New York: Dover, 1960, 4 vols. in 2.

No current history, but short summaries of all papers from 1693 to 1899 (Studnička) concerned with determinant theory in their chronological order of publication. Useful chronological lists of writings.

1238. ———. *Contributions to the History of Determinants 1900-1920*. London, Glasgow: Blackie and Son, 1930.

Short summaries of writings published between 1900 and 1920 (with lists of authors whose writings are reported on) and subject index.

1239. Studnička, Franz Joseph. "A.L. Cauchy als formaler Begründer der Determinantentheorie; eine literarisch-historische Studie." *Abhandlungen der königlichen bömischen Gesellschaft der Wissenschaften von Jahre 1875 und 1876, Abhandlungen der mathematisch-naturwissenschaftlichen Classe* (Prague), VIII (1877). Also printed separately, Prague: n.p., 1876, 39 pp.

The paper tries to prove that Cauchy was the formal founder of determinant theory and discusses Cauchy's relevant papers as well as those of his predecessors.

ANALYSIS

1240. Ayoub, R. "Euler and the Zeta Function." *American Mathematical Monthly* 81 (1974), 1067-1086.

Beginning with a short biography of Euler, goes on to consider Euler's attempts to evaluate what is today called the Riemann zeta function. Although unsuccessful, Euler did discover its functional equation for integral values of the independent variable.

1241. Bachmakova, Isabella G. "Les méthodes différentielles d'Archimède." *Archive for History of Exact Sciences* 2 (1964), 87-107.

The author argues that Archimedes' method of finding tangents, as in his "On Spirals," influenced geometers of the 17th century (e.g., Torricelli) in laying the basis for differential calculus.

1242. Baron, M. *The Origins of the Infintesimal Calculus*. Oxford: Pergamon Press, 1969.

See item 775.

1243. Bernkopf, Michael. "The Development of Function Spaces with Particular Reference to their Origins in Integral Equation Theory." *Archive for History of Exact Sciences* 3 (1966), 1-96.

See items 1069, 1137.

1244. ———. "A History of Infinite Matrices. A Study of Denumerably Infinite Linear Systems as the First Step in the History of Operations Defined on a Function Space." *Archive for History of Exact Sciences* 4 (1967/1968), 308-358.

See items 1070, 1138.

1245. Birkhoff, Garrett. *A Source Book in Classical Analysis*. Cambridge, Mass.: Harvard University Press, 1973.

See items 52 and 966.

1246. Bochner, S. "The Rise of Functions." In *Complex Analysis*. (Rice University Studies 56.) Houston, Tex.: William Marsh Rice University, 1970, 3-21.

Beginning with antiquity, the author surveys the subsequent development of concepts, including continuity, piecewise analytic functions, trigonometric series, orthogonal systems, and analytic continuation. A short but good general introduction.

1247. ———. "Singularities and Discontinuities." *Complex Analysis*, Vol. II. (Rice University Studies, 59.) Houston, Tex.: William Marsh Rice University, 1973, 21-41.

An eclectic study of important work on singularities in the 19th and 20th centuries with links to 18th-century mathematics and a glance back to ancient Greece. Bochner sees an anomaly in that Greek mathematics lacked a theory of continuity, while Greek philosophy *had* such a concept. Discusses Cauchy's residue formula plus Hartog's discovery of the importance of non-singularities for the theory of complex variables.

1248. Bohlman, G. "Uebersicht ueber die wichtigsten Lehrbücher der Infinitesimal-Rechnung von Euler bis auf die heutige Zeit." *Jahresbericht der Deutschen Mathematiker-Vereinigung* 6 II (1899), 91-110.

Dated, but still very useful for anyone interested in pedagogical traditions. The works are divided on a thematic rather than a strictly chronological basis. The arithmetization of analysis is followed from Euler through Lagrange, Cauchy, and finally Weierstrass. Other aspects include the naive approach, the systematic-arithmetic approach leading to Peano's work, and the philosophical and physical traditions.

1249. Borel, E. *Notice sur les travaux scientifiques de M. Emile Borel*. 2nd ed. Paris: Gauthier-Villars, 1921.

First issued in 1912, with supplements in 1918 and 1921. Written by Borel, this detailed self-evaluation of his work (71 pages) begins with a general introduction, followed by five chapters: I. Functions of a real variable; II. Functions of a complex variable; III. Entire functions and meromorphic functions; IV. Arithmetic, algebra, differential equations; V. Geometry, probability, and statistical mechanics. Bibliographies also included. The original *Notice* and supplements are reprinted in E. Borel, *Oeuvres de Emile Borel* (Paris: Editions du Centre National de la Recherche Scientifiques, 1972, in 4 vols.), Vol. 1, 119-201.

1250. Bos, Hendrik J.M. "Differentials, Higher Differentials and the Derivative in the Leibnizian Calculus." *Archive for History of Exact Sciences* 14 (1974), 1-90.

See items 776, 926.

1251. Bottazzini, Umberto. "Le funzioni a periodi multipli nella corrispondenza tra Hermite e Casorati." *Archive for History of Exact Sciences* 18 (1977), 39-88.

See item 1087.

1252. Bourbaki, Nicolas. *Eléments d'histoire des mathématiques.* Paris: Hermann, 1966.

See items 69, 984, 1103.

1253. Boyer, Carl B. *History of the Calculus and Its Conceptual Development.* New York: Dover, 1959.

See items 777, 860, 2230.

1254. ———. *A History of Mathematics.* New York: Wiley, 1967.

See items 70, 985.

1255. Brill, A., and M. Noether. "Die Entwicklung der Theorie der algebraischen Funktionen in älterer und neuer Zeit." *Jahresbericht der Deutschen Mathematiker-Vereinigung* 3 (1894), 107-566.

See item 1079.

1256. Burkhardt, Heinrich. "Trigonometrische Reihen und Integrale bis etwa 1850." *Enzyklopädie der mathematischen Wissenschaften*, II (A12). Leipzig: B.G. Teubner, 1914-1915, 819-1354.

A vast study with very thorough references. Discusses development of analytic functions via trigonometric series; also nonharmonic representations, Fourier integrals, integration of partial differential equations in two or more variables, and numerous applications.

1257. ———. "Entwicklungen nach oscillierenden Functionen und Integration der Differentialgleichungen der mathematischen Physik." *Jahresbericht der Deutschen Mathematiker-Vereinigung* 10 (1901-1908), viii + 1804 pp. This was also issued separately in 2 vols., Leipzig: B.G. Teubner, 1908, xii + xii + 1800 pp. Reprinted New York: Johnson Reprint Corp., 1960.

See items 1418, 1679.

1258. ———. "Ueber den Gebrauch divergenter Reihen in der Zeit 1750-1860." *Mathematische Annalen* 70 (1911), 189-206.

An attempt to understand how 18th- and early 19th-century mathematicians dealt with infinite series before the advent of rigorous convergence tests and the modern theory of limits. The author suggests that this branch of analysis has diverse roots, but was largely unrelated to the one field in which rigor had been well established, namely, geometry.

1259. Cajori, Florian. *A History of the Conceptions of Limits and Fluxions in Great Britain from Newton to Woodhouse.* Chicago and London: Open Court, 1919.

See item 927.

1260. ———. "The History of Notations of the Calculus." *Annals of Mathematics* 25 (1923), 1-46.

Considers development of notation from Newton and Leibniz to Peano and W.H. Young. Cajori contends that new notations have usually only been adopted when the need became imperative, while on the other hand a lack of suitable notation has often arrested progress in the subject.

1261. ———. *History of Mathematical Notations*. Chicago: Open Court, 1928-1929. 2 vols.

See items 73, 643, 986, 1821.

1262. Cleave, J.P. "The Concept of 'Variable' in 19th-Century Analysis." *British Journal for the Philosophy of Science* 30 (1979), 266-278.

Cleave defends the position of Robinson and Lakatos, who argue that the foundations of Cauchy's calculus can best be understood as an anticipation of the approach of non-standard analysis. This interpretation was questioned in item 1280.

1263. Crowe, Michael J. *A History of Vector Analysis*. London: University of Notre Dame Press, 1967.

See items 1020, 1195.

1264. Dauben, Joseph Warren. *Georg Cantor*. Cambridge, Mass.: Harvard University Press, 1978.

See items 1097, 2102, 2240.

1265. ———. "The Trigonometric Background to Georg Cantor's Theory of Sets." *Archive for History of Exact Sciences* 7 (1971), 181-216.

A detailed examination of how Cantor's early work on Fourier series led him to his revolutionary investigations of the real number line and ultimately the theory of sets.

1266. Davis, P.J. "Leonard Euler's Integral: A Historical Profile of the Gamma Function." *American Mathematical Monthly* 66 (1959), 849-869.

Describes Euler's early work, its generalization to the complex plane by Gauss, infinite factorizations by Weierstrass, the function-theoretic investigations of Hadamard, and the culminating characterization of the gamma function as given by the Bohr-Mollerup-Artin theorem.

1267. Deakin, Michael A.B. "The Development of the Laplace Transform, 1737-1937, I. Euler to Spitzer, 1737-1800." *Archive for History of Exact Sciences* 25 (1981), 343-390.

An attempt to provide a more balanced narrative history of the development of the Laplace transform, this study also clearly establishes Euler's priority in its discovery. See also item 1681.

1268. Denjoy, A. "Arnaud Denjoy: évolution de l'homme et de l'oeuvre." *Astérique*. Edited by G. Choquet. Paris: Société Mathématique de France, 1975, 28-29.

Henri Cartan provides the biographical appreciation. Denjoy's own "Mon oeuvre mathématique: sa genèse et sa philosophie" and "Le mécanisme des opérations mentales chez les mathématiciens" are also included, as is a bibliography of Denjoy's works.

1269. Dieudonné, Jean. "The Work of Nicolas Bourbaki." *American Mathematical Monthly* 77 (1970), 134-145.

See item 1163.

1270. ———. *Abrégé d'histoire des mathématiques, 1700-1900*. Paris: Hermann, 1978. 2 vols.

Written by numerous scholars under Dieudonné's supervision. Volume I contains chapters on 18th-century analysis, complex variable theory, and foundations of analysis. Volume II has chapters on elliptic functions and Abelian integrals, functional analysis, and integration and measure theory.

1271. Dubbey, John M. "Cauchy's Contribution to the Establishment of the Calculus." *Annals of Science* 22 (1966), 61-67.

Cauchy's definition of "limit" for the foundation of calculus is criticized as not being analytic and for not giving a proper arithmetical procedure for finding limits.

1272. Dugac, Pierre. *Histoire du théorème des accroissments finis*. Paris: Université Pierre et Marie Curie, 1979.

A study of concepts and related results leading to the proof of the Mean Value Theorem. Dugac argues that Dini was the first to give a rigorous proof; he also documents that Weierstrass was directly influenced by Bolzano by making use of the former's unpublished lecture notes and other materials. Related results include Rolle's Theorem, the Intermediate Value Theorem, Cauchy's criterion, and the Bolzano-Weierstrass Theorem.

1273. ———. *Limite, point d'accumulation, compact*. Paris: Université Pierre et Marie Curie, 1980.

Of the three topics given in the title, limits receive the most attention. Dugac begins by a consideration of the zeros of an analytic function. The work of Bolzano and Cantor are featured, making ample use of unpublished materials.

1274. ———. "Eléments d'analyse de Karl Weierstrass." *Archive for History of Exact Sciences* 10 (1973), 41-176.

See item 1080.

1275. ———. *Sur les fondements de l'analyse au XIXe siècle*. Louvain: Université Catholique de Louvain, 1980.

A history of the foundations of analysis from d'Alembert to Cantor and Dedekind. Includes bibliography and topics for examinations.

1276. ———. "Notes et documents sur la vie et l'oeuvre de René Baire." *Archive for History of Exact Sciences* 15 (1976), 297-383.

See item 1111.

1277. ———. "Charles Méray (1835-1911) et la notion de limite." *Revue d'histoire des sciences et leurs applications* 23 (1970), 330-350.

Analyzes Méray's major contribution to mathematics, namely, his work dating from 1869 on the theory of irrational numbers.

1278. Duren, W.L. "The Development of Sufficient Conditions in the Calculus of Variations." *University of Chicago Contributions to the Calculus of Variations*. Vol. 1. Chicago: Chicago University Press, 1930, 245-349.

Gives background from Euler to Jacobi leading up to the first sufficiency proofs of Weierstrass and Scheefer. Discusses the stronger results obtained by Kneser, Hilbert, Osgood, Carathéodory, Bolza, and Tonelli, as well as work on the Lagrange and Mayer problems and various isoperimetric problems. Related works by Mayer, Kneser, Bliss, Hahn, and Morse handle the problem of obtaining sufficient conditions when endpoints are variable.

1279. Enneper, A. *Elliptische Funktionen: Theorie und Geschichte*. 2nd ed. Halle: L. Nebert, 1890.

A seven-page historical introduction followed by over 500 pages of text. Still this work will be useful for researchers who want *detailed* information about who did what in this important field of early 19th-century analysis. The treatment is topical and the author cites his references very thoroughly.

1280. Fisher, G. "Cauchy's Variables and Orders of the Infinitely Small." *British Journal for the Philosophy of Science* 30 (1979), 261-265.

Fisher criticizes the non-standard analysis interpretation of Cauchy's work in the foundations of calculus as given by Robinson, Lakatos, and Cleave. He argues that Cauchy's use of variable quantities and infinitesimals is fundamentally unclear. Cleave offers a rebuttal in item 1262.

1281. ———. "The Infinite and Infinitesimal Quantities of du Bois-Reymond and Their Reception." *Archive for History of Exact Sciences* 24 (1981), 101-163.

Discusses reactions of Cantor, Dedekind, Peano, Russell, Pringsheim, Stolz, Borel, Hardy, Hausdorff, et al. Issues center around the validity and/or utility of theories of the infinitely large and small. Fisher cites this as a case study indicating the shortcomings of Lakatos's theory in *Proofs and Refutations*, ed. J. Worrall and E.G. Zahar (Cambridge, Eng.: Cambridge University Press, 1976).

1282. Flett, T.M. "Some Historical Notes and Speculations Concerning the Mean-Value Theorems of the Differential Calculus." *Bulletin of the Institute of Mathematics and Its Applications* 10 (1974), 66-72.

Explains why the attempts of Lagrange, Cauchy, and Jordan to prove the Mean-Value Theorem failed.

1283. Fréchet, M. "La vie et l'oeuvre de Emile Borel." *L'Enseignement Mathématique* 11 (1965), 1-94.

A biography of 20 pages followed by a lengthy discussion of the general character of Borel's work. Consideration is given to his work in number theory, infinite series, set theory, measure theory, real variables, complex variables, differential equations, geometry, and applied mathematics.

1284. ———. *Notice sur les travaux scientifiques de M. Maurice Fréchet*. Paris: Hermann, 1933.

Begins with a list of university degrees, teaching positions, and scientific awards and distinctions. Mathematical publications (pp. 7-25) are divided into seven categories: applied mathematics (7 publications); probability and errors (28 publications); geometry (29 publications); classical analysis (40 publications); and general themes (including pedagogy) (4 papers); pages 27-101 constitute Fréchet's own "Notice sur les travaux scientifiques."

1285. Freudenthal, H. "Did Cauchy Plagiarize Bolzano?" *Archive for History of Exact Sciences* 7 (1971), 375-392.

Freudenthal answers no, and in so doing, he attempts to refute Grattan-Guinness's argument in item 1293.

1286. Funk, P. "Bolzano als Mathematiker." *Sitzungsberichte Öesterreicher Akademie der Wissenschaften* (Vienna) 252, Teil 5 (1967), 121-134.

1287. Gibson, G.A. "James Gregory's Mathematical Work." *Proceedings of the Edinburgh Mathematical Society* (1922/1923), 2-35.

Drawn primarily from Rigaud's *Correspondence of Scientific Men of the Seventeenth Century*, vol. 2, and the *Commercium Epistolicum*, compiled by John Collins. Attempts to determine Gregory's role in the early development of calculus.

1288. Goldstine, Herman H. *A History of Numerical Analysis from the 16th through the 19th Century*. New York, Heidelberg, Berlin: Springer, 1977.

See items 798, 863, 1947.

1289. ———. *A History of the Calculus of Variations from the 17th through the 19th Century*. New York, Heidelberg, Berlin: Springer, 1980.

See items 779, 864.

1290. Grabiner, J.V. "The Origins of Cauchy's Theory of the Derivative." *Historia Mathematica* 5 (1978), 379-409.

See item 1071.

1291. ———. *The Origins of Cauchy's Rigorous Calculus*. Cambridge, Mass.: MIT Press, 1981.

See item 903.

1292. Grattan-Guinness, Ivor. "Berkeley's Criticism of the Calculus as a Study in the Theory of Limits." *Janus* 56 (1969), 215-227.

The explication of Berkeley's criticism of the calculus is related to his theological views and philosophical Idealism. Mention is also made of the reaction of contemporaries to Berkeley's *The Analyst*.

1293. ———. "Bolzano, Cauchy and the 'New Analysis' of the Early Nineteenth Century." *Archive for History of Exact Sciences* 6 (1970), 372-400.

Argues that Cauchy plagiarized Bolzano based on the coincidence of ideas in the latter's 1817 pamphlet and in the *Cours d'analyse*, the social scene surrounding Parisian mathematics, and Cauchy's personality. This argument has been rebutted in item 1285.

1294. ———. *The Development of the Foundations of Mathematical Analysis from Euler to Riemann*. Cambridge, Mass.: MIT Press, 1970.

See items 865, 1072.

1295. ———. "Preliminary Notes on the Historical Significance of Quantification and of the Axioms of Choice in the Development of Mathematical Analysis." *Historia Mathematica* 2 (1975), 475-488.

A programmatic sketch of topics for future study and their significance for the history of mathematics.

1296. Hadamard, J. *Notice sur les travaux scientifiques de M. Jacques Hadamard*. Vol. 1. Paris: Gauthier-Villars, 1901. Reprinted, with vol. 2, Paris: Hermann, 1912.

Volume 1 covers Hadamard's works from 1884 to 1901, with a bibliography on pp. 7-11; volume 2 includes publications from 1901 to 1912 and has a bibliography on pp. 7-13.

1297. Hall, A.R. *Philosophers at War: The Quarrel Between Newton and Leibniz*. Cambridge: Cambridge University Press, 1980.

See item 831.

1298. Hankins, Thomas. *Jean d'Alembert*. Oxford: Clarendon Press, 1970.

An intellectual biography stressing d'Alembert's work in mechanics and his place within the *philosophes'* circle. Argues convincingly that d'Alembert was a Cartesian rationalist whose mathematics was rooted in the geometrization of nature. Some discussion of the wave equation and the foundation of his mathematics.

1299. ———. *Sir William Rowan Hamilton*. Baltimore, London: Johns Hopkins University Press, 1980.

Hankins skillfully weaves together personal, social, intellectual, and scientific aspects of Hamilton's life. His chapter on Hamilton's principle and the optical-mechanical analogy discusses the importance this had for Schrödinger's work in wave mechanics. The background to Hamilton's discovery of the quaternions is also very interesting.

1300. Hawkins, Thomas. *Lebesgue's Theory of Integration. Its Origins and Development.* Madison: University of Wisconsin Press, 1970.

See items 1074, 1147, 2113.

1301. ———. "Non-Euclidean Geometry and Weierstrassian Mathematics: The Background to Killing's Work on Lie Algebras." *Historia Mathematica* 7 (1980), 289-342.

A groundbreaking study of Killing's work as it emerged from his training in the Weierstrassian approach to the foundations of mathematics. The Berlin tradition is contrasted with the physical-mathematical approach that dominated Göttingen, and whose leading spokesman was Klein. Hawkins gives a close-up look at how these traditions affected 19th-century outlooks and research.

1302. Hellinger, E. "Hilbert's Arbeiten ueber Integralgleichungen und unendliche Gleichungssysteme." In D. Hilbert's *Gesammelte Abhandlungen.* Vol. 3. Berlin: Springer, 1935, 94-145.

A short introduction discusses the work of Sturm and Liouville, the integral equation theory of Volterra and Fredholm, and the work of Poincaré and Hill on infinite determinants. This is followed by a thorough examination of Hilbert's work and related efforts by the author, Toeplitz, Riesz, Weyl, von Neumann, Friedrichs, Wintner, Banach, Kneser, Courant, et al.

1303. Hewitt, E., and R. Hewitt. "The Gibbs-Wilbraham Phenomenon: An Episode in Fourier Analysis." *Archive for History of Exact Sciences* 21 (1979), 129-160.

Discusses the mathematics as well as the history behind this curious subject in the history of Fourier analysis.

1304. Hofmann, Joseph E. *Leibniz in Paris, 1672-1676.* Translated by A. Prag and D.T. Whiteside. Cambridge: Cambridge University Press, 1974.

See item 832.

1305. ———. "Johann Bernoulli, der Propagator der Infinitesimalmethoden." *Praxis Mathematica* 9 (1967), 209-212.

See item 880.

1306. ———. "Ueber Jakob Bernoulli's Beiträge zur Infinitesimalmathematik." *L'Enseignement Mathématique* 2 (1956), 61-171.

See items 805, 879.

1307. ———. "Aus der Frühzeit der Infinitesimalmethoden: Auseinandersetzung um der algebraischen Quadratur algebraischer Kurven in der zweiten Hälfte des 17. Jahrhunderts." *Archive for History of Exact Sciences* 2 (1965), 271-343.

Traces the early history of the problem of determining conditions for the integration of an algebraic function by means of algebraic functions. Gives detailed analysis of the contributions of Leibniz, Tschirnhaus, Johann Bernoulli, Craig, et al.

1308. Huke, A. "An Historical and Critical Study of the Fundamental Lemma in the Calculus of Variations." *University of Chicago Contributions to the Calculus of Variations*, Vol. 1. Chicago: Chicago University Press, 1930, 45-160.

Discusses the lemma and early proofs (1823-1870) attempted by Dirksen, Sarrus, Todhunter, et al. The first modern proof was by Heine (1870), followed by du Bois-Reymond, Weierstrass, Zermelo, Kneser, Landau, et al. The problem was modified by du Bois-Reymond and proved in this form by Hilbert and several others. Analogues of the fundamental lemma were established by Haar and Schauder.

1309. Jarník, V. *Bolzano and the Foundations of Mathematical Analysis*. Prague: Society of Czechoslovak Mathematicians and Physicists, 1981.

Translations of papers on Bolzano's contributions to the foundations of analysis, function theory, and his definition of a continuous, nowhere-differentiable function. Illustrated and with an introduction by J. Folta.

1310. Jensen, C. "Pierre Fermat's Method of Determining Tangents of Curves and Its Applications to the Conchoid and the Quadratrix." *Centaurus* 14 (1969), 72-85.

An attempt to reconstruct Fermat's incomplete solution to the problem of constructing tangents to these curves.

1311. Jourdain, P.E.B. "The Theory of Functions with Cauchy and Gauss." *Biblioteca Mathematica* 3 (1905), 190-207.

Considers Cauchy's 1814 memoir on definite integrals with regard to the origin of his treatment of continuity, imaginary numbers, and the theory of complex integration. This work is contrasted with Gauss's investigations on the integration of functions of a complex variable.

1312. ———. "The Origins of Cauchy's Conception of the Definite Integral and of the Continuity of a Function." *Isis* 1 (1913), 661-703.

The evolution of the function concept and continuity is traced from the vibrating string problem, through Fourier series, to Cauchy's work. A broad, non-technical account that discusses the contributions of Bolzano and De Morgan, and stresses Fourier's work as paving the way to a rigorous, pure mathematics.

1313. Juškevič, Adolf P. "Gottfried Wilhelm Leibniz und die Grundlagen der Infinitesimalrechnung." *Studia Leibnitiana* 2 (1969), 1-19.

Emphasis is given to Leibniz's facility in developing useful symbols, as well as to his views on infinitesimals. A French version appears in *Organon* 5 (1968), 153-168.

1314. Kahane, J.-P. "Leopold Fejér et l'analyse mathématique au début du XXe siècle." *Cahiers du Séminaire d'Histoire des Mathématiques* 2 (1981), 67-84.

Evaluates the significance of Fejér's work and provides a bibliography.

1315. Kitcher, P. "Fluxions, Limits, and Infinite Littleness: A Study of Newton's Presentation of the Calculus." *Isis* 64 (1973), 33-49.

Argues that Newton's use of fluxions, infinitesimals, and the method of prime and ultimate ratios, should be understood as responses to distinct mathematical needs. Thus the universality of Newton's genius can be seen as reflecting the complexities of the mathematics of his day.

1316. Klein, F. *Vorlesungen über die Entwicklung der Mathematik im 19. Jahrhundert*. Berlin: Springer Verlag, 1926-1927. 2 vols. Reprinted in 1 vol. New York: Chelsea, 1967.

This classic work is by now rather dated, but still offers insights into the period that give it an enduring value. The chapter on Gauss borders on hagiography, but is at the same time highly suggestive as an interpretive key for the development of mathematics throughout the 19th century. The treatment of work by Cauchy, Dirichlet, Abel, and Jacobi is somewhat brief, whereas Riemann and Weierstrass are explored in detail. Klein's geometric outlook predominates throughout, and this has somewhat influenced his choice of subject matter. Thus analysis is largely subordinated to the development of mechanics and mathematical physics. Some topics (e.g., elliptic and automorphic functions) do, however, receive considerable attention. See also items 989, 1570, 2086.

1317. Kline, Morris. *Mathematical Thought from Ancient to Modern Times*. New York: Oxford University Press, 1972.

No topic receives as much attention in this book as does analysis. The chapters on Mathematization of Science and The Creation of Calculus go hand in hand. Ordinary and Partial Differential Equations occupy four chapters (one each for both the 18th and 19th centuries), while The Calculus of Variations comprises two (again one each for the 18th and 19th centuries). Other chapters include Complex Variables, Integral Equations, Real Variables, and Functional Analysis. See also items 82, 867, 990, 1106, 1571, 2145.

1318. Kolman, A. *Bernard Bolzano*. Berlin: Akademie Verlag, 1963.

Originally published as Ernest Kol'man, *Bernard Bol'tsano* (Moscow: Akademii Nauk CCCR, 1955, 223 pp.), with illustrations and portraits. Includes bibliographies.

1319. Koppelman, Elaine. "The Calculus of Operations and the Rise of Abstract Algebra." *Archive for History of Exact Sciences* 8 (1971), 155-242.

See items 1042, 1207.

1320. Langer, R.E. "Fourier Series: The Genesis and Evolution of a Theory." *American Mathematical Monthly* 54, Part 2 (1947), 1-81.

A leisurely exposition of the subject's development beginning with the vibrating string problem, through Fourier's study of heat diffusion, to the study of characteristic values, orthogonality conditions, Green's function, and the formal Fourier series representation of an arbitrary function.

1321. Laugwitz, D. "Zur Entwicklung der Mathematik des Infinitesimalen und Infiniten." *Jahrbuch Ueberblicke Mathematik* (1975), 45-50.

Brief history of the development of infinitesimals and infinities from Euler to Riemann, with emphasis on Euler's work on divergent series. A sequel to an article in the same journal by W.A.J. Luxemburg (pp. 31-44).

1322. Lebesgue, Henri. "À propos de quelques travaux mathématiques récents." *L'Enseignement Mathématique* 17 (1971), 1-48.

Based on unpublished notes from 1905. Discusses the development of real variable theory including contributions of Euler, Fourier, Cauchy, Peano, Cantor, Dini, Lipschitz, Arzela, Osgood, Jordan, Baire, and du Bois-Reymond.

1323. ———. *Notice sur les travaux scientifiques de M. Henri Lebesgue*. Toulouse: Edouard Privat, 1922.

A short summary of the development of real variable theory during the 19th century is followed by chapters detailing Lebesgue's work on measure theory and integration, representation of functions by infinite series, the problems of Dirichlet and Plateau in the calculus of variations, topology, geometry, and miscellaneous other areas.

1324. Lévy, P., et al. "La vie et l'oeuvre de J. Hadamard." *L'Enseignement Mathématique* 13 (1967), 1-72.

Articles covering Hadamard's work in functional analysis (Lévy), function theory (Mandelbrojt), number theory (Mandelbrojt), partial differential equations (Malgrange), and geometry (Malliavin), plus a complete bibliography.

1325. Lützen, Jesper. "Heaviside's Operational Calculus and the Attempts to Rigorise It." *Archive for History of Exact Sciences* 21 (1979), 161-200.

See items 1075, 1435, 1498.

1326. Mackey, G.W. "Harmonic Analysis as the Exploitation of Symmetry: A Historical Survey." *Bulletin of the American Mathematical Society* 3 (1980), 543-698.

See item 1148.

1327. Manning, K.R. "The Emergence of the Weierstrassian Approach to Complex Analysis." *Archive for History of Exact Sciences* 14 (1975), 297-383.

See item 1085.

1328. Markusewitsch, A.I. *Skizzen zur Geschichte der analytischen Funktionen*. Berlin: VEB Deutscher Verlag der Wissenschaften, 1955.

See item 936.

1329. ———. "Emotions of a Young Mathematician (a Letter from N.N. Luzin to M. Ja. Vygodskiĭ." *Matematika v Skole* 6 (1976), 25-32. In Russian.

Luzin characterized his letter as a "psychological document of the state of mind at an early stage of development." It gives an image of mathematics at Moscow University in the early 20th century, while discussing certain implications of the function and set theory of Bolzano and Weierstrass.

1330. Mathews, Jerold. "William Rowan Hamilton's Paper of 1837 on the Arithmetization of Analysis." *Archive for History of Exact Sciences* 19 (1978), 177-200.

Examines the content of Hamilton's paper and offers several reasons for its lack of influence on mathematicians of his day. See also item 1015.

1331. McHugh, J.A.M. "An Historical Survey of Ordinary Differential Equations with a Large Parameter and Turning Points." *Archive for History of Exact Sciences* 7 (1971), 277-324.

See item 1437.

1332. McShane, E.J. "Recent Developments in the Calculus of Variations." *American Mathematical Society Semicentennial Publications*. Vol. 2. New York: American Mathematical Society, 1938, 69-97.

A discussion of important developments during the period 1913-1938. Topics include the Bolza problem, Morse theory, work of Hilbert, Bliss, Tonelli, Carathéodory, et al.

1333. Medvedev, F.A., ed. *Frantsuzskaya Shkola Teorii Funksii i Mnozhestv, na Rubezhe XIX-XX VV*. Moscow: Nauka, 1976.

The French School of the Theory of Functions and Sets at the Turn of the XIX-XX Centuries centers around the work of Borel, Baire, and Lebesgue. To a lesser extent it also considers Fréchet, Denjoy, and others. The impact of the French school on mathematics in Italy, England, Russia, and other countries is also considered.

1334. ———. "Henri Lebesgue's Works on the Theory of Functions (On the Hundredth Anniversary of His Birth)." *Uspehi Matematičeskii Nauk* 30 (1975), 227-238. In Russian.

An historical examination of Lebesgue's theory of the integral, set in the mathematical context of the late 19th century.

1335. Mittag-Leffler, G. "An Introduction to the Theory of Elliptic Functions." *Annals of Mathematics* 24 (1922-1923), 271-351.

A useful introduction to the subject as it developed after the fundamental discoveries of Abel and Jacobi. The lion's share of the attention, however, is devoted to Weierstrass's work.

1336. Monna, Antonie Frans. "The Concept of Function in the 19th and 20th Centuries, in Particular with Regard to Discussions between Baire, Borel and Lebesgue." *Archive for History of Exact Sciences* 9 (1972), 57-84.

See item 1077.

1337. Oberschelp, A. "Die Entwicklung der Leibnizschen Idee der unendlich kleinen Grössen in der modernen Mathematik." *Studia Leibnitiana Supplementa* 2 (1969), 27-33.

Emphasizes Leibniz's contention that infinitesimals cannot be compared with finite numbers, and relates this to a consideration of A. Robinson's non-standard analysis.

1338. Ovaert, J.L. "La thèse de Lagrange et la transformation de l'analyse." *Philosophie et calcul de l'infini*. Paris: Maspero, 1976, 157-200.

Lagrange's formal power series approach to analysis is compared with the work of his predecessors, L'Hospital and Maclaurin. Includes appendices outlining the contents of their calculus texts and reproducing the sections pertaining to the foundations of these three systems.

1339. Parkinson, G.H.R. "Science and Metaphysics in the Leibniz-Newton Controversy." *Studia Leibnitiana Supplementa* 2 (1969), 79-112.

Stresses that the differences between Newton and Leibniz were primarily philosophical. Whereas Newton proceeded by induction, Leibniz developed his own ideas as early as 1703-1705 in conjunction with objections to Locke's arguments in his *New Essays*.

1340. Pesin, I.M. *Classical and Modern Integration Theories*. New York: Academic Press, 1970.

A useful reference, giving a bird's-eye view of developments from Cauchy to Lebesgue and beyond. Generally expository, but with occasional proofs, the work of Borel, Lebesgue, Young, Stieltjes, Denjoy and Khinchin, Perron, and Daniell form the main body of the book.

1341. Pierpont, J. "Mathematical Rigor, Past and Present." *Bulletin of the American Mathematical Society* 34 (1928), 23-53.

Illustrates the relativism of mathematical rigor by surveying accepted standards in different periods: calculus in the 17th and 18th centuries, Cauchy and Weierstrass on analysis, Kronecker on algebra, Cantor, Dedekind, and Poincaré on set theory, and Peano, Hilbert, Russell, and Frege on mathematical logic are among the examples considered.

1342. Plancherel, M. "Le développement de la théorie des séries trigonométriques dans le dernier quart de siècle." *L'Enseignement Mathématique* 24 (1924-1925), 19-58.

An update on progress in this field supplementing Burkhardt's (see item 1256). Topics include Riesz-Fischer Theorem, plus work of Lebesgue, Hardy and Littlewood, Cesaro, Fejér, W.H. Young, de la Vallée-Poussin, et al.

1343. Porter, T.I. "A History of the Classical Isoperimetric Problem." *University of Chicago Contributions to the Calculus of Variations* 2 (1933), 475-517.

Begins with origins and early proofs in antiquity. Geometric proofs of the isoperimetric property of the circle were accomplished by Steiner, Blaschke, and Bonneson. Hurwitz gave a proof using Fourier series, and Weierstrass, Bolza, Tonelli, and Bonneson gave proofs using the calculus of variations. A bibliography of work on the problem is included.

1344. Reiff, R.A. *Geschichte der unendlichen Reihen*. H. Lauppsche Buchhandlung, 1889. Reprinted Wiesbaden: Martin Sändig, 1969.

See items 784, 935.

1345. Riesz, F. "L'évolution de la notion d'integrale depuis Lebesgue." *Annales de l'Institut Fourier* 1 (1949), 29-42.

This short paper reflects a lecture Riesz delivered in Paris, and also at Grenoble, in the same year. He traces the development of the integral from Lebesgue, including developments made by Fatou, Stieltjes, Young, Perron, Lusin, Egoroff, Denjoy, de la Vallée-Poussin, Daniell, Carathéodory, and Stone, among others. No bibliography.

1346. Ross, B. "The Development of Fractional Calculus 1695-1900." *Historia Mathematica* 4 (1977), 75-89.

Beginning with the work of L'Hospital and Leibniz, the author proceeds to consider the contributions of Euler, Laplace, Abel, Liouville, Riemann, Sarim, and Letnikov, among others. He ends with a remark on Heaviside's operators.

1347. Sachse, A. "Essai historique sur la représentation d'une fonction arbitraire d'une seule variable par une série trigonométrique." *Bulletin des Sciences Mathématiques* 4 (1880), 43-64, 83-112.

Discusses work of Riemann, Dirichlet, du Bois-Reymond, Weierstrass, Heine, Lipschitz, Schwarz, et al.

1348. Schlesinger, L. "Bericht ueber die Entwicklung der Theorie der linearen Differentialgleichungen seit 1865." *Jahresbericht der Deutschen Mathematiker-Vereinigung* 18 (1909), 133-266.

See item 1095.

1349. Schlissel, Arthur. "The Development of Asymptotic Solutions of Linear Ordinary Differential Equations, 1817-1920." *Archive for History of Exact Sciences* 16 (1976/1977), 307-378.

Considers the origins of the theory in work of Carlini, Liouville, Green, Stokes, and Hankel. Poincaré is the central figure, as his contributions made this subject a distinct new branch of modern mathematics. His work is examined along with that of Birkhoff, Horn, Schlesinger, Tamarkin, and Debye. See also items 1440, 1697.

1350. Schneider, Ivo. "Der Mathematiker Abraham de Moivre (1667-1754)." *Archive for History of Exact Sciences* 5 (1968/1969), 177-317.

See items 890, 2040.

1351. Scott, J.F. *The Mathematical Work of John Wallis*. Oxford: Oxford University Press, 1938.

See item 858.

1352. Scriba, Christoph J. "The Inverse Method of Tangents: A Dialogue Between Leibniz and Newton (1675-1677)." *Archive for History of Exact Sciences* 2 (1964), 113-137.

The author stresses that in this early correspondence the geometric view still predominates. The use of integral tables for the quadrature of curves is recognized by both parties as an important new tool for analysis, but the motivation still comes from classical rather than analytic geometry.

1353. ———. "Neue Dokumente zur Entstehungsgeschichte des Prioritätsstreits zwischen Leibniz und Newton um die Erfindung des Infinitesimalrechnung." *Studia Leibnitiana Supplementa* 2 (1969), 69-78.

On the basis of manuscripts and letters by Gregory and Wallis, it is shown that Wallis and Gregory precipitated the dispute with Leibniz by accusing him of having stolen the calculus from Newton. Wallis was motivated by his anger over an anonymous review written by Leibniz of his *Opera Mathematica* (1695).

1354. Sinaceur, H. "Cauchy et Bolzano." *Revue d'histoire des sciences* 26 (1973), 97-112.

The author disputes the claim of Grattan-Guinness that Cauchy plagiarized Bolzano (see item 1293). Instead he attempts to show that Cauchy and Bolzano represent two distinct mathematical traditions: the former being an exponent of mainstream geometry, the latter a precursor of Weierstrassian analytic rigor.

1355. Smail, L.L. *History and Synopsis of the Theory of Summable Infinite Processes*. Eugene: University of Oregon Press, 1925.

This study presents a comprehensive survey of results (to about 1920) on the theory of summability of non-convergent series and related forms. Beginning with a brief history of the early use of divergent series (pp. 1-3), a summary is given of each memoir, paper, or work dealing with the subject, including important definitions and theorems (without proof), arranged in chronological order from Frobenius (1880) to Takenaka (1923), pp. 4-173. An index of names of authors (pp. 174-175) and an index to important topics of the subject matter are also provided.

1356. Stanton, R.J., and R.O. Wells, eds. *History of Analysis*. (Rice University Studies, 64, nos. 2 & 3.) Houston, Tex.: William Marsh Rice University, 1978.

Proceedings of a Conference on the History of Analysis. Papers on "Mathematics and Society--A Historical View" (F.E. Browder), "The Emergence of Analysis in the Renaissance and After" (S. Bochner), "The Creation of the Theory of Group Characters" (T. Hawkins, see item 1030), and "Harmonic Analysis as the Exploitation of Symmetry" (G.W. Mackey).

1357. Stolz, O. "B. Bolzano's Bedeutung in der Geschichte der Infinitesimalrechnung." *Mathematische Annalen* 18 (1881), 255-279.

Compares approaches of Bolzano and Cauchy, and argues that not only did Bolzano have priority but in important respects he went beyond him. Also shows the connection between Bolzano's ideas and subsequent developments.

1358. Strφmholm, P. "Fermat's Methods of Maxima and Minima and of Tangents. A Reconstruction." *Archive for History of Exact Sciences* 5 (1968), 47-69.

Attempts to rehabilitate Wieleitner's thesis that Fermat had no single "method" for finding tangents to curves.

1359. Taton, René. *L'oeuvre scientifique de Monge*. Paris: Presses Universitaires de France, 1951, 441 pp.

See items 923, 1018, 1587.

1360. Taylor, A.E. "The Differential: Nineteenth and Twentieth Century Developments." *Archive for History of Exact Sciences* 12 (1974), 355-383.

See item 1157.

1361. Todhunter, Isaac. *A History of the Calculus of Variations during the Nineteenth Century*. New York: Chelsea, 1962, reprint of the 1861 edition.

Beginning approximately where Woodhouse (item 933) leaves off. The first half is devoted to the work of Lagrange, Lacroix, Dirksen, Ohm, Gauss, Poisson, Ostrogradsky, Delaunay, Sarrus, Cauchy, Legendre, Brunacci, and Jacobi. The second deals with the numerous expositors and commentators of the mid-19th century. See Goldstine, items 779, 864, for a more recent account.

1362. Toeplitz, O. *Die Entwicklung der Infinitesimalrechnung. Eine Einleitung in der Infinitesimalrechnung nach der genetischen Methode*. Berlin: Springer, 1949.

Based on a lecture course given in 1926, this work was prepared for publication by Gottfried Köthe from an unfinished manuscript that Toeplitz left behind at the time of his death in 1940. The genetic method refers to the analysis of a modern subject by tracing its various roots in the past. This is therefore not intended as a complete and unbiased account of the development of calculus, but rather as an attempt to sketch how the subject came to acquire its present form.

1363. Tucciarone, J. "The Development of the Theory of Summable Divergent Series from 1880 to 1925." *Archive for History of Exact Sciences* 10 (1973), 1-40.

Analyzes the early work of Frobenius, Hölder, Cesàro, and Borel from 1880 to 1900 leading to Knopp's recognition of the essential criteria for a theory of divergent series. Further developments by Riesz, Hardy, Toeplitz, Hausdorff, Tauber, and Wiener are considered, and a substantial bibliography is included.

1364. Turnbull, H.W. *The Mathematical Discoveries of Newton*. London, Glasgow: Blackie & Son, 1945.

A short but useful sketch of Newton's work including chapters on: Early Influences (Wallis, Barrow, et al.), The Binomial Theorem, The Method of Fluxions, The *De Quadratura*, The Solid of Least Resistance and the Curve of Quickest Descent, Interpolations and Finite Differences, The *Arithmetica Universalis*, Cubic Curves, and The Geometry in the *Principia*.

1365. ———. *James Gregory Tercentenary Memorial Volume*. Edinburgh: Royal Society of Edinburgh, 1939.

See item 820.

1366. Voss, A. "Differential- und Integralrechnung." *Enzyklopädie der mathematischen Wissenschaften*, II (A2). Leipzig: B.G. Teubner, 1899, 54-134.

Useful for its extensive references to research of the 18th and 19th centuries. Applications receive more attention than pure theory; contains an illustrated appendix on planimeters and integrators.

1367. Walker, E. *A Study of the Traité des indivisibles of Gilles Personne de Roberval*. New York: Columbia University Press, 1932.

See item 850.

1368. Westfall, Richard S. *Never at Rest. A Biography of Isaac Newton.* Cambridge: Cambridge University Press, 1980, xviii + 908 pp.

Westfall's monumental biography is especially strong on Newton's mathematics, as he has made extensive use of Whiteside's recent work. Chapter 4 deals with Newton's early mathematical discoveries. The scientific work of 1665-1666 is the subject of Chapter 5, whereas Chapter 6 discusses his work while Lucasian professor at Cambridge. Chapter 10 examines the mathematics of the *Principia*, and Chapter 14 the priority dispute with Leibniz.

1369. Whiteside, D.T. "Patterns of Mathematical Thought in the Later Seventeenth Century." *Archive for History of Exact Sciences* 1 (1960/1962), 179-388.

See item 773.

1370. Woodhouse, R. *A History of the Calculus of Variations in the Eighteenth Century.* New York: Chelsea, 1964, reprint of the 1810 edition.

See item 933.

1371. Youschkevitch, Adolf P. "The Concept of Function up to the Middle of the Nineteenth Century." *Archive for History of Exact Sciences* 16 (1977), 37-85.

See item 932.

1372. ———. "Lazare Carnot and the Competition of the Berlin Academy in 1786 on the Mathematical Theory of the Infinite." In *Lazare Carnot Savant.* Edited by C.C. Gillispie. Princeton: Princeton University Press, 1971.

See item 883.

1373. ———. "Euler und Lagrange ueber die Grundlagen der Analysis." *Sammelband der zu ehren des 200 Geburtstages Leonhard Eulers.* Berlin: Deutschen Akademie der Wissenschaften zu Berlin, 1959, 224-244.

See item 928.

1374. ———. *Die Entwicklung des Funktionsbegriff.* Translated by K. Reich. Munich: Deutsches Museum, 1972.

A 31-page translation of the original Russian.

1375. ———. "J.A. da Cunha et les fondements de l'analyse infinitésimale." *Revue d'histoire des sciences et de leurs applications* 26 (1973), 3-22.

J.A. da Cunha (1744-1787) was a Portuguese mathematician whose book, *Principrios Mathematicos* (published posthumously in 1790, translated into French in 1811), is an early attempt to base the calculus on rigorous foundations before either Bolzano or Cauchy. He anticipated the latter in developing a criterion for sequential convergence and in defining the differential of a function.

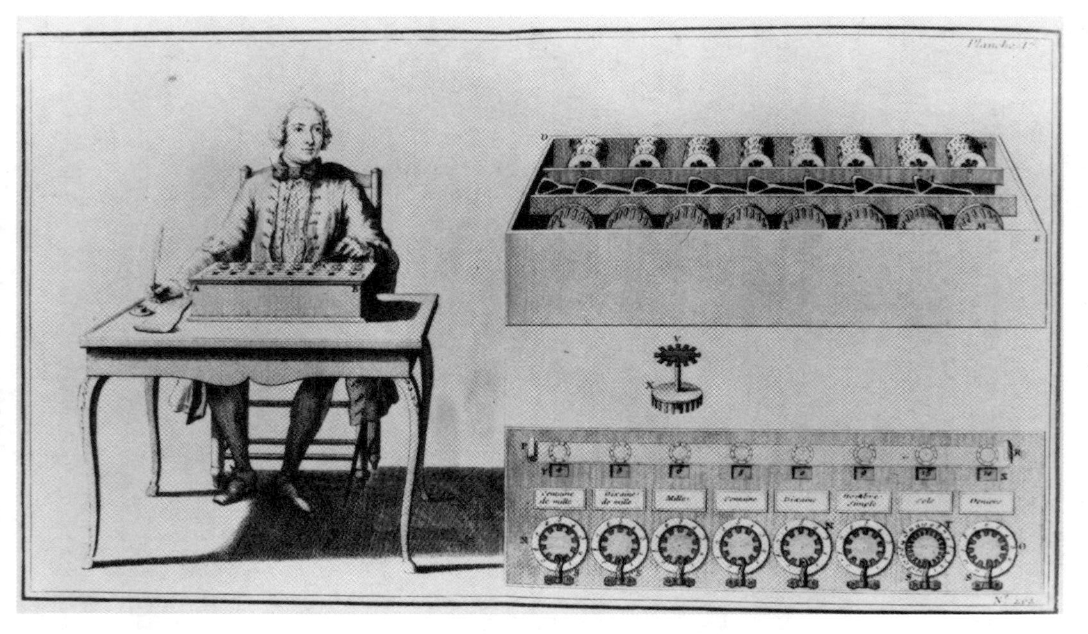

Figure 10. This illustration of Pascal's machine shows both the construction of the cogwheels and use of the machine. From *Machines et inventions approuvées par l'Académie royale des sciences* (Paris, 1735).

COMPUTING

This bibliography concentrates on the period from Babbage to the invention of the stored program electonic digital computer in the mid-1940s, but also contains items relating to earlier work on mechanical calculation and on devices incorporating sequence control mechanisms. Works have been emphasized here that deal specifically with issues relevant to the history of mathematics and its connections with the history of computing and computers.

1376. *Scientific Computing Service Limited: A Description of Its Activities, Equipment and Staff*. London: Scientific Computing Service Ltd., 1946, 23 pp.

An interesting brochure, which includes a brief summary of L.J. Comrie's career, and a bibliography of publications by Comrie, and by the Scientific Computing Service.

1377. Aiken, H.H., and the staff of the Computation Laboratory. "A Manual of Operation for the Automatic Sequence Controlled Calculator." *Annals of the Computation Laboratory of Harvard University*. Vol. 1. Cambridge, Mass.: Harvard University Press, 1946, 561 pp.

The first chapter "Historical Introduction" is a very useful account of the development of calculating machines, difference engines, and Babbage's work. It includes mention of Müller and Torres y Quevedo.

1378. Apokin, I.A., and L.E. Maĭstrov. *Razvitie Vychislitel'nykh Mashin*. Moscow: Nauka, 1974. 400 pp.

An account of the history of digital computing from the earliest aids to calculation to the modern computer. Chapter 1 includes discussion of the abacus in China, Europe, and Russia, while Chapter 2, on mechanical calculators, discusses the work of Jakobson, Chebyshev, and Odhner, as well as Schickard, Pascal, and Leibniz. Chapter 3 covers tabulating machines and electromechanical desk calculators. The next chapter "The Birth of Electronic Computing," which names M.A. Bonch-Bruevich as having invented an electronic trigger circuit in 1918, one year before the independent work of Eccles and Jordan, describes such projects as the Harvard Mark I, the Atanasoff-Berry Computer, ENIAC, EDVAC, etc. Chapter 5 describes early stored program computers, and states that the first Russian computers were the MESM and BESM. The final four chapters discuss transistorized and integrated circuit computers, computer applications, and the future of computer technology. A large number of references are listed, to both Russian- and English-language sources.

1379. Babbage, Charles. *The Ninth Bridgewater Treatise: A Fragment*. London: John Murray, 1837.

Makes numerous references to his work on calculating engines, and includes a short appendix describing the history of his efforts to produce first a difference and then an analytical engine.

1380. ———. *Passages from the Life of a Philosopher*. London: Longman, Green, Longman, Roberts and Green, 1864, 496 pp. Reprinted New York: Augustus M. Kelley, 1969.

Contains several chapters relating to the difference engines and the analytical engine (which have been reprinted in H.P. Babbage [item 1381] and Morrison and Morrison [item 1406]. A fascinating book, which throws much light on the strange character of Charles Babbage.

1381. Babbage, H.P., ed. *Babbage's Calculating Engines: Being a Collection of Papers Relating to Them, Their History, and Construction*. London: E. and F.N. Spon, 1889, 342 pp. Reprinted Los Angeles: Tomash Publishers, 1983.

A reprinting of the writings of Babbage and others, on both the difference and the analytical engines, edited by his son. Much of the material has been reprinted in Morrison and Morrison (item 1406).

1382. Bauer, F.L. *Between Zuse and Rutishauser--The Early Development of Digital Computing in Central Europe*. Item 1404, pp. 505-524.

A very interesting and well-illustrated paper, covering both computer hardware and programming languages and techniques. The relationship between Zuse's work on computers and his ideas for the Plankalkül is explored, and the influence of his work on such people as Rutishauser is discussed.

1383. ———, and H. Wössner. "The 'Plankalkül' of Konrad Zuse: A Forerunner of Today's Programming Languages." *Communications of the Association for Computing Machinery* 15 (7) (1972), 678-685.

A fairly detailed account of the language, and an analysis of it compared to modern programming languages.

1384. Bowden, B.V., ed. *Faster Than Thought*. London: Sir Isaac Pitman and Sons, 1953, 416 pp.

Contains much material on Babbage, including a reprint of the Lady Lovelace translation of Ménabréa's article. The major part of the book consists of chapters on the then-current British computing machine projects.

1385. Burks, A.W., H.H. Goldstine, and J. von Neumann. *Preliminary Discussion of the Logical Design of an Electronic Computing Instrument*. Vol. 1, Part 1. Princeton: Institute for Advanced Study, 28 June 1946; 2nd ed. 2 September 1947. Reprinted in item 1414; extracts reprinted in Randell, item 1408.

The famous report, on the design of what is now known as the "von Neumann"-style machine. Gives a detailed description of the plans for the parallel binary I.A.S. computer, including great detail on the arithmetic unit.

1386. Collier, B. "The Little Machines that Could've: The Calculating Machines of Charles Babbage." Dissertation, Harvard University, 1970.

A valuable detailed study of Babbage's machines, and of the circumstances surrounding their development. Based on the Babbage correspondence in the British Museum, the Babbage sketchbooks and drawings in the Science Museum, and the Buxton collection of Babbage manuscripts in the Museum of the History of Science at Oxford.

1387. Comrie, L.J. "The Application of the Hollerith Tabulating Machine to Brown's Tables of the Moon." *Monthly Notices, Royal Astronomical Society*, 92 (7) (1932), 694-707.

Describes the facilities and operation of then current punches, tabulators, and sorters, and the use of these for calculating E.W. Brown's Tables of the Moon. States that Hollerith equipment was first used at the Nautical Almanac Office in 1929.

1388. ———. "Modern Babbage Machines." *Bulletin, Office Machinery Users Assoc. Ltd.* (London) (1932), 29 pp.

An article on difference engines and techniques.

1389. ———. "The Application of Commercial Calculating Machines to Scientific Computing." *Mathematical Tables and Other Aids to Computation* 2 (1946), 149-159.

Extended survey of the use of desk calculators, adding machines, and punched card machinery for scientific calculations.

1390. ———. "Babbage's Dream Comes True." *Nature* 158 (1946), 567-568.

Review of Aiken, item 1377.

1391. Eckert, J.P. "A Preview of a Digital Computing Machine." *Theory and Techniques for the Design of Electronic Digital Computers. Lectures delivered 8 July-31 Aug. 1946*. Edited by C.C. Chambers. Philadelphia: Moore School of Electrical Engineering, Univ. of Pennsylvania, 1947, 10.1-10.26.

An account of the plans for EDVAC, and a description of its intended order code. Discusses how the decision to provide a single common memory for constants, variables, and instructions led to the concept of a stored program, and the implications of this concept on computer design.

1392. Fleck, G., ed. *A Computer Perspective* ("By the Office of Charles and Ray Eames"). Cambridge, Mass.: Harvard University Press, 1973, 175 pp.

A profusely illustrated book based on an I.B.M.-sponsored exhibition. Although aimed at a popular audience, it contains a vast amount of information, relating directly or indirectly to the origins of computers, that is not readily available.

1393. Goldstine, Herman H. *The Computer from Pascal to von Neumann*. Princeton, N.J.: Princeton University Press, 1972, 378 pp.

The first part of this book discusses the history of digital and analogue calculating devices and concurrent developments in mathematics. However, the main purpose of the book is to give an extensive account, from the viewpoint of a particular participant, of the ENIAC, EDVAC, and I.A.S. projects. The account makes available for the first time a wealth of material taken from contemporary documents. Particular attention is paid to the work of von Neumann and to his role in the EDVAC and I.A.S. projects.

1394. Gravelaar, N.L.W.A. "John Napier's Werken." *Verhandelingen der Koninklijke Akademie van Wetenschappen to Amsterdam*. Eerste Sectie, Deel VI, No. 6. Amsterdam: Johannes Müller, 1899, 159 pp. In Dutch.

The fullest analysis available of John Napier's work on logarithms, calculating aids, arithmetic, algebra, etc. Almost the only publication to point out that Napier's "Local Arithmetic" (Appendix to his *Rabdologiae*, 1617; see item 1407 below) was nothing other than arithmetic in binary notation.

1395. Hawkins, W.F. "The First Calculating Machine (John Napier, 1617)." *The New Zealand Mathematical Society Newsletter* 16 (December 1979), Supplement, 1-23.

Complete translation (with commentary) of John Napier's specification of his Promptuary for multiplication (in *Rabdologiae*, 1617; see item 1407 below), which was a major advance from abacus-type devices towards fully mechanical calculating machines. A working Promptuary has been constructed from this translation.

1396. Hinsley, F.H., with E.E. Thomas, C.F.G. Ransom, and R.C. Knight. *British Intelligence in the Second World War: Its Influence on Strategy and Operations*. Vol. 1. London: H.M.S.O., 1979, 601 pp.

The first official account of this subject, written with unrestricted access to wartime files, including files which are unlikely ever to be made public. This first of three volumes covers the period until the summer of 1941. Contains much authoritative information regarding the extent to which the various types of Enigma messages were deciphered, but little on the techniques used or the individuals concerned. Appendix 1 (pp. 487-495), entitled "Breaking the Enigma: Polish, French and British Contributions," indicates that the Polish cryptanalysts produced their first "cryptographic Bombe" in 1937, and that the first of the British Bombes, "which were of quite different design from the Polish, and much more powerful," was delivered by May 1940.

1397. Horsburgh, E.M., ed. *Napier Tercentenary Celebration: Handbook of the Exhibition*. Edinburgh: Royal Society of Edinburgh, 1914. Also published as *Modern Instruments and Methods of Calculation: A Handbook of the Napier Tercentenary Celebration Exhibition*. London: G. Bell and Sons, 1914. Reprinted Los Angeles: Tomash Publishers, 1983.

Contains, in addition to Percy Ludgate's article on "Automatic Calculating Machines," good descriptions of various then current calculating instruments and machines: Archimedes, Colt's Calculator, Brunsviga, Burroughs Adding and Listing machine, Comptometer, Layton's Arithmometer, Mercedes-Euklid Arithmometer, Millionaire, and the Thomas Arithmometer.

1398. Kepler, Johannes. *Opera Omnia*. Vol. 7. Edited by Ch. Frisch. Frankfurt: Heyder & Zimmer, 1858.

Wilhelm Schickard's first report of his calculating machine is quoted briefly on p. 300 (cf. Kepler, 1718, p. 683, item 1400).

1399. ———. *Gesammelte Werke*. Vol. 18. Edited by Max Caspar. Munich: C.H. Beck, 1959.

Prints letters from Wilhelm Schickard to Kepler of 20 December 1623 and 25 February 1624, describing his calculating machine. See also item 827.

1400. ———. *Epistolae ad Joannem Kepplerum*. Edited by Michael Gottlieb Hanschius. Leipzig: n.p., 1718.

Wilhelm Schickard's letter to Kepler of 20 December 1623, in which he described very briefly his calculating machine, is printed on p. 683. That account was cited by D. Stewart and W. Minto (1787, item 1409, p. 39), and by C. Frisch (Kepler, 1858, item 1398, p. 300).

1401. Lavington, S. *Early British Computers*. Manchester: Manchester University Press, 1980, 139 pp.

Computing in Great Britain, from the 1930s to 1960. Brief illustrated accounts of COLOSSUS, ACE, EDSAC, Manchester Mark 1, DEUCE, PEGASUS, etc.

1402. Lovelace, Augusta Ada, Countess of. "Sketch of the Analytical Engine Invented by Charles Babbage, by L.F. Ménabréa of Turin, Officer of the Military Engineers, with Notes upon the Memoir by the Translator." *Taylor's Scientific Memoirs* 3 (1843) Article 29, 666-731. Reprinted in Bowden, item 1384 above, Morrison and Morrison, item 1406 below, and H.P. Babbage, item 1381 above.

The single most important paper published on Babbage's analytical engine. Lady Lovelace's notes are more lengthy than the original Ménebréa paper, which is itself a good description of the basic principles of the analytical engine.

1403. Lyndon, R.C. "The Zuse Computer." *Mathematical Tables and Other Aids to Computation* 2 (20) (1947), 355-359.

Describes the then incomplete Z4 computer. At this date the memory held 16 numbers--the projected size being 1024.

1404. Metropolis, N., et al., eds. *A History of Computing in the Twentieth Century*. New York: Academic Press, 1980.

The proceedings of a conference at Los Alamos in 1978, at which most of the pioneers of computing presented their accounts of early computing. A major source for the history of computing.

1405. Moore, D.L. *Ada, Countess of Lovelace: Byron's Legitimate Daughter*. London: John Murray, 1977, 397 pp.

A carefully researched biography, which provides a very good account of the remarkable yet tragic life of Lady Lovelace. Her friendship and collaboration with Charles Babbage are covered at some length, though there is comparatively little technical detail concerning their correspondence and discussions about the analytical engine.

1406. Morrison, P., and E. Morrison, eds. *Charles Babbage and his Calculating Engines: Selected Writings by Charles Babbage and Others*. New York: Dover Publications, 1961, 400 pp.

A valuable selection of material on Babbage's engines, taken mainly from *Passages from the Life of a Philosopher* (item 1380) and from *Babbage's Calculating Engines* (item 1381).

1407. Napier, John. *Rabdologiae*. Edinburgh: Andreas Hart, 1617, 154 pp. In Latin. Later editions: Leyden 1626, 1628; Italian translation *Rabdologia*, Verona, 1623; Dutch translation *Eerste Deel Vande Nievwe Telkonst*, Gouda, 1626. Facsimile edition Osnabruch: O. Zeller, 1966.

John Napier described his numbering rods ("Napier's Bones") for multiplication and division, with special rods for square and cube roots and with applications to mensuration. The appendix on Local Arithmetic was the first full publication of binary arithmetic, as far as square root extraction. Napier stressed that the most important part of the book is the Appendix on the Promptuary for multiplication, which can be regarded as the first calculating machine (cf. W.F. Hawkins, 1979, item 1395).

1408. Randell, B. *The Origins of Digital Computers: Selected Papers*. 3rd ed. Berlin: Springer Verlag, 1982, 580 pp.

A set of 34 original papers and manuscripts relating to the origins of digital computers. Introductory and linking text is provided in order to place the work of the various pioneers into perspective, and to cover such topics as early calculating machines and sequence-control mechanisms and the development of electromagnetic and electronic digital calculating devices. There is a valuable annotated bibliography of over 850 items.

1409. Stewart, D. (Earl of Buchan), and W. Minto. *An Account of the Life, Writings and Inventions of John Napier, of Merchiston*. Perth: R. Morison, 1787, 142 pp.

A valuable book about Napier, with clear accounts of the computing devices in *Rabdologiae* (Napier, 1617, item 1407). A brief historical survey of calculating machines after the Promptuary begins (p. 39) with Schickard's first report of his calculating machine, in his 1623 letter to Kepler (cf. Kepler, 1718, item 1400, p. 683).

1410. Travis, I. "The History of Computing Devices." *Theory and Techniques for the Design of Electronic Digital Computers. Lectures Delivered 8 July-31 Aug. 1946*. Edited by C.C. Chambers. Philadelphia, Pa.: Moore School of Electrical Engineering, University of Pennsylvania, 1947, 2.1-2.3.

A very brief account of calculating techniques, and of the development of analogue and digital calculating devices.

1411. Turing, Alan M. "On Computable Numbers, with an Application to the Entscheidungsproblem." *Proceedings of the London Mathematics Society* 42 (1936), 230-267.

The famous paper which introduced the concept of a "universal" computing machine. See also A.M. Turing, "Correction to 'On Computable Numbers,'" *Proceedings of the London Mathematical Society* 43 (1937), 544-548.

1412. ———. *Proposals for Development in the Mathematics Division of an Automatic Computing Engine (A.C.E.)*. Report E.882. Teddington, Middlesex, England: Executive Committee, National Physical Laboratory, 1945. Reprinted as NPL Report, Computer Science 57, April 1972.

A report, which also carried the title *Proposed Electronic Calculator*, giving detailed plans for the ACE, known to have been written some time during 1945.

1413. Turing, S. *Alan M. Turing*. Cambridge: W. Heffner and Sons, 1959, 157 pp.

Biography of Alan Turing, by his mother. States that Turing submitted a proposal to the British Government for the construction of a computer, and that it was on the basis of this that he joined N.P.L. in October 1945. "Before the war he had already begun to build a computer of his own with wider scope, he hoped, than those then in operation." He visited America during the war and "probably saw something of the progress of computing machinery in the States."

1414. Von Neumann, J. *Collected Works*. Edited by A.H. Taub. Oxford: Pergamon, 1963. 6 vols.

Includes the very important papers on computers, written jointly with A.W. Burks and H.H. Goldstine (1946, item 1385), and with H.H. Goldstine in 1947 and 1948.

1415. Wilkes, M.V. *Automatic Digital Calculators*. London: Methuen, 1956, 305 pp.

Contains one of the best early surveys and detailed discussions of the origins of computers. After a detailed discussion of the Babbage analytical engine, Ludgate is mentioned, and Torres y Quevedo's work is described very briefly. It then gives quite a lot of detail on the Harvard Mark I, ENIAC, EDVAC, the Bell Laboratories Computer Models V and VI, and the Harvard Mark II. (The early part of the book is based on the Cantor Lectures given by Wilkes in 1951.)

1416. Wilkinson, J.H. "The Pilot ACE at the National Physical Laboratory." *Radio and Electronics Engineer* 45 (7) (July 1975), 336-340.

Provides a detailed account of the trials and tribulations involved in developing the Pilot ACE Computer, whose first successful public demonstration was in December 1950. It describes how the design was based on that of the "Test Assembly," a prototype design by a team led by Harry Huskey, which was in turn based on what Turing described as Version V of his original 1945 proposal for an Automatic Computing Engine. It states that the term "Engine" was chosen "in recognition of the pioneering work of Babbage on his Analytical Engine," and that Turing was already at work on Version V in May 1946 when the author joined N.P.L.

1417. Zuse, K. *Der Computer--Mein Lebenswerk*. Munich: Verlag Moderne Industrie, 1970.

An autobiography, with many technical details about his work, starting in 1934, on the design of program-controlled calculators. The work of his collaborator Schreyer, who investigated the design of an electronic version of the Z3, and of Dirks, who developed a magnetic drum store, is also covered.

DIFFERENTIAL EQUATIONS

1418. Burkhardt, Heinrich F.K.L. "Entwicklungen nach oscillirenden Functionen und Integration der Differentialgleichungen der mathematischen Physik." *Jahresbericht der Deutschen Mathematiker-Vereinigung* 10 (1901-1908), viii + 1804 pp. Also issued separately in 2 vols. Leipzig: B.G. Teubner, 1908, xii + xii + 1800 pp. Reprinted New York: Johnson Reprint Corp., 1960.

Discusses in great detail differential equations originating in mathematical physics, with special emphasis on expansion methods for their solution. Covers almost all ramifications of this huge subject from its beginning in the 18th century through the end of the 19th century, and is extremely helpful as a guide to primary sources. See also item 1679.

1419. Demidov, Sergei S. "K istorii teorii differentsialnykh uravnenii s chastnymi proizvodnymi." *Istoriko-Matematicheskie Issledovaniia* 18 (1973), 181-202.

Considers the attempts to reduce second order partial differential equations to simple standard forms from Euler (1770) to du Bois-Reymond (1889), and the changing programs toward a general theory of such equations.

1420. ———. "Razvitie issledovanii po uravnenijam s chastnymi proizvodnymi pervogo porjadka v XVIII-XIV vv." *Istoriko-Matematicheskie Issledovaniia* 25 (1980), 71-103.

Surveys the history of first order partial differential equations in the 18th and 19th centuries. Distinguishes four periods in the development of this theory: a formal-analytical period up to 1770 (Euler, d'Alembert), a geometrical period (Lagrange, Monge), the period of Jacobi's second method, and the period of Lie's general theory.

1421. ———. "Vozniknovenie teorii differentsialnykh uravnenii s chastnymi proizvodnymi." *Istoriko-Matematicheskie Issledovaniia* 20 (1975), 204-220.

See items 938, 1682.

1421a. ———. "Differentsialnye uravneniya s chastnymi proizvodnymi v rabotakh zh. dalambera." *Istoriko-Matematicheskie Issledovaniia* 19 (1974), 94-124.

See item 937.

1422. Dobrovolski, W.A. "Sur l'histoire de la classification des points singuliers des équations différentielles." *Revue d'Histoire des Sciences* 25 (1972), 3-11.

Surveys 19th-century studies about singular points of differential equations and analytical functions. Contends that the classification theory of singularities originates in the work of N.E. Joukovsky (1876).

1423. ———. "Contribution à l'histoire du théorème fondamental des équations différentielles." *Archives internationales d'Histoire des Sciences* 22 (1969), 223-234.

Traces the origins and principal methods for the proof of existence and uniqueness of solutions to differential equations in the work of Cauchy and Weierstrass.

1424. *Encyklopädie der mathematischen Wissenschaften. Vol II.1.1. Analysis.* Leipzig: B.G. Teubner (issued in three parts), 1899-1916.

The *Encyklopädie* was set up as a compendium of the state of knowledge around 1900. The articles are very detailed and cover narrow subdisciplines. They are a useful guide to the primary literature and frequently contain historical remarks or judgments. Volume II.1.1. contains items 1425-1431, which are relevant to the theory of differential equations. See also items 5, 967.

1425. Painlevé, P. "Gewöhnliche Differentialgleichungen; Existenz der Lösungen," article nr. II (A4a), 189-229.

1426. Vessiot, E. "Gewöhnliche Differentialgleichungen; Elementare Integrationsmethoden," article nr. II (A4b), 230-293.

1427. Weber, E. von. "Partielle Differentialgleichungen," article nr. II (A5), 294-399.

1428. Maurer, L., and H. Burkhardt. "Kontinuierliche Transformationsgruppen," article nr. II (A6), 401-436.

This article focuses almost exclusively on the ideas of S. Lie.

1429. Bôcher, Maxime. "Randwertaufgaben bei gewöhnlichen Differentialgleichungen," article nr. II (A7a), 437-463.

1430. Burkhardt, Heinrich, and W.F. Meyer. "Potentialtheorie (Theorie der Laplace-Poisson'schen Differentialgleichung)," article nr. II (A7b), 464-503.

See also item 1980.

1431. Sommerfeld, A. "Randwertaufgaben in der Theorie der partiellen Differentialgleichungen," article nr. II (A7c), 504-570.

1432. Engelsman, Steven B. "Lagrange's Early Contributions to the Theory of First Order Partial Differential Equations." *Historia Mathematica* 7 (1980), 7-23.

See items 939, 1091.

1433. Gilain, Christian. "La théorie géométrique des équations différentielles de Poincaré et l'histoire de l'analyse." Dissertation, Université de Paris I, 1977, 145 pp.

Characterizes Poincaré's article "Sur les courbes définies par une équation différentielle" (1880-1886) as a turning point in the study of differential equations, because it breaks with the dominance of complex-function methods, initiates a global study of differential equations, and introduces topology as a method of research. See also item 1092.

1434. Hofmann, Joseph E. "Über Auftauchen und Behandlung von Differentialgleichungen im 17. Jahrhundert." *Humanismus und Technik* 15 (Part 3), (1972), 1-40.

Argues that Debeaune's differential equation (1638) finally inspired Leibniz's formal treatment of linear first order differential equations and Johann Bernoulli's method of solving differential equations by power series and variation of constants. See also item 782.

1435. Lützen, Jesper. "Heaviside's Operational Calculus and the Attempts to Rigorise It." *Archive for History of Exact Sciences* 21 (1979/1980), 161-200.

Argues that there were two approaches to make Heaviside's operational calculus rigorous: one based on integral transformations and the other leading to abstract algebraical formulation. See also items 1075, 1498.

1436. ———. "The Prehistory of the Theory of Distributions." Dissertation, University of Aarhus, Denmark, November 1979, vii + 389 pp. Also published in the series *Studies in the History of Mathematics and Physical Sciences*, 7. New York: Springer, 1982.

Distinguishes four independent fields which inspired the theory of distributions: generalized solutions to differential equations, generalized Fourier transforms, generalized functions, and de Rham-currents. Argues that Schwartz's desire for unification and his understanding of duality were instrumental for the birth of distribution theory.

1437. McHugh, J.A.M. "An Historical Survey of Ordinary Linear Differential Equations with a Large Parameter and Turning Points." *Archive for History of Exact Sciences* 7 (1970/1971), 277-324.

This is a highly technical survey article, describing mathematical techniques and main results of this theory between 1830 and 1970. Extensive bibliography of primary sources.

1438. Ravetz, Jerome R. "Vibrating Strings and Arbitrary Functions." *The Logic of Personal Knowledge: Essays Presented to M. Polanyi on his 70th Birthday*. London: Routledge and Kegan Paul, 1961, 71-88.

See item 931.

1439. Rothenberg, Siegfried. *Geschichtliche Darstellung der Entwicklung der Theorie der singulären Lösungen totaler Differentialgleichungen von der ersten Ordnung mit zwei variablen Grössen*. Dissertation, University of Munich; Leipzig: Teubner, 1908.

Divides the history of singular solutions into seven periods, and argues that a paradox in Lagrange's theory was the prime motivation for 19th-century attempts to understand the nature of such solutions.

1440. Schlissel, Arthur. "The Development of Asymptotic Solutions of Linear Ordinary Differential Equations, 1817-1920." *Archive for History of Exact Sciences* 16 (1976/1977), 307-378.

Describes the development through generalization and ramification of the notion of asymptotic solutions from Carlini (1817) to Debye's "saddle point method." See also items 1349, 1697.

1441. ———. "The Initial Development of the WKB Solutions of Linear Second Order Ordinary Differential Equations and Their Use in the Connection Problem." *Historia Mathematica* 4 (1977), 183-204.

Traces the idea of WKB solution back to the work of George Green (1837) and analyzes the attempts to rigorize it. See also items 1096, 1155, 1698.

1442. Simonov, N.I. "Sur les recherches d'Euler dans le domaine des équations différentielles." *Revue d'histoire des sciences et de leurs applications* 21 (1968), 131-156.

See item 940.

1443. Truesdell, Clifford A. *The Rational Mechanics of Flexible or Elastic Bodies 1638-1788. (Leonhardi Euleri Opera Omnia series 2 volumen XI/2.)* Zurich: Orell Füssli, 1960, 435 pp.

"Editor's Introduction" to volumes 12 and 13 of Euler's *Opera Omnia*, Series 2. See also items 960, 1722.

1444. Wallner, C.R. "Totale und partielle Differentialgleichungen; Differenzen und Summenrechnung; Variationsrechnung." *Vorlesungen über Geschichte der Mathematik. Vierter Band; von 1759 bis 1799*. Edited by Moritz Cantor. Leipzig: Teubner, 1908, 871-1074.

This section in the classic by Cantor is still of value as an introduction to the primary literature of ordinary and partial differential equations in the second half of the 18th century.

ELECTRICITY AND MAGNETISM

The sciences of electricity and magnetism have a very long history about which much has been written. What follows is not intended to be a comprehensive bibliography of this literature but for the most part is confined to works dealing with these subjects in their mathematical mode. Studies treating the earlier, non-mathematical stages in their development have in general been omitted, as have those which, though they deal with more recent events, are chiefly concerned with experimental matters or with technological applications. Only in the case of Faraday has an exception been made here: even though Faraday used no mathematics, the importance of his contributions to the rise of electromagnetic field theory was such that two major studies of his work, by Bence Jones, item 1490, in the 19th century, and L.P. Williams, item 1539, in the 20th, have been included. These are, however, only two of many works devoted to one or another aspect of Faraday's electrical and magnetic researches; no attempt has been made to encompass the remainder. Finally, studies dealing with 20th-century developments in relativity and quantum theory have also been omitted. In other words, the intention has been to survey what has been written on the history of the classical mathematical theories of electromagnetism. The list is not restricted, however, to works in which the mathematical content is overt. Rather, the aim has been to provide a reasonably comprehensive coverage (especially of the English-language literature) of works dealing in one way or another with the evolution of electrical and magnetic theory in the period when that theory had become mathematical in form, whether or not those works are themselves expressed in technical mathematical language. A few studies have been included that treat 18th- and early 19th-century steps towards mathematizing the subject, but the majority refer to developments during the heyday of classical physics, the period between 1850 and 1905. Most have been published within the past few years, for it is only recently that a significant number of historians of science has taken up this area for investigation.

Bibliographies

1445. Ekelöf, Stig, ed. *Catalogue of Books and Papers Relating to the History of Electricity in the Library of the Institute for Theoretical Electricity, Chalmers University of Technology*. Gothenberg: Chalmers University Books, 1964-1966, 109 pp.; 111 pp. 2 vols.

The catalogue of the valuable collection that has been formed over many years by Professor Ekelöf. Vol. I deals with works published prior to 1820, Vol. II with those published after that date. The collection is rich in Continental and especially

Scandinavian material. It does not aspire to completeness; the aim has been to assemble "not all works, but all *important* works."

1446. Frost, A.J., ed. *Catalogue of Books and Papers relating to Electricity, Magnetism, the Electric Telegraph, etc., including the Ronalds Library*. London and New York: E. and F. Spon, 1880, xxvii + 564 pp.

A catalogue initially compiled over many years by the famous telegraph engineer, Sir Francis Ronalds, F.R.S., up to his death in 1873. The catalogue includes not only titles of the works in Ronalds's own remarkable collection of materials on electricity, magnetism, and related subjects, but also all other works on the same subjects which came to his notice. Entries are listed alphabetically by author. The Ronalds collection itself, particularly rich in Continental publications, is now housed at the Institute of Electrical Engineers, London, where it is supplemented by the smaller but also very valuable Silvanus P. Thompson collection.

1447. Mottelay, Paul Fleury. *Bibliographical History of Electricity and Magnetism, Chronologically Arranged*.... London: Charles Griffin & Co., 1922, xx + 673 pp.

A monumental work, chronologically arranged, advancing from earliest times up to the age of Faraday. Entries amount to brief accounts of the publications, discoveries, or inventions attributed to the authors noted at each date, to which are appended more or less extensive lists of authorities consulted.

1448. Weaver, William D., ed. *Catalogue of the Wheeler Gift of Books, Pamphlets and Periodicals in the Library of the American Institute of Electrical Engineers*. Introduction, descriptive and critical notes by Brother Potamian. New York: American Institute of Electrical Engineers, 1909, viii + 504 pp.; 475 pp. 2 vols.

An annotated catalogue of the celebrated Latimer Clark collection, of which it was claimed at the time it was sold that it held "practically every known publication in the English language previous to 1886, on magnetism, electricity, galvanism, the lodestone, mariner's compass, etc." A number of other items are included that were published after that date, and many in other languages. The order of entries in the catalogue is chronological, with a separate chronological listing of "Excerpts from Periodicals-Miscellanea." There are an index of authors and an elaborate system of cross-referencing of entries. There is also a separate subject index for the very extensive section of the collection that relates to the telegraph.

1449. Wilson, David B., ed. *Catalogue of the Manuscript Collections of Sir George Gabriel Stokes and Sir William Thomson, Baron Kelvin of Largs, in the Cambridge University Library*. Cambridge: Cambridge University Library, 1976, 589 pp.; 363 pp. 2 vols.

An invaluable guide to two large and important collections of manuscripts.

Historical Monographs and Articles

1450. *Aepinus's Essay on the Theory of Electricity and Magnetism.* Introductory monograph and notes by R.W. Home. Translation by P.J. Connor. Princeton, N.J.: Princeton University Press, 1979, xiv + 514 pp.

Includes a full English translation of Aepinus's *Essay* (1759), together with an annotated bibliography of his published writings and a 224-page introductory monograph comprising a biographical outline and an extended discussion of the eighteenth-century sciences of electricity and magnetism and the place of Aepinus's work therein. It is argued that even though Aepinus was not prepared to assume any particular form for the laws of electrical and magnetic action, his work nevertheless constitutes the beginning of the mathematical sciences of electricity and magnetism, and thereby an important step in the mathematization of physics more generally.

1451. Agostino, Salvatore d'. "I vortici dell'etere nella teoria del campo elettromagnetico di Maxwell: La funzione del modello nella construzione della teoria." *Physis* 10 (1968), 188-202.

Discusses the different roles played by mechanical models of the aether in Maxwell's various papers on electrodynamics.

1452. ———. "Il pensiero scientifico di Maxwell e lo sviluppo della teoria del campo elettromagnetico nella memoria 'On Faraday's Lines of Force.'" *Scientia* 103 (1968), 291-301. French translation in supplement, pp. 155-164.

Examines Maxwell's ideas on the nature of physical theories and on methodology, as set out in his first memoir on electromagnetism and elsewhere. Focuses particularly on his advocacy of theoretical pluralism, and on his search for mechanical models for physical theories.

1453. ———. "Hertz's Researches on Electromagnetic Waves." *Historical Studies in the Physical Sciences* 6 (1975), 261-323.

An extended discussion of the development during the 1880s of Hertz's theoretical conceptions in electrodynamics leading to his famous experiments on electric waves. The importance for Hertz of Helmholtz's theory of polarization is emphasized, but so too is his ultimate acceptance of a purer Maxwellian outlook.

1454. ———. "La scoperta di una velocità quasi uguale alla velocità della luce nell'elettrodinamica di Wilhelm Weber (1804-1891)." *Physis* 18 (1976), 297-318.

Describes how a constant with the dimensions of a velocity appeared in Weber's fundamental law of electrodynamic action, and discusses the implications of the discovery of Weber and Kohlrausch that its value approximated that of the velocity of light. Argues that Weber did not attach any special significance to this result, and in this connection contrasts Weber's research program, seen as aiming to reduce electrodynamics to mechanics, with Maxwell's program based on contiguous action and the notion of a field.

1455. Berkson, William. *Fields of Force: The Development of a World View from Faraday to Einstein*. London: Routledge and Kegan Paul, 1974, xiv + 370 pp.

A lively survey of the development of the concept of a field from Faraday to Einstein. Is particularly concerned to delineate the "problem situation" within which each person discussed (most notably Faraday, Maxwell, Hertz, Lorentz, and Einstein) was working.

1456. Bork, Alfred M. "Maxwell, Displacement Current, and Symmetry." *American Journal of Physics* 31 (1963), 854-859.

Argues against the statement often made in physics texts that Maxwell introduced the displacement current to make his equations symmetrical. Shows that symmetry was brought in as a consideration by Heaviside.

1457. ———. "Maxwell and the Electromagnetic Wave Equation." *American Journal of Physics* 35 (1967), 844-849.

Discusses and illustrates by flow charts Maxwell's three derivations of the wave equation from the basic electromagnetic equations, in the papers of 1865 and 1868 and in the *Treatise*. See also item 1953.

1458. ———. "Maxwell and the Vector Potential." *Isis* 58 (1967), 210-222.

Presents detailed evidence to show that, in contrast to a commonly expressed modern view, for Maxwell the vector potential represented not a mere mathematical construction, but a real physical quantity which had its conceptual origin in Faraday's notion of the "electrotonic state."

1459. Bowley, R.M., et al. *George Green: Miller, Snienton*. Nottingham: Nottingham Castle, 1976, 96 pp.

Contains useful biographical information in essays on "George Green: His Family and Background" by Frieda M. Wilkins-Jones and "George Green: His Academic Career" by David Phillips. The latter essay is valuable due to the inclusion of a number of letters which passed among Green, Sir Edward Ffrench Bromhead, and William Whewell. A 7-page introduction on "George Green: His Achievements and Place in Science" is worthless. See also item 1979.

1460. Bromberg, Joan L. "Maxwell's Displacement Current and His Theory of Light." *Archive for History of Exact Sciences* 4 (1967), 218-234.

Rejects, on the basis of a detailed analysis of the relevant parts of Maxwell's published papers, the accounts usually given of the introduction by him of the displacement current, arguing that he introduced it not on aesthetic grounds or in order to obtain a consistent set of equations but for pragmatic reasons, as a plausible method of advancing his calculations. Points to errors and serious confusions in Maxwell's initial presentation of the idea, and shows how these were gradually eliminated in his later publications on electromagnetism. See also item 1956.

1461. ———. "Maxwell's Electrostatics." *American Journal of Physics* 36 (1968), 142-151.

Argues that there is "a real and fundamental ambiguity, contradiction, or obscurity" in each of Maxwell's major discussions of electrostatics, but a different one in each. Finally, in the *Treatise*, Maxwell's equations mask his physical ideas. Displacement was initially introduced as a synonym for dielectric displacement. In later versions of the theory, however, Maxwell altered the signs appearing in his equations in such a way as to render those governing displacement incompatible with his earlier conception. As a result, displacement came to be seen by his successors as a conception original with him, and his intended meaning became lost.

1462. Buchwald, Jed Z. "William Thomson and the Mathematization of Faraday's Electrostatics." *Historical Studies in the Physical Sciences* 8 (1977), 101-136.

Argues that Thomson's application of Fourier's heat diffusion equation to electrostatics did not arise from an attempt to express Faraday's ideas about electric force in mathematical terms. Rather, his primary purpose initially was to use Laplace's theory of attraction to solve a problem in heat theory. Only subsequently, when he became aware of the difficulties associated with Poisson's conception of a physical layer of electric fluid on the surface of a conductor, did he abandon the fluid theory of electricity and seek to reconcile Green's analyses based on the notion of potential with Faraday's doctrines.

1463. ———. "The Hall Effect and Maxwellian Electrodynamics in the 1880's: Part I, The Discovery of a New Electric Field." *Centaurus* 23 (1979-1980), 51-99; and "Part II, The Unification of Theory, 1881-1893." *Centaurus* 23 (1979-1980), 118-162.

A detailed account of Hall's discovery, based on his manuscript notebooks and setting it securely within a tradition of electromagnetic theorizing grounded on Maxwell's *Treatise* in which electrical conduction was regarded as an entirely secondary field phenomenon. Shows, too, how the Hall effect was absorbed into the Maxwellian theory, yet at the same time highlighted difficulties in the treatment of conductivity within that theory.

1464. Campbell, Lewis, and William Garnett. *The Life of James Clerk Maxwell*. London: Macmillan, 1882, xvi + 662 pp. 2nd ed., abridged and rev. London: Macmillan, 1884. 1st ed. reprinted with a preface by Robert H. Kargon. New York and London: Johnson Reprint Corp., 1969.

Contains (1) a narrative *Life*, which includes many letters to Campbell and others, by Campbell, a boyhood friend and lifelong correspondent of Maxwell; (2) an account of Maxwell's scientific work, by Garnett, one of his students at Cambridge;

(3) a selection of his poetry. In the second edition, Garnett's contribution was greatly condensed and distributed throughout the biography, and seven new letters (including four to Faraday) added. These have also been included in the reprint edition.

1465. Caneva, Kenneth L. "From Galvanism to Electrodynamics: The Transformation of German Physics and Its Social Context." *Historical Studies in the Physical Sciences* 9 (1978), 63-159.

On the basis of German writings on electricity and magnetism from the first half of the 19th century, delineates two very different styles in German physics during this period, namely, "concretizing science," which was qualitative in character and asserted experience as the direct source of theory, and "abstracting science," which was quantitative, mathematical, and hypothetico-deductive in structure. Argues that these were maintained by two different generational groups, and links the change in scientific outlook from one generation to the next with wider changes in German culture and society following the Napoleonic invasions.

1466. Chalmers, A.F. "The Electromagnetic Theory of James Clerk Maxwell and Some Aspects of Its Subsequent Development." Dissertation, University of London, 1971.

Provides the basis for items 1467 and 1468. Includes in addition a chapter on "The Subsequent Extension of Maxwell's Theory," in which Helmholtz's theory of a polarizable ether, seen as a compromise between the action at a distance and continuous field theories, is presented as the crucial link between the work of Maxwell and that of Hertz and Lorentz.

1467. ———. "The Limitations of Maxwell's Electromagnetic Theory." *Isis* 64 (1973), 469-483.

Identifies several different limitations of Maxwell's theory, including (1) Maxwell's conception that light was a mechanical state of the ether arising from a mechanical interaction between the matter of the source and the surrounding ether, and not from *electrical* disturbances, meant that he failed to recognize the possibility of electromagnetic radiation; (2) his conception of charge was vague and unsatisfactory; yet (3) despite its vagueness, it was precise enough to lead to some falsifiable (and ultimately falsified) conclusions. These limitations are attributed to the theory's being, ironically, too much a theory about mechanisms in the aether, as a result of which it lacked important elements provided by the rival action-at-a-distance theory.

1468. ———. "Maxwell's Methodology and His Application of It to Electromagnetism." *Studies in History and Philosophy of Science* 4 (1973), 107-164.

Argues that Maxwell's innovations in electromagnetism were achieved in spite of the methodology to which he purportedly ascribed: in particular, *contra* Duhem, Maxwell was led to the concept of displacement and hence the idea that light was an electromagnetic phenomenon, not by his various attempts to reduce electricity and magnetism to the principles of mechanics, but by arguments arising within the science of electricity itself. Analyzes a number of persisting (and connected) difficulties in Maxwell's notions of displacement and charge, arguing that Maxwell was actually hindered from resolving these by his methodological views.

1469. Cuvaj, Camillo. "Henri Poincaré's Mathematical Contributions to Relativity and the Poincaré Stresses." *American Journal of Physics* 36 (1968), 1102-1113.

Gives a summary account of some of the main achievements in Poincaré's major paper (published 1906) on the theory of the electron. An addendum, with a list of corrections, was published in *American Journal of Physics* 38 (1970), 774-775.

1470. Doran, B.G. "Origins and Consolidation of Field Theory in Nineteenth-Century Britain: From the Mechanical to the Electromagnetic View of Nature." *Historical Studies in the Physical Sciences* 6 (1975), 132-260.

A general and non-mathematical account in which Larmor's work in the 1890s is portrayed as the culmination of a long tradition in British physics of non-mechanical theories of the ether. The argument rests, however, on a number of dubious re-interpretations of earlier workers.

1471. Duhem, Pierre. *Les théories électriques de J. Clerk Maxwell: Etude historique et critique*. Paris: A. Hermann, 1902.

A polemical critique of Maxwell's ideas.

1472. Everitt, C.W.F. *James Clerk Maxwell: Physicist and Natural Philosopher*. New York: Charles Scribner's Sons, 1975.

An expanded version of the "Maxwell" entry in the *Dictionary of Scientific Biography*, item 10. An excellent survey of Maxwell's life and scientific work.

1473. Gillispie, Charles Coulston. *The Edge of Objectivity: An Essay in the History of Scientific Ideas*. Princeton, N.J.: Princeton University Press, 1960, 562 pp.

A justly esteemed work which includes (pp. 458-492) a section on Maxwell in which he is presented as "the ultimate impresario of classical physics, who brought the chief characters, the atom and the ether, to the center of the stage, and there left them all exposed to the winds of criticism blowing up out of positivism." See also item 862.

1474. Gillmor, C. Stewart. *Coulomb and the Evolution of Physics and Engineering in 18th-Century France*. Princeton, N.J.: Princeton University Press, 1971, xvii + 328 pp.

A well-researched biography, together with analyses of Coulomb's various contributions to physics and engineering. Argues that mathematical physics emerged in late 18th-century France from an amalgamation of three previously separate traditions, namely, rational mechanics, experimental physics, and practical engineering, and that Coulomb, with his strong background in engineering, played a seminal role in this. See also item 1689.

1475. Goldberg, Stanley. "The Abraham Theory of the Electron: The Symbiosis of Experiment and Theory." *Archive for History of Exact Sciences* 7 (1970), 7-25.

Shows how Abraham (1875-1922) attempted to derive a wholly electrodynamic basis for mechanics on the assumption of a rigid electron, Maxwell's equations, and an absolute frame of reference determined by the ether. Contrasts Abraham's views with Lorentz's theory based on a deformable electron, assesses the relationship between his work and Kaufmann's experimental investigations, and discusses the limitations of his approach.

1476. ———. "The Lorentz Theory of Electrons and Einstein's Theory of Relativity." *American Journal of Physics* 37 (1969), 982-994.

Reviews the development of Lorentz's theory of electrons insofar as it relates to the electrodynamics of moving bodies. Argues that the principle of relativity did not play an important role in Lorentz's theory, and that though Lorentz eventually realized the distinctions between his own work and that of Einstein, he was unwilling to embrace Einstein's formulation completely and thereby to reject the ether.

1477. ———. "In Defense of Ether: The British Response to Einstein's Special Theory of Relativity, 1905-1911." *Historical Studies in the Physical Sciences* 2 (1970), 89-125.

Shows that in Britain in these years Einstein's theory was largely neglected, and the concept of the ether generally maintained. Attributes British slowness to come to terms with Einstein to the fact that most British theoreticians were trained at Cambridge, and that through the Tripos examinations their training was directed chiefly at questions of ether mechanics.

1478. Green, H. Gwynedd. "A Biography of George Green, Mathematical Physicist of Nottingham and Cambridge, 1793-1841." In *Studies and Essays in History of Science and Learning Offered in Homage to George Sarton....* Edited by M.F. Ashley Montagu. New York: Henry Schuman, 1946, 549-594.

A brief (18-page) account of Green's career, unfortunately entirely lacking in technical discussion of his work, together with various documents relating to his life. Also includes extracts from Kelvin's correspondence concerning his re-discovery of Green's subsequently renowned *Essay on the Application of Mathematical Analysis to the Theories of Electricity and Magnetism* (1828).

1479. Haas-Lorentz, G.L. de, ed. *H.A. Lorentz--Impressions of His Life and Work*. Amsterdam: North-Holland, 1957, 172 pp.

Reminiscences of Lorentz by his daughter and others who knew him, with a chapter on "The Scientific Work," without any mathematical details, by A.D. Fokker.

1480. Heilbron, John L. *Electricity in the 17th and 18th Centuries: A Study of Early Modern Physics*. Berkeley, Los Angeles, London: University of California Press, 1979, xiv + 606 pp.

An outstanding and comprehensive history of electricity to about the year 1800, including perceptive discussions of the conceptual difficulties confronting those who wished to quantify the subject and of the mathematizing activities of Aepinus, Cavendish, Coulomb, and Poisson.

1481. Heimann, P.M. "Maxwell and the Modes of Consistent Representation." *Archive for History of Exact Sciences* 6 (1969-1970), 171-213.

Argues that there was a "fundamental dichotomy" in Maxwell's thinking on electromagnetism, whereby on some occasions he took lines of force to be the basic entities of the theory and on others he sought to reduce these to states of polarization of particles of matter and ether.

1482. ———. "Maxwell, Hertz, and the Nature of Electricity." *Isis* 62 (1971), 149-157.

Emphasizes the contradictions in Maxwell's discussions of the nature of electricity in his *Treatise* and elsewhere, and argues that Hertz's desire to eliminate these was fundamental to his reformulation of Maxwell's theory.

1483. Hesse, Mary B. *Forces and Fields: The Concept of Action at a Distance in the History of Physics*. London: Thomas Nelson & Sons, 1961, x + 318 pp.

Chap. VIII (pp. 189-225) is devoted to "The Field Theories," and includes a brief summary of the tension in 19th-century electromagnetic theory between action-at-a-distance and field conceptions.

1484. ———. "Logic of Discovery in Maxwell's Electromagnetic Theory." In *Foundations of Scientific Method: The Nineteenth Century*. Edited by Ronald N. Giere and Richard S. Westfall. Bloomington, London: Indiana University Press, 1973, 86-114.

Investigates Maxwell's explicit discussions of physical method and their application in his electromagnetic theory. Argues that, in his mature theory, Maxwell attempted to justify his introduction of the displacement current by a generalized method of induction and analogy, and by no means regarded the idea as a hypothetical concept.

1485. Hirosige, Tetu. "Electrodynamics before the Theory of Relativity, 1890-1905." *Japanese Studies in the History of Science* 5 (1966), 1-49.

Presents three alternative streams of thought guiding elecrodynamics during this period, namely, (1) Hertz's axiomatic approach, (2) Larmor's etherial dynamics, and (3) Lorentz's and Wiechert's theory of electrons. Describes how Lorentz's theory came to be widely (but by no means universally) accepted shortly after 1900, and shows how this carried with it the view that the ether was merely the seat of the electromagnetic field, and not, after all, a mechanical substance. See also item 2081.

1486. ———. "Origins of Lorentz' Theory of Electrons and the Concept of the Electromagnetic Field." *Historical Studies in the Physical Sciences* 1 (1969), 151-209.

An excellent analysis of the evolution of Lorentz's ideas on electromagnetism, from his early work on an electromagnetic theory of optics based on an action-at-a-distance conception, up to his major presentation of the electron theory in 1895. Concentrates particularly on Lorentz's gradual conceptualization of the electromagnetic field as a dynamical state of a stationary aether devoid of all mechanical qualities, rather than as a mechanical system in the ether as it was for Maxwell. See also item 2082.

1487. ———. "The Ether Problem, the Mechanistic Worldview, and the Origins of the Theory of Relativity." *Historical Studies in the Physical Sciences* 7 (1976), 3-82.

A comprehensive survey of the nineteenth-century background to the work of Lorentz, Poincaré, and Einstein on the "ether problem" and relativity. Argues that while both Lorentz and Poincaré were working within a traditional problem situation which tried to reduce electromagnetism to mechanics (or vice versa), Einstein was not. On the contrary, under the influence of Mach's critique of the mechanistic world view--seen here as much more important for Einstein than his criticisms of absolute space and time--Einstein was seeking a unification of electromagnetism and mechanics "at a higher level" where the two theories were considered to be of equal standing. See also item 2083.

1488. Hoppe, Edmund. *Geschichte der Elektrizität*. Leipzig: J.A. Barth, 1884, xx + 622 pp.

A thorough and very comprehensive general history of electricity up to the discovery of conservation of energy and Weber's electric force "law." Concludes with a substantial section on 19th-century technical applications.

1489. Jammer, Max. *Concepts of Mass in Classical and Modern Physics*. Cambridge, Mass.: Harvard University Press, 1961, 230 pp.

Contains a chapter (pp. 136-153) on "The Electromagnetic Concept of Mass," in which the work of Heaviside and Abraham in particular is described.

1490. Jones, H. Bence. *Life and Letters of Faraday*. London: Longmans, Green and Co., 1870, x + 427 pp.; viii + 499 pp. 2 vols.

A "biography" constructed almost entirely out of Faraday's own letters and journals, with short linking passages.

1491. Kargon, Robert. "Model and Analogy in Victorian Science: Maxwell's Critique of the French Physicists." *Journal of the History of Ideas* 30 (1969), 423-436.

Includes a section (pp. 431-436) on "Maxwell and the Electromagnetic Field" in which Maxwell's attitude towards his mechanical models is discussed.

1492. Kastler, Alfred. "Ampère et les lois de l'électrodynamique." *Revue d'histoire des sciences et de leurs applications* 30 (1977), 143-157.

An exposition of Ampère's route to his law of force between current elements, summarizing Ampère's own account in his *Mémoire sur la théorie mathématique des phénomènes électrodynamiques, uniquement déduite de l'expérience*.

1493. Knudsen, Ole. "From Lord Kelvin's Notebook: Ether Speculations." *Centaurus* 16 (1971), 41-53.

Publishes a brief extract from one of Kelvin's notebooks, together with a brief commentary and explanatory notes. The passage in question is dated January 6, 1859, and Kelvin appears to have written it out in order to clarify his own thinking. In it, he sets out his notion of the ether as an ideal elastic substance which he wishes to substitute for the traditional conception of the ether as made up of discrete particles exerting forces on each other at a distance. The Faraday effect figures prominently in his discussion.

1494. ———. "The Faraday Effect and Physical Theory." *Archive for History of Exact Sciences* 15 (1976), 235-281.

Emphasizes the role played by the Faraday effect in the development of Maxwell's electromagnetic theory, namely, that it convinced him that the magnetic field was constituted of aetherial vortices. Argues that it was this conviction which lay behind Maxwell's well-known difficulties concerning the nature of electricity which, it is shown, vitiated even his detailed analysis of the Faraday effect itself. Contrasts Maxwell's problems here with the power of the Continental action-at-a-distance approach displayed in Carl Neumann's analysis of the same effect.

1495. ———. "Electric Displacement and the Development of Optics after Maxwell." *Centaurus* 22 (1978), 53-60.

Argues that Maxwell's concept of displacement was of a real physical quantity representing the polarization of the ether and with no specific connection with matter. Gibbs in his papers on electromagnetic optics followed this, but Lorentz, following Helmholtz, regarded displacement as a composite with distinct components of polarization in both matter and the ether. The separation of the two helped pave the way for the electron theory.

1496. ———. "19th Century Views on Induction in Moving Conductors." *Centaurus* 24 (1980), 346-360.

Takes as his starting point the opening paragraph of Einstein's 1905 relativity paper containing his famous comments about an asymmetry in the classical treatment of electromagnetic induction. Shows that questions of symmetry and invariance in connection with electromagnetic induction had been discussed by a number of 19th-century authors, and points to a general feature of these discussions, namely, their linking of the problem of induction with the question of the relationship between the motions of matter and the aether.

1497. Koenigsberger, Leo. *Hermann von Helmholtz*. Braunschweig: Friedrich Vieweg und Sohn, 1902-1903. 3 vols. A 1-vol. English translation and abridgement by Frances A. Welby, Oxford: Clarendon Press, 1906.

The standard biography, including long extracts from Helmholtz's papers and correspondence. (These are considerably abbreviated in the English edition.)

1498. Lützen, Jesper. "Heaviside's Operational Calculus and the Attempts to Rigorise It." *Archive for History of Exact Sciences* 21 (1979-1980), 161-200.

Examines several examples of Heaviside's non-rigorous use of the basics of operational calculus in electrical calculations, showing his extensive reliance on physical intuition. Traces the efforts of later mathematicians to render these techniques more rigorous. See also items 1075, 1435.

1499. McCormmach, Russell. "The Electrical Researches of Henry Cavendish." Dissertation, Case Western Reserve University, 1967.

A comprehensive analysis of Cavendish's electrical investigations, based on a thorough study of Cavendish's manuscripts, as well as on Maxwell's published edition of the electrical papers.

1500. ———. "J.J. Thomson and the Structure of Light." *British Journal for the History of Science* 3 (1967), 362-387.

Describes Thomson's speculations concerning a discontinuous structure in the electromagnetic field, based upon the presumed discreteness of Faraday-style tubes of force and leading to the view that light was granular in character. Discusses the relationship between Thomson's ideas and Einstein's notion of light quanta, suggesting that even though Thomson resolutely opposed the quantum theory, the familiarity of British physicists with his own ideas about light helped reconcile them to it.

1501. ———. "H.A. Lorentz and the Electromagnetic View of Nature." *Isis* 61 (1970), 458-497.

An authoritative study of Lorentz's development of the electron theory as the foundation of a universal purely electromagnetic physics. The inherently non-mechanical character of the theory is emphasized, as is the authoritative position it came to occupy in physics, especially in Germany, around 1900.

1502. ———. "Einstein, Lorentz, and the Electron Theory." *Historical Studies in the Physical Sciences* 2 (1970), 41-87.

An exceptionally clear account of the evolution of Einstein's thought during the first ten years of the 20th century. Emphasizes the central place of electrodynamics in the problem situations with which Einstein was concerned, and the importance of Lorentz's work to him, even in those situations where he deviated most sharply from Lorentz's conceptions. See also item 2089.

1503. McGuire, J.E. "Forces, Powers, Aethers, and Fields." In *Methodological and Historical Essays in the Natural and Social Sciences*. Edited by Robert S. Cohen and Marx W. Wartofsky. (Boston Studies in the Philosophy of Science, XIV.) Dordrecht, Boston: D. Reidel, 1974, 119-159.

An attempt to isolate "some of the turning points in the history of the emergence of field concepts as a prolegomenon to understanding the dynamics of conceptual change involved." Includes a discussion (which draws heavily on Heimann's work, item 1481) of the ideas of Maxwell, and, much more briefly, an account of the ideas of Poynting, J.J. Thomson, Larmor, and Lorentz.

1504. Merleau-Ponty, Jacques. *Leçons sur la genèse des théories physiques: Galilée, Ampère, Einstein*. Paris: Vrin, 1974, 172 pp.

Includes (pp. 69-112) an interesting discussion of Ampère's *Théorie mathématique des phénomènes électro-dynamiques* (1827), viewing this as illustrative of one of the characteristic stages in the formation of modern physical theory in which the Galilean objective of mathematizing experimental knowledge is accepted, but in which a new problem arises where, with this objective in mind, a choice has to be made between different modes of conceptualizing the situation.

1505. Miller, Arthur I. "A Study of Henri Poincaré's 'Sur la dynamique de l'électron.'" *Archive for History of Exact Sciences* 10 (1973), 207-328.

A detailed discussion, in its historical context, of Poincaré's notable attempt to formulate a purely electromagnetic theory of a deformable electron. The paper includes extended analyses of earlier theories of the electron due to Abraham and Lorentz, and evidence for the influence of Poincaré's work despite its being soon overtaken by Einstein's. Emphasizes Poincaré's adherence to an electromagnetic world-picture in which the principle of relativity was a law open to experimental verification.

1506. ———. *Albert Einstein's Special Theory of Relativity: Emergence (1905) and Early Interpretation (1905-1911)*. Reading, Mass.: Addison-Wesley, 1981, xxviii + 466 pp.

A "biography" of Einstein's 1905 relativity paper that includes a long introductory chapter (pp. 11-121), with considerable mathematical detail, on "Electrodynamics: 1890-1905." This surveys the theoretical contributions of Lorentz, Poincaré, and Abraham and their interaction with the experimental work of Kaufmann. See also item 2089a.

1507. Miller, John David. "Rowland and the Nature of Electric Currents." *Isis* 63 (1972), 5-27.

An account, based on then recently discovered manuscript sources, of Rowland's lifelong concern with experimental investigations of the nature of electric currents, especially through his efforts to demonstrate magnetic effects due to moving electric charges.

1508. ———. "Rowland's Magnetic Analogy to Ohm's Law." *Isis* 66 (1975), 230-241.

Describes how Rowland attempted to translate into mathematical form Faraday's analogy between a magnet and its surrounding force field and a voltaic battery immersed in water, arriving ultimately at the conclusion that magnetic induction is related to the magnetic potential between two points by a law exactly analogous to Ohm's law for electrical currents.

1509. Molella, Arthur Philip. "Philosophy and Nineteenth-Century German Electrodynamics: The Problem of Atomic Action at a Distance." Dissertation, Cornell University, 1972, 263 pp.

Focuses on the interaction between Weber's theory of electrodynamic action at a distance between "atoms" of electricity and contemporary German philosophical atomism, especially as expounded by G.T. Fechner and J.F.C. Zöllner.

1510. Moyer, Donald Franklin. "Energy, Dynamics, Hidden Machinery: Rankine, Thomson and Tait, Maxwell." *Studies in History and Philosophy of Science* 8 (1977), 251-268.

Discusses the role of generalized equations of motion in Maxwell's electrodynamics (following earlier suggestions by Rankine and Thomson and Tait) as a method of generating mechanical explanations even though the underlying machinery remains hidden.

1511. ———. "Continuum Mechanics and Field Theory: Thomson and Maxwell." *Studies in History and Philosophy of Science* 9 (1978), 35-50.

Briefly discusses some nineteenth-century developments in continuum mechanics, especially those due to William Thomson, and shows how Maxwell used a generalized form of Thomson's line of reasoning to construct the electromagnetic theory set out in his *Treatise*. See also item 1713.

1512. Olesko, Kathryn Mary. "The Emergence of Theoretical Physics in Germany: Franz Neumann and the Königsberg School of Physics, 1830-1890." Dissertation, Cornell University, 1980, 545 pp.

An outstanding thesis based on extensive research on previously untapped archival material. Includes some discussion of Neumann's contribution to electromagnetic theory, but the emphasis throughout is on wider questions concerning the conceptualization of physics as a mathematical science and the institutionalization of the new approach.

1513. Pihl, Mogens. *Der Physiker L.V. Lorenz.* Copenhagen: Einar Munksgaard, 1939, 128 pp.

A University of Copenhagen physics doctoral thesis. Includes a brief (3 pages) biographical outline of Lorenz and a bibliography of his writings on physical subjects, together with critical analyses of his work on optics, the conductivity of metals, and kinetic and elasticity theory. Seeks throughout to clarify Lorenz's ideas by translating them into less complicated and more satisfactory modern mathematical form.

1514. ———. "The Scientific Achievements of L.V. Lorenz." *Centaurus* 17 (1972), 83-94.

A brief account, in English, of Lorenz's principal contributions to mathematical physics, especially in optics, the electromagnetic theory of light, conductivity, and the theory of telephone cables.

1515. Pyenson, Lewis. "Physics in the Shadow of Mathematics: The Göttingen Electron-Theory Seminar of 1905." *Archive for History of Exact Sciences* 21 (1979), 55-89.

Describes the material studied at the seminar as a summary of immediately pre-Einsteinian electron theory--mainly the work of Hertz, Abraham, Schwarzschild, Sommerfeld, and, above all, Lorentz--and argues that the failure of the group to resolve in a satisfactory manner the outstanding problems of electromagnetic theory was due to their over-emphasizing the purely mathematical techniques involved, at the expense of physical theory. See also item 2093.

1516. Rosenfeld, L. "The Velocity of Light and the Evolution of Electrodynamics." *Nuovo Cimento. Supplemento* (Società italiana di fisica) 4 (1956), 1630-1669.

Discusses the developments leading up to Maxwell's identification of light as an electromagnetic phenomenon, and also the less familiar story of Lorenz's independently arriving at the same conclusion by a different route. Stresses the importance for both men of a dynamical conception of nature which a few years later, in Hertz's day, was on the wane.

1517. Schaffner, Kenneth F. "The Lorentz Electron Theory and Relativity." *American Journal of Physics* 37 (1969), 498-513.

Traces Lorentz's work on the electrodynamics of moving bodies from 1887 to 1909. Discusses the evolving role played in Lorentz's theory by the contraction hypothesis, and the type of support this enjoyed within the theory. Emphasizes the nonreciprocal character of the transformation equations in Lorentz's theory, in which the ether continues to provide a privileged reference frame.

1518. ———, ed. *Nineteenth-Century Aether Theories*. With a commentary by K.F. Schaffner. Oxford, New York: Pergamon Press, 1972, ix + 278 pp.

A volume in the Commonwealth and International Library series, "Selected Readings in Physics." Comprises an excellent 121-page introduction by Schaffner, together with extracts from the writings of Fresnel, Stokes, Michelson and Morley, Green, MacCullagh, W. Thomson, Fitzgerald, Heaviside, Larmor, and Lorentz.

1519. ———. "Outlines of a Logic of Comparative Theory Evaluation with Special Attention to Pre- and Post-Relativistic Electrodynamics." *Historical and Philosophical Perspectives of Science*. Edited by R.H. Stuewer. (Minnesota Studies in the Philosophy of Science, Vol. 5.) Minneapolis: University of Minnesota Press, 1970, 311-354.

Uses the logic to assess the relative standing in 1905 of Lorentz's and Einstein's electrodynamic theories, concluding that the former ranked higher in "theoretical context sufficiency" and the latter in simplicity considerations.

1520. Siegel, Daniel M. "Completeness as a Goal in Maxwell's Electromagnetic Theory." *Isis* 66 (1975), 361-368.

Suggests that Maxwell's electromagnetic theory, and in particular his modification of Ampère's law by the inclusion of the displacement current, should be seen as "a shining example of systematic and goal-oriented theoretical endeavor rewarded," where the goal was theoretical completeness, that is, a theory which enabled one to calculate effects "in the limiting cases where the known formulae are inapplicable."

1521. ———. "Classical-Electromagnetic and Relativistic Approaches to the Problem of Nonintegral Atomic Masses." *Historical Studies in the Physical Sciences* 9 (1978), 323-360.

Discusses the competing answers provided by classical electromagnetic theory (in which, following Lorentz, all mass is regarded as electromagnetic) and relativity theory, to the question of why atomic masses are not integral multiples of the mass of the hydrogen atom. These answers at first developed independently, but from 1916 they interacted. Not until the 1930s did the relativistic approach prevail. It is emphasized, following McCormmach, that in the early days of relativity theory, Lorentz's theory "was not some hoary predecessor, but rather a near contemporary, just as new and full of promise."

1522. Simpson, Thomas K. "Maxwell and the Direct Experimental Test of His Electromagnetic Theory." *Isis* 57 (1966), 411-432.

Reports the results of an unsuccessful effort to trace any of Maxwell's speculations concerning the possibility of a direct experimental verification of his theory of electromagnetic propagation through a medium. Shows that many of the materials required for such direct experimentation were available in Maxwell's day, and takes his silence concerning them as evidence that his preoccupations were with different questions, and in particular with the nature of light rather than with electromagnetic phenomena for their own sake.

1523. ———. "A Critical Study of Maxwell's Dynamical Theory of the Electromagnetic Field in the *Treatise on Electricity and Magnetism*." Dissertation, Johns Hopkins University, 1968, 604 pp.

Interprets Maxwell's *Treatise* as a systematic attempt to articulate Faraday's insights in the language of formal mathematical physics. Delineates two different phases in Maxwell's presentation of the dynamical theory in the *Treatise*, one inductive, the second deductive, with the dynamical theory proper appearing in the latter. This two-part form is seen to correspond to Maxwell's wider philosophy of science; the first phase is intended to uncover *a priori* intuitions of fundamental ideas from which the subsequent deductive phase can proceed.

1524. ———. "Some Observations on Maxwell's *Treatise on Electricity and Magnetism*." *Studies in History and Philosophy of Science* 1 (1970), 249-263.

An article abstracted from the author's doctoral thesis (see previous entry). Discusses the relationship between Maxwell's metaphysics and his use of Lagrangian methods in developing his dynamical theory of the electromagnetic field. Argues that, for Maxwell, "the Lagrangian mode of the dynamical theory is not ... simply a convenience for an imperfect stage of a science; it is the appropriate mode for human knowledge of nature, which is essentially relative."

1525. Spitzer, Paul Georg. "Joseph John Thomson: An Unfinished Social and Intellectual Biography." Dissertation, Johns Hopkins University, 1970, 244 pp.

Discusses the social and intellectual milieu in which Thomson rose to prominence, with much detail concerning the style and social position of physics in late Victorian Cambridge, though without reference to Thomson's private papers and correspondence or to other unpublished material. Analyzes, without giving mathematical details, Thomson's publications up to the late 1880s. These are seen as very conventional in character.

1526. Stine, Wilbur Morris. *The Contributions of H.F.E. Lenz to Electromagnetism*. Philadelphia: The Acorn Press, 1923, 157 pp.

Includes a very sketchy biographical chapter, together with a bibliography of Lenz's scientific publications and an account, unfortunately historically exceedingly naive and with little mathematical detail, of his major papers on electromagnetism.

1527. Thompson, Silvanus P. *The Life of William Thomson, Baron Kelvin of Largs*. London: Macmillan, 1910, xx + 1,297 pp. 2 vols.

A typical "life and letters" in the Victorian style and on a grand scale, with generous quotations throughout from Thomson's letters. Appendices list his various academic and other distinctions, his publications (661 items), and patents granted to him.

1528. Topper, David Roy. "J.J. Thomson and Maxwell's Electromagnetic Theory." Dissertation, Case Western Reserve University, 1970, 184 pp.

Describes Maxwell's work in electromagnetism, emphasizing his use in his later papers of the Lagrangian formulation in order to ground the theory satisfactorily on dynamical principles while yet remaining ignorant of the actual mechanical systems presumed to be involved. Shows how Thomson extended this approach in order to show, through the use of cyclic coordinates, that potential energy is formally equivalent to the kinetic energy of hidden motions; and how on this basis he banished forces from his dynamical theory in favor of etherial vortex motions and the like. Argues that Thomson, far from upholding an electromagnetic theory of matter, remained unfailingly committed to the classical program of reducing electromagnetism to dynamical principles.

1529. ———. "Commitment to Mechanism: J.J. Thomson, the Early Years." *Archive for History of Exact Sciences* 7 (1971), 393-410.

An article abstracted from the author's doctoral thesis (see item 1528). Argues that Thomson was committed from the outset of his career to giving a complete mechanical explanation of electromagnetic phenomena, and shows how he accomplished this in a mathematical proof involving the use of cyclic coordinates in Lagrange's equations in order to reduce the potential energy term (and hence the concept of force) to kinetic energy, interpreted as the kinetic energy of hidden motions.

1530. Tricker, R.A.R. *Early Electrodynamics: The First Law of Circulation*. Oxford: Pergamon, 1965, x + 217 pp.

Deals with the development of Ampère's theory of the electrodynamics of steady currents. Includes extensive extracts from Ampère's writings and those of Biot and Savart, shorter ones from papers by Oersted (1820) and Grassmann (1845), two short chapters setting the historical stage, and an 89-page commentary. The latter is not purely historical, but is also concerned with elucidating the logical status of the theory.

1531. ———. *The Contributions of Faraday and Maxwell to Electrical Science*. Oxford: Pergamon, 1966, x + 289 pp.

Includes biographical sketches of both Faraday and Maxwell, straightforward accounts of their work on electromagnetism, and extracts from their chief writings on the subject. Also has an interesting chapter on "The Logical Status of the Law of Electromagnetic Induction."

1532. Turner, Joseph. "A Note on Maxwell's Interpretation of Some Attempts at Dynamical Explanation." *Annals of Science* 11 (1955), 238-245.

Discusses the constraints that Maxwell imposed on the invention of dynamical explanations in physics, especially his requirements (1) that the mechanism proposed be a "consistent representation" in the sense that it is consistent with the fundamental principles of dynamics and (2) that there be some independent evidence for it.

1533. ———. "Maxwell on the Logic of Dynamical Explanation." *Philosophy of Science* 23 (1956), 36-47.

A brief but exceptionally clear exposition of Maxwell's attitude towards his theories, drawing for this purpose the following useful distinctions: (1) a *physical analogy* is a relation between a branch of one science and a branch of another, such that both branches possess the same mathematical form; (2) a *dynamical analogy* is a physical analogy in which one of the branches of science involved is a branch of dynamics; (3) a *dynamical explanation* is a dynamical analogy taken literally. In these terms, Maxwell's 1861 paper is seen as an attempt to provide a dynamical explanation of electromagnetism, while his 1864 paper has the more modest aim of providing a dynamical analogy.

1534. ———. "Maxwell on the Method of Physical Analogy." *British Journal for the Philosophy of Science* 6 (1956), 226-238.

An elaboration of Maxwell's views on the nature and usefulness of physical analogies.

1535. Wangerin, Albert. *Franz Neumann und sein Wirken als Forscher und Lehrer*. Braunschweig: F. Vieweg und Sohn, 1907, x + 185 pp.

Comprises an outline of Neumann's career as "the first exponent of theoretical physics in Germany"; a systematic survey of his published work on crystallography, heat, optics (especially the optical behavior of crystals), and electromagnetism, and in pure mathematics; brief summaries of his lectures as published by his students; a 30-page history of his renowned Königsberg seminar on theoretical physics; and a brief account of his lifelong struggle to have a physics laboratory erected at Königsberg.

1536. Whittaker, Sir Edmund Taylor. *A History of the Theories of Aether and Electricity*. Vol. I: *The Classical Theories*. Vol. II: *The Modern Theories 1900-1926*. London: Thomas Nelson & Sons, 1951-1953. Reprinted New York: Harper Torchbooks, 1960, 434 pp.; 310 pp. 2 vols.

Volume I is a revised edition of a work first published in 1910 under the title *A History of the Theories of Aether and Electricity, from the Age of Descartes to the Close of the Nineteenth Century*. Still the standard history of the subject, and likely to remain so even though recent historiography has challenged some of its premises and recent scholarship some of its conclusions. The treatment of Einstein in Volume II is, however, notoriously unfair. See also items 1972, 2095.

1537. Wiederkehr, Karl Heinrich. "Wilhelm Webers Stellung in der Entwicklung der Elektrizitätslehre." Dissertation, Universität Hamburg, 1960, 254 pp.

A detailed account of Weber's career and scientific work. Emphasizes the importance for Weber of his early collaboration with Gauss.

1538. ———. *Wilhelm Eduard Weber: Erforscher der Wellenbewegung und der Elektrizität 1804-1891.* (Grosse Naturforscher, Bd. 32.) Stuttgart: Wissenschaftliche Verlagsgesellschaft, 1967, 227 pp.

A straightforward account, without technical detail, of Weber's life and scientific work. Based on the author's doctoral dissertation, item 1537.

1539. Williams, L. Pearce. *Michael Faraday: A Biography*. London: Chapman and Hall; New York: Basic Books, 1965, xvi + 531 pp.

An outstanding intellectual biography that carefully traces the development, in interaction with his experimental investigations, of Faraday's theoretical conceptions concerning electricity and magnetism. Some of Williams's opinions, especially concerning the influence of Boscovich's ideas on Faraday, have proved controversial, but in general he presents a plausible account of a highly original thinker.

1540. ———. *The Origins of Field Theory*. New York: Random House, 1966, xii + 148 pp.

Intended as a college text, this work draws heavily upon the same author's much larger intellectual biography of Faraday (see item 1539), but also devotes two chapters to setting the scientific background to Faraday's work, and a concluding chapter to Maxwell's mathematization of Faraday's qualitatively expressed ideas. An excellent introduction to the subject.

1541. Wise, Matthew N. "The Flow Analogy to Electricity and Magnetism: Kelvin and Maxwell." Dissertation, Princeton University, 1977, 303 pp.

Focuses on the use by Kelvin and Maxwell (in his early work) of a flow analogy for electromagnetic forces in order to bring the mathematical techniques developed by Fourier to bear on Faraday's experimentally based conceptions. On this basis presents an interpretation of the origins of electromagnetic field theory which emphasizes the role of mathematical techniques in producing conceptual change.

1542. Woodruff, A.E. "Action at a Distance in Nineteenth Century Electrodynamics." *Isis* 53 (1962), 439-459.

Describes the developments leading up to Weber's formulation of his well-known expression for the force between two moving electric charges, and also the principal difficulties that others, especially Helmholtz and Maxwell, found in Weber's conception.

1543. ———. "The Contributions of Hermann von Helmholtz to Electrodynamics." *Isis* 59 (1968), 300-311.

A good clear account which brings out the relationship of Helmholtz's work to that of his predecessors (Neumann, Weber, Maxwell) and of Hertz.

GEOMETRY

Like the subject itself, histories of geometry rapidly become specialized. The reader should consult other sections, especially for articles on Greek, Moslem, Arabic, Indian, and Chinese geometry. The references given here pertain almost exclusively to Western mathematics since 1600. Even so, some articles are listed elsewhere, under certain centuries (as, for example, a number of references to Gauss). Much recent work on the history of geometry will be found in the biographical entries in the *Dictionary of Scientific Biography*, item 10. However, this bibliography is selective and not every article known has been listed; those given should provide a starting point.

1544. Bachelard, S. *La représentation géométrique des quantités, imaginaires au début du XIX siècle*. Paris: Université de Paris, 1967.

Traces the history of the representation of complex numbers in terms of real numbers or as points in the plane, leading eventually to the formal conception of alternative ways of viewing complex quantities. An emphasis is placed on the large number of mathematicians who arrived at the idea of the geometric representation of complex numbers at about the same time, and the reaction of others to such new ideas.

1545. Bonola, R. *Non-Euclidean Geometry. A Critical and Historical Study of Its Development*. Translated by H.S. Carslaw. New York: Dover, 1955.

See item 1063.

1546. Bos, Hendrik J.M. "On the Representation of Curves in Descartes' *Géométrie*." *Archive for History of Exact Sciences* 24 (1981), 295-338.

Bos discusses Descartes's work with regard to his introduction of new mechanical methods for the production of curves that were considered to be inadmissable in Greek geometry. He attempts to show the connection between this work and Descartes's general mathematical program. See also item 810.

1547. Boyer, Carl B. *History of Analytic Geometry*. New York: Scripta Mathematica, 1956.

See item 785.

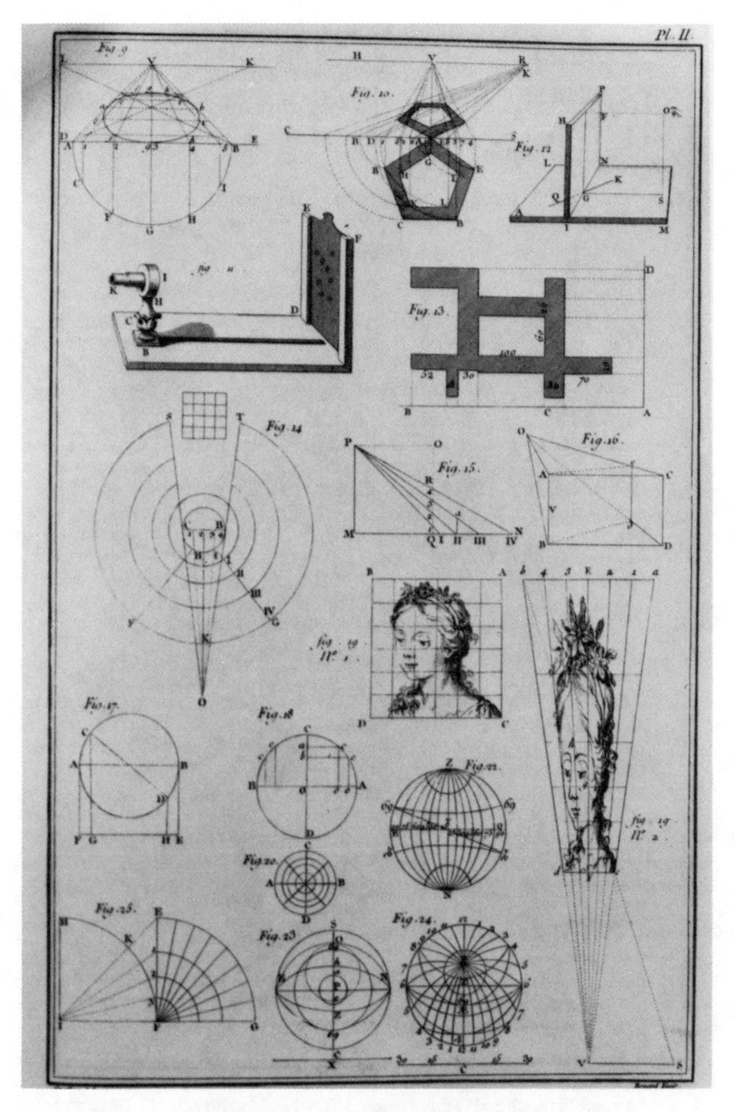

Figure 11. This plate from Diderot's and D'Alembert's famous *Encyclopédie* (1767) illustrates the geometry of perspective, including the principles of projections and vanishing points, and mechanical instruments for demonstrating and establishing the empirical principles of perspective.

1548. Brill, A., and Noether, M. "Die Entwicklung der Theorie der algebraischen Funktionen in älterer und neuer Zeit." *Jahresbericht der Deutschen Mathematiker-Vereinigung* 3 (1894), 107-566.

See item 1079.

1549. Chasles, Michel. *Aperçu historique sur l'origine et le développement des méthodes en géométrie*. 2nd ed. Paris, 1875.

See item 942.

1550. Clagett, Marshall, ed. *Nicole Oresme and the Medieval Geometry of Qualities and Motions. A Treatise on the Uniformity and Difformity of Intensities Known as Tractatus de configurationibus qualitatum*. Madison: University of Wisconsin Press, 1968.

Contains an introduction, English translation, and commentary to Oresme's *Tractatus de configurationibus qualitatum et motuum.* In addition to the geometrical treatment of motion, there are substantial philosophical discourses on aesthetics, music, and magic. This volume also contains a (partial) translation of Oresme's *Quaestiones supes geometriam Euclidis*. See also item 495.

1551. Contro, W.S. "Von Pasch zu Hilbert." *Archive for History of Exact Sciences* 15 (1975-1976), 283-295.

Pasch's axiomatization of projective geometry was part of two distinct lines of work: the axiomatic approach of Peano in Italy and others in Germany, and the empirical approach to the study of axioms. These two lines were eventually brought together by Hilbert to form modern geometric axiomatics.

1552. Coolidge, J.L. *A History of the Conic Sections and Quadric Surfaces*. Oxford: Oxford University Press, 1945.

1553. ———. *A History of Geometrical Methods*. Oxford: Oxford University Press, 1940.

See item 786.

1554. Dąmbska, I. "*An Essay on the Foundations of Geometry* de B. Russell et la critique de ce livre en France dans les années 1898-1900." *Organon* 10 (1974), 145-153.

This paper summarizes Russell's publication (1897) of his Fellowship Examination at Trinity College, Cambridge. Couturat's review and the discussion following in *Revue de Métaphysique et de Morale* are considered. The English original with a foreword by Morris Kline is also available (New York: Dover, 1956).

1555. Dieudonné, Jean. *Cours de géométrie algébrique I*. Paris: Presses Universitaires de France, 1974.

Designed as an introduction to a course in modern algebraic geometry, this work is unfailingly stimulating in its presentation of projective geometry, Riemann and the birational geometry of curves, generalization to higher dimensions, the Italian school, and the rigorization of their work by Zariski and Weil, and culminates in Grothendieck's edifice.

1556. Dou, A.M. "Logical and Historical Remarks on Saccheri's Geometry." *Notre Dame Journal of Symbolic Logic* 11 (1970), 385-415.

Discusses Saccheri's *Euclides ab omni maevo vindicatus* (1733), in which he attempts to prove Euclid's fifth postulate (on parallels). It is now known that Saccheri's axioms may lead to elliptic and hyperbolic as well as Euclidean geometries. The author gives a commentary on Saccheri's methods, and shows their influence on Lambert, Gauss, Bolyai, and Lobachevsky.

1557. Engel, F., and P. Staeckel. *Die Theorie der Parallellinean von Euklid bis auf Gauss*. Leipzig: Teubner, 1895. 2 vols.

Despite the title, the book passes from Euclid to Wallis with no mention of intermediate developments. Prime focus is on the work of Saccheri, Lambert, Gauss, Schweikart, and Taurinus. See also item 1064.

1558. ———. *Urkunden zur Geschichte der nichteuklidische Geometrie*. Leipzig: Teubner, 1899-1913. 2 vols. Reprinted New York and London: Johnson Reprint Corporation, 1972.

Volume 1 concerns Lobachevsky; volume 2, Wolfgang and Johann Bolyai. Each volume includes a frontispiece, the first portraying Lobachevsky, the second, Bolyai. Engel edited volume 1, which provides German translations of two geometric treatises originally written in Russian (235 pp.). Engel then offers detailed notes (pp. 237-344), followed by a lengthy study of Lobachevsky's life and works (pp. 349-445), along with a bibliography and indexes. Volume 2 was edited by Stäckel. It begins with a lengthy, detailed biographical and critical history of the Bolyai family, Wolfgang's work as a professor of mathematics, his son Johann's discovery of "absolute geometry," imaginary numbers, connections with Lobachevsky's works, the theory of parallel lines, etc. The rest of volume 2 provides German translations of selected portions of the writings of both Bolyais.

1559. Fano, G. "Gegensatz von synthetischer und analytischer Geometrie in seiner historische Entwicklung im XIX. Jahrhundert." *Enzyklopädie der mathematischen Wissenschaften*, III. AB4a. Leipzig: B.G. Teubner, 1907-1910, 221-288.

The key actors are Monge, Poncelet, Möbius, Steiner, and Chasles in synthetic geometry, and Möbius, Plücker, v. Staudt in analytic geometry. Algebraic geometry is also discussed through the works of Grassmann, Clebsch, et al., as are the contributions to differential geometry made by Monge, Dupin, Gauss, and Lie.

1560. Freudenthal, H. "Geometry." *Encyclopedia Britannica* 10 (1969), 186-195.

An overview of Greek, projective, and non-Euclidean geometries, the axiomatic method, philosophical questions, geometrical transformations, and differential geometry; concludes with a brief look at topology.

1561. ———. "The Main Trends in the Foundations of Geometry in the 19th Century." *Logic, Methodology, and Philosophy of Science.* Edited by E. Nagel, P. Suppes, and A. Tarski. Stanford: Stanford University Press, 1962.

A penetrating account of the origins and motivation for late 19th-century work on the foundations of geometry. The investigations of Helmholtz, Riemann, Klein, and the recognition of non-Euclidean geometry led to work of Pasch and the Italian school, and culminated in Hilbert's *Grundlagen der Geometrie.*

1562. Galuzzi, Massimo. "Il problema delle tangenti nella *Géométrie* di Descartes." *Archive for History of Exact Sciences*, 22 (1980), 37-51.

See *Isis Critical Bibliography* (1981), item 1738.

1563. Gray, J.J. *Ideas of Space.* Oxford: Clarendon Press, 1979.

See item 1066.

1564. ———, and L. Tilling. "Johann Heinrich Lambert, Mathematician and Scientist, 1728-1777." *Historia Mathematica* 6 (1979), 236-258.

1565. Hawkins, Thomas. "Non-Euclidean Geometry and Weierstrassian Mathematics: The Background to Killing's Work on Lie Algebras." *Historia Mathematica* 7 (1980), 289-342.

See item 1301.

1566. Itard, J. *La Géométrie de Descartes.* (Les Conférences du Palais de la Découverte, Sér. D, Histoire des Sciences, No. 39.) Paris: Université de Paris, 1956, 14 pp.

1567. Jammer, M. *Concepts of Space.* Cambridge, Mass.: Harvard University Press, 1954.

Stresses the philosophy and history behind the concept, especially the theological implications of space independent of matter. Greater emphasis is placed on the ancient and medieval periods.

1568. Kagan, V. *Lobachevsky and His Contribution to Science.* Moscow: Foreign Languages Publishing House, 1957.

Brief (91-page) general appreciation of Lobachevsky, with illustrations, in the series "Men of Russian Science." Shortened translation of *Lobachevskii i ego mesto v nauke* (Moscow, 1955, 301 pp.). For a biographical appreciation of Kagan, see A.M. Lopshits and P.K. Rashevskii, "Venianin Fedorovich Kagan (1869-1953)," *Zamechatel'nye uchenye Moskovskogo Universiteta* 39 (1969), 42 pp., with portrait.

1569. Klein, F. *Elementary Mathematics from an Advanced Standpoint.* Part 2, Geometry. Translated by E.R. Hedrick and C.A. Noble. New York: Dover, 1939; translation of the 1925 Springer 3rd edition.

There is probably no better single introduction to Klein's intuitive, geometric approach than this best seller based on lectures primarily intended for prospective Gymnasium teachers. The "advanced standpoint" particularly refers to the outlook of Klein's Erlangen Program. Offers many key insights into work of Möbius, Plücker, Lie, and others, including of course Klein himself. Applications are stressed throughout.

1570. ———. *Vorlesungen über die Entwicklung der Mathematik im 19. Jahrhundert.* Berlin: Springer Verlag, 1926-1927. 2 vols. Reprinted in 1 vol. New York: Chelsea, 1967.

This great classic thoroughly reflects the author's decidedly geometric outlook. Not only does geometry itself receive a disproportionate amount of attention (Monge and his school, Chasles and his, Möbius, Plücker, Steiner, Staudt, Cayley, etc.), but most of the other subjects are presented in connection with their applicability to geometry. Thus function theory and Riemann surfaces, group theory and crystallography, algebra and geometric curve theory are but a few of the many connections Klein emphasizes. The second volume illustrates the utility of the Erlangen Program and general invariant theory in mathematical physics. See also items 989, 1316, 2086.

1571. Kline, Morris. *Mathematical Thought from Ancient to Modern Times.* New York: Oxford University Press, 1972.

Kline's work has a great deal of coverage on geometry with chapters on Projective Geometry, Coordinate Geometry, Analytic and Differential Geometry in the 18th Century, Revival of Projective Geometry, Non-Euclidean Geometry, Differential Geometry of Gauss and Riemann, Projective and Metric Geometry, Algebraic Geometry, Foundations of Geometry, Tensor Analysis and Differential Geometry, and many other related topics. See also items 82, 867, 990, 1106, 1317, 2145.

1572. Lanczos, C. *Space through the Ages. The Evolution of Geometrical Ideas from Pythagoras to Hilbert and Einstein.* New York: Academic Press, 1970.

Despite the title only one-third history, two-thirds differential geometry, leading up to Einstein's theory of gravitation.

1573. Loria, Gino. *Die hauptsächlichsten Theorien der Geometrie in ihrer früheren und heutigen Entwicklung.* Translated by F. Schütte. Leipzig: B.G. Teubner, 1888.

German translation of *Il passato ed il presente delle principali teorie geometriche.* Excellent for its extensive references to mid-19th-century geometric researches.

1574. ———. "Da Descartes e Fermat a Monge e Lagrange. Contributo alla storia della geometrica analitica." *Reale Accademica dei*

Lincei. Atti. Memoirie della classe di scienze fisiche, mathematiche e naturali, Series 5, 14 (1923), 777-845.

Covers a vast array of work beginning with background in antiquity, fundamental contributions of Descartes and Fermat, the coordinatization of the plane and 3-space, relation to the development of calculus, and the theory of algebraic curves.

1575. ———. "Perfectionnements, evolution, metamorphoses du concept de coordonnées; Contribution à l'histoire de la géométrie analytique." *Mathematica* XVIII (1942), 125-145; XX (1944), 1--2; XXI (1945), 66-83.

1576. Mainzer, K. *Geschichte der Geometrie*. Mannheim: Bibliografisches Institut, 1980.

See item 788.

1577. Molland, A.G. "Shifting the Foundations: Descartes' Transformation of Ancient Geometry." *Historia Mathematica* 3 (1976), 21-49.

Molland discusses Descartes's *Géométrie* with particular reference to his introduction of new mechanical methods for the production of curves in opposition to Greek ideas.

1578. ———. "An Examination of Bradwardine's Geometry." *Archive for History of Exact Sciences* 19 (1978), 113-175.

Examines Bradwardine's *Geometria Speculativa* as an example of the medieval geometric outlook. The study is based on the edition of this work prepared in the author's Ph.D. dissertation (Cambridge University, 1967). See also item 577.

1579. Neuenschwander, E. "Der Nachlass von Casorati (1835-1890) in Pavia." *Archive for History of Exact Sciences* 19 (1978), 1-89.

Describes the contents and significance of Casorati's nearly untouched Nachlass, including his correspondence with a number of foreign mathematicians and his notes on conversations with Weierstrass and Kronecker. See also item 1089.

1580. Reich, K. "Die Geschichte der Differentialgeometrie von Gauss bis Riemann (1828-1868)." *Archive for History of Exact Sciences* 11 (1973), 273-382.

See items 1062, 1717.

1581. Richards, J. "The Evolution of Empiricism: Hermann von Helmholtz and the Foundations of Geometry." *British Journal for the Philosophy of Science* 28 (1977), 235-253.

Stresses the empirical component in Helmholtz's geometric theories, and shows how his researches were integrated around problems in the physiology of perception. This led him to a variety of investigations, and in particular to the geometry of physical space. Also discusses influences of Kant and Riemann.

1582. Scholz, Erhard. *Geschichte des Mannigfaltigkeitsbegriffs von Riemann bis Poincaré*. Boston: Birkhäuser, 1980.

Photo-printed from a typed manuscript, this work was originally produced as a dissertation at the Mathematical Institute of Bonn University, Germany, in 1979. It opens with a discussion of n-dimensional geometry in the first half of the 19th century, and then proceeds to a detailed discussion of Riemann's contributions, those of Beltrami, Helmholtz, Klein, connections with the foundations of geometry, topology of surfaces, complex function theory, Poincaré's work on the topology of higher dimensional spaces, and the connections with the history of "Mannigfaltigkeiten" and modern mathematics. Three appendixes discuss connections with automorphic functions, birational geometry of algebraic structures and Poincaré's study of the second Betti number of algebraic surfaces. Bibliography, but *no* index.

1583. Scott, J.F. *The Scientific Work of René Descartes (1596-1650)*. London: Taylor and Francis, 1952, vii + 211 pp. Reprinted 1976.

Discusses *Discours de La Méthode*, *La Dioptrique*, *Les Météores*, and the *Principia Philosophiae*, but the bulk of the book is an analysis of the contents of *La Géométrie*. See also item 813.

1584. Sommerville, D.M.Y. *Bibliography of Non-Euclidean Geometry*. New York: Chelsea, 1970.

See item 1068.

1585. Staeckel, P. "Gauss als Geometer." In *Materialien für eine Wissenschaftliche Biographie von Gauss*. Edited by F. Klein, M. Brendel, and L. Schlesinger. *Nachrichten der K. Gesellschaft der Wissenschaften zu Göttingen, mathem.-physik. Klasse* 5 (1917). Reprinted in C.F. Gauss, *Werke* 10 (2) (4), Göttingen: Gesellschaft der Wissenschaften, 1922-1933, 1-123.

Chapters on Gauss's work in foundations of geometry, geometry of position, complex numbers and their relation to geometry, elementary and analytic geometry, and the general theory of surface curvature. Includes a bibliography of his work in these areas.

1586. Struik, Dirk J. "Outline of a History of Differential Geometry." *Isis* 19 (1933), 92-120; 20 (1933), 161-191.

Briefly discusses pre-Leibnizian contributions, the work of Euler, Clairaut, Monge and his school, Gauss, the French school of the 1840s, Riemann, Italian contributions, modern developments, sources, and external influences.

1587. Taton, René. *L'oeuvre scientifique de Monge*. Paris: Presses Universitaires de France, 1951, 441 pp.

A brief biography followed by a discussion of Monge's work in descriptive geometry, analytic and differential geometry, and as a precursor of modern geometry, mathematical analysis, and other scientific areas. See also items 923, 1018.

1588. ———. *L'oeuvre mathématique de G. Desargues*. Paris: Presses Universitaires de France, 1951, 232 pp.

Examines Desargues's life and work and its significance. The centerpiece is the text of Desargues's "Brouillon Project." Letters to Mersenne and from Descartes and Beaugrand are included, along with a bibliography of Desargues's work. See also item 809.

1589. Torretti, R. *Philosophy of Geometry from Riemann to Poincaré*. Dordrecht: Reidel, 1978, xiii + 459 pp.

Despite its title, 250 of its 459 pages are devoted to a chiefly historical account of non-Euclidean geometries, including the development of manifolds (following Gauss and Riemann), projective geometry (up to Klein), the Helmholtz-Lie space problem, and axiomatics (Pasch, Peano, and Hilbert). See also items 1582, 2271.

LOGIC

This section covers logic from ancient to modern times, although it excludes from its lists survey articles on very recent developments in a given area. In addition, literature concerned with the philosophy of logic is cited only if it has a noticeably historical character. The term "logic" has been interpreted historically to refer to whatever bodies of knowledge were regarded as "logical" at the time; but choice has been exercised so as to concentrate on works which deal with the bearing of "logic" on mathematics.

General Histories

1590. Bocheński, I.M. *A History of Formal Logic*. Notre Dame, Ind.: University of Notre Dame Press, 1961, xxii + 567 pp. A revised translation of the German original, *Formale Logik*. Freiburg, Munich: Verlag Karl Alber, 1956. Reprinted New York: Chelsea, 1970.

Mainly a selection of quotations from primary sources (all periods). Not a source from which to *learn* the history of logic, but useful passages and extensive bibliography. Divides logic into four varieties: Greek, Scholastic, Mathematical (one-third of text), and Indian. It consists of short texts, in English, and a minimum of connective commentary. Requires some sophistication in logic. Complements Kneale, item 1593, nicely, but both essentially end with *Principia Mathematica* (1910-1913). Contains an extensive bibliography.

1591. Dumitriu, A. *History of Logic*. Tunbridge Wells: Abacus, 1977. 4 vols.

Rather breathless scamper over all periods, and also a wide range of applications of logic.

1592. Jϕrgenson, J. *A Treatise of Formal Logic*. Copenhagen, London: Levin and Munksgaard, 1931. Reprinted New York: Russell and Russell, 1962. 3 vols.

Has some historical information, though not impressive over all.

1593. Kneale, William, and Martha Kneale. *The Development of Logic.* Oxford: Clarendon Press, 1962, viii + 762 pp.

The most complete and best survey of the history of logic from the Greeks to the early years of this century. Half of the book deals with the nineteenth century. It does treat set theory and the foundations of mathematics, and is quite readable by those with very little knowledge of logic.

1594. Prantl, C. *Geschichte der Logik in Abendlände.* Leipzig: S. Hirzel, 1855-1870. Reprinted Leipzig: G. Fock, 1927; Graz: Akademische Druck- und Verlagsanstalt, 1955. 4 vols.

Of interest for the extraordinary *history of* the history of logic, where the decline of logic in the late Renaissance left Prantl knowing absolutely *less* logic than did some of his primary figures.

1595. Prior, A.N., et al. "Logic, History of." *The Encyclopedia of Philosophy.* Vol. 4. Edited by Paul Edwards. New York: Macmillan, 1967, 513-571.

A good introduction to the whole of the history of logic.

1596. Scholz, H. *Geschichte der Logik.* Berlin: Junker und Dünnhaupt, 1931. English translation, *Concise History of Logic*, by K.F. Leidecker. New York: Philosophical Library, 1961.

1597. Styazhkin, N.I. *From Leibniz to Peano....* Cambridge, Mass.: MIT Press, 1969.

1598. Ueberweg, F. *System der Logik und Geschichte der logischen Lehren.* Bonn, 1857, and later editions.

Journals

Articles on the history of logic appear occasionally in journals for the history of science or philosophy, in some journals for logic (especially the *Notre Dame Journal of Formal Logic*), and in the new journal *History and Philosophy of Logic.* Writings on the subject are usually reviewed in the *Journal of Symbolic Logic.*

Bibliographies

1599. Church, Alonzo. "A Bibliography of Symbolic Logic." *The Journal of Symbolic Logic* 1 (1936), 121-218. "Additions and Corrections," 3 (1938), 178-212. Reprinted Princeton: Princeton University Press, 1938.

A comprehensive bibliography of symbolic logic for the period 1666-1935. Items of "especial interest or importance" have been indicated with an asterisk, a judgment which has stood the test of time. The "Additions and Corrections" contains a very fine-grained and extensive subject index. Subsequent volumes of the journal have reviewed nearly all publications in, and closely related to, symbolic logic (including set theory, which was excluded from the original bibliography), and listed the remainder. Volume 26 (1961) is devoted to indices of contributed papers, abstracts, reviews by author, reviews by reviewer, and an extensive index of reviews by subject. Unfortunately, for future historians and current logicians, this comprehensive reviewing policy had to be terminated in 1976 for financial reasons.

1600. Moss, J.M.B., and D. Scott. *A Bibliography of Books on Symbolic Logic*. Oxford: Clarendon Press, to appear.

1601. Risse, W. *Bibliographia Logica III. Verzeichnis der Zeitschriftenartikel zur Logik*. Hildesheim, New York: Olms, 1979.

Classified bibliography of primary literature in logic (and some aspects of the history of logic). Most literature cited is post-World War II. Curiously patchy in coverage, though exhaustive when good.

Editions

Among the mathematicians whose editions were listed in Section II, pp. 14-21, those who were significantly interested in logic included Bolzano, Boole, Brouwer, Hilbert, Leibniz, Peano, and Weyl. In addition, there are the following editions for logicians and philosophers of note:

1602. Frege, G. *Kleine Schriften*. Edited by I. Angelelli. Hildesheim: Olms, 1967, viii + 434 pp.

1603. ———. *Nachgelassene Schriften*. Edited by H. Hermes and others. Hamburg: Meiner, 1969; 2nd ed. 1983, viii + 388 pp. Partial English edition *Posthumous Writings*. Oxford: Blackwell, 1979.

1604. ———. *Wissenschaftlicher Briefwechsel*. Edited by H. Hermes and others. Hamburg: Meiner, 1976, xvi + 310 pp. Partial English edition *Philosophical and Mathematical Correspondence*. Oxford: Blackwell, 1980, ix + 214 pp.

See item 2246.

1605. Gentzen, Gerhard. *The Collected Papers of Gerhard Gentzen*. Edited by M.E. Szabo. (Studies in Logic.) Amsterdam: North-Holland, 1969, xiv + 338 pp.

See item 1114.

1606. Herbrand, Jacques. *Logical Writings*. Edited by Warren D. Goldfarb, translated by Jean van Heijenoort. Cambridge, Mass.: Harvard University Press; Dordrecht, Holland: D. Reidel, 1971, x + 312 pp. From the French edition, *Ecrits logiques*. Edited by J. van Heijenoort. Paris: Presses Universitaires, 1968.

Includes a preface by J. van Heijenoort and a biographical notice by C. Chevalley and A. Lautmann, as well as a note on Herbrand's thoughts by C. Chevalley.

1607. Husserl, E. *Husserliana*. The Hague: Nijhoff, 1950-.

1608. Leśniewski, S. (Edition to appear from Nijhoff.)

1609. Łukasiewicz, Jan. *Jan Łukasiewicz. Selected Works*. Edited by L. Borkowski. (Studies in Logic.) Amsterdam: North-Holland, 1970, xii + 405 pp.

1610. Mostowski, Andrzej. *Foundational Studies. Selected Works*. Edited by K. Kuratowski et al. (Studies in Logic.) Amsterdam: North-Holland, 1979. 2 vols.

1611. Peirce, C.S. *Collected Papers*. Edited by P. Weiss, P. Hartshorne, and A. Burks. Cambridge, Mass.: Harvard University Press, 1931-1958. 8 vols.

Misleading title: astonishingly incomplete. Logic mostly in volumes 3 and 4.

1612. ———. *Writings of Charles S. Peirce. A Chronological Edition*. Editor-in-chief M. Fisch. Indianpolis: Indiana University Press, 1982-.

Logical manuscripts are to appear in some volumes. Volume 1 has just been published. Others are now in preparation.

1613. Ramsey, F.P. *The Foundations of Mathematics*. Edited by R.B. Braithwaite. London: Routledge & Kegan Paul, 1931.

1614. ———. *Foundations*. Edited by D.H. Mellor. London: Routledge & Kegan Paul, 1978, viii + 287 pp.

A different edition from its predecessor; in particular, differently incomplete!

1615. Robinson, Abraham. *Selected Papers*. Edited by H.J. Keisler, S. Körner, W.A.J. Luxemburg, and A.D. Young. New Haven: Yale University Press, 1979. 3 vols.

Volume 1 is devoted to Model Theory and Algebra; Volume 2 to Nonstandard Analysis and Philosophy; Volume 3 contains Robinson's work in Aeronautics. Each volume is preceded by George Seligman's detailed "Biography of Abraham Robinson," and contains critical overviews by the editors of the significance of Robinson's various contributions to the many fields in which he worked. See also item 1128.

1616. Russell, Bertrand A.W. *The Collected Papers*. Edited by K. Blackwell and others. London: Allen and Unwin, 1983-. 28 vols.

Logical and mathematical papers to appear mostly in volumes 2-6.

1617. Skolem, Thoralf. *Selected Works in Logic*. Edited by Jens Erik Fenstad. Oslo: Universitetsforlaget, 1970, 732 pp.

See item 2269.

1618. Tarski, Alfred. *Logic, Semantics, Metamathematics: Papers from 1923 to 1938*. Edited by J.H. Woodger. Oxford: Clarendon Press, 1956, xiv + 472 pp. Rev. ed. by J. Corcoran. Indianapolis: Hackett, 1983, xxx + 506 pp.

See item 2270.

Source Books

1619. Davis, Martin. *The Undecidable. Basic Papers on Undecidable Propositions, Unsolvable Problems and Computable Functions*. Hewlett, N.Y.: Raven Press, 1965, vi + 440 pp.

An anthology of seminal papers on recursion theory by Gödel, Church, Turing, Rosser, Kleene, and Post.

1620. Van Heijenoort, Jean. *From Frege to Gödel. A Source Book in Mathematical Logic, 1879-1931*. Cambridge, Mass.: Harvard University Press, 1967, xi + 660 pp.

A source book par excellence. English versions of the most important papers of mathematical logic up to and including Gödel's work, with excellent introductions by the editor. See also item 2272.

Ancient Logic

1621. Corcoran, J., ed. *Ancient Logic and Its Modern Interpretations*. Dordrecht: Reidel, 1974.

1622. Łukasiewicz, J. *Aristotle's Syllogistic*. 2nd ed. Oxford: Oxford University Press, 1957.

1623. Mates, B. *Stoic Logic*. Berkeley, Los Angeles: University of California Press, 1953.

Medieval and Renaissance

1624. Ashworth, Earline Jennifer. *Language and Logic in the Post-Medieval Period*. (Synthèse Historical Library 12.) Dordrecht, Holland; Boston, Mass.: D. Reidel Publishing Company, 1974, 304 pp.

Contains a systematic exposition of doctrines, e.g., supposition, consequences. Concentrates on the survival of medieval doctrines. Modern figures, such as Descartes and Leibniz, are excluded.

1625. ———. *The Tradition of Medieval Logic and Speculative Grammar from Anselm to the End of the Seventeenth Century. A Bibliography from 1836 Onwards*. (Subsidia Mediaevalia 9.) Toronto: Pontifical Institute of Mediaeval Studies, 1978, 111 pp.

Concentrates on formal logic and semantics. Modern figures, such as Descartes and Leibniz, are excluded. Helpful indices of names, original texts, texts in translation, subjects.

1626. Boehner, Philotheus. *Medieval Logic. An Outline of Its Development from 1250-c. 1400*. Manchester: Manchester University Press, 1952, 130 pp.

1627. ———. *Collected Articles on Ockham*. Edited by Eligius M. Buytaert. St. Bonaventure; New York; Louvain; Paderborn: The Franciscan Institute, 1958, 482 pp.

1628. Maierù, Alfonso. *Terminologia logica della tarda scolastica*. Rome: Edizioni dell'Ateneo, 1972, 687 pp.

Exhaustive account of different uses of a small number of terms including *appellatio*, *copulatio*, *confusio*, and *propositio modalis*.

1629. Moody, Ernest Addison. *Truth and Consequence in Mediaeval Logic*. Amsterdam: North-Holland Publishing Company, 1953, 113 pp.

Classic and stimulating account of formal inferences and semantic paradoxes, but to be used with care: based on a limited number of sources, and the use of modern techniques and concepts can be misleading.

1630. ———. "Medieval Logic" under "Logic, History of." *The Encyclopedia of Philosophy*. Vol. 4. Edited by Paul Edwards. New York, London: Macmillan and Free Press, 1967, 528-534.

1631. Pinborg, Jan. *Logik und Semantik im Mittelalter. Ein Ueberblick*. Stuttgart, Bad Cannstatt: Friedrich Frommann Verlag--Günther Holzboog KG, 1972, 216 pp.

Not so much a history as a brief look at a series of authors and themes. Its only fault is its brevity.

1632. ———, ed. *The Logic of John Buridan*. (Opuscula Graecolatina; Supplementa Musei Tusculani, Vol. 9.) Copenhagen: Museum Tusculanum, 1976, 165 pp.

A very useful collection of papers, exemplifying recent scholarship in medieval logic.

1633. Rijk, Lambertus Marie de. *Logica Modernorum. A Contribution to the History of Early Terminist Logic. Vol. 1. On the Twelfth Century Theories of Fallacy*. Assen: Van Gorcum, 1962, 674 pp. *Vol. 2. Parts One and Two. The Origin and Early Development of the Theory of Supposition*. Assen: Van Gorcum, 1967, 615 pp.; 909 pp.

A monumental compilation of material, particularly valuable for the printing of hitherto unknown texts.

1634. Risse, Wilhelm. *Die Logik der Neuzeit. 1. Band. 1500-1640*. Stuttgart, Bad Cannstatt: Friedrich Frommann Verlag--Günther Holzboog, 1964, 573 pp.; *2. Band. 1640-1780*. Stuttgart, Bad Cannstatt: Friedrich Frommann Verlag--Günther Holzboog, 1970, 748 pp.

Organized according to schools of thought. Very scholarly, but marred by author's ignorance of formal logic.

1635. ———. *Bibliographia Logica. Verzeichnis der Druckschriften zur Logik mit Angabe ihrer Fundorte. Band I. 1472-1800.* Hildesheim: Georg Olms, 1965, 293 pp.

Organized chronologically, but helped by author and very brief subject indices. Includes a lot of non-logic.

1636. ———. *Bibliographia Logica. Band IV. Verzeichnis der Handschriften zur Logik.* Hildesheim, New York: Georg Olms, 1979, 390 pp.

List of logic manuscripts from the 7th to the 18th century. Two main sections, medieval authors and more recent authors, organized alphabetically. Good indices.

1637. Spade, Paul Vincent. "Recent Research on Medieval Logic." *Synthèse* 40 (1979), 3-18.

Deals mainly with work published since 1960.

1638. Thomas, Ivo. "Interregnum" under "Logic, History of." *The Encyclopedia of Philosophy*. Vol. 4. Edited by Paul Edwards. New York, London: Macmillan and Free Press, 1967, 534-537.

Useful brief introduction to the period between the Middle Ages and Leibniz.

17th and 18th Centuries

1639. de Condillac, E. *Logique. Logic.* Translated and edited by W.R. Albury. New York: Abaris, 1980.

A translation *en face*; a photoreproduction of the 1798 original, together with an excellent editor's introduction.

1640. Wolters, G. *Basis und Deduktion. Studien zur Entstehung und Bedeutung der Theorie der axiomatischen Methode bei J.H. Lambert (1728-1777).* Berlin: de Gruyter, 1980.

Boolean Algebra and Algebraic Logic

1641. Barone, F. "Peirce e Schröder." *Filosofia* 17 (1966), 181-224.

1642. Boole, M.E. *Collected Works*. London: Daniel, 1931. 4 vols.

Amidst much crankiness, useful and probably authoritative comments on Boole's conception of his logic (see L.M. Laita, "Boolean Algebra and Its Extra-Logical Sources: The Testimony of Mary Everest Boole," *History and Philosophy of Logic* 1 [1980], 37-60).

1643. Grattan-Guinness, Ivor. "Wiener on the Logics of Russell and Schröder." *Annals of Science* 32 (1975), 103-132.

See item 1144.

1644. Rosado Haddock, G.E. "Edmund Husserls Philosophie der Logik und Mathematik in Lichte der gegenwärtigen Logik und Grundlagenforschung." Dissertation, University of Bonn, 1973.

1645. Smith, B., ed. *Parts and Moments. Studies in Logic and Formal Ontology*. Munich: Philosophia, 1982, 564 pp.

Includes extensive bibliography of post-Boolean developments in part-whole logic.

1646. Smith, G., ed. *The Boole-De Morgan Correspondence 1842-1864*. Oxford: Clarendon Press, 1982, 156 pp.

Mathematical Logic

1647. Bunn, R. "Developments in the Foundations of Mathematics." In *From the Calculus to Set Theory, 1630-1910*. Edited by I. Grattan-Guinness. London: Duckworth, 1980, 220-255.

1648. Chihara, C.S. *Ontology and the Vicious-Circle Principle*. London, Ithaca, N.Y.: Cornell University Press, 1973.

See item 2237.

1649. Church, Alonzo. *Introduction to Mathematical Logic*. Vol. I. Princeton, N.J.: Princeton University Press, 1956, x + 378 pp.

This highly respected monograph contains numerous insightful historical comments on its main themes: propositional, first, and second order logic. See also item 2238.

1650. Curry, Haskell B. *Foundations of Mathematical Logic*. New York: McGraw-Hill, 1963. Reprinted New York: Dover, 1977, viii + 408 pp.

A high-level textbook with numerous historical remarks.

1651. Grattan-Guinness, Ivor. *Dear Russell-Dear Jourdain: A Commentary on Russell's Logic, Based on His Correspondence with Philip Jourdain*. New York: Columbia University Press; London: Duckworth, 1977, vi + 234 pp.

Deals with the period 1902-1919 when Russell was writing *Principia Mathematica* and doing his most important work in logic. Extensive editorial comments connect the excerpts from the letters. Treats Jourdain's obsession with well ordering and his (somewhat disappointing) notes on the *Principia*. A very useful 25-page bibliography is included. See also items 2110, 2248.

1652. Guillaume, M. "Axiomatique et logique." In *Abrégé d'histoire des mathématiques 1700-1900*. Vol. 2. Edited by J. Dieudonné. Paris: Hermann, 1978, 315-430.

Despite the title of the book, the article has a substantial section on 20th-century mathematical logic. The Dieudonné volumes are also discussed above in item 1270.

1653. Hermes, H. "Zur Geschichte der mathematischen Logik und Grundlagenforschung in den letzten fünfundsiebzig Jahren." *Jahresbericht der Deutschen Mathematiker-Vereinigung* 68 (1966), 75-96.

Considers the history of mathematical logic and foundational studies since 1890.

1654. Jourdain, Philip E.B. "The Development of the Theories of Mathematical Logic and the Principles of Mathematics." *Quarterly Journal of Pure and Applied Mathematics* 41 (1910), 324-352; 43 (1912), 219-314; 44 (1913), 113-128.

Includes annotations from logicians to whom Jourdain submitted his manuscript on their work for comments.

1655. Quine, W.V.O. *Set Theory and Its Logic*. 2nd ed. Cambridge, Mass.: Harvard University Press, 1969, xiv + 361 pp.

Useful footnotes and remarks on the bearing of logic on the axiomatizations of set theory.

1656. Thiel, C. *Sinn und Bedeutung in der Logik Gottlob Freges*. Meisenheim am Glan: Hain, 1965. English translation, *Sense and Reference in Frege's Logic*. Dordrecht: Reidel, 1968, ix + 172 pp.

20th-Century Logic (General)

1657. Chang, C.C. "Model Theory 1945-1971." *Proceedings of the Tarski Symposium*. Edited by Leon Henkin et al. (Proceedings of Symposia in Pure Mathematics, Vol. 25.) Providence, R.I.: American Mathematical Society, 1974, 173-186.

Presents a diagram of the main areas of research in model theory and a diagram of those topics which influenced later work.

1658. Goldfarb, W. "Logic in the Twenties: The Interpretation of the Quantifier." *Journal of Symbolic Logic* 44 (1979), 351-368.

1659. Grattan-Guinness, Ivor. "On the Development of Logics Between the Two World Wars." *American Mathematical Monthly* 88 (1981), 495-509.

Covers not only the principal philosophies of mathematics but also first- versus higher-order logics and the status of infinitary logics, recursion and computability, non-classical logics, Polish contributions to logic, and the profession of logic.

1660. Kleene, Stephen C. "The Work of Kurt Gödel." *The Journal of Symbolic Logic* 41 (1976), 761-778.

A high level, but quite understandable, summary and evaluation of the work of Kurt Gödel (1906-1978), undoubtedly the most important logician of the twentieth century. For additional biographical information as well as a list of Gödel's 25 publications, see Hao Wang, "Kurt Gödel's Intellectual Development," *The Mathematical Intelligencer* 1 (1978), 182-185.

1661. Mangione, C. "La logica del ventesimo secolo." In *Storia del pensiero filosofico e scientifico*. Edited by L. Geymonat. Milan: Garzanti, Vol. 6, 1972, 469-682; Vol. 7, 1976, 299-433.

1662. Moore, G.H. "Beyond First Order Logic: The Historical Interplay between Mathematical Logic and Axiomatic Set Theory." *History and Philosophy of Logic* 1 (1980), 95-137.

See item 1149.

1663. Mostowski, Andrzej. *Thirty Years of Foundational Studies. Lectures on the Development of Mathematical Logic and the Study of the Foundations of Mathematics in 1930-1964*. Helsinki: Acta Philosophica Fennica, no. 17, 1965; New York: Barnes and Noble, 1966, 180 pp.

An expository work on the development of logic and the foundations of mathematics by an important participant who is willing to state his own judgments. See also item 1151.

1664. Vaught, R.L. "Model Theory before 1945." *Proceedings of the Tarski Symposium*. (Proceedings of Symposia in Pure Mathematics, Vol. 25.) Edited by Leon Henkin et al. Providence, R.I.: American Mathematical Society, 1974, 153-172.

Model theory here refers to the study of the relationships between sets of sentences in a formal language and the structures which satisfy them. This is a careful treatment of the history of the deepest results in model theory: the Löwenheim-Skolem theorem, Gödel's completeness theorem for first-order logic, and the seminal work of Tarski, especially his definition of truth. A bibliography of original papers is appended.

Polish Logic

1665. Jordan, Zbigniew A. "The Development of Mathematical Logic in Poland between the Two Wars." In *Polish Logic 1920-1939. Papers by Ajdukiewicz, Chwistek, Jaśkowski, Jordan, Leśniewski, Łukasiewicz, Słupecki, Sobociński, and Wajsberg*. Oxford: Clarendon Press, 1967, 346-397.

Poland's significant contributions to logic, some of which are published here in English for the first time, are outlined in this essay which was first published in 1944.

1666. MacCall, S., ed. *Polish Logic 1920-1939*. Oxford: Clarendon Press, 1967.

Selection of some Polish writings. See Review by W.A. Pogorzelski in *Journal of Symbolic Logic* 35 (1970), 442-446.

1667. Rickey, W.F. *An Annotated Leśniewski Bibliography*. Bowling Green: Bowling Green State University, 1972. Supplement 1976.

Non-Classical Logics

1668. Lewis, C.I. *A Survey of Symbolic Logic*. Berkeley, California: University of California Press, 1918, vi + 406 pp. Reprinted New York: Dover, 1960, xi + 327 pp. (Chapters V and VI omitted).

Valuable for 19th-century logic. Excellent bibliography. See also item 2256.

1669. Rescher, N. *Many-Valued Logic*. New York: McGraw-Hill, 1969.

Contains extensive bibliography.

1670. Rutz, P. *Zweiwertige und mehrwertige Logik. Ein Beitrag zur Geschichte und Einheit der Logik*. Munich: Ehrenwirth, 1973.

Philosophy of Logic

1671. Bowne, G.D. *The Philosophy of Logic 1880-1908*. The Hague: Mouton, 1966.

1672. Haack, S. *Philosophy of Logics*. Cambridge: Cambridge University Press, 1978, xvi + 276 pp.

1673. Rostand, F. *Sur la clarté des démonstrations mathématiques*. Paris: Vrin, 1962, 166 pp.

Fine, little-known study of errors in proofs, including some logical errors.

1674. Ruzavin, G.I. *O prorode matematicheskogo znaniya....* Moscow: Misl'., 1968.

MATHEMATICAL PHYSICS

The name "mathematical physics" can be extended to cover almost all of mathematics and physics, as well as major sections of every other science, such as engineering, biomechanics, geophysics, physical chemistry. This bibliography restricts the term to the following five areas: mechanics, elasticity, fluid dynamics, heat conduction, and sound. To these five has been added a general bibliography of works covering more than one field or works of outstanding importance. Each section has its own introduction.

No attempt has been made at a comprehensive listing here as the field is too vast; but the choice of histories, texts, and reviews, with a leavening of original sources, has been made so that no major idea or publication will be overlooked after diligent consultation of those actually listed. The five topics are of unequal weight: Fluid Mechanics and Elasticity outweigh the other three since mechanics, heat, and sound form part of these topics.

To most of the sections have been added histories of major mathematical methods used in that topic; for example, differential geometry is used in elasticity theory. Again, some collateral topics have been mingled with each main topic; for example, crystallography has been combined with heat conduction. Further, some national histories have been sprinkled over the six sections, with the location of these histories depending on which of the topics features in the history.

The period most intensively covered is 1700-1900, with some protrusion at either end. Many original sources have been deliberately though reluctantly omitted in order to keep the bibliography within reasonable bounds; since almost all of the originals are featured in Truesdell's various works, not too much pain will be occasioned by this.

General Studies

Into this section are collected all the general methods of mathematical physics: vector analysis, series, transforms, differential equations, general histories of mathematical physics, and related technology. The mathematical methods concentrated on here are twofold: those for solution of differential equations, with Burkhardt on series solutions, Deakin on the Laplace transform solution, Schlissel on asymptotic solutions, and Demidov on general partial differential equations; and those of vector analysis, in the texts of Gibbs and Weatherburn.

There are several histories: Auerbach, Dugas, Grigor'yan, Rosenberger, and Truesdell on mechanics or physics in general, and that of Singer on later 19th-century technology; to these could be added the history by Szabó listed under Mechanics. The several texts--Auerbach-Hort, Courant-Hilbert, Flügge, Gibbs-Wilson, Kneser, Thomson and Tait, Weatherburn--are mixed with several general histories of wide ranging topics and a few major sources, such as Green, Helmholtz. While most of these could find homes in the various sections, almost all of them cover three or more sections or are unique of their kind--the articles by Burkhardt and Deakin, for example.

1675. Abro, A. d'. *The Rise of the New Physics: Its Mathematical and Physical Theories*. New York: Dover, 1951, xi + 1-426 pp. + 24 portraits and v + 427-982 pp. + 12 portraits. 2 vols. Originally published as *Decline of Mechanism*. New York: Van Nostrand, 1939.

Volume 1 covers the classical period to circa 1900, while volume 2 deals with the "new" theories of relativity, quantum mechanics, and the associated models. The emphasis is on sketching the physics, with a not-too-heavily mathematical treatment. There are almost no references, and while useful as background, this is not a direct reference for the history of mathematics or physics.

1676. Auerbach, Felix. *Entwicklungsgeschichte der modernen Physik, zugleich eine Uebersicht ihrer Tatsachen, Gesetze und Theorien*. Berlin: Springer, 1923, viii + 344 pp.

This is an exposition of physics rather than a history. There are two parts to the book: nine general chapters deal with the development of physics from space, time, matter, and energy

nil impediret, recta pergeret ad *c*, (per leg. 1.) deſcribens lineam *Bc* æqualem ipſi *AB*; adeo ut radiis *AS*, *BS*, *cS* ad centrum actis, confectæ forent æquales areæ *ASB*, *BSc*. Verum ubi corpus venit ad *B*, agat vis centripeta impulſu unico ſed magno, efficiatque ut corpus de recta *Bc* declinet & pergat in recta *BC*. Ipſi *BS* parallela agatur *cC*, occurrens *BC* in *C*; & completa ſecunda temporis parte, corpus (per legum corol. 1.) reperietur in *C*, in eodem plano cum triangulo *ASB*. Junge *SC*; & triangulum *SBC*, ob

LIBER PRIMUS.

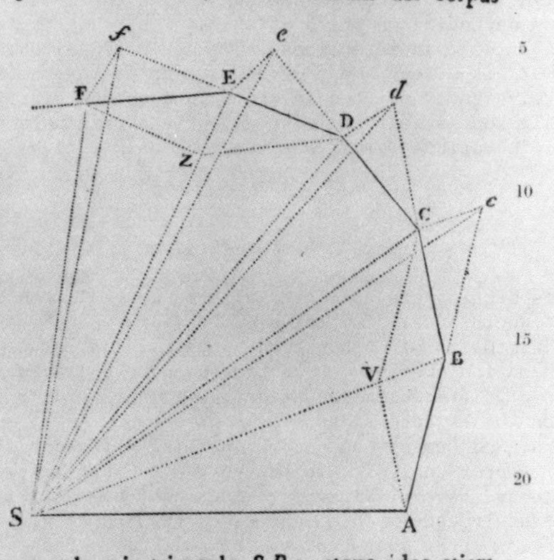

parallelas *SB*, *Cc*, æquale erit triangulo *SBc*, atque ideo etiam triangulo *SAB*. Simili argumento ſi vis centripeta ſucceſſive agat in *C*, *D*, *E*, &c. faciens ut corpus ſingulis temporis particulis ſingulas deſcribat rectas *CD*, *DE*, *EF*, &c. jacebunt hæ omnes in eodem plano; & triangulum *SCD* triangulo *SBC*, & *SDE* ipſi *SCD*, & *SEF* ipſi *SDE* æquale erit. Æqualibus igitur tempori-

Figure 12. This is the diagram accompanying Proposition I, Theorem I of Section II of Newton's *Principia*. Newton applied the infinitesimal calculus in proving the proposition, by allowing the triangles ASB, BSC, etc., to increase in number while their breadth diminished "in infinitum." The proposition itself is related to the determination of centripetal forces, and reads as follows (not shown on this page): "The areas which revolving bodies describe by radii drawn to an immovable center of force lie in the same immovable planes, and are proportional to the times in which they are described."

through classical mechanics and thermodynamics, then vibrations, wave motion, heat conduction, and radiation through to black-body radiation, quantum mechanics, and relativity. Then seven special chapters deal with individual topics of mathematical physics: rigid bodies, elastic solids, fluids, sound, heat, electromagnetism, and light. The author was at the center of developments in physics that he describes.

1677. ———, and Wilhelm Hort. *Handbuch der physikalischen und technischen Mechanik*. Leipzig: Johann Ambrosius Barth, 1926-1931. 7 bände.

This is typical of all such handbooks and encyclopedias. The seven volumes run to a thousand or so pages each, and mix engineering, technology, mathematics, and physics, with articles by many different authors. Korn, Nemenyi, Lichtenstein, and others feature in its pages, and the articles are well referenced in their footnotes as well as at their ends.

1678. Bikermann, Jacob Joseph. "Capillarity before Laplace: Clairaut, Segner, Monge, Young." *Archive for History of Exact Sciences* 18 (1978), 103-122; "Theories of Capillary Attraction." *Centaurus* 19 (1975), 182-206.

The first of these articles relates the work of the authors in the title to that of Laplace; it is mainly polemical and about priority, and has 30 references (at the end). The second article covers Mariotte, Laplace, Young, Poisson, Franz Neumann, and Dupré, and chides physicists about the lack of modern experimental work done on the hypotheses of these people; there are 17 references.

1679. Burkhardt, Heinrich Friedrich Karl Ludwig. "Entwicklungen nach oscillirenden Funktionen und Integration der Differentialgleichungen der mathematischen Physik." *Jahresbericht der Deutschen Mathematiker-Vereinigung* 10 (1901-1908), viii + 1804 pp. This was also issued separately in two volumes, Leipzig: B.G. Teubner, 1908, xii + xii + 1800 pp. Reprinted New York: Johnson Reprint Corp., 1960.

This wonderful report with several thousand references, mainly and unfortunately in the 9,000 footnotes, concentrates on analysis rather than mathematical physics, although vibrating bodies and other applications are discussed at length. It covers this field to 1850, with only a small number after this date, and is a report in the tradition of Hilbert, Brill-Noether, and Minkowski in this series. A somewhat condensed version of this review appears in the *Encyklopädie der mathematischen Wissenschaften mit Einschluss ihrer Anwendungen*, as two articles, again by Burkhardt, "Trigonometrische Interpolation (Mathematische Behandlung periodischer Naturerscheinungen)," in volume II (A9a), pp. 642-694, and again "Trigonometrische Reihen und Integrale bis etwa 1850," in volume II (A12), pp. 819-1354; see item 1256. The latter review is extended by that of Emil Hilb and Marcel Riesz in Part 3, "Neuere Untersuchungen über trigonometrische Reihen" (1922), pp. 1189-1228, where the story is taken as far as Reimann; other articles in Part 3 of Volume II give further information on this and related topics. For literature up to 1920, see Maurice Marie Albert Lecat, *Bibliographie des séries trigonométriques, avec un appendice sur le calcul des variations* (Louvain: n.p., 1921, viii + 167 pp.). There is a *Compléments* (Louvain, 1924, 15 pp.). See also item 1418.

1680. Courant, Richard, and David Hilbert. *Methoden der mathematischen Physik*. Berlin: Springer, Band I, 1924; Band II, 1937. There is an English translation, *Methods of Mathematical Physics*. New York: Interscience, 1953.

A good description of the writing of these volumes is given in Constance Reid's *Courant in Göttingen and New York: The Story of an Improbable Mathematician* (New York: Springer, 1976, pp. 198-199), where Courant's preference for the by-then-dated "classical" approach to integral equations and existence theorems for partial differential equations is indicated. The material and style of these volumes dominated the world stage for a generation, as "the" approach to the solution of linear (second-order) partial differential equations; cf. also M.S. Berger, *Bulletin of the American Mathematical Society*, n.s. 4 (1981), 362-368.

1681. Deakin, Michael A.B. "The Development of the Laplace Transform, 1737-1937. I. Euler to Spitzer, 1737-1880"; "The Development of the Laplace Transform, 1737-1937. II. Poincaré to Doetsch, 1880-1937." *Archive for History of Exact Sciences* 25 (1981), 343-390; 26 (1982), 351-381.

These two articles present a thoroughly documented (over 200 references) and balanced picture of the mathematical and engineering history of this important theoretical and practical tool. See also item 1267.

1682. Demidov, Sergei S. "Vozniknovenie teorii differentisial'nykh uravnenii s chastnymi proizvodnymi" (Origins of the theory of partial differential equations). *Istoriko-Matematicheskie Issledovaniia* 20 (1975), 204-220.

A short introduction to Euler and d'Alembert on partial derivatives in differential equations, with fifteen references. See also item 938.

1683. Dugas, René. *Histoire de la mécanique*. Preface de Louis de Broglie. Paris: Dunod; Neuchatel: Griffon, 1950, 649 pp. There is an English translation by J.R. Maddox, *A History of Mechanics*. New York: Central Book Company, 1955, 671 pp.

This major history is a series of accounts of the work of individuals, in five general periods: to the Renaissance; the 17th century, the century of formation; the 18th, century of organization; then Lagrange and after; and finally the 19th to 20th centuries. It is impossible that the treatment on such a scale be fully balanced; see the reviews quoted.

Reviews: Cohen, I. Bernard, *Isis* 42 (1951), 271-272.
Truesdell, Clifford, *Mathematical Reviews* 14 (1953), 341-343.

1684. Duhem, Pierre Maurice Marie. *La théorie physique: son objet et sa structure*. Paris: Chevalier and Rivière, 1906. 2nd ed. as *La théorie physique: son objet--sa structure*. Revue et augmentée. Paris: Rivière, 1914, viii + 514 pp.

This history and philosophy of physics describes the place of mechanical models, mathematical deduction, and experiment in theoretical physics; it argues that the nature of physical laws is such that a whole theory, not just a single law, is falsified or destroyed by experimental evidence to the contrary. See also his *Evolution de la mécanique* (Paris: Joanin, 1903, 348 pp.).

1685. Flügge, S., ed. *Encyclopedia of Physics--Handbuch der Physik*. Band III, Teile 1 & 3. Berlin: Springer, 1960 (Part I); 1965 (Part 3).

This volume deals with mechanics, and the articles of most importance are the following: from Part 1, "The Classical Field Theories," pp. 226-793, by C. Truesdell and R.A. Toupin, and the appendix to this article, "Tensor Fields," pp. 794-858, by J.L. Ericksen. From Part 3, "The Non-Linear Field Theories of Mechanics," by C. Truesdell and W. Noll, which occupies all viii + 602 pages of this Part.

With nearly 2,000 references, and an overview of the whole area of mathematical physics, these three give a résumé, with historical foundations, of the geometry, kinematics, balance equations, and response-functions to 1960, as well as the mechanics of fluids, elastic solids, and other more exotic materials to 1965. Thermodynamics and electromagnetism are not neglected. The references to the older literature, back to 1680, complement the three reviews by Truesdell in the *Leonhardi Euleri Opera Omnia* and the *Mechanical Foundations*, and provide rapid and accurate entry into the literature.

1686. Frankel, Eugene. "Jean-Baptiste Biot: The Career of a Physicist in 19th-Century France." Dissertation, Princeton University, 1972, vii + 404 pp.

This history depicts Biot, Laplace, and Arago in the period 1795-1830. It details some of the research done by Biot and Arago, and argues that science became a profession to be followed, with second-rate workers appearing in some of the roles. It is very good for a general overview of the period and of the force towards the mathematicization of physics under Poisson, Laplace, Biot, Ampère, and others. There are 300 references at the end of the thesis.

1687. ———. "J.-B. Biot and the Mathematicization of Experimental Physics in Napoleonic France." *Historical Studies in the Physical Sciences* 8 (1977), 33-72.

The period is reviewed for trends in mathematicization rather than details. For details one should consult the work of Ivor

Grattan-Guinness, Rod Home, John Clark, and others quoted by them. The 100 or so references lie in the footnotes.

1688. Gibbs, Josiah W., and Edwin Bidwell Wilson. *Vector Analysis: A Text-book for the Use of Students of Mathematics and Physics, Founded upon the Lectures of J. Willard Gibbs*. New York: Charles Scribner's Sons, 1901. 2nd ed. 1909, xviii + 436 pp. Reprinted New York: Dover, 1960.

Geometry, algebra, calculus (both differential and integral), and application of vectors appear in a text devoted solely to vector analysis. There are only three references: to Gibbs's original pamphlet of 1880-1881 (reprinted in *The Scientific Papers of J. Willard Gibbs*, Vol. 2, pp. 17-90), to Heaviside's *Electromagnetic Theory*, and to Föppl's lectures on *Die Maxwell'-sche Theorie der Electrizität* (Leipzig and Berlin: B.G. Teubner, 1912). The applications to gravity, electromagnetism, elasticity, relativity, and optics became the standard prescription for texts on vector and tensor calculus by Weatherburn, McConnell, Levi-Civita, and others.

1689. Gillmor, C. Stewart. *Coulomb and the Evolution of Physics and Engineering in 18th-Century France*. Princeton, N.J.: Princeton University Press, 1971, xvii + 328 pp.

With nearly 300 references listed at its end, this book is an invaluable source for the work of Coulomb and others on friction, torsion, electrostatics, and civil engineering in this period. Coulomb was one of the first to use mathematics in manpower planning (he published a description of this work in 1799). As with many other biographies of scientific men, Coulomb's life is here written separately from the report of his work: the force driving him to create is not tied to the creation. The dilemma facing all writers of such biographies is the readership of his work: does the writer aim for a general audience that understands men but not much science, or for the restricted one that has some understanding of both? See also item 1474.

1690. Grattan-Guinness, Ivor. "Mathematical Physics in France, 1800-1840: Knowledge, Activity and Historiography." In *Mathematical Perspectives: Essays on Mathematics and Its Historical Development*. Edited by J. Dauben. New York: Academic Press, 1981.

This work covers the whole range of mathematical physics in France: research, education, use, and application, with approximately 100 references listed at the end together with several, very informative tables.

1691. Green, George. "On the Laws of Reflexion and Refraction of Light at the Common Surface of Two Non-Crystallized Media." *Transactions of the Cambridge Philosophical Society* 7 (1842), 1-24.

This marvellous piece of what Saint-Venant disparagingly called "pure analysis," contains (1) the Lagrangian approach to mechanics via d'Alembert's principle, (2) expansions in Eulerian homogeneous functions (polynomials), (3) the Cauchy-Green tensor for elastic deformations, (4) infinitesimal Lie-group generators of the isotropic symmetry group, (5) the Stokes elasticity equations via the elastic potential, and all this in eight pages.

1692. Grigor'yan, Ashot Tigranovich, and Iosif Benediktovich Pogrebysskii. *Istoriya Mekhaniki. S drevneishikh vremen do kontsa XVIII veka* (History of mechanics. From earliest times to the end of the 18th century). Moscow: Nauka, 1971, 300 pp., and *Istoriya Mekhaniki. S kontsa XVIII veka do serediny XX veka* (History of mechanics. From the end of the 18th century to the middle of the 20th). Moscow: Nauka, 1972, 416 pp.

This history, a sequence of articles by a score of authors, has thousands of references but very little mathematics. The volumes have compressed bibliographies for each topic and an index of persons. The meaning of mechanics extends to all of the usual topics in classical mathematical physics; thus these two volumes are useful as a general guide but are not detailed enough to be the leader in this field.

1693. Helmholtz, Hermann von. *Ueber die Erhaltung der Kraft: eine physikalische Abhandlung, vorgetragen in der sitzung der physikalischen Gesellschaft zu Berlin am 23. Juli 1847.* Berlin: Reimer, 1847, 71 pp. Reprinted as *Ostwalds Klassiker*, Nr. 1, Leipzig: Wilhelm Engelmann, 1902, 60 pp., and in Helmholtz's *Wissenschaftliche Abhandlungen*. Leipzig: J.A. Barth, 1822-1825, Bd. I, S., pp. 12-75. Both these reprints contain *Zusätze* added by Helmholtz in 1881.

This wide ranging paper, covering as it does mechanical, thermal, electrical, and electromagnetic forces and energy, is the foundation paper for the principle of conservation and equivalence of the various forms of energy, despite Mayer.

1694. Kneser, Adolf. *Die Integralgleichungen und ihre Anwendungen in der mathematischen Physik: Vorlesungen an der Universität zu Breslau.* Braunschweig: Friedrich Vieweg, 1911, viii + 243 pp.

Integral equations from Fredholm are applied to heat conduction, vibrations, in one-, two-, and three-dimensions, as well as to the Dirichlet problem. Fifty references are listed at the end. This book completely rewrote this section of mathematical physics before Courant-Hilbert.

1695. Minkowski, Hermann. "Kaplillarität." *Encyklopädie der mathematischen Wissenschaften mit Einschluss ihrer Anwendungen*, V.9. Leipzig: B.G. Teubner, 1907, 558-613.

This lucid article gives a solid historical and mathematical introduction to capillarity as well as some experimental detail; it provides a mathematical counterweight for the articles by Bikermann. There are about 100 references, mostly in the footnotes.

1696. Rosenberger, Ferdinand. *Die Geschichte der Physik in Grundzügen, mit synchronistischen Tabellen der Mathematik, der Chemie und beschreibenden Naturwissenschaften, sowie der allgemeinen Geschichte.* Braunschweig: Friedrich Vieweg. Band I, 1882, ix + 175 pp. Band II, 1884, vii + 407 pp. Band III, 1887-1890, xiii + 826 pp. Reprinted Hildesheim: Georg Olms, 1965.

This is a general history without much mathematics. The volumes cover, respectively, the history of physics from earliest times to the Middle Ages; then from 1600 to 1780; and finally the next hundred years to 1880.

1697. Schlissel, Arthur. "The Development of Asymptotic Solutions of Linear Ordinary Differential Equations, 1817-1920." *Archive for History of Exact Sciences* 16 (1976/1977), 307-378.

With 100 references to all the masters, this is the key article on asymptotic methods for differential equations with large parameters. The references relate to applications as well as theory. See also items 1349, 1440.

1698. ———. "The Initial Development of the WKB Solutions of Linear Second Order Ordinary Differential Equations and Their Use in the Connection Problem." *Historia Mathematica* 4 (1977), 183-204.

The development of a method for obtaining approximate solutions to ordinary differential equations whose coefficients are not necessarily analytic is sketched for the period 1840-1930. The method is ascribed to Wentzel, Kramers, and Leon Brillouin, all dealing with quantum theory and wave theory of light, but many others contributed from various fields of application: Green (water and light waves), Rayleigh (waves), Gans (light), Stokes (sun spectra), Jeffreys (geology), and these things are indicated in the 40 or so references. See also items 1096, 1155, 1441.

1699. Singer, Charles Joseph, et al. *A History of Technology. Vol. 5: The Late Nineteenth Century, c. 1850 to c. 1900*. Oxford: Oxford University Press, 1958, xxxviii + 888 pp. + 44 pls.

This volume is good for applications of mathematical solutions to technology and engineering in the period although it does not itself contain any mathematics.

1700. Smith, Crosbie W. "'Mechanical Philosophy' and the Emergence of Physics in Britain: 1800-1850." *Annals of Science* 33 (1976), 3-29.

The article details the slow mathematicization of physics in Britain, i.e., the transition from observation to theoretical explanation as in Thomson, Rankine, Tait, and Maxwell. The 100 or so references are in the footnotes. The article is descriptive rather than mathematical.

1701. Thomson, Sir William, and Peter Guthrie Tait. *Treatise on Natural Philosophy*. Oxford: At The Clarendon Press, 1867, xxiii + 727 pp. The first and only published volume of a projected four. 2nd ed. published in two parts, Part 1 in 1879, xvii + 508 pp., and Part 2 in 1883, xxvii + 527 pp.

The British bible of mathematical physics, it was very widely used and translated. Despite this, it was cut to just the "physics" and published as the *Elements of Natural Philosophy* in 1873 and 1879, to suit British students who lacked adequate mathematical background. The book gave the first accessible proof of Stokes's theorem, and was one of the first texts on hydrodynamics and elasticity.

1702. Truesdell, Clifford A. "The Mechanical Foundations of Elasticity and Fluid Dynamics." *Journal of Rational Mechanics and Analysis* 1 (1952), 125-300; with corrections, 2 (1953), 593-616. Reprinted as *Continuum Mechanics I: The Mechanical Foundations of Elasticity and Fluid Dynamics*. New York: Gordon and Breach, 1966, xvi + 218 pp.

This classic article is a general exposition of deformable masses, solids, liquids, and others, as this theory stood in 1949, with emphasis on work just prior to that date. The exposition is firmly based on the original sources, and many of the 600 references are dated prior to 1900; these, together with the detailed author index and the appendices to the 1966 reissue, make this an invaluable source for the history of elasticity and fluid dynamics from 1600 to 1950.

1703. ———. *Essays in the History of Mechanics*. Berlin: Springer Verlag, 1968, x + 384 pp.

Several chapters of this book deal with the history of particle and rigid-body mechanics as opposed to that of continuum mechanics dealt with elsewhere in the book. It presents the history of the principle of moment of momentum for both types of mechanics, and treats of pre-Cauchy continuum mechanics. The style is lively, the author is an expert on his subject, and one meets the primary sources and their authors.
Review: Brush, S.G., *Isis* 61 (1970), 115-118.

1704. Weatherburn, Charles Ernest. *Advanced Vector Analysis with Application to Mathematical Physics*. London: G. Bell, 1924, xvi + 222 pp.

J.H. Michell and C.E. Weatherburn popularized the use of vector analysis and tensor calculus in Australia as soon as they obtained a copy of Gibbs-Wilson. They attempted to introduce it into British applied mathematics but failed. This book and Weatherburn's *Elementary Vector Analysis* (London: Bell, 1921), became standard texts for much too long a period. True, the work was based on all the good authors: Beltrami, Darboux, Gibbs, Love, Maxwell, and Poincaré; but it was never modernized to the standard of, say, Schouten's *Der Ricci-Kalkül*.

Elasticity

The history of elasticity predates Galileo, but the modern theory may be said to start with him. The histories by Todhunter and Pearson and Truesdell cover the whole of our period, 1660-1960, and can be used as a standard to correct that by Saint-Venant for the period 1750-1850. The major mathematical topics listed here are differential geometry in the histories by Reich, Phillips, and Vincensini, to which could be added René Taton's biography of Gaspard Monge, *L'oeuvre scientifique de Monge* (Paris: Presses Universitaires de France, 1951);

and integral theorems, those of Green, Ostrogradski, and Stokes, used both in elasticity and in fluid mechanics as well as in other fields of mathematical physics. See the articles by Katz and Stolze.

The prominent place of differential geometry in this theory is made clear by one of Saint-Venant's appendices and by the treatise from the Cosserat brothers. The text by Schouten given here could be multiplied many times over from the works of Gibbs, Weatherburn, Levi-Civita, and many others indicated in Truesdell's extensive works. The major original sources listed are the articles by Cauchy, Maxwell, and Poisson; these, and articles by Stokes (1849) given under Fluid Mechanics, item 1747 below, and by Green (1842) given under General Studies, item 1691 above, form the tinder for Saint-Venant's fiery polemic in his fifth appendix. As extinguishers for this fire one may use Love's text where the elastic constants are derived in detail, or Flügge as listed under General Studies, item 1685 above.

1705. Cauchy, Augustin-Louis. "De la pression ou tension dans un corps solide" (1822); "Addition à l'article précédent"; "Sur la condensation et la dilatation des corps solides" (1822). *Exercices de mathématique, Seconde Année*. Paris: De Bure, Frères, 1827, 60-78; 79-81; 82-93. "Sur les équations qui expriment les conditions d'équilibre ou les lois de mouvement intérieur d'un corps solide élastique ou non élastique" (1822); "Sur l'équilibre et le mouvement d'un système de points matériels sollicités par des forces d'attraction ou de repulsion" (1828); "De la pression ou tension dans un système de points matériels" (1828). *Exercices de mathématique, Troisième Année*. Paris: De Bure, Frères, 1828, 195-226; 227-252; 253-277.

The first three articles deal with analyses of stress, particularly principal stresses and the symmetry of the stress (tensor). The last three articles have two forms of simplistic linear stress-strain relations which reduce to one, possibly two, the unknown elastic coefficients; the equations of motion derived from a "molecular" hypothesis, ending with nine, then six, then three elastic constants in general; and then the reduction to one constant in the isotropic case, as before. The dating of the articles is taken from the *Comptes rendus*.

1706. Cosserat, Eugène Maurice Pierre, and François Cosserat. *Théorie des corps déformables*. Paris: Hermann, 1909, vi + 227 pp.

This treats of the deformation (kinematics) of lines, surfaces, and three-dimensional media, and it uses a potential and a Lagrange-type integral. It is one of the classics, and with the moving trihedron or local frame, it makes one of the first applications of this facet of differential geometry to this area.

1707. Euler, Leonhard. *Leonhardi Euleri Opera Omnia, series secunda, Volumen X. Commentationes mechanicae ad theorium corporum flexibilium et elasticorum pertinentes, Volumen prius*. Edited by F. Stüssi and H. Favre. Zurich: Orell Füssli, 1947, ix + 451 pp.

This contains 19 papers on elasticity of one-dimensional bodies, including bars, columns, and strings, for the period 1728-1767. The papers present the basic steps in the formulation and solution of partial differential equations.

1708. ———. *Leonhardi Euleri Opera Omnia, series secunda, Volumen XI, Sectio prima. Commentationes mechanicae ad theoriam corporum flexibilium et elasticorum pertinentes, Volumen posterius*. Edited by F. Stüssi and E. Trost. Zurich: Orell Füssli, 1957, x + 383 pp.

This contains twenty papers on the vibration of strings and elastic laminae for 1768-1784, including two posthumous ones.

1709. Katz, Victor J. "The History of Stokes' Theorem." *Mathematics Magazine* 52, No. 3 (1979), 146-156.

This incomplete history of Stokes's theorem in the wide, modern sense concentrates on Ostrogradski in particular, and on the "vector" forms of the equations, leading to the modern differential-form statement. The 28 references give a good but only partial introduction to this area; they could be supplemented by the works of Chasles, Liouville, Thomson, Lamé, and Despeyrous, as well as the reviews given under the bibliography on potential theory.

1710. ———. "The History of Differential Forms from Clairaut to Poincaré." *Historia Mathematica* 8 (1981), 161-188.

This concentrates on the work of Clairaut, Ostrogradski, Betti, Poincaré, and Volterra; it omits the work of Euler and Fontaine. Further, the author suppresses all physical arguments and applications as they appear in Clairaut, Thomson, and Stokes in particular. There are 53 references listed at the end.

1711. Love, Augustus Edward Hugh. *A Treatise on the Mathematical Theory of Elasticity*. Cambridge: At the University Press, Volume I in 1892 and Volume 2 in 1893. Later editions to 1927 had only one volume. Reprinted New York: Dover, 1952, from the fourth edition, xviii + 643 pp.

The reprint text dates from 1906 with revisions to 1926. It predates the revolution in concept and scope that came in the period 1940-1966 with Truesdell, Oldroyd, and Noll. The several hundred references are buried in the footnotes scattered around the text. The major problems in elasticity theory--the several geometries, the nonlinear response of the materials, the problem of expressing the properties of materials by functions or integrals--all of these are mentioned, but in the classical 19th-century way, in notations obscuring the features of the model. The clear notation of Gibbs-Wilson with its vectors and dyadics (linear transformations) was not to penetrate the British fog for another 30 years or more.

1712. Maxwell, James Clerk. "On the Equilibrium of Elastic Solids." *Transactions of the Royal Society of Edinburgh* 20 (1853), 87-120.

This is a masterly derivation of small displacement and temperature-change elasticity equations, based on a reading of the relevant French theoretical literature--Navier, Poisson, Lamé, Clapeyron, and Cauchy are cited--and the experimental evidence of Wertheim and Oersted among many others mentioned. Of course Stokes's equations are obtained. These equations are supported by the solution of fourteen worked examples in tension, compression, torsion, bending, and heating, with reference again to experiment. This is a tour-de-force neither understood nor appreciated by Saint-Venant fifteen years later.

1713. Moyer, Donald Franklin. "Continuum Mechanics and Field Theory: Thomson and Maxwell." *Studies in History and Philosophy of Science* 9 (1978), 35-50.

This details the mathematical development of vector fields in these areas with nearly 150 references to Maxwell, Thomson, and others. See also item 1511.

1714. Oravas, Gunhard, and Leslie McLean. "Historical Development of Energetical Principles in Elastomechanics. I: From Heraclitos to Maxwell"; "Historical Development of Energetical Principles in Elastomechanics. II. From Cotterill to Prange." *Applied Mechanics Reviews* 19 (1966), 647-658; 919-933.

The first review covers the period 1640-1860, the second 1860-1930. The references listed at the ends of the articles total about 160 or so. The treatment is uneven and not always accurate: *Poisson* extended Lagrange's idea of the potential to electrostatics and elasticity in 1811-1814, and these papers were read by George Green who named and used the concept. Despite such faults, these articles are welcome antidotes to the surfeit of nonmathematical papers on this topic.

1715. Phillips, Esther R. "Karl M. Peterson: The Earliest Derivation of the Mainardi-Codazzi Equations and the Fundamental Theorem of Surface Theory." *Historia Mathematica* 6 (1979), 137-163.

This article continues the development of an aspect of the history of differential geometry "after" Riemann, i.e., the period 1853-1880. It concentrates on the relations between the coefficients of the first and second fundamental forms for surfaces in three-dimensional space. There are two dozen references. See also item 1060.

1716. Poisson, Simeon-Denis. "Mémoire sur les surfaces élastiques." *Mémoires de la Classe des Sciences mathématiques et physiques de L'Institut, année 1812, Partie* 2 (1816), 167-225.

This article was read on the 1st of August, 1814, while an Académie prize competition on its topic was current, but it was not an entry for the prize. It introduced into the French tradition a molecular hypothesis for forces between particles in elastic solids as well as an elastic potential from which the forces were derived. This form of the potential certainly gave Lagrange's biharmonic equation for the vibrations of the surface; but it opened a Pandora's box of elastic constants, and the resulting confusion and controversy was to ruin French engineering education for fifty years.

1717. Reich, Karin. "Die Geschichte der Differentialgeometrie von Gauss bis Riemann (1828-1868)." *Archive for History of Exact Sciences* 11 (1973), 273-382.

This is a magnificent history of the second period in *differential* geometry. Marvellous, detailed references (800 at least) are given, as well as an index of authors. One misses the flavor of the originals in that the applications are mainly suppressed, e.g., the heat conduction of Riemann and Lamé, the physical arguments of Thomson and Chasles, the conformal mapping problem of geodesy from Gauss and Jacobi. The result is a wonderful description of the *pure* mathematics in the text of this article with the applications and motivation easily mined from the references. See also item 1062.

1718. Saint-Venant, Adhémar Jean Claude Barré de, and Claude Louis Marie Henri Navier. *Résumé des leçons données à l'Ecole des ponts et chaussées sur l'application de la mecanique à l'établissement des constructions et des machines. Première partie contenant les leçons sur la résistance des matériaux et sur l'établissement des constructions en terre, en maçonnerie et en charpente. Première section. De la Résistance des Corps Solides, par Navier. Troisième édition avec des notes et des appendices, par M. Barré de Saint-Venant. Tome premier. Fascicule I.* Paris: Dunod, 1864, i-cccxi pp. (Saint-Venant) + 1-509 pp. (Navier) + 510-852 pp. (Saint-Venant) + 1 pl.

This is invaluable as both text and history. The Riemann curvature tensor is used in linear approximation in an appendix, a forerunner of modern theories due to Eckart and others. The long historical introduction by Saint-Venant and the controversial fifth appendix (also by Saint-Venant) attest to the persistence of accepted positions rather than to the virtue of research in Paris in 1840-1870. The articles of Green, Stokes, and Maxwell cited here and in the bibliography of fluid mechanics come in for severe but unwarranted criticism.

1719. Schouten, Jan Arnoldus. *Der Ricci-Kalkül: eine Einführung in die neueren Methoden und Probleme der mehrdimensionalen Differentialgeometrie*. Berlin: Springer, 1924. Last reprinted 1978, x + 312 pp.

This often-reprinted work is one of the classic texts in differential geometry of other-than-three dimensional Euclidean space. There are over 200 references to literature up to 1923 collected at the end. This book consolidated the new geometries in all but English-speaking countries and led to wider applications of differential geometry.

1720. Stolze, Charles H. "A History of the Divergence Theorem." *Historia Mathematica* 5 (1978), 437-442.

This brief history discusses the work of Green, Gauss, and Ostrogradski on the original divergence theorem, and of Heaviside and Gibbs on the vector form. The undoubted priority for the surface-volume form of the theorem lies with Ostrogradski; the dozen references are adequate for this end but not for the topic.

1721. Todhunter, Isaac, and Karl Pearson. *A History of the Theory of Elasticity and of the Strength of Materials from Galilei to the Present Time. Volume I, 1639-1850, Galilei to Saint-Venant. Volume II, Parts I and II, Saint-Venant to Lord Kelvin.* Cambridge: At the University Press, Vol. I in 1886, xvi + 924 pp.; Vol. II posthumously for Todhunter in 1893, 726 pp. and 546 + 12 pp.

This history is in the usual Todhunter style: it presents a summary of each article, with little evaluation and no synthesis. It should be used in conjunction with the originals and with other histories, e.g., those of Saint-Venant and Truesdell.

1722. Truesdell, Clifford A. *The Rational Mechanics of Flexible or Elastic Bodies, 1638-1788: Introduction to Leonhardi Euleri Opera Omnia, series secunda, Volumina X et XI. Leonhardi Euleri Opera Omnia, series secunda, Volumen XI, sectio secunda.* Zurich: Orell Füssli, 1960, 435 pp.

This work (volume 11, part 2, of the second series of Euler's *Opera Omnia*) contains long quotations and extensive summaries of the results due to Huygens, Hooke, Leibniz, Mariotte, the Bernoullis, Euler, d'Alembert, Lagrange, and others on vibrating strings and the general properties of elastic bars. Proper credit is given to d'Alembert (1746) as first formulator and solver of the wave equation. See also item 960.

1723. Vincensini, P. "La géométrie différentielle au XIXe siècle (avec quelques réflexions générales sur les mathématiques)." *Scientia* 107 (1972), 617-660. English translation, 661-696.

This is a general review of the whole of the 19th century's differential geometry with 111 references listed at the end.

1724. Vizgin, V.P. "Vzaimosvyaz' fiziki i matematiki v XIX veke" (The relationship between physics and mathematics in the 19th century). *Istoriko-Matematicheskie Issledovaniia* 22 (1977), 111-126.

The essence of this article is distilled into four pages of chronological tables for the century, listing dates, authors, physics topics, and mathematics.

Fluid Mechanics

Three major topics fall under this heading: the foundations of fluid mechanics proper; aerodynamics and boundary-layer theory; and hydrology-hydraulics-meteorology. The mathematical method indicated is complex function theory, with Stokes's and the divergence theorem being tied to elasticity.

So here we have several reviews--Beyer, Brillouin, Hicks, Love, Stokes, and Truesdell--and some histories--Durand, Euler, Rouse and Ince, and Tokaty. To these are added a few texts, some minor historical articles, and several original sources. The articles by d'Alembert and Michell detail some of the uses of complex analysis in two-dimensional fluid flow, and more can be gathered for the texts by Lamb, Lichtenstein, and Serrin.

The major guides to the early (1650-1800) history is the pair of reviews by Truesdell in the *Leonhardi Euleri Opera Omnia*, items 1732, 1733, and Neményi's history, item 1743. For the 19th century, the reviews by Stokes, Hicks, Brillouin, and Love are fairly complete. The 20th century saw remarkable developments in high-speed flow, beginning with the work of Prandtl on boundary-layers; these are reported in the reviews by Dryden and Tani, covering the period to 1970. For types of materials other than viscous and non-viscous fluids, the seminar by Markowitz forms an easy introduction to the more formal work done by Truesdell and others.

1725. Alembert, Jean-le-Rond d'. *Essai d'une nouvelle théorie de la résistance des fluides*. Paris: David, 1752. Reprinted Brussels: Culture et Civilisation, 1966, xlvi + 212 pp. + 2 pls.

On pp. 60-70 d'Alembert derives the so-called Cauchy-Riemann equations, lays the foundation for the application of complex function theory to two-dimensional fluid mechanics, and suggests the potential and stream functions for such and many other applications.

1726. Bernoulli, Daniel. *Hydrodynamica, sive de viribus et motibus fluidorum commentarii*. Strassburg: Johannes Reinholdus Dulseckerus, 1738, 4 pp. + 304 pp. + 12 pls.

This fundamental work appears in English translation as part of *Hydrodynamics, by Daniel Bernoulli* and *Hydraulics, by Johann Bernoulli*, translated from the Latin by Thomas Carmody and Helmut Kobus, preface by Hunter Rouse (New York: Dover, 1968). Both of these works are extensively treated by Truesdell and Szabó, since they lay the foundation for Eulerian (modern) fluid dynamics.

1727. Beyer, Robert T., ed. *Foundations of High Speed Aerodynamics. Facsimiles of Nineteen Fundamental Studies as They Were Originally Reported in the Scientific Journals*. New York: Dover, 1951, xi + 286 pp.

This covers the period 1870-1950, and has an extensive bibliography compiled by one of the foremost workers in the area at the time, George F. Carrier.

1728. Biswas, Asit K. "Beginning of Quantitative Hydrology." *Journal of the Hydraulics Division, Proceedings of the American Society of Civil Engineers* 94 (1968), 1299-1316.

This article covers the contributions of Perrault, Mariotte, and Halley to experimental quantitative hydraulics and the hydrological cycle. There are 34 references at the end of the article.

1729. Brillouin, Marcel. "Questions d'hydrodynamique." *Annales de la Faculté des Sciences de Toulouse pour les sciences mathêmatiques et les sciences physiques* 1 (1887), 1-80.

Each article in the early volumes of this new series of these *Annales* was paged separately, and this article was the first of a series of survey articles planned by the new editors. References were collected at the end of the article, a revolution in style for the time. Here on pp. 73-80, we have 160 of them, covering the period 1820-1886 on vortex theory (with wonderful illustrations in the text), on jets and free surfaces before Michell, as well as viscosity. This is undoubtedly the major review of fluid flow between Stokes (1846) and Love (1901); it betrays no French bias, and deals with viscosity only briefly.

1730. Dryden, Hugh L. "Fifty Years of Boundary-Layer Theory and Experiment." *Science* 121 (1955), 375-380.

This is a brief review of boundary-layer theory from Prandtl's introduction of the concept (1904) to 1954, with 150 references listed at the end of the article.

1731. Durand, William Franck, ed. *Aerodynamic Theory*. Berlin: Springer, 1934-1936. 6 vols. Reprinted New York: Dover, 1963, and bound as 6 vols. in 3.

The twenty sections of this work cover mathematics, fluid mechanics, the history of aeronautics, and the whole of aerodynamics to date of publication. The historical sketch by R. Giacomelli and E. Pistolesi, "Historical Sketch of Aviation Theory," on pp. 305-395 of Volume 1, does not have a complete bibliography but there is a substantial one at the end of Volume 2, with references to other, exhaustive bibliographies; many other sections contain further bibliographies.

1732. Euler, Leonhard. *Leonhardi Euleri Opera Omnia, series secunda, Volumen XII. Commentationes mechanicae ad theoriam corporum fluidorum pertinentes, Volumen prius*. Edited by C. Truesdell. Zurich: Orell Füssli, 1954, cxxv + 288 pp.

This volume contains, on pp. vii-cxxv, Truesdell's famous review of *Rational Fluid Mechanics, 1687-1765*, continued below; it features the work of Newton, Huygens, Clairaut, Euler, the Bernoullis, d'Alembert, and Lagrange. Major parts of this review paraphrase or summarize the five foundation papers by Euler which appear on pp. 1-168 in this volume; these papers date from 1755 to 1757. Statements about Johann Bernoulli on pp. xxxii and xxxvii should be read in the light of Truesdell's later comments given in his review of Szabo's book: "An Essay Review of Geschichte der mechanischen Prinzipien und ihrer wichtigsten Anwendungen," *Centaurus* 23 (1980), 163-175, esp. p. 168.

1733. ———. *Leonhardi Euleri Opera Onmia, series secunda, Volumen XIII. Commentationes mechanicae ad theoriam corporum fluidorum pertinentes. Volumen posterius*. Edited by C. Truesdell. Zurich: Orell Füssli, 1955, cxviii + 375 pp.

This contains an editor's introduction, a review *The Theory of Aerial Sound, 1687-1788*, and the completion of the above review, *Rational Fluid Mechanics, 1765-1788*. The five papers of Euler in this volume cover the period 1768-1777 and deal with equilibrium and motion of fluids, motion of fluids in tubes, and sound tubes.

1734. Frisinger, H. Howard. "Mathematicians in the History of Meteorology: The Pressure-Height Problem from Pascal to Laplace." *Historia Mathematica* 1 (1974), 263-286.

This review covers 17th- and 18th-century beginnings of mathematical meteorology, with a bibliography of 50 fundamental references.

1735. Helmholtz, Hermann von. *Wissenschaftliche Abhandlungen von Hermann Helmholtz*. Edited by Arthur König. Leipzig: Johann Ambrosius Barth. Band I, 1882; viii + 938 pp. + 2 pls.; Band II, 1883, vi + 1021 pp. + 5 pls.; Band III, posthumously, 1895, xxxviii + 654 pp.

The first volume contains his papers on energy, hydrodynamics, acoustics, electrodynamics, and galvanism; the second covers physiology, including optics and acoustics, and physical optics as well; the third contains the remainder of his papers, a bibliography, and an evaluation of his scientific work by G. Wiedemann. In particular, Band 1 contains the two papers, "Ueber Integrale der hydrodynamischen Gleichungen, welche den Wirbelbewegungen entsprechen," *Journal für die reine und angewandte Mathematik* 55 (1858), 25-55 (here 101-134) and "Ueber discontinuirliche Flüssigkeitsbewegungen," *Monatsberichte der Königlichen Akademie der Wissenschaften zu Berlin* (1868), 215-228 (here 146-157). The first of these introduced vorticity, expanded by his friend Thomson, while the second introduced free surfaces in fluid flow, exploited by Kirchhoff, Schwarz, Christoffel, and Michell.

1736. Hicks, William Mitchinson. "Report on Recent Progress in Hydrodynamics. Part I. General Theory." *Report of the Fifty-First Meeting of the British Association for the Advancement of Science, 1881*. London: John Murray, 1882, 57-88.

This review covers the forty years from Stokes's report of 1846, and contains: the general equations of motion as handled by Clebsch, Thomson, and Tait; vortex theory, Helmholtz, Thomson; free surfaces, Helmholtz, Rayleigh, Kirchhoff, Christoffel; motion of bodies through fluids, by most of the above authors as well as Lamb and Bjerknes; viscous fluids, Meyer, Maxwell, Bobylev; waves, Stokes, Boussinesq, Rankine, Reynolds. The many references are given in the footnotes.

1737. ———. "Report on Recent Progress in Hydrodynamics. Part II. Special Problems." *Report of the Fifty-Second Meeting of the British Association for the Advancement of Science, 1882*. London: John Murray, 1883, 39-70.

This review of special problems continues the previous item and deals with solutions of Laplace's equation under various boundary conditions. The authors of the hundreds of footnoted references are, in the main, those given in the *General Theory* report. The main special problems reviewed are: two-dimensional motions, using the (complex) potential; three-dimensional motions using the potential and the (electro)magnetic analogy; viscous fluids. Solutions for cylinders, spheres, ellipsoids, all moving through fluids, or for fluids moving through pipes and various other configurations, are reviewed.

1738. Lamb, Horace. *Hydrodynamics*. 6th ed. Cambridge: The University Press, 1932, xv + 738 pp. Reprinted many times, including New York: Dover, 1952. The first edition was in 1879, as a *Treatise on the Mathematical Theory of the Motion of Fluids*, with subsequent editions in 1895, 1906, 1916, and 1924.

This text is a classic, much used and often quoted. Its several hundred references are, as usual, interred in the footnotes. Over half of the text is devoted to waves, and there is a chapter on rotating masses of liquid. Despite its being last revised in 1932, this book seems timeless, covering many "modern" topics such as solitary waves, and geophysical and atmospheric flows.

1739. Lichtenstein, Leon. *Grundlagen der Hydromechanik*. Berlin: Springer, 1928, xvi + 506 pp. Reprinted in *Die Grundlehren der mathematischen Wissenschaften in Einzeldarstellungen*, Band 30. Berlin: Springer, 1968.

This text lies in the modern tradition where Lamb lies in the past: Lichtenstein is solidly mathematical, paying attention to the separation of geometry from physics, to questions of existence, and to a general overview rather than the simple plethora of particular and bewildering examples of the mainstream British tradition.

1740. Love, Augustus Edward Hugh. "Hydrodynamik: Physikalische Grundlegung" and "Hydrodynamik: Theoretische Ausführungen." *Encyklopädie der mathematischen Wissenschaften mit Einschluss ihrer Anwendungen*, IV.15. and IV.16. Leipzig: B.G. Teubner, 1901, 48-83; 84-147.

These reviews were revised and extended by Love, Paul Appell, H. Bergelin, and H. Villat in the French translation of them, in the *Encyclopédie des sciences mathématiques pures et appliquées*, Tome IV, Partie 5 (1912), 61-101 and (1914) 102-208. All these have largely been superseded by Truesdell's *Encyclopedia of Physics* articles, but they do offer concentration on fluid dynamics. As usual, the many hundred references are given in the footnotes.

1741. Markowitz, Herschel. "The Emergence of Rheology." *Physics Today* 21, No. 4 (1968), 23-30.

This review is a good general introduction to rheology and its main features and founders, written in an informal style.

1742. Michell, John Henry. "On the Theory of Free Stream Lines." *Philosophical Transactions of the Royal Society of London* A 181 (1890), 389-431.

This paper bridges the gap between the work of Schwarz, Christoffel, and Kirchhoff (1865-1870) and J.J. Thomson (1900); with a new technique for handling the transformations, it links together methods used in electrostatics, fluid dynamics, and heat conduction, as well as theoretical work on the Riemann mapping theorem.

1743. Neményi, Paul F. "The Main Concepts and Ideas of Fluid Dynamics in Their Historical Development." *Archive for History of Exact Sciences* 2 (1962), 52-86.

The emphasis of this posthumous article lies on a description of the work of Leonardo da Vinci, Galileo, Torricelli, and Mariotte as leading to the work of Newton. There are several short sections dealing with the Bernoullis, Euler, d'Alembert, Bossut, and de Borda, while the "moderns" are briefly dismissed. Experiment is to the forefront, but this is one of the few references which tackle pre-Eulerian fluids. See also his "Wasserbauliche Strömungslehre" in *Handbuch der Physikalischen und Technischen Mechanik, Band 5, Mechanik der Flüssigkeiten nebst technischen Anwendungsgebieten* (Leipzig: Johann Ambrosius Barth, 1931), 967-1145.

1744. Prandtl, Ludwig. *Fundamentals of Hydro- and Aeromechanics* and *Applied Hydro- and Aeromechanics*. New York: United Engineering Trustees, 1934, viii + 270 pp.; xvi + 311 pp. Taken from his lectures by O.G. Tietjens. Reprinted New York: Dover, 1957.

These texts are an edited presentation of the theory and applications in these areas, particularly boundary-layer theory. All detail and wealth of example is eliminated in order to make the outline of the theory presented very clear and to make the instructive examples stand out. The major reference to his work on boundary-layer theory is: "Ueber Flüssigkeitsbewegung bei sehr kleiner Reibung," *Verhandlungen der III. Internationalen Mathematiker-Kongress, Heidelberg 1904* (Leipzig: Teubner, 1905), 484-491. This is reprinted in his *Gesammelte Abhandlungen zur angewandten Mechanik, Hydro- und Aerodynamik*, herausgegeben ... von Walter Tollmien, Hermann Schlichting, Henry Görtler, und F.W. Riegels, in drei Teilen (Berlin: Springer, 1961), xix + 1620 pp. The boundary-layer article appears on pp. 575-584 of the second volume, which contains Prandtl's contributions to boundary-layer theory, turbulence, and gas dynamics; volume 1 covers elasticity, plasticity, rheology, buckling, aerodynamics, and contains a complete bibliography of his work; the third volume contains meteorology, model testing, and miscellanea.

1745. Rouse, Hunter, and Simon Ince. *History of Hydraulics*. Iowa City: State University of Iowa, 1957, xii + 269 pp. Reprinted with corrections from supplements to *La Houille Blanche*, 1954-1956. Reprinted with corrections New York: Dover, 1963.

This is a good introduction to the concepts of hydraulics, with the work of Euler and Johann Bernoulli somewhat neglected while that of d'Alembert is overblown. About one hundred references are located at the ends of the chapters. The history covers early antiquity to mid 20th century, with about two-thirds from the 17th century onwards. There is an even spread on engineering, mathematics, and personalities, but the level of reliability is not so even, and sources should be checked: "Little is known about [Ernst Heinrich and Wilhelm Eduard Weber]," say pp. 144-145, "except that they were Professors at Leipzig and Halle respectively." This "little known" Wilhelm Eduard Weber was at the center of one of the largest political storms of the 19th century in Germany, was the driving force in Gauss's work on the electric telegraph and electrodynamics and on terrestrial magnetism, and later did much electrodynamics on his own.

Review: Truesdell, Clifford, *Isis* 50 (1959), 69-71.

1746. Serrin, James. "Mathematical Principles of Classical Fluid Mechanics." *Encyclopedia of Physics/Handbuch der Physik.* Edited by S. Flügge. Vol. VIII, Part 1, Fluid Dynamics I. Berlin: Springer, 1959, 125-263.

Being a mathematical presentation of classical fluid mechanics, this text refers to the original works in its extensive footnotes, and becomes a valuable historical source. Fifty or so major references are cited at the end and extended bibliographies are indicated.

1747. Stokes, Sir George Gabriel. "On the Theories of the Internal Friction of Fluids in Motion, and of the Equilibrium and Motion of Elastic Solids." *Transactions of the Cambridge Philosophical Society* 8 (1849), 287-319.

Offprints of this classic paper appeared in 1845, but the actual volume of the *Transactions* appeared late (not until 1849). It exercised a strong and lasting influence on fluid dynamics through Maxwell and the "Notes on Hydrodynamics" written by Stokes and Thomson for students (see their collected papers). It also laid the base for Stokes's next work on pendulums.

1748. ———. "On the Effect of the Internal Friction of Fluids on the Motion of Pendulums." *Transactions of the Cambridge Philosophical Society* 9 (1856), 8-106.

Every physics student has seen Stokes's formula for the terminal velocity of objects falling through viscous fluids, but this long article is devoted to the very important practical and theoretical problem of the length of a seconds pendulum moving in air at a certain temperature and density and the formula is a byproduct. Applications of these results lie in gravity measurements and astronomy, as is attested by the authors cited: Bessel, Biot, du Buat, Baily, and Coulomb on the experimental side, for half the paper is a comparison of this new theory with old experiment.

1749. ———. "Report on Recent Researches in Hydrodynamics." *Report of the Fifteenth Meeting of the British Association for the Advancement of Science, 1846*. London: John Murray, 1847, Part I, 1-46.

This is one of the earliest reviews of fluid dynamics, written by the man who began solving problems based on Laplace's equation rather than just laying down theory or solving one- and two-dimensional simple cases. It was extended by that of Hicks and complemented by that of Brillouin.

1750. Tani, Itiro. "History of Boundary-Layer Theory." *Annual Review of Fluid Mechanics* 9 (1977), 87-111.

This excellent historical review of boundary-layer theory has 30 references to the mathematics but is itself not heavily mathematical; it covers the period 1845-1970 in fluid mechanics, and the period 1905-1970 on boundary-layer theory in detail. It covers the spread of Prandtl's ideas, so vital in aerodynamics, but also gives credit to those who discovered these ideas independently and contemporaneously. Each volume in this series of reviews begins with a short personal or institutional history of about 10-20 pages.

1751. Tokaty, Grigori Alexandrovich. *A History and Philosophy of Fluid Mechanics*. Henly-on-Thames: Foulis, 1971, ix + 241 pp.

This complete history, from earliest times to the present, has about 400 references in its footnotes. It includes Russian contributions (19th to 20th century) not included in the reviews listed here, as well as German (19th to 20th century) and French (18th to 19th century). There is a good balance of mathematics, engineering, and history.

1752. Truesdell, Clifford A. "Notes on the History of the General Equations of Hydrodynamics." *American Mathematical Monthly* 60 (1953), 445-458.

This is an introductory, descriptive history covering Newton, Daniel Bernoulli, d'Alembert, Euler, Navier, Cauchy, Poisson, and Stokes, 1687-1845, and forms a preliminary to the great reviews, q.v.

1753. ———. *The Kinematics of Vorticity*. (Indiana University Publications in Science Series, No. 19.) Bloomington: Indiana University Press, 1954, xviii + 232 pp.

This thorough study, historically based on the reviews in the *Leonhardi Euleri Opera Omnia*, is fully documented with over 300 items in its bibliography. It is (relatively) easy to read, mathematically speaking, and concentrates on the geometry and kinematics of rotational motion and the vector fields and theorems attaching thereto.

Heat Conduction

Heat has been associated with problems in energy and its conservation and transformation, and with the development of the theory of crystals and their groups. The problem of heat conduction led to developments in differential geometry by Lamé and Riemann (see Reich's article under Elasticity), in series by Fourier, and in quantum theory by Planck, who also developed the theory of heat radiation.

The major histories of thermodynamics listed here are those by Mach, Bachelard, and Truesdell. The texts by Planck and Carslaw and Jaeger give good coverage to problems and applications. The original source is Fourier, and the two works by Herivel and Grattan-Guinness show its practical and theoretical importance. For the principle of conservation of energy and its history one should consult the book by Elkana, particularly its appendix. Finally, the only original source for crystallography given here is the collection of articles by Fedorov, but the reviews by Burkhardt, Wiederkehr, and Wigner give good coverage.

1754. Bachelard, Gaston. *Etude sur l'évolution d'un problème de physique: la propagation thermique dans les solides*. Paris: J. Vrin, 1927, 184 pp. Reprinted 1973, v + 183 pp.

This is a discussion of the history of the nature of heat, as a fluid or as movement; it discusses the work of Biot, Fourier and Comte, Duhamel, Lamé, and Boussinesq, in the 19th century. The mathematics tends to be treated rather than the models of heat conduction and of heat conduction through crystals; there are about 40 references.

1755. Burkhardt, J.J. "Zur Geschichte der Entdeckung der 230 Raumgruppen." *Archive for History of Exact Sciences* 4 (1967), 235-246.

This details priority investigations for Schoenflies, Barlow, Fedorov, and Sohnke, and has a bibliography of 80 titles.

1756. Carslaw, Horatio Scott. *Introduction to the Theory of Fourier's Series and Integrals and the Mathematical Theory of Heat Conduction*. London: Macmillan, 1906, xvii + 434 pp. After several editions this was split into two books, one of which is the following.

1757. ———, and John Conrad Jaeger. *Conduction of Heat in Solids*. Oxford: Clarendon Press, 1946, with a second edition in 1959, x + 510 pp.

The first edition ran to 187 pages on series and integrals and formed an excellent introduction to functions of one and two real variables, and to 220 pages on heat conduction in solids. Very soon the separate parts became standard references. After World War II, the theory of heat conduction was issued separately after being completely rewritten by Jaeger. In the process the basic motivation from Carslaw's research with Sommerfeld on Riemann surfaces was totally suppressed for the practical engineering style demanded in Australia at the time.

1758. Elkana, Yehuda. *The Discovery of the Conservation of Energy*. Cambridge, Mass.: Harvard University Press, 1974, x + 213 pp.

This is a philosophical rather than a mathematical history. It deals with the work of Mayer, Joule, Thomson, Rankine, Helmholtz, and Carnot in an appendix on priority in the conservation of energy. There are nearly 300 references listed at the end. See also Erwin N. Hiebert, *Historical Roots of the Principle of Conservation of Energy* (Madison: The State Historical Society of Wisconsin, 1962, 118 pp.).

1759. Fedorov, Evgraf Stepanovich. *Symmetry of Crystals*. N.p.: American Crystallographic Association, 1971, x + 315 pp. Translated by David and Katherine Harker from the Russian, *Simmetriya i struktura kristallov*. Moscow: Akademii͡a Nauk SSSR, 1949.

This translation of Fedorov's five great papers on crystallography makes his work easily accessible. The work of later Russian authors and of others appears in the journals *Kristallografiya* and *Acta Crystallographia*, in which many survey and somewhat historical articles appear. See also the history by Ilarion Ilarionovich Shafranovskii, *Istoriya Kristallografii: S drevneishikh vremen do nachala XIX stoletiya* (The history of crystallography from earliest times to the beginning of the 19th century) (Leningrad: Nauka, 1978, 293 pp.).

1760. Fourier, Jean-Baptiste-Joseph. *Théorie analytique de la chaleur*. Paris: Didot, 1822. This was reprinted in *Oeuvres de Fourier*. Edited by Gaston Darboux. *Tome premier: Théorie analytique de la chaleur*. Paris: Gauthier-Villars, 1888, xxviii + 563 pp.; *Tome second: Mémoires publiées dans divers récueils*. Paris: Gauthier-Villars, 1890, xiii + 636 pp.

This is the classic publication of the material contained in his prize essay of 1807; it was translated into English by Alexander Freeman as *The Analytical Theory of Heat by Joseph Fourier* (Cambridge: At the University Press, 1878, xxiii + 466 pp.), and reprinted many times. The manuscript of the 1807 prize memoir was not printed in the collected works, even though the second volume of these covers heat radiation, cooling of the earth, temperature of the planets, along with the numerical solution of equations, and one memoir on virtual velocities; that text is contained in the next item.

1761. Grattan-Guinness, Ivor, in collaboration with J.R. Ravetz. *Joseph Fourier, 1768-1830: A Survey of His Life and Work, Based on a Critical Edition of His Monograph on the Propagation of Heat, Presented to the Institut de France in 1807*. Cambridge, Mass.: MIT Press, 1972, xii + 516 pp.

This reproduces the text of the 1807 essay with interspersed comments on the work. The survey fills about one-fifth of the book, and there are nearly 400 references in the footnotes and the bibliography. See also item 1073. This item needs to be supplemented by the next one.

1762. Herivel, John. *Joseph Fourier: The Man and the Physicist*. Oxford: Clarendon Press, 1975, xii + 350 pp.

This work is mainly biographical although two large chapters are devoted to mathematical-physical material. There is a substantial appendix giving English translations of 28 letters, and there are 100 references in the bibliography. The importance of these last three items lies in the fact that the thread linking Daniel Bernoulli, Fourier, Dirichlet, and Riemann has been transformed into a striking tapestry of partial differential equations and differential geometry, to say nothing of the work of Cantor in cardinality.

1763. Lee, William H.K., ed. *Terrestrial Heat Flow*. (Geophysical Monograph Series, no. 8.) Washington, D.C.: American Geophysical Union, 1965.

The first two articles, by Edward C. Bullard, "Historical Introduction to Terrestrial Heat Flow," pp. 1-6, and by John C. Jaeger, "Application of the Theory of Heat Conduction to Geothermal Measurements," pp. 7-23, contain 120 references to the history, theory, and applications for this topic in the period 1850-1960.

1764. Mach, Ernst. *Die Prinzipien der Wärmelehre. Historisch-kritisch entwickelt*. Leipzig: Johann Ambrosius Barth, 1896, viii + 472 pp. Reprinted in three further editions to 1923. There does not seem to be a complete translation into English.

This has the usual virtues of and suffers from the usual defects of Mach's work. Two references which may be used to guide one through the Mach-Planck controversy are: Ulrich Hoyer, "Von Boltzmann zu Planck," *Archive for History of Exact Science* 23 (1980), 47-86 (on entropy, with 40 references); and Hans Georg Schöpf, *Von Kirchhoff bis Planck: Theorie der Wärmestrahlung in historisch-kritischer Darstellung* (Braunschweig: Vieweg, 1978, 199 pp.).

1765. Planck, Max Karl Ernst Ludwig. *Vorlesungen über die Theorie der Wärmestrahlung*. Leipzig: Johann Ambrosius Barth, 1906. Translated into English by Morton Masius as *The Theory of Heat Radiation*. Philadelphia: P. Blakiston's Son, c. 1914, xiv + 225 pp. To this item should be added: *Vorlesungen über Thermodynamik*. Leipzig: Veit, 1897, vi + 248 pp. Translated into English by Alexander Ogg as *Treatise on Thermodynamics*. London: Longmans, Green and Co., 1903, xiii + 272 pp.

The first of these treats of radiation, entropy, various oscillators, and probability (for quantum theory). The second has the usual thermodynamics as well as chemical thermodynamics (missing from Mach), and introduces the Planck thermodynamic potential, this being the Gibbs-Duhem potential divided by the negative of the absolute temperature. References are few in both treatises, totalling about 20.

1766. Riemann, Georg Friedrich Bernhard. *Bernhard Riemann's Gesammelte mathematische Werke und wissenschaftliche Nachlass. Herausgegeben unter Mitwirkung von R. Dedekind von H. Weber*. Leipzig: Teubner, 1876, viii + 576 pp. Reprinted, with addition, from the 2nd German ed. of 1892, New York: Dover, 1953.

The Paris Académie sent the original of Riemann's *Commentatio mathematica, qua respondere tentatur quaestioni ab Ill*[us-trissi]*ma Academia Parisiensi propositae: "Trouver quel doit être l'état calorifique d'un corps solide ..."* to Weber for him to make a copy and publish it in the collected works (pp. 370-383, with Weber's comments on pp. 384-399). This is the first publication of this important derivation of metric and curvature tensors; this curvature was stated to be invariant under coordinate transformations in the appropriate way.

1767. Truesdell, Clifford A. *The Tragicomical History of Thermodynamics, 1822-1854*. New York: Springer, 1980, xii + 372 pp.

This is the full development of Truesdell's short course, delivered in three lectures at the International Centre of Mechanical Sciences in Udine, Italy, in 1971: *Tragicomedy of Classical Thermodynamics* (Courses and Lectures No. 70, Vienna and New York: Springer-Verlag, 1973, 41 pp.). The original "tragicomedy" had Fourier, Carnot, Clapeyron, Helmholtz, Clausius (on thermodynamics), and Gibbs (on thermostatics) in the leading roles of what was but a sketch. This history portrays the period 1800-1860, with emphasis placed on that in the title. The work of Fourier, Carnot, Clausius, the British School (Joule, Rankine, Thomson, and Helmholtz), and Reech is presented in detail. There are over 200 references listed as "sources" and there are several hundred others given in these sources and in the footnotes. The treatment is both historical and mathematical. See also item 2140.

1768. Wiederkehr, Karl Heinrich. "Das Weiterwirken der Haüyschen Idee von der Polyedergestalt der Moleküle in der Chemie, die Umgestaltung der Haüyschen Strukturtheorie durch Seeber und Delafosse, und Bravais' Entdeckung der Gittertypen." *Centaurus* 22 (1978), 177-186.

This is a good introduction to pre-group theoretic crystal theory.

1769. Wigner, Eugene P. "Symmetry Principles in Old and New Physics." *Bulletin of the American Mathematical Society* 74 (1968), 793-815.

This review deals with symmetry and the invariance it induces, as seen in crystallography and in quantum mechanics. There are about 100 references at the end.

1770. Wise, M. Norton. "William Thomson's Mathematical Route to Energy Conservation: A Case Study of the Role of Mathematics in Concept Formation." *Historical Studies in the Physical Sciences* 10 (1979), 49-83.

The analogy of heat, fluid flow, magnetism, and mechanics based on the Laplace and Poisson equations is described and used to gently illustrate the development of concepts via mathematics. However, the physical side of Thomson's thinking is not played down. See also Crosbie W. Smith, "Natural Philosophy and Thermodynamics: William Thomson and 'the Dynamical Theory of Heat,'" *British Journal of the History of Science* 9 (1976), 293-319, as this is the nonmathematical complement to the paper by Wise.

Sound and Vibrations

This theory begins with the papers by d'Alembert, and rapidly develops with work by Euler, Daniel Bernoulli, and Lagrange; the developments at the turn of the century due to Sophie Germain and Poisson, particularly in elasticity theory, lay the foundations for Rayleigh's treatise. On the practical side, there are corrections to Newton's results by

Laplace, many developments in bells, bars, and instruments, and some of these are indicated.

The major text is Rayleigh's *Theory of Sound*, and the original sources listed are d'Alembert, Euler, and Lagrange. To these have been added the entertaining contributions by Bucciarelli and Dworsky, Lehr, and Miller, and the review by Lindsay. Further material can be found under Elasticity and Fluid Mechanics.

1771. Alembert, Jean-le-Rond d'. "Recherches sur la courbe qui forme une corde tendue mise en vibration"; "Suite des recherches sur la courbe qui forme une corde tendue mise en vibration." *Histoire de l'Académie royale des sciences et belles-lettres de Berlin pour 1747* (1749), 214-219; 220-249. Reproduced by Librairie Hachette, with an introduction by René Taton, in microfiche 7972 (deux pièces) Paris: Microéditions Hachette, 1979.

These papers introduce d'Alembert's famous solutions to the wave equation, as well as the equation itself, together with the individual exponential solutions, although these are not rewritten in the sine and cosine forms.

1772. Bucciarelli, Louis L., and Nancy Dworsky. *Sophie Germain: An Essay in the History of the Theory of Elasticity*. Dordrecht, Boston: D. Reidel, 1980, xi + 147 pp.

The title says elasticity but the subject is sound and vibrations based on elasticity; it stems from Chladni's patterns on vibrating circular discs, and from the attempts to form a theory to account for them come Lagrange's biharmonic equation and the molecular elasticity theory of Poisson (1814). The elasticity goes on to the efforts of Navier (1820) and Cauchy (1828). Additional references may be found in reviews by I. Grattan-Guinness, *Annals of Science* 38 (1981), 663-690, especially 670-671, and by J.J. Cross, *Annals of Science* 39 (1982), 85-88, where further references may be found. See also item 2361.

1773. Dostrovsky, Sigalla. "The Origins of Vibration Theory: The Scientific Revolution and the Nature of Music." Dissertation, Princeton University, 1969, v + 270 pp. Microfilm or Xerox copy of typescript.

1774. Dyment, S.A. "The Laplace Correction." *Science Progress* 26 (1931), 231-240.

This is a short history of the propagation of sound in air from 1660 to 1820, with 25 references.

1775. Euler, Leonhard. *Leonhardi Euleri Opera Omnia, series quarta A, Commercium epistolicum, Volumen V. Commercium cum A.C. Clairaut, J. d'Alembert et J.L. Lagrange*. Edited by A. Pavolic and R. Taton. Basel: Birkhäuser, 1980, vii + 611 pp.

These letters, especially those involving Lagrange in 1759-1761, have some of the seeds of the theory of sound developed by Euler and Lagrange in those years.

1776. Hunt, Frederick Vinton. *Origins in Acoustics: The Science of Sound from Antiquity to the Age of Newton*. New Haven: Yale University Press, 1978, xv + 196 pp.

This history gives a thorough introduction to pre-Euler-Lagrange practice in acoustics, with about 400 references in the notes which are collected at the end of the book.

1777. Lagrange, Joseph-Louis. "Nouvelles recherches sur la nature et la propagation du son." *Mélanges de philosophie et mathématique de la Societé royale de Turin pour les années 1760-1761, Miscellanea Taurinensia* 2 (1760-1761), 11-172.

This article, besides being a key reference in the history of sound, includes one of the first attempts to use a Fourier integral transform for a partial differential equation; the transformation includes surface integrals involving a rectangular box as boundary, and Lagrange makes some vague hand-waving gestures about integrals on general two-dimensional surfaces.

1778. Lehr, André. "On Vibration Patterns before Chladni." *Janus* 52 (1965), 113-120.

The tuning of bells in the Netherlands in the 17th century was done by means of bars of the correct tone. These bars were held fixed by bolts through holes drilled at nodes. The nodes were found by sprinkling sand on the bars, by then striking these bars, and by marking the places where the sand heaped together. The bells were sounded and fixed bars, sprinkled with sand, resonated, with good tuning making the sand grains leap and jump. This is a descriptive paper with accurate translations of the quotations from the sources.

1779. Lindsay, Robert Bruce. "The Story of Acoustics." *Journal of the Acoustical Society of America* 39 (1966), 629-644.

This is a general review of the history of sound from 1600 to 1960, with good references to the literature; it is not mathematical in itself.

1780. Miller, Dayton Clarence. *Anecdotal History of the Science of Sound to the Beginning of the 20th Century*. New York: Macmillan, 1935, xii + 114 pp.

This history is deceptively titled with the word "anecdotal," as it gives a good coverage of the topic, particularly of the 18th and 19th centuries. Further, it gives good comments on other histories and has 15 plates scattered throughout the text and 100 references at the end.

1781. Strutt, John William, Third Baron Rayleigh. *The Theory of Sound*. London: Macmillan, Volume I in 1877 and Volume II in 1878. Reprinted New York: Dover, 1945, with an historical introduction by Robert Bruce Lindsay, from one of the later editions.

This is a complete theory of sound, useful as an historical source because of the multitudinous references in the footnotes.

It begins with a general theory of vibrations, and then deals with the vibration of strings, bars, membranes, and plates; the second volume deals with sound in air, air in chambers of various shapes and sizes.

Mechanics

This classical topic has three faces; potential theory and gravity (see Poincaré here and Potential Theory elsewhere), celestial mechanics (Poincaré, Wintner), and the mechanics of particles and rigid bodies (Kovalevskaya, Lagrange, Laplace, and Szabó). The General section above contains further references. Many classics due to Routh, Whittaker, Mathieu, and the well-known textbook writers of the last century have been omitted; the names of these writers are mentioned in these two sections and that on Potential Theory. Some judgment on these may be gleaned from Felix Klein's history of the development of mathematics in the 19th century.

1782. Fierz, Markus. *Vorlesungen zur Entwicklungsgeschichte der Mechanik*. (Lecture Notes in Physics, No. 15.) Berlin: Springer, 1972, 97 pp.

This history covers the whole period from Plato to Newton, with emphasis on the 16th and 17th centuries: Copernicus, Kepler, Galileo, Huygens, and Newton. The two dozen references are not really sufficient; the treatment is not mathematical.

1783. Grigor'yan, Ashot, Tigranovich. *Ocherki istorii mekhaniki v Rossii* (Notes on the history of mechanics in Russia). Moscow: Akademiia Nauk SSSR, 1961, 292 pp.

This history covers Russian contributions to mechanics, dynamics, stability, aerodynamics and general hydrodynamics, and elasticity. The coverage is mainly 19th and 20th centuries, and there are 600 references at the end. See also his *Ocherki istorii matematiki i mekhaniki: sbornik statei* (Notes on the history of mathematics and mechanics: a collection of articles) (Moscow: Akademiia Nauk SSSR, 1963, 272 pp.).

1784. ———. "On the Development of Variational Principles of Mechanics." *Archives Internationales d'histoire des Sciences* 18 (1965), 23-35.

This article concentrates on the Russian school using minimum principles in mechanics, particularly by Ostrogradski, Sludski, Joukovski, Talysin, Chaplygin, and Suslov, in the period 1850-1920. It is mathematical, with a dozen references.

1785. Kovalevskaya, Sophie. "Sur le problème de la rotation d'un corps solide autour d'un point fixe." *Acta Mathematica* 12 (1889), 177-232.

This prize essay crowned by the Paris Académie in 1888 gives a third case where these equations of motion can be integrated by theta functions of two variables; the cases due to Euler-Poisson and to Lagrange are rather special, i.e., linear with respect to the time variable. While this is the only mention of the elliptic functions and Abelian integrals in this section, these "modern" techniques were pervasive in mathematical physics, as attested by the chapters in E.T. Whittaker and G.N. Watson's *A Course of Modern Analysis* (Cambridge: At the University Press, 1902), and many later editions and reprintings.

1786. Lagrange, Joseph-Louis. *Mécanique analytique*. Paris: Veuve Desaint, 1788, xii + 512 pp. 2nd ed. 1808-1815; 3rd ed. 1853-1855 (edited by. J. Bertrand); 4th ed. 1888-1889 (edited by Gaston Darboux) in the *Oeuvres de Lagrange*, J.A. Serret and Gaston Darboux, editors, Paris: Gauthier-Villars, Tome XI, 1888, xxii + 502 pp., and Tome XII, 1889, viii + 391 pp.

This classic contains his general theory of mechanics, celestial mechanics, and the elements of hydrodynamics.

1787. Laplace, Pierre Simon de. *Traité de mécanique céleste*. Paris: various publishers, 1799-1827. In detail, Tome I, edited by J.B.M. Duprat, 1799, xxxii + 368 pp.; II, J.B.M. Duprat, 1799, 382 pp.; III, J.B.M. Duprat, 1802, xxiv + 304 pp. + Supplément, 1808, 24 pp.; IV, Courcier, 1805, xl + 348 pp. + deux Suppléments, 65 + 79 pp.; V, Bachelier, 1823-1825, viii + 420 pp. + Supplément, 1827, 35 pp. There is an English translation of the first four volumes by Nathaniel Bowditch, *Celestial Mechanics by the Marquis de Laplace*. Boston: Hillard, Gray, Little and Wilkins, 1829-1839. Reprinted New York: Chelsea, 1966. 4 vols. See also the *Oeuvres* de Laplace (1843-1847) and the *Oeuvres complètes de Laplace*. Paris: Gauthier-Villars, 1878-1912, and the article on Laplace in Volume XV of the *Dictionary of Scientific Biography* (item 10, pp. 273-403) by Charles C. Gillispie, Robert Fox, and Ivor Grattan-Guinness.

This classic contains mechanics, celestial mechanics, capillarity, and fluid mechanics (for planetary atmospheres).

1788. Mach, Ernst. *Die Mechanik in ihrer Entwicklung. Historisch-kritisch dargestellt*. Leipzig: Brockhaus, 1883, x + 483 pp., and eight other editions to 1933. Translated into English in six editions as *The Science of Mechanics. A Critical and Historical Exposition of its Principles*, first done by Thomas J. McCormack. Chicago: Open Court, 1893, xiv + 534 pp., and five other editions to 1960.

This evergreen classic must be read carefully and checked against the original sources, especially for Galileo and the Bernouillis.

1789. Pogrebysskii, Iosif Benediktovich. *Ot Lagranzha k Einshteinu: Klassicheskaya mekhanika XIX veka* (From Lagrange to Einstein: Classical mechanics in the 19th century). Moscow: Nauka, 1966, 326 pp.

This is a set of notes with 200 references at the very end. The notes cover the statics of Lagrange, Carnot, and Monge, as well as the dynamics of Lagrange; elasticity and hydrodynamics in the molecular style of Laplace, including the work of Poinsot as well; the analytical methods of Hamilton, Jacobi, and Ostrogradski; and the geometrization of mechanics under Darboux, Beltrami, and Lipschitz.

1790. Poincaré, Jules Henri. *Théorie de Potentiel Newtonien: Leçons professées à la Sorbonne pendant le premier semestre 1894-1895 par H. Poincaré, redigées par Edouard Le Roy ... et Georges Vincent*. Paris: Georges Carré et C. Naud, 1899, 366 pp.

This is not a text as it covers the research current at the time; all of the latest researches, including results published only "yesterday," appeared in these lectures in the form they took in the literature. The topics begin classically as in Riemann: Laplace's equation, series expansions and solutions, Green's formulae, Poisson's integral. And then they march into the present: attracting surfaces, Green's function, Dirichlet's problem, Harnack's theorem, double layers, balayage, Neumann's method, simply connected but not necessarily convex spaces, and the fundamental functions (singular solutions).

1791. ———. *Leçons de mécanique céleste professées à la Sorbonne*. Paris: Gauthier-Villars, Tome I, 1905, vi + 368 pp.; Tome II, partie 1, 1907, 168 pp.; Tome II, partie 2, 1909, 138 pp.; Tome III, 1910, 472 pp. + 2 pls.

Félix Tisserand (1845-1896), in his *Traité de mécanique céleste* (Paris: Gauthier-Villars, 1889-1896), reproduces the thought and argument of the founders of celestial mechanics. Poincaré in *Les méthodes nouvelles* took this as known and expounded new methods to replace the old. These *Leçons* fall in between, treating the problems *de novo* but with less emphasis on analytical rigor.

1792. ———. *Les méthodes nouvelles de la mécanique céleste*. Paris: Gauthier-Villars, Tome I, 1892, 385 pp.; Tome II, 1893, viii + 479 pp.; Tome III, 1899, 414 pp.

This is the classic exposition of periodic solutions, asymtotic solutions, invariant integrals, perturbation; it has the flavor of a geometer seeking analytic rigor. See also J. Kovalevsky, "Problèmes de mécanique céleste," *Revue des questions scientifiques* 144 (1973), 189-210.

1793. Sharlin, Harold Issadore. "William Thomson's Dynamical Theory: An Insight into a Scientist's Thinking." *Annals of Science* 33 (1975), 133-147.

This article supplies some of the physical thinking so lacking in histories of mathematics. It gives very useful indications of Thomson's mode of argument in the vital period 1846-1851.

1794. Szabó, István. *Geschichte der mechanischen Prinzipien und ihrer wichtigsten Anwendungen*. Basel, Stuttgart: Birkhäuser, 1977, xvi + 491 pp.

This history attempts to cover the period 1600-1975. All of the hundreds of references appear in the footnotes, and the indexes are brief. There is a severe concentration on the classical and the linear, and the list of topics is necessarily highly selective: rigid-body mechanics of the 18th century; the mechanics of Eulerian fluids, viscous fluids and gas dynamics; linear, homogeneous, isotropic, elastic solid materials; impact. See the favorable review by Truesdell in *Centaurus* 23 (1980), 163-175. See also item 958.

1795. Truesdell, Clifford. "Whence the Law of Moment of Momentum?" *Mélanges Alexandre Koyré. Tome I, L'aventure de la science.* Edited by I.B. Cohen and R. Taton. Paris: Hermann, 1964, 588-612.

This is a brief "working-historian's" view of the separateness of this principle (from the principle of linear momentum), its basis and development, its use and meaning, and of the *Eulerian* equations of motion. There are about 30 references in the footnotes. This is reprinted with additions and illustrations in Truesdell's *Essays*, item 1703.

1796. Wintner, Aurel. *The Analytical Foundations of Celestial Mechanics*. Princeton: Princeton University Press, 1941, xii + 448 pp.

The three- and n-body problems are treated in some detail. This is the last of the analytical treatises before the topological flood; it summarizes the tradition of Hill, Poincaré, Levi-Civita, and G.D. Birkhoff. The historical part lies in the notes, pp. 411-443, where all the references are given for the sections of the text. See also Alfred Gautier's *Essai historique sur le problème des trois corps; ou, Dissertation sur la théorie des mouvements de la lune et des planètes, abstraction faite de leur figure* (Paris: Veuve Courcier, 1817, xii + 283 pp. + planche).

MATHEMATICS AND NAVIGATION

This mathematical area has been one of great practical importance for five centuries, but the attention paid to it by historians has been patchy, perhaps because of the specialized technical knowledge and interest required, perhaps because no great scientific ideas were seen to be involved, and perhaps because it has been regarded as being away from the mainstream of mathematical development. With a few exceptions it has not attracted the continued attention of mathematicians of the first rank. (The distinguished algebraist, William Burnside, FRS [1852-1927], was regarded as strange when he left Cambridge to take up an appointment as Professor of Mathematics at the Royal Naval College, Greenwich, England, which had been set up as the "technical university" of the navy. The traditional view is expressed by the equally distinguished pure mathematican, J.E. Littlewood, FRS [1885-1977], in his *A Mathematician's Miscellany* [London: Methuen, 1953]: "I wasted time on optics and astronomy [*not* worth knowing] ...," [p. 70, his emphasis]). Nevertheless, the work of service officers, "mathematical practitioners" and amateurs must not be discounted; the first-named group were among the comparatively few people in public office in the eighteenth and nineteenth centuries who had had the opportunity of a technical education.

The modern reader who requires a thorough knowledge of many of the developments put forward will have to go back to primary sources and to the many unexpected and scattered sources which are listed in the historical accounts given in this bibliography.

1797. *Admiralty Manual of Navigation*. London: Her Majesty's Stationery Office, 1955. 3 vols.

Volume 3 is the mathematical volume, going beyond the requirements of executive officers. It indicates the professional requirements of knowledge of its date.

1798. Bowditch, Nathaniel. *The New American Practical Navigator*. Newport: William R. Wilder, 1802, iii + 589 pp., 8 pls.

The first edition of the standard North American work. The "real" first edition appeared in 1799 under the name of J.M. Moore. Numerous subsequent editions issued. An epitome of navigation, with tables.

1799. Cotter, Charles H. *A History of Nautical Astronomy*. London: Hollis and Carter, 1968, xii + 387 pp.

Extends over the whole period from the earliest times; fullest on later techniques. The 375-item bibliography includes a large number of primary sources. Many mathematical methods given in the text; others are referred to. Further detailed information is given in the same author's "A History of Nautical Astronomical Tables" (unpublished Ph.D. dissertation, University of London, 1975, 821 pp.), which contains copious references to work in many countries. Issued on microfiche, London: Mansell, 1977.

1800. Forbes, Eric Gray. *Tobias Mayer (1723-62) Pioneer of Enlightened Science in Germany*. Göttingen: Vandenhoeck & Ruprecht, 1980, 248 pp. + 16 pls.

This scientific biography is listed because of Mayer's contributions to the study of atmospheric refraction and to the important theory of lunar distances, applicable to the longitude problem. The bibliography gives full references to Mayer's technical work and accounts thereof. Forbes has also edited both Mayer's published and unpublished works, as well as the Euler-Mayer correspondence of 1751-1755.

1801. Inman, James. *Nautical Tables Designed for the Use of British Seamen*. London: J.D. Potter, 1957, xxxi + 505 pp. + Supplement ii + XXXIV.

The first edition is dated 1821. Later revisions were made by William Hall and H.B. Goodwin. A standard work up to recent times. Other tables, such as Brown's and Norie's, were favored by the merchant marine; this volume is, according to its title page, "Sanctioned for Use in the Royal Navy."

1802. *Journal of the Institute of Navigation*. London, 1948-1971; subsequently *Journal of Navigation*, 1972-.

An important quarterly journal publishing articles and reviews on modern developments. Occasional historical contributions.

1803. Melluish, R.K. *An Introduction to the Mathematics of Map Projections*. Cambridge: University Press, 1931, viii + 145 pp.

A mathematical treatment of the various projections and their properties. Some historical information is given incidentally.

1804. Michel, Henri. *Traité de l'astrolabe*. Paris: Gauthier-Villars, 1947, viii + 202 pp., 24 pls. Reissued Paris: Alain Brieux, 1967, with notes, corrections, and preface by F. Maddison.

An account of the various types of astrolabes, with an explanation of their design, construction, and use. (The impression given that conformality is needed in their design is incorrect.) See also item 382.

1805. *The Nautical Almanac for the Year 1967*. 200th anniversary ed. London and Washington, D.C.: Her Majesty's Stationery Office and the United States Government Printing Office, 1966, 276 + xxxv pp. + an historical section (pp. 3b-3n) by W.A. Scott on "The Nautical Almanac and Astronomical Ephemeris 1767 to 1967."

The almanac is published annually, but the London and Washington editions are printed separately. A related official work is *Sight Reduction Tables for Air Navigation* (Washington, D.C.: U.S. Hydrographic Office, 1951-1952; London: H.M. Stationery Office, 1952-1953), of which volumes 2 and 3 are permanent; volume 1 is issued for selected star epochs of about 5 years. Despite their name, the simplicity of these tables has made them attractive to nautical navigators.

1806. Neugebauer, Otto. *A History of Ancient Mathematical Astronomy*. Berlin, Heidelberg, New York: Springer-Verlag, 1975, xxi + 1457 pp., 9 pls. 3 vols.

Neugebauer in this classic work gives a mathematical account, with references, of map projections at pp. 879-892. For early mathematical geography, see pp. 733-736, and for Ptolemy's *Geography* ("few books have exercised such a profound influence on human thought and civilization") see pp. 934-940. Ptolemy's *Planisphaerium* (pp. 857-876) really comes under Astronomy, and the *Analemma* (pp. 839-843) under Geometry. See also items 135, 246.

1807. Shirley, John W. *Thomas Harriot: A Biography*. Oxford: Clarendon Press, 1983, xii + 508 pp.

Harriot was the leading contributor of his time to the mathematical, astronomical, and optical bases of navigation and map-making. Comprehensive bibliography.

1808. Smart, W.M. *Text-Book on Spherical Astronomy*. Cambridge: University Press, 1965, xii + 430 pp.

A standard work first published in 1931. Chapter 13 introduces mathematical elements of the determination of position at sea. Chapter 1 on spherical trigonometry may be supplemented by J.G. Leathem's revision of I. Todhunter's *Spherical Trigonometry* (London: Macmillan, 1901), which has historical references that are still of value. A useful work on the earlier period is Mary C. Zeller's *The Development of Trigonometry from Regiomontanus to Pitiscus* (Ann Arbor: Edwards Bros., 1946).

1809. Taylor, Eva G.R. *The Mathematical Practitioners of Tudor and Stuart England, 1485-1714*. Cambridge: University Press for the Institute of Navigation, 1954, xi + 443 pp., 12 pls.

See items 630, 801, 2216.

1810. ———. *The Mathematical Practitioners of Hanoverian England, 1714-1840*. Cambridge: University Press for the Institute of Navigation, 1966, xvi + 503 pp., 12 pls., including portraits.

Despite lack of detailed references for much of their material, these are essential accounts and handbooks for their periods. Both volumes contain general chapters followed by lengthy sections of biographies and works. See also items 995, 2217.

1811. Waters, David W. *The Art of Navigation in England in Elizabethan and Early Stuart Times*. London: Her Majesty's Stationery Office, 1978, xl + 696 pp., 87 pls. 2nd ed. in 3 vols.

Touches also on earlier Spanish and Portuguese work. The extensive bibliographical and analytical content, both of primary and secondary sources, makes this an essential reference, rendering unnecessary the listing of many other important works.

NUMBERS AND NUMBER THEORY

This section of the bibliography canvasses a number of diverse but related topics, including the origins of counting, numbers and number systems, the history of arithmetic, combinatorics, elementary number theory and more advanced topics involving algebraic and analytic number theory, which begin to overlap with other sections of the bibliography, in particular with the sections on algebra and analysis. The history of numerical analysis is covered in a shorter section of its own following this one. Cross-references to other sections of the bibliography have been used whenever possible to accommodate the interests of economy without sacrificing as wide a sampling as possible of the subjects considered here. Even so, except for the origins of counting and number systems as they evolved from ancient times to the present, this section of the bibliography concentrates primarily on aspects of number and number theory dating roughly from the Renaissance.

Numbers

1812. Ascher, M., and R. Ascher. "Numbers and Relations from Ancient Andean Quipus." *Archive for History of Exact Sciences* 8 (1972), 288-320.

Quipus--knotted cords, whose precise use has been much debated--are discussed, with an analytic description of nine quipus given in the appendix.

1813. ———. *Code of the Quipu. A Study in Media, Mathematics and Culture.* Ann Arbor: University of Michigan Press, 1981.

A detailed, in-depth analysis of the quipu.

1814. Audisio, F. "Il numero π." *Periodico Matematico* 11 (1931), 11-42 and 149-150.

Brief account of the history of *pi.*

1815. Bortolotti, E. "Sul numero π." *Periodico Matematico* 11 (1931), 110-113.

A companion piece to item 1814.

1816. Bosteels, G. *La vie des nombres.* Namur: Ad. Wesmael-Charlier, 1960.

Discusses the history of the numerals from one to twenty, as well as basic arithmetic operations.

1817. Boyer, Carl B. "Fundamental steps in the development of numeration." *Isis* 35 (1944), 153-168.

A general survey, including five figures.

1818. ———. "Zero: the Symbol, the Concept, the Number." *National Mathematics Magazine* 18 (1944), 323-330.

1819. Brainerd, C.J. *The Origins of the Number Concept.* New York: Praeger, 1979.

Concerns the cultural evolution of numbers, logical theories of number, and how numerical concepts are grasped in the course of human psychological development. This book tries to integrate all these subjects as part of a related whole. Chapters 6-11 are devoted to what the author terms "ontogenetic changes" in number concepts. Some of the material here appeared previously in serials like the *Journal of General Psychology*, etc.

1820. Cajori, Florian. "The Controversy of the Origin of Our Numerals." *Scientific Monthly* 9 (1919), 458-464.

Reviews recent theories, especially those of Kaye and Carra de Vaux. Cajori favors a Hindu origin, but admits reliable evidence can be found no earlier than the 9th century. The earliest record of use of zero is an inscription in India dated to 867 A.D.

1821. ———. *A History of Mathematical Notations*. Chicago: Open Court, 1928-1929. 2 vols.

Volume 1 is concerned with notations in elementary mathematics; Volume 2 is devoted to notations mainly in higher mathematics. The second volume ends with a discussion of empirical generalizations on the growth of mathematical notations. Both volumes are well illustrated. See also items 73, 643, 986.

1822. Dantzig, T. *Number, the Language of Science. A Critical Survey Written for the Cultured Non-Mathematician*. Rev. ed. London: Allen and Unwin, 1962.

This is a standard popularization, first written in 1930, and issued in numerous editions since. G. Sarton reviewed the book in *Isis* 16 (1931), 455-459, as well as subsequent editions in 20 (1934), 592, and 31 (1940), 475-476. There is also a French translation of the original edition published in Paris (Payot, 1931). See also item 75.

1823. Dhombres, J. *Etude epistemologique et historique des idêes de nombre de mesure et de continu*. Nantes: U.E.R. de Mathématiques, 1976.

Duplicated manuscript. Discusses numeration, the number concept, real numbers, transfinite numbers, with an interesting chapter on "parallel" developments in Chinese mathematics.

1824. Gānguli, Sāradākānta. "The Indian Origin of the Modern Place-Value Arithmetical Notation." *American Mathematical Monthly* 39 (1932), 251-256.

See item 417.

1825. Glathe, A. *Die chinesische Zahlen*. Tokyo: Deutsche Gesellschaft für natur- und völkerkunde Ostasiens, 1932.

On Chinese numbers; with 15 plates and 3 drawings.

1826. Green, D.R. "The Historical Development of Complex Numbers." *Mathematical Gazette* 60 (411) (1976), 99-107.

Brief survey, with references to Heron, Diophantos, Mahavira, Bhascara, Bar Chiia Abraham, Pacioli, Chuquet, Cardano, Bembelli, Descartes, Leibniz, Harriot, Cotes, de Moivre, Euler, Wallis, Argand, Wessel, Gauss, d'Alembert, Cauchy.

1827. Hartner, W. "Zahlen und Zahlensysteme bei Primitiv- und Hochkulturvölkern." *Paideuma* 2 (1943), 268-326.

A detailed, illustrated study of number-words and symbols in the languages and writing of many cultures, both primitive and highly developed.

1828. Hughes, B.B. "The Earliest Known Record of California Indian Numbers." *Historia Mathematica* 1 (1974), 79-82.

Discusses numeration in the 18th century among American Californian Indians.

1829. Kavett, H., and P. Kavett. "The Eye of Horus is upon You." *The Mathematics Teacher* 68 (1975), 390-394.

A short discussion of Egyptian numeration, including hieroglyphic and hieratic scripts. Fractions are also discussed.

1830. Kennedy, H.C. "Peano's Concept of Number." *Historia Mathematica* 1 (1974), 387-408.

Discusses the number concept, as well as Peano's postulates and the foundations of mathematics.

1831. Larsen, H.D. *Arithmetic for Colleges*. New York: Macmillan, 1950.

Offers an introduction to the history of numbers and numerology, with historical notes.

1832. Laugwitz, D. "Bemerkungen zu Bolzanos Grössenlehre." *Archive for History of Exact Sciences* 2 (1965), 398-409.

Argues that the analysis of Bolzano's work given by Rychlik, item 1839, and van Rootsellar, item 1838, are inadequate. Laugwitz argues that the major deficiency in Bolzano's work concerns a faulty definition, and furthermore, upon interpreting Bolzano's theory in terms of the infinitesimals he and Schmieden introduced in 1958 in "Eine Erweiterung der Infinitesimalrechnung," *Mathematische Zeitschrift* 69 (1958), 1-39, Bolzano's theory can be made entirely consistent and rigorous.

1833. Mazaheri, A. "Formes 'sounites' et formes 'Chi'ites' des chiffres arabes au les avatars de chiffres indiens en Islam." *Proceedings of the XIIth International Congress for the History of Science (1971)*. Moscow: Editions "Nauka," 1974, Section IV, 60-63.

Considers Sunnite and Shiite forms of Arabic numerals and the avatars of Indian numerals in Islam, concentrating on the 11th to the 15th centuries.

1834. Menninger, K. *Zahlwort und Ziffer. Eine Kulturgeschichte der Zahl*. Göttingen: Verlag Vandenhoeck & Ruprecht, Vol. 1: Zahlreihe und Zahlsprache, 1957; Vol. 2: Zahlschrift und Rechnen, 1958. 2 vols.

First published in 1934, Menninger's classic work on the subject of number and numerals was translated into English in 1969 (see item 1835). Volume 1 concerns number sequences and the words or language of counting. Volume 2 is devoted to the writing of numbers and to the development of computation and various arithmetic operations. With 170 illustrations.
Review: Fehr, H.F., *American Mathematical Monthly* 66 (1959), 437.
Hofmann, J.E., *Archives internationales d'Histoire des Sciences* 10 (1957), 255; 11 (1958), 199-200.

1835. ———. *Number Words and Number Symbols; a Cultural History of Numbers*. Translated by P. Broneer. Cambridge, Mass.: MIT Press, 1969.

See item 1834.

1836. Miller, G.A. "Historical Note on Negative Numbers." *American Mathematical Monthly* 40 (1933), 4-5.

1837. Mitchell, U.G., and M. Strain. "The Number *e*." *Osiris* 1 (1936), 176-196.

1838. Rootselaar, B. van. "Bolzano's Theory of Real Numbers." *Archive for History of Exact Sciences* 2 (1964), 168-180.

Rootselaar offers a different view of Bolzano's theory of real numbers and its deficiencies than those presented by Rychlik, item 1839. Above all, he opposes the view that Bolzano's theory can be refined to make it rigorous by showing that it is inherently inconsistent. Virtually no bibliography. See also Laugwitz, item 1832.

1839. Rychlik, K. *Theorie der reellen Zahlen in Bolzanos handscriftlichem Nachlasse*. Prague: Rieger, 1962.

Presents Bolzano's theory of real numbers based primarily upon manuscripts and other original sources. See also items 1832 and 1838.

1840. Sarton, George. "Decimal Systems Early and Late." *Osiris* 9 (1950), 581-601.

Includes two figures.

1841. Scriba, C.J. *The Concept of Number*. Manheim: Hochschultaschenbücher. Bibliographisches Institut, 1968.

A fundamental study that provides a basic introduction to the development of the number concept.

1842. ———. "Number." *Dictionary of the History of Ideas* 3 (1973), 399-407.

Discusses development of the number concept and numeration; for a more detailed study, see item 1841.

1843. Sergescu, P. "Histoire du nombre." *Conférences du Palais de la Découverte, 23*. Paris: Université de Paris, 1953.

Review: Kraitchik, M., *Archives internationales d'Histoire des Sciences* 9 (1955), 78.

1844. Smeltzer, D. *Man and Number. An Account of the Development of Man's Use of Number Through the Ages*. New York: Emerson Books, 1958.

A short, general introduction with bibliography.

1845. Smith, David Eugene. "The Roman Numerals." *Scientia* 40 (1926), 1-8 and 69-78.

Part 1 considers the origins of Roman numerals; Part 2 discusses other problems associated with their history.

1846. Solomon, B.S. "'One is no number' in China and the West." *Harvard Journal of Asiatic Studies* 17 (1954), 253-260.

1847. Thureau-Dangin, François. "Sketch of a History of the Sexagesimal System." Translated by S. Gandz. *Osiris* 7 (1939), 95-141.

Revised version of the author's *Esquisse d'une histoire du système sexagesimal* (Paris: Geuthner, 1932).

1848. Van der Waerden, Bartel Leendert. "Hamilton's Discovery of Quaternions." *Mathematics Magazine* 49 (1976), 227-234.

Discusses the origins of quaternions, as well as division algebras, octonians, and complex numbers in general, with references to Hamilton's notes and letters.

1849. Whitrow, G.J. "Continuity and Irrational Number." *Mathematical Gazette* 17 (1933), 151-157.

An historical discussion, briefly, of the subject from Pythagoras to Cantor.

1850. Wieleitner, H. "Zur Frühgeschichte des Imaginären." *Jahresbericht der Deutschen Mathematiker-Vereinigung* 36 (1927), 74-88.

Survey of contributions, especially those of Cardano and Bombelli, with special reference to the work of Bortolotti on the subject.

Number Systems

1851. Glaser, A. *History of Binary and Other Nondecimal Numeration.* Southampton, Pa.: A. Glaser, 1971; Los Angeles: Tomash Publishers, 1981.

Based on the author's dissertation at Temple University, with the title "History of Modern Numeration Systems." It opens with consideration of the work of Harriot, and continues to the present. Discusses the question of which base is best, and offers bibliographies of primary and secondary sources, chapter by chapter. The appendix to the 1981 edition includes a photo-reproduced copy of Fontennelle's article, "Nouvelle arithmétique binaire," 1703.

1852. Guitel, T. *Histoire comparée des numeration ecrite*. Paris: Flammarion, 1975.

Covers such topics as numeration in general, Egyptian, Aztec, Greek, Roman, Arabic, Hebrew, Ethiopian, Sumerian, Babylonian, Mayan, Chinese, and Indian numeration, including discussion of zero. Numerous illustrations and tables, with classification of 25 numeration systems, are also included.

1853. Laki, K. "The Number System Based on Six in the Proto Finno-Ugric Language." *Journal of the Washington Academy of Sciences* 50 (1960), 1-11.

1854. Petrosyan, G.B. "Ob alfavitnykh sistemakh schisleniya." *Istoriko-Matematicheskie Issledovaniia* (1978), 144-155.

Figure 13. The goddess "Arithmetica" has had a long and glorious reign in the history of mathematics. Here she is shown overseeing a contest for speed between two of her earliest disciples—Pythagoras (on the right), whose cult worshipped the divinity of number in all things, and Boethius (on the left), the leading arithmetical authority of the Middle Ages. From the expression on the latter's face, the competition between the old counting board and the "new math" of Hindu-Arabic numbers (and the base-ten calculations they facilitated) was not much of a contest.

Discusses alphabetic numeration systems and how to carry out arithmetic operations in such systems.

1855. Powell, Marvin A. "The Antecedents of Old Babylonian Place Notation and the Early History of Babylonian Mathematics." *Historia Mathematica* 3 (1976), 417-439.

Argues for an earlier origin than has been claimed previously. See also item 155.

1856. Shirley, John W. "Binary Numeration before Leibniz." *American Journal of Physics* 19 (1952), 452-454.

1857. Shrader-Frechette, M. "Complementary Rational Numbers." *Mathematics Magazine* 51 (1978), 90-98.

Discusses the history of studies concerning patterns of digits found in repeating decimals.

Counting and the Abacus

1858. Barnard, F.P. *The Casting-Counter and the Counting Board. A Chapter in the History of Numismatics and Early Arithmetic.* Oxford: Clarendon Press, 1916.

Somewhat out of date on the subject of arithmetic, but still useful on the subject of counting boards.

1859. Brooke, M. "How the Shona Count." *Journal of Recreational Mathematics* 6 (1973), 296-298.

Brief, but provides information on finger counting, names of the days of the week, the months, and basic measurements of the Shona of Rhodesia.

1860. Closs, M.P. "The Nature of the Maya Chronological Count." *American Antiquity* 42 (1977), 18-27.

Covers Mayan mathematics, numeration, and aspects of time.

1861. Evans, Gillian R. "*Duc oculum.* Aids to Understanding in Some Mediaeval Treatises on the Abacus." *Centaurus* 19 (1976), 252-263.

Surveys 11th- and 12th-century treatises on arithmetic, especially those used in education. Also covers numeration in general and fractions. See also item 557.

1862. ———. "From Abacus to Algorism: Theory and Practice in Medieval Arithmetic." *British Journal for the History of Science* 10 (1977), 114-131.

See item 558.

1863. Kimble, G.W. "From Pebbles to Position." *Historia Mathematica* 6 (1979), 323-325.

Note on positional counting systems.

1864. Maistrov, L.E. "Rol alfabitnykh sistem numeratsii." *Istoriko-Matematicheskie Issledovaniia* 19 (1974), 39-49.

Discusses various opinions concerning the origin and role of alphabetical systems in numeration, particularly in Greek mathematics.

1865. Marschak, A. *The Roots of Civilization. The Cognitive Beginnings of Man's First Art, Symbol and Notation.* New York: McGraw-Hill, 1972.

Investigates prehistoric artifacts in connection largely with numeration and calendars. Many illustrations, with a good index, bibliography, and sources.

1866. Moon, P. *The Abacus.* New York: Gordon and Breach, 1971.

Although history comprises only a small portion of this book, descriptions of counting boards, as well as Chinese, Japanese, Roman, and Russian abacuses make this a useful source.

1867. Seidenberg, A. "The Diffusion of Counting Practices." *University of California Publications in Mathematics.* Item 2308.

1868. ———. "The Ritual Origin of Counting." *Archive for History of Exact Sciences* 2 (1962), 1-40.

This article begins with a discussion of number-mysticism and ritual, considers the evidence of counting rituals, tabus on counting, stones, taxation, money, ordinal names, etc. The basic hypothesis is that counting was invented as a means of calling participants in a ritual onto the ritual scene. Extensive bibliography. See also item 413.

Arithmetic

1869. Auluck, F.C., ed. "Proceedings of the Symposium on al-Bīrūnī and Indian Sciences held at New Delhi on November 8-9, 1971." *Indian Journal of the History of Science* 19 (2) (1975), 89-276.

Twenty-five papers, including Brahmagupta and Bag on Bīrūnī and Indian arithmetic.

1870. Detlefsen, M., D. Erlandson, H.J. Clark, and C. Young. "Computation with Roman Numerals." *Archive for History of Exact Sciences* 15 (2) (1976), 141-148.

Simple procedures for adding and multiplying Roman numerals are given to show that computation with Roman numerals is not so difficult--or impossible--as is often held.

1871. Eganyan, A.M. *Grecheskaya logistika.* Yerevan: Erevanskii Armyanskii Gosudarstvennyi pedagogicheskii institut imeni Kh. Abovyana, 1972.

A critical review of secondary literature on Greek arithmetic from Nesselmann (1842) through Vogel's classic study (1936), and more recent work as well. Analysis of ancient and medieval sources is also included.

1872. Fettweis, E. "Streitfragen aus der Geschichte der Arithmetik in ethnologische Beleuchtung." *Scientia* 88 (1953), 235-249.

1873. Gillings, R.J. "Tests of Divisibility." *Scripta Mathematica* 22 (1956), 294-296.

1874. Hermelink, H. "The Earliest Reckoning Books Existing in the Persian Language." *Historia Mathematica* 2 (1975), 299-303.

Discusses 11th-century Persian arithmetic, especially that of Abū Ja'far Muh. b. Ayyūb Tabarī.

1875. Itard, J. *Arithmétique et théorie des nombres*. Paris: Presses Universitaires de France, 1963.

Review: Taton, R., *Revue d'Histoire des Sciences* 17 (1964), 168-169.

1876. Karpinski, Louis C. *The History of Arithmetic*. New York: Rand McNally, 1925.

Review: Cajori, F., *Isis* 8 (1926), 231-232.

1877. Lam, Lay Yong. "The *Jih yung suan fa*: An Elementary Arithmetic Textbook of the Thirteenth Century." *Isis* 63 (1972), 370-383.

1878. Nordgaard, M.A. "The Origin and Development of Our Present Method of Extracting the Square and Cube Roots of Numbers." *Mathematics Teacher* 17 (1924), 223-238.

1879. Saidan, A.S., ed. *Arabic Arithmetic. The Arithmetic of Abū al-Wafā' al-Būzajānī, 10th Century, Mss. Or. 103 Leiden and 40 M Cairo*. Amman, Jordan: Jordanian University, 1971.

Entirely in Arabic, with introduction, commentaries, and ample references to the arithmetic of al-Karaji. Discusses Arab mathematics, arithmetic, Babylonian mathematics, Greek mathematics, sexagesimal systems, finger reckoning. See also item 287.
Review: King, D., *Isis* 64 (1976), 123-125.

1880. ———. "The Arithmetic of Abū'l-Wafā'." *Isis* 65 (1974), 367-375.

A shorter version of the author's *Arabic Arithmetic*, annotated above in item 1879.

1881. Smith, C.S. "A Seventeenth-Century Octonary Arithmetic." *Isis* 66 (1975), 390-394.

Reproduces in facsimile two pages of notes in a contemporary volume.

1882. Taton, René. *Histoire du calcul*. (Collection Que Sais-Je?, No. 198.) Paris: Presses Universitaires, 1946.

Review: Sergescu, P., *Archives internationales d'Histoire des Sciences* 1 (1947), 182.

1883. ———. "L'évolution d'une operation: la multiplication." *La Nature* 76 (1948), 268-271.

Combinatorics

1884. Biggs, Norman L. "The Roots of Combinatorics." *Historia Mathematica* 6 (1979), 109-136.

History to 1650.

1885. Knobloch, E. "Die mathematischen Studien von G.W. Leibniz zur Kombinatorik, auf Grund fast ausschliesslich handschriftlicher Aufzeichnungen dargelegt und Kommentiert." *Studia Leibnitiana Supplementa* 11 (1973).

This study of Leibniz's contributions to combinatorics was followed several years later by a volume of texts, *Studia Leibnitiana Supplementa* 16 (1976), 339 pp. See also items 833 and 834.

1886. ———. "Marin Mersennes Beiträge zur Kombinatorik." *Sudhoffs Archive* 58 (1974), 356-379.

Contents and probable origins of Mersenne's contributions to combinatorics.

1887. ———. "Die mathematischen Studien von G.W. Leibniz zur Kombinatorik." *Janus* 63 (1976), 1-26.

Discusses Leibniz, his manuscripts, and his followers, including Weingartner, Stern, Hindenburg, and Boscovitch.

Logarithms

1888. Adams, C.W. "When Was Logarithmic Paper First Used? Answer to Query No. 80. *Isis* 30 (1939), 95-6." *Isis* 31 (1940), 429-430.

Answers were also given by G. Sarton, *Isis* 30 (1939), 95-96, and W.F. Durand, *Isis* 32 (1947), 117. Durand actually had a hand in the development of logarithmic paper.

1889. Bruins, E. "Computation of Logarithms by Huygens." *Janus* 65 (1978), 97-104.

Historical discussion of Huygens's method of 1661, including references to Bürgi, Napier, Briggs, and Decker.

1890. Eccarius, W. "Alexander von Humboldts Bemühungen um Additionslogarithmen." *NTM-Schriftenreihe für Geschichte der Naturwissenschaften, Technik, und Medizin* 16 (1979), 30-42.

Transcription with notes of two early manuscripts on logarithms by von Humboldt.

Number Theory

1891. Archibald, Raymond Clare. "Perfect Numbers." *American Mathematical Monthly* 28 (1921), 140-153.

1892. ———. "Quadratic Diophantine Equations." *Scripta Mathematica* 2 (1933), 27-33.

1893. ———. "Goldbach's Theorem." *Scripta Mathematica* 3 (1935), 44-50.

1894. Baker, A. "Some Historical Remarks on Number Theory." *Historia Mathematica* 2 (1975), 549-553.

Discusses primarily the Catalan Conjecture.

1895. Barnes, C.W. "The Representation of Primes of the Form 4n + 1 as the Sum of Two Squares." *L'Enseignement Mathématique* 18 (1972), 289-299.

Discusses primarily the work of Legendre on representing a prime as the sum of two squares. Continued fraction constructions are also discussed.

1896. Bateman, P., and H. Diamond. "John E. Littlewood (1885-1977). An Informal Obituary." *Mathematical Intelligencer* 1 (1978), 28-33.

Discusses Littlewood's contributions to analytic number theory, as well as to analysis. Collaborations with Hardy and others are also discussed.

1897. Berndt, B.C. "Ramanujan's Notebooks." *Mathematics Magazine* 51 (1978), 147-164.

Offers detailed descriptions of the more interesting entries in Ramanujan's notebooks, along with a history touching on number theory, special functions, and series, as well as general biographical information and general content of the notebooks.

1898. Collison, M.J. "The Origins of the Cubic and Biquadratic Reciprocity Laws." *Archive for History of Exact Sciences* 17 (1977), 63-69.

The author shows that Gauss developed new proofs of the quadratic reciprocity law in order to find methods applicable to the theory of cubic and biquadratic residues. The dispute between Eisenstein and Jacobi over priority of proofs of the cubic and biquadratic reciprocity laws is described.

1899. Dickson, L.E. *History of the Theory of Numbers*. Washington, D.C.: The Carnegie Institution, 1919-1923. Reprinted New York: Stechert, 1934, and New York: Chelsea, 1952, 1971, xii + 486, xxii + 803, iv + 313 pp. 3 vols.

Ore described this work as "a complete, encyclopedic account of the history of the discoveries in number theory up to 1918." Volume 1 is devoted to divisibility and primality; volume 2 covers Diophantine analysis; volume 3 takes up quadratic and higher forms, with a chapter on the class number by G.H. Cresse. The work was reviewed as each volume appeared. See also item 1048. *Reviews:* Child, J.M., *Isis* 3 (1920), 446-448; *Isis* 4 (1921), 107-109; *Isis* 6 (1924), 96-98.

1900. ———. "Perfect and Amicable Numbers." *Scientific Monthly* 12 (1921), 349-354.

1901. Drake, R.C. "A Developmental Study of Ideal Theory." Dissertation, American University, *Dissertation Abstracts* 37 (1976), 797-B.

Treats development of Ideal Theory from its origins in Fermat's Last Theorem and algebraic number theory, as well as in certain polynomial rings and algebraic geometry. Emphasis is given to Kummer, Dedekind, and Kronecker.

1902. Dress, F. "Théorie additive des nombres, problème de Waring et théorème de Hilbert." *L'Enseignement Mathématique* 18 (1972), 175-190.

Treats additive number theory, Waring's problem, Hilbert's Theorem, sums of powers of integers, etc. Errata on pp. 301-302.

1903. Edwards, H.M. *Riemann's Zeta Function.* New York: Academic Press, 1974.

History of the prime number theorem, the Riemann hypothesis, and study of the Riemann zeta function, including analysis of the works of Chebyshev, Hadamard, Hardy, Mangoldt, Stieltjes, Littlewood, and others. A translation of Riemann's famous paper of 1859, "On the Number of Primes Less Than a Given Magnitude," is also included.

1904. Gendrikhson, N.N. "O nekotorykh rabotakh Gaussa po teorii algebraicheskikh chisel." *Problemy istorii matematiki i mekhaniki* 1 (1972), 56-60.

Discusses Gauss's work on the theory of algebraic numbers, including cyclotonic fields. The extent to which Gauss's unpublished work anticipated later developments, including Kummer's work and the concept of ideal factors, is explored.

1905. Goodstein, L.J. "A History of the Prime Number Theorem." *American Mathematical Monthly* 80 (1973), 599-615.

Concentrates on developments from Euler to Hadamard and La Vallée-Poussin, with selected bibliography, samples, and corrections of Gauss's tables, as well as a translation of a letter from Gauss to Encke of December 24, 1849. See also item 1906.

1906. ———. "Correction to 'A History of the Prime Number Theorem.'" *American Mathematical Monthly* 80 (1973), 1115.

Corrects an earlier assertion that the sieve of Eratosthenes had appeared in Euclid's *Elements*.

1907. Halasz, G. "The Number-Theoretic Work of Paul Turan." *Acta Arithmetica* 37 (1980), 9-19.

Includes number theory, power sum method, zeta function, partitions, and uniform distributions.

1908. Hardy, G.H. *Ramanujan: Twelve Lectures on Subjects Suggested by His Life and Work*. Cambridge, England: Cambridge University Press, 1940.

This book opens with a biographical essay, "The Indian Mathematician Ramanujan," pp. 1-21. The eleven lectures which follow are meant to demonstrate the actual and original achievements of Ramanujan, especially on the theory of prime numbers, the analytic theory of numbers, partitions, representation of numbers as sums of squares, and Ramanujan's function.

1909. Hofmann, Joseph E. "Über zahlentheoretische Methoden Fermats und Eulers, ihre Zusammenhänge und ihre Bedeutung." *Archive for History of Exact Sciences* 1 (1961), 122-159.

This article offers a detailed discussion of the number-theoretic results of Fermat and Euler, compares their methods, investigates specific problems like Pythagorean triples, the four-cube problem $x^3 + y^3 = a^3 + b^3$, rational points on curves, etc. Extensive footnotes and bibliographic sources are provided, as well as a complete name index. See also item 792.

1910. Holzer, L. "Eulers Forschungen in seiner Anleitung zur Algebra vom Standpunkt der modernen Zahlentheorie." *Sammelband zu Ehren des 250. Geburtstages Leonhard Eulers*. Berlin: Akademie-Verlag, 1959, 209-223.

Review: Ore, O., *Mathematical Reviews* 23 (1962), 4, A 36.

1911. Lee, E.J., and J.S. Madachy. "The History and Discovery of Amicable Numbers." *Journal of Recreational Mathematics* 5 (1972), 77-93.

This article is written in several parts, and discusses the work of Ore and amicable numbers. Part 1 discusses chronology of discovery and rediscovery of pairs of amicable numbers, while parts 2 and 3 offer tables of the 1,107 known amicable pairs (as of 1972), with a classification and a table listed by discoverer and type.

1912. LeLionnais, F., ed. *Great Currents of Mathematical Thought*. Translated by R.A. Hall and H.G. Bergmann. New York: Dover, 1971. 2 vols.

Volume I includes essays by various distinguished French mathematicians on diverse subjects, including the number concept, Fermat's Last Theorem, Kummer's ideal numbers, transfinite numbers, cardinal numbers, ordinal numbers, pi, etc. Volume II deals with mathematics in the arts and sciences.

1913. Matvievskaya, G.P. "Teoriya kvadratichnykh irratsialnostei i teoriya otnoshenii v evrope do XVII v" (The theory of quadratic irrationals and the theory of ratio in Europe to the XVIIth century). *Proceedings of the XIIIth International Congress for the History of Science*, 1971, 4. Moscow: Nauka, 1974, 77-80.

Theories of irrational numbers, ratio, and proportion, concentrating largely on medieval mathematics.

1914. Melnikov, I.G. "Voprosy teorii chisel v tvorchestve Ferma i Eilera." *Istoriko-Matematicheskie Issledovaniia* 19 (1974), 9-38.

Discusses number theory in the works of Fermat and Euler.

1915. Mirsky, L. "In Memory of Edmund Landau, Glimpses from the Panorama of Number Theory and Analysis." *The Mathematical Scientist* 2 (1977), 1-26.

Sketch of Landau's life and outline of his achievements as discoverer and teacher in both number theory and analysis, with some indication of his influence on others.

1916. Neumann, O. "Bemerkungen aus heutigen Sicht über Gauss' Beiträge zu Zahlentheorie, Algebra und Funktionentheorie." *NTM-Schriftenreihe für Geschichte der Naturwissenschaften, Technik und Medizin* 16 (1979), 22-39.

Comments on contemporary assessments of Gauss's contributions to number theory, algebra, and the theory of functions.

1917. Ore, Øystein. *Number Theory and Its History*. New York, Toronto, London: McGraw-Hill, 1948, x + 370 pp.

Based on a lecture course on the theory of numbers and its history given over several years at Yale University. Presents "the principal ideas and methods of number theory as well as their historical background and development through the centuries." Includes chapters on counting and the recording of numbers, numerology and various number systems, Euclid, prime numbers, perfect and amicable numbers, Diophantine problems, congruences, theorems in number theory due to Fermat and Euler, as well as a chapter on the converse of Fermat's theorem and the classical construction problems, with emphasis on regular polygons, as well as the theory of decimal expansions.

1918. Paiow, M. "Die mathematische Staatsstelle. II." *Archive for History of Exact Sciences* 12 (1974), 174-185.

Considers the passage in Plato's *Republic* on the subjects of nuptual and geometric numbers.

1919. Pieper, H. *Variationen über ein zahlentheoretisches Thema von Carl Friedrich Gauss*. Basel: Birkhäuser, 1978.

Discusses Gauss's quadratic reciprocity theorem, with 14 variations on his results, including those by Lejeune-Dirichlet, Legendre, König, Eisenstein, Frobenius, Hensel, Mirimanoff, and Kronecker.

1920. Ross, B. "Euler's Letter to Goldbach Announcing the Discovery of an Integral Representation for X!" *Ganita Bharati: Bulletin of the Indian Society for History of Mathematics* 1 (1979), 9-12.

Translates Euler's letter of January 8, 1730, into English.

1921. Saidan, A.S. "Number Theory and Series Summations in Two Arabic Texts." *Proceedings of the First International Symposium for the History of Arabic Science*. Vol. II. (Papers in European Languages.) Edited by A.Y. al-Hassan, et al. Aleppo, Syria: Institute for the History of Arabic Science, 1978, 145-163.

Concentrates on the 11th and 14th centuries.

1922. Schwarz, W. "Zum zahlentheoretischen Werk von Enrico Bombieri." *Jahrbuch Überblicke Mathematik* (1975), 15-151.

On Bombieri's work in number theory.

1923. Scriba, Christoph J. "Zur Entwicklung der additiven Zahlentheorie von Fermat bis Jacobi." *Jahresbericht der Deutschen Mathematiker-Vereinigung* 72 (1970), 122-142.

Discusses primarily Fermat, Euler, and Jacobi.

1924. Shedlovskii, A.B. "O rabotakh P.L. Chebysceva po teorii chisel." *Problemy istorii matematiki i mekhaniki* 1 (1972), 4-9.

Discusses Chebyshev's work on number theory, including prime numbers and the prime number theorem.

1925. Smith, David Eugene. *A Source Book in Mathematics*. (Source Books in the History of the Sciences, edited by Gregory D. Walcott.) New York: McGraw-Hill, 1929.

The first 200 pages reproduce in English translations snippets from 24 authors ranging from Treviso's first printed *Arithmetic* of 1478 to selections from Recorde, Steven, Napier, Wallis, Pascal, Euler, Hermite, Gauss, Kummer, Chebyshev, etc. Frequent illustrations.

1926. Steinig, J. "On Euler's Idoneal Numbers." *Elemente der Mathematik* 21 (1966), 73-88.

1927. Stroyes, J. "Survey of the Arab Contributions to the Theory of Numbers." *Proceedings of the First International Symposium for the History of Arabic Science*. Vol. II. (Papers in European Languages.) Edited by A.Y. al-Hassan, et al. Aleppo, Syria: Institute for the History of Arabic Science, 1978, 173-181.

1928. Uhler, H.S. "A Brief History of the Investigations on Mersenne Numbers and the Latest Immense Primes." *Scripta Mathematica* 18 (1952), 122-131.

1929. Weil, A. "Two Lectures on Number Theory, Past and Present." *L'Enseignement Mathématique* 20 (1974), 87-110.

Two Ritt Lectures delivered at Columbia University in 1972. These "talks" have been transcribed from tape recordings, and hence have the appearance of rambling presentations rather than polished lectures. History of number theory over the past 300 years is the general subject. The first lecture deals with number theory and elliptic functions; the second with the zeta function and related developments like the Riemann hypothesis and Weil's own conjectures of 1948 extending Riemann's hypothesis to algebraic varieties of arbitrary dimension over finite fields--conjectures established by Deligne. Authors given special attention include Fermat, Euler, Lagrange, Goldbach, Gauss, Riemann, Dedekind, and several twentieth-century mathematicians.

1930. ———. "Sur les sommes de trois et quatre carrés." *L'Enseignement Mathématique* 20 (1974), 215-222.

A very brief review of the sums of three and four squares, mentioning work of Lagrange, Euler, Jacobi, Dirichlet, and Kronecker.

1931. ———. "La cyclotomie jadis et naguere." *L'Enseignement Mathématique* 20 (1974), 247-263.

Discusses cyclotomy from Greek geometry to algebraic methods of Lagrange, Gauss, Jacobi, Cauchy, Eisenstein, Kummer, Stickelberger, Davenport, Hasse, Weil, Selberg, and Chowbe.

1932. ———. *Essais historiques sur la théorie des nombres*. Geneva: Université, 1975. Also issued as part of *L'Enseignement Mathématique* 22 (1975).

A collection, under one cover, of Weil's four papers in items 1929, 1930, and 1931. The two papers described in item 1929 above, despite the title of item 1932, are still in English.

1933. Willard, D. "Husserl's Essay 'On the Concept of Number' with a Translation of 'On the Concept of Number: Psychological Analysis,' by Edmund Husserl (1887)." *Philosophia Mathematica* 9 (1972-1973), 40-52; 10 (1973), 37-87.

The introductory essay is only four pages long, but the translation of Husserl's "Habilitationsschrift" runs to 50 pages.

1934. Willerding, M.F. "Figurate Numbers." *School Science and Mathematics* 72 (1972), 151-158.

Illustrated with tables and diagrams, this article offers an introduction to plane figurate numbers.

1935. Yates, S. "Prime Divisors of Repunits." *Journal of Recreational Mathematics* 8 (1975), 33-37.

Discusses the history, period lengths, and provides a table from $n = 2$ to 100 of prime divisors of $R_n = (10^n - 1)/9$.

1936. Zassenhaus, H., and L. Rüdenberg. *Hermann Minkowski. Briefe an David Hilbert*. Berlin: Springer-Verlag, 1973.

This collection includes letters from Minkowski to Hilbert from 1885 to 1908. Rüdenberg provides reminiscences of her father (Minkowski), and Zassenhaus provides historical background on the Hilbert-Minkowski report on number theory. Well illustrated with portraits of Minkowski's contemporaries, and an annotated index of names.

Fermat's Last Theorem

1937. Edwards, H. "The Background of Kummer's Proof of Fermat's Last Theorem for Regular Primes." *Archive for History of Exact Sciences* 14 (1975), 219-236.

This article questions the tradition that Kummer's proof developed out of an earlier fallacious proof; a previously unknown letter from Liouville to Dirichlet related to this history is given. The author also shows that Kummer's theory of "ideal complex numbers" contained a serious flaw that went unnoticed for ten years. See also the postscript to this paper in the same journal, 17 (1977), 381-394, in which the full text of a manuscript by Kummer, withdrawn from publication in 1844, is transcribed. See also items 1052, 1053.

1938. ———. *Fermat's Last Theorem. A Genetic Introduction to Algebraic Number Theory*. New York: Springer-Verlag, 1977.

Although this book offers a good deal of historical information, it is meant to be a textbook. Emphasis here is upon Euler, Fermat, Kummer, Kummer's theory of ideal factors, Fermat's Last Theorem for regular primes, determination of the class number, divisor theory for quadratic integers, Gauss's theory of binary quadratic forms, and Dirichlet's class number formula. See also item 1054.

1939. Ferguson, R.P. "On Fermat's Last Theorem." *Journal of Undergraduate Mathematics* 6 (1974), 1-14 and 85-97.

Includes tables of Bernoulli numbers and discussion of Fermat's Last Theorem.

1940. Hofmann, Joseph E. "Über eine zahlentheoretische Aufgabe Fermats." *Centaurus* 16 (1972), 169-202.

Determination of Pythagorean triangles whose legs and hypotenuse are square numbers.

1941. Mahoney, Michael Sean. "Fermat's Mathematics: Proofs and Conjectures." *Science* 178 (1972), 30-36.

Fermat's Last Theorem was probably not his last, but it summarized his mathematical career. Concentrates on Fermat's methods in number theory, especially that of infinite descent to prove the "last theorem."

1942. ———. *The Mathematical Career of Pierre de Fermat, 1601-1665*. Princeton, N.J.: Princeton University Press, 1973.

A detailed biography of Fermat, with considerable attention given to number theory, as well as other aspects of Fermat's mathematics, his life, and times. Chapter VI is especially relevant for the history of number theory, and is entitled "Between Traditions." It includes such topics as "Numbers, Perfect and Not so Perfect," "Triangles and Squares," "Infinite Descent and the 'Last Theorem,'" etc. See pp. 280-348. See also items 627, 818.
Reviews: Boyer, C., *Science* (July 13, 1973), 152-153;
Weil, A., *Bulletin of the American Mathematical Society* 79 (1973), 1138-1149.

1943. Mazur, B. "Review of *Ernst Edward Kummer, Collected Papers* (Edited by Andre Weil)." *Bulletin of the American Mathematical Society* 83 (1971), 976-988.

The review summarizes the main contributions of Kummer to the study of Fermat's Last Theorem, higher reciprocity laws, ideal complex numbers, congruences of p-adic integers, etc.

1944. Noguès, R. *Théorème de Fermat. Son histoire.* Paris: Vuibert, 1932.

Review: Revue Générale des Sciences 43 (1932), 482.

1945. Ribenboim, P. "The Early History of Fermat's Last Theorem." *The Mathematical Intelligencer* 2 (1976), 7-21.

NUMERICAL ANALYSIS

1946. Cajori, Florian. "Horner's Method of Approximation Anticipated by Ruffini." *Bulletin of the American Mathematical Society* 17 (1911), 409-414.

1947. Goldstine, Herman H. *A History of Numerical Analysis from the 16th through the 19th Century.* New York, Heidelberg, Berlin: Springer-Verlag, 1977, xiv + 348 pp.

This book is the standard work on the subject, and gives comprehensive treatment to the technical development of numerical analysis during the period when the foundations of the subject were being laid. Beginning with the work of Napier, Briggs, Bürgi, Viète, and Kepler, the early study of logarithms and interpolation leads to a chapter on the Age of Newton, followed by detailed analysis of the contributions of Euler and Lagrange, Laplace, Legendre, and Gauss, with a final chapter on 19th-century figures including Jacobi, Cauchy, and Hermite. See also items 798, 863.
Review: Bruins, E.M., *Janus* 65 (1978), 303-312.

1948. Hofmann, Joseph E. "Nicolaus Mercators Logarithmotechnica (1668)." *Deutsche Mathematik* 3 (1938), 446-466.

For other works dealing with the history of logarithms, see items 1888, 1889, 1890.

1949. ———. "Weiterbildung der logarithmischen Reihe Mercators in England." *Deutsche Mathematik* 3 (1938), 598-605.

1950. Lohne, J.A. "Thomas Harriot als Mathematiker." *Centaurus* 11 (1966), 19-45.

Includes discussion of Harriot's work related to numerical analysis. See also item 821.

1951. Naux, C. *Histoire des logarithmes de Neper à Euler*. Paris: A. Blanchard, 1966.

See item 783.

1952. Turnbull, H.W. "A Study in the Early History of Interpolation." *Proceedings of the Edinburgh Mathematical Society* 3 (1932-1933), 150-172.

MATHEMATICAL OPTICS

The works listed below fall into two general categories, the history of geometrical optics and the history of physical optics to the extent that it involved quantitative considerations. The evolution of optical theory is stressed but not to the exclusion of applications. Included are general histories of optics as well as samplings of the literature bearing on developments in each of the major historical periods from antiquity to modern times. Accounts of the most recent developments (holography and fiber optics, for example) will not be found in the studies cited.

1953. Bork, Alfred M. "Maxwell and the Electromagnetic Wave Equation." *American Journal of Physics* 35 (1967), 844-849.

Exhibits by way of logical flow charts using modern vectorial notation Maxwell's three mathematical derivations of the wave equations in electromagnetic theory. See also item 1457.

1954. Boyer, Carl B. *The Rainbow from Myth to Mathematics*. New York and London: Thomas Yoseloff, 1959, 376 pp.

An exhaustive historical survey of the subject from antiquity to the 20th century based on a careful analysis of primary sources. While not neglecting literary-mythical treatments of the rainbow, the author gives major emphasis to the unending quest to devise a mathematical theory accounting for all aspects of this complex phenomenon.

1955. Broglie, Louis de. *Matter and Light. The New Physics*. Translated by W.H. Johnston. London: C. Allen and Unwin; New York: W.W. Norton and Co., 1939. Reprinted New York: Dover Publications, 1946, 1951, 300 pp.

A collection of essays by the Nobel Laureat surveying the 20th-century revolution in physics and its philosophical implications. Focus on the quest for a synthetic view of matter and light consistent with the evidence for both discreteness and continuity at the basis of physical processes. A slightly updated version of the original French edition of 1937.

1956. Bromberg, Joan L. "Maxwell's Displacement Current and His Theory of Light." *Archive for History of Exact Sciences* 4 (1967), 218-234.

Shows how Maxwell was led to a generalization of Ampère's law assigning magnetic effects to displacement as well as to conduction currents. The electromagnetic theory of Maxwell had its origins in this mathematical investigation. See also item 1460.

1957. Cantor, G.N. "Berkeley, Reid, and the Mathematization of Mid-Eighteenth-Century Optics." *Journal of the History of Ideas* 38 (1977), 429-448.

Characterizes the epistemological debate of the period on the question of whether visual perception of magnitudes and distances rests on an innate geometry of the mind or on experience alone. Initiated by Berkeley, who rejected the geometrical model in favor of perception as a "visual language," the debate bore on the larger issue of how far optics might be reduced to mathematics.

1958. Chappert, André. *Etienne Louis Malus (1775-1812) et la théorie corpusculaire de la lumière*. Paris: Vrin, 1977, 277 pp.

Credits Malus with a major role in the progress of optics despite the fact that shortly after his death the corpuscular theory, on which he based all his optical studies, was superseded by the wave theory. The main support for the argument comes from a consideration of Malus's enduring contributions to geometrical optics, most notably the theorem that bears his name.

1959. Edgerton, Samuel Y., Jr. *The Renaissance Rediscovery of Linear Perspective*. New York: Basic Books, 1975, 206 pp.

A well-written, informative exposition of the achievement of Brunelleschi and Alberti set against a broad cultural-historical background. The cartographic technique of Ptolemy's *Geography* is shown to have been a major influence in the application of linear perspective to pictorial representation during the Renaissance.

1960. Klein, Martin J. "Einstein's First Paper on Quanta." *The Natural Philosopher* 2 (1963), 59-86.

A critical analysis of the revolutionary paper of 1905 proposing the photon conception of light. Klein argues that the light quantum hypothesis was rooted, not in Planck's theory of black-body radiation, but rather in Einstein's own profound studies of thermodynamics and statistical mechanics.

1961. Lejeune, Albert. *Euclide et Ptolémée. Deux stades de l'optique géométrique grecque*. Louvain: Bibliothèque de l'université, 1948, 196 pp.

Focuses on the theory of vision set out by Ptolemy in the first three books of the *Optics*. With Ptolemy, Lejeune argues, optics found its proper method, abandoning the strictly geometrical, axiomatic procedure of Euclid in favor of a more broadly based approach that included physical and psychological as well as mathematical components.

1962. ———. "Recherches sur la catoptrique grecque d'après les sources antiques et médiévales." *Académie royale de Belgique, Classe des lettres et des sciences morales et politiques, Mémoires* (Brussels), 52 (2) (1957), 199 pp.

Sequel to the author's *Euclide et Ptolémée*, in which the principal subject is the treatment of reflection and refraction in the later books of Ptolemy's *Optics*. For a fuller understanding of Ptolemy's work Lejeune discusses contributions to catoptrics by other ancient Greeks, Euclid, Archimedes, Hero of Alexandria, and the pseudo-Euclid.

1963. Lindberg, David C. *Theories of Vision from al-Kindi to Kepler*. Chicago and London: University of Chicago Press, 1977, 324 pp.

Argues that Kepler's theory of the retinal image, the first successful solution to the problem of vision, was the culminating achievement of the medieval *perspectiva* tradition. The tradition combined two major approaches to optics carried over from antiquity, the mathematical (Euclid) and the physical and psychological (Aristotle and Galen). Extensive bibliography.

1964. Mach, Ernst. *The Principles of Physical Optics: An Historical and Philosophical Treatment*. Translated by John S. Anderson and A.F.A. Young. New York: E.P. Dutton and Co., 1925; London: Methuen and Co., 1926. Reprinted New York: Dover Publications, c. 1953, 324 pp.

Sets forth in a topical arrangement the fundamental concepts of optics and the historical threads in their development. Mathematical derivations and experimental results bulk large in an account that proposes to free optical principles of their "metaphysical ballast." Published originally in German in 1921, the work excludes "radiation, the decline of the emission theory of light, Maxwell's theory, together with relativity," subjects Mach intended to treat in a sequel.

1965. Ronchi, Vasco. "Classical Optics Is a Mathematical Science." *Archive for History of Exact Sciences* 1 (1960-1962), 160-171.

Argues that the modern study of images produced by lenses and mirrors is not, as is generally thought, a physical science based on observation and experiment but rather a mathematical science grounded on inferences logically drawn from an assumed hypothesis, to wit, Kepler's conception of the eye's distance-measuring triangle. The failure to recognize Kepler's idea for what it is, a mere working hypothesis to be modified or rejected as experiments dictate, has drawn attention away from serious discrepancies between the theory of images and observed phenomena.

1966. ———. *The Nature of Light. An Historical Survey.* Translated by V. Barocas. Cambridge, Mass.: Harvard University Press, 1970, 288 pp.

Surveys ideas on the nature of light from Greek antiquity to the middle of the 19th century. Calls attention to the difficulties experienced historically in distinguishing between light as an external reality and light as a subjective sensation and to the attendant problems in developing a proper science of photometrics. An expansion of the original Italian edition of 1939.

1967. Sabra, A.I. *Theories of Light from Descartes to Newton.* London: Oldbourne, 1967, 363 pp.

Not a survey but a study of "problems and controversies ... particularly important in the development of seventeenth-century theories about the nature of light and its properties." In this period, which embraced the contributions of Fermat, Huygens, Hooke, and Pardies as well as those of Descartes and Newton, a mathematical-experimental investigation of refraction (and its application to dispersion by Newton) was a central concern.

1968. Steffens, Henry John. *The Development of Newtonian Optics in England.* New York: Science History Publications, 1977, 190 pp.

Characterizes Newtonian optics as a coherent, comprehensive system of scientific description and explanation distinct both from Newton's optics, which inspired it, and from the wave theory of the early 19th century, which supplanted it. The Newtonians approached optics as a quantitative study of the dynamics of light corpuscles.

1969. Synge, J.L. *Geometrical Optics. An Introduction to Hamilton's Method.* Cambridge: Cambridge University Press, 1962, 110 pp.

A concise exposition of Hamilton's generalized mathematical approach to optics by one of the editors of his optical papers. This little text first appeared in 1937.

1970. Verdet, Emile. Introduction to *Oeuvres complètes d'Augustin Fresnel.* Edited by Henri de Senarmont, Emile Verdet, and Léonor Fresnel. Paris: Imprimerie Impériale, 1866-1870. 3 vols.

A detailed, critical analysis of the work of the French physicist, whose mathematical skills played a major role in the establishment of the wave theory of light and its triumph over the corpuscular theory.

1971. Westfall, Richard S. "The Development of Newton's Theory of Color." *Isis* 53 (1962), 339-358.

Argues that Newton's theory of color, although it presupposed a mechanistic framework, was a revolutionary break with the views of Aristotelians and mechanists alike, both schools regarding colors essentially as "modifications of pure light by the admixture of darkness." By associating specific colors with specific kinds of rays, each displaying its own characteristic degree of refrangibility, Newton made the study of colors a mathematical science for the first time.

1972. Whittaker, Sir Edmund Taylor. *A History of the Theories of Aether and Electricity*. London: Thomas Nelson & Sons, 1951-1953. Reprinted New York: Harper & Brothers, Harper Torchbooks, 1960, 434 pp.; 319 pp. 2 vols.

A detailed, authoritative work, broader in subject matter than its title suggests and covering the period from Descartes and Newton through the first quarter of the twentieth century. Relative to mathematical optics it is particularly valuable for its treatment of the many models of the "luminiferous aether" called forth by the wave theory. See also items 1536, 2095.

1973. Wilde, Emil. *Geschichte der Optik vom Ursprunge dieser Wissenschaft bis auf die gegenwärtige Zeit*. 1838-1843. 2 vols. Reprinted Wiesbaden: Dr. Martin Sändig, 1968, 352 pp.; 407 pp. 2 vols. in 1.

Provides useful, if dated, accounts of the most important writers on optics from Greek antiquity through the 18th century. Almost 250 pages devoted to Newton and his critics, among whom Goethe receives the most extensive treatment. Mathematical derivations and proofs of the optical laws under discussion. A separate chapter on caustics, rarely considered in standard histories of optics.

POTENTIAL THEORY

The entries listed below are intended to provide access to the whole range of literature on potential theory from 1680 to 1950, with emphasis on the classical period, 1800 to 1900. No attempt has been made at a comprehensive listing here, but the choice of histories (Todhunter, Bacharach, Burkhardt and Meyer, Sologub) and of texts (Betti, Boussinesq, Clausius, Kellogg, La Vallée Poussin, Mathieu, Neumann, Riemann) has been guided by the desire to provide such a listing indirectly. The other principle behind this selection is the need for correction: the older histories and texts omit material, they repeat erroneous claims, they cannot be updated as they would be in the world of Orwell's *1984*. So a sprinkling of important articles by Chasles, La Vallée-Poussin, and Poincaré has been included in the spirit of Kenneth May.

1974. Bacharach, Max. *Abriss der Geschichte der Potentialtheorie*. Wurzburg: Sturtz Druckerei, 1883, 78 pp.

This doctoral thesis is a summary of the history of potential theory and is equally as valuable as Todhunter. There are over 350 references; all the main issues are treated (not always correctly: see the Gauss-Chasles-Poincaré controversy); a long list of contemporary applications, mainly in electricity and magnetism, are cited though not always evaluated. The work of George Green is featured, probably through Carl Neumann's influence, and is given equal prominence with that of Gauss. The best coverage is for the period 1825 to 1880.

1975. Betti, Enrico. *Teorica della forze che agiscono secondo la legge di Newton, e sua applicazione alla elettricità statica.* Pisa, 1865, 144 pp. Estratto dal *Nuovo cimento*, vols. 18, 19, 20.

1976. ———. *Teorica delle Forze Newtoniane e sue applicazioni all'elettrostatica e al magnetismo.* Pisa: T. Nistri, 1879, viii + 359 pp.

1977. ———. *Lehrbuch der Potentialtheorie und ihrer Anwendungen auf Elektrostatik und Magnetismus, von Enrico Betti.* Autorisirte deutsche Aufgabe. Besorgt, und mit Zusätzen, sowie einen Vorwort versehen von W. Franz Meyer. Stuttgart: W. Kohlhammer, 1885, xv + 434 pp.

Originally three articles in the *Nuovo cimento* in 1863-1864, then a greatly expanded book quickly translated into German, this material on gravity, electrostatics, magnetism, and their potentials became very influential on the Continent in the period 1880-1900, since it was well presented, easy to read, and accurate. The original articles are reprinted in the *Opere matematiche di Enrico Betti*, t. 2 (Milan: Hoepli, 1913).

1978. Boussinesq, Valentin Joseph. *Application des potentiels à l'étude de l'équilibre et du mouvement des solides élastiques, avec des notes étendues sur divers points de physique mathématique et d'analyse.* Paris: Gauthier-Villars, 1885, 722 pp.

Applications are in elasticity, fluid mechanics, heat, waves, in two and three dimensions, including bending of plates. This complements Bacharach and Mathieu, as it lays emphasis on French contributions, almost exclusively *not* in gravitation, electricity, and magnetism (for which, see Mathieu, item 1993).

1979. Bowley, R.M., et al. *George Green: Miller, Snienton.* Nottingham: Nottingham Castle, 1976, 96 pp.

This small book gives an excellent life of Green, as well as a valuable set of letters spanning his academic career. See also item 1459.

1980. Burkhardt, Heinrich, and W.F. Meyer. "Potentialtheorie." *Encyklopädie der mathematischen Wissenschaften mit Einschluss ihrer Anwendungen*, II.A7b. Leipzig: B.G. Teubner, 1900, 464-503.

This article gives a synopsis of potential theory from Lagrange to Hilbert. There are over 300 references. The presentation is systematic and critical as well as historico-didactic. It covers definitions of the potential, plane and three-dimensional potentials, the contributions of Green, Gauss, and Thomson, and the various principles and problems; it concludes with expansions, approximation methods, and convergence proofs.

The German contribution is not overemphasized. The best coverage is for the period 1820 to 1900.

1981. Chasles, Michel. "Enoncé de deux théorèmes généraux sur l'attraction des corps et la théorie de la chaleur." *Comptes rendus de l'Académie des Sciences de Paris* 8 (1839), 209-211.

The appearance of this paper caused the deluge of articles on potential theory from 1840 to 1895. It announced the conjectured theorem that a given amount of mass may be spread on a given surface in infinitely many ways such that the potential is constant inside the surface and equal to a certain potential outside the surface, i.e., that various distributions can produce the same potential. This theorem was seemingly refuted by Gauss in 1840, but was proved by Poincaré in 1896 and Frostman in 1935.

1982. Clausius, Rudolf Julius Emmanuel. *Die Potentialfunction und das Potential: ein Beitrag zur mathematischen Physik.* Leipzig: Barth, 1859, vi + 108 pp. 2nd ed., 1877, x + 178 pp.

This small, oft-quoted text represents the watershed between the older presentations of Franz Neumann and Dirichlet (following Gauss) and the new age of Carl Neumann and Riemann (following Green). References are few. Apart from the usual theorems and the calculation of potential functions for bodies of various special shapes, the last third of the book deals with the potential of systems of masses, or their (potential) energy, and the relation of energy to motion (1877 edition).

1983. Gårding, Lars. "The Dirichlet Problem." *The Mathematical Intelligencer* 2, No. 1 (1979), 43-53.

An expanded version of a talk for a student audience at Lund which covers the period from Poisson to Frostman in modern terminology.

1984. Gauss, Carl Friedrich. "Theoria attractionis corporum sphero̎idicorum ellipticorum homogeneorum methodo nova tractata." *Commentationes recentiores Societatis regiae scientiarum Gottingensis* 2 (1813), 1-24, bound in the volume as number 9 or 10. Read 18 Mart. 1813.

This classic paper contains a trivial version of the divergence theorem involving no derivatives in the surface integral, all of which integrals are zero, so that no volume integral is involved. It also contains results on the attractions of ellipsoids already derived by Ivory (1809), where the attraction is calculated by reducing the problem for internal and external points for the given ellipsoid to the problem of finding the attraction of a point on the surface of another ellipsoid. Involved are integrals with discontinuous values as to whether a point is inside, on, or outside a surface.

1985. ———. "Allgemeine Lehrsätze in Beziehung auf die in verkehrten Verhältnisse des Quadrats der Entfernung wirkenden Anziehungs- und Abstossungs-kräfte." *Resultate aus den Beobachtungen des magnetischen Vereins im Jahre 1839.* Herausgegeben von

C.F. Gauss und W. Weber. Leipzig: Weidmannsche Buchhandlung und Verlag, 1840, 1-51.

The major theorem of this paper, the reason why it was written, and a theorem repeatedly stated by Gauss to be true and to be proved rigorously, directly contradicts Chasles's conjecture of 1839. However, Gauss's theorem is false; he attempted to prove it by varying the mass distribution. Despite this error, this is an extraordinarily rich paper, containing surface distributions (not just thin layers) for the first time. This presentation, somewhat pre-Green in treatment, is repeated by Dirichlet, Clausius, Riemann, and Franz Neumann. The change away from it comes with Carl Neumann and his extensive use of Green's work.

1986. Green, George. *Mathematical Papers of George Green.* Edited by N.M. Ferrers. London: Macmillan, 1871. Reprinted New York: Chelsea Publishing Co., 1970, xii + 336 pp.

This volume contains all Green's published works including the *Essay* (pp. 1-115). Many other treasures await the diligent reader: prepotentials, applied Lie groups, the theory of elasticity with a potential, etc.

1987. ———. *An Essay on the Application of Mathematical Analysis to the Theories of Electricity and Magnetism.* Nottingham: Printed for the author by T. Wheelhouse, 1828, ix + 72 pp.

This *Essay* really needs no comment, since its contents are well known--the potential, the various formulae (but not the so-called Green's theorem for the plane), the mean value theorem for harmonic functions, the Green's function, the interior and exterior forms of the so-called Dirichlet problem for one or more closed surfaces. First referred to in other literature in 1835, it was brought into prominence by Thomson in 1845, then reprinted by Crelle in three parts in 1850-1855, and thereafter universally developed and applied.

1988. Hilbert, David. "Ueber das Dirichlet'sche Princip." *Jahresbericht der Deutschen Mathematiker-Vereinigung* 8 (1899), 184-188.

This note announces a new *hope* for proving the Dirichlet principle in general, but contains no proof; the essence lies in two simple examples of the principle which is enunciated in a very general form.

1989. ———. "Ueber das Dirichlet'sche Princip." *Festschrift zur Feier des 150-jährigen Bestehens der Königlichen Gesellschaft der Wissenschaften zu Göttingen. Abhandlungen der mathematisch-physikalischen Klasse*. Berlin: Weidmannsche Buchhandlung, 1901, 1-27.

The proof of the Dirichlet principle in general form, for Riemann surfaces, with the appropriate, requisite boundary values. Reprinted in *Mathematische Annalen* 59 (1904), 161-186.

1990. Kellogg, Oliver Dimon. *Foundations of Potential Theory*. Berlin: Springer, 1929. Reprinted New York: Dover, 1953, ix + 384 pp.

This classic text is not couched in measure theory terms and is one of the last in the Green-Dirichlet-Riemann tradition. It is a comprehensive, detailed, and documented work with several hundred references in its footnotes.

1991. La Vallée-Poussin, Charles-Jean-Gustave-Nicolas de. *Le potentiel logarithmique. Balayage et représentation conforme*. Louvain: Librairie universitaire; Paris: Gauthier-Villars, 1949, xii + 452 pp.

This text is a work born ten years after its time; it is limited to extending the author's prewar results in potential theory, together with a systematic and historical treatment of the topic. References are few. Certainly there are new ideas: extensions of the ideas of distribution and capacitary potential, and complete sets. But the war prevented any new research results from flowing into Belgium after 1939; so, a picture of potential theory as it was, with a little glimpse towards the future. It is one of the first texts couched in terms of measure theory.

1992. ———. "Gauss et la théorie du potentiel." *Revue des questions scientifiques* 133 (1962), 314-330.

This seminar was delivered in Brussels on 18 November 1939. It displays all the author's virtues: it is clear, honest, and balanced. The paper analyzes Gauss's 1840 paper in the *Resultate*, praising its virtues, stating its several defects and errors, pointing to their correction, and finally it gives a detailed picture of some of the priorities of authorship of principles, problems, and errors in potential theory and boundary value problems for elliptic differential equations.

1993. Mathieu, Emile Léonard. *Théorie du potentiel et ses applications à l'électrostatique et au magnétisme*. Première partie: Théorie du potentiel. Seconde partie: Électrostatique et magnétisme. Paris: Gauthier-Villars, 1885 and 1886. *Traité de physique mathématique*, t. III et IV, vi + 179 pp. and vi + 235 pp., respectively.

These two texts cover that half of potential theory and its applications not covered by Boussinesq. References are few. The first part contains the theorems of potential theory, expressed in terms of both attraction between masses and heat conduction; one-third of this part is devoted to bodies composed of thin layers whose boundary surfaces are level surfaces for the potential. The second part applies the potential to electro- and magnetostatics.

1994. Monna, Antonie Frans. *Dirichlet's Principle: A Mathematical Comedy of Errors and Its Influence on the Development of Analysis*. Utrecht: Oosthoek, Scheltema en Holkema, 1975, vii + 138 pp.

An uneven history of the Dirichlet principle as applied to potential theory and complex function theory. It relies on German sources for the evaluation of Gauss and Green; it is unsatisfactory in its treatment of the French school from Chasles to Poincaré. But it gives an outline, from an expert, of the principle's purpose, development, and importance. An informal type of history, with 123 references, covering 1800-1940. See also item 1076.

1995. Neumann, Carl Gottfried. *Das Dirichlet'sche Prinzip in seiner Anwendung auf die Riemann'schen Flächen*. Leipzig: Teubner, 1865, 80 pp.

This, together with his *Vorlesungen über Riemann's Theorie der Abel'schen Integrale* (Leipzig: Teubner, 1865), marks the beginning of Neumann's long *affaire* with the Dirichlet principle, the Dirichlet problem, and their proofs. Here functions of a complex variable are used to map the Riemann surface or sphere to the plane, under various boundary conditions including prescribed discontinuities on the boundary of the polygonal region corresponding to the surface. The modern version of this work is in Hermann Weyl's *Der Idee der Riemannschen Fläche* (Leipzig, Berlin: Teubner, 1913).

1996. ———. *Untersuchungen über das logarithmische und Newton'sche Potential*. Leipzig: Teubner, 1877, xvi + 368 pp.

The most quoted work in the field codifies nearly 20 years of his work. References do not abound, because most of the work is new, generalizing all his own previous work as well as that of others. It is characterized by order, discipline, and system, attempting to *prove* where Green, Gauss, and Dirichlet had only *sketched*, to fill in gaps in arguments or to create new methods of proof. The detailed contents are now classical: elementary potential theory; distributions of mass, charge; double distributions or doublets; the use of the method of arithmetic means to establish the existence of a solution to the first boundary value problem for *convex* surfaces (avoiding the Dirichlet principle error). His attempts, from 1860 to 1887, to prove Dirichlet's principle for *general* surfaces, were fruitless, much to his despair.

1997. ———. "Ueber die Integration der partiellen Differentialgleichung: $\partial^2\Phi/\partial x^2 + \partial^2\Phi/\partial^2 = 0$." *Journal für die reine und angewandte Mathematik* 59 (1861), 335-366.

This paper initiated the strict solution of the first boundary value problem for the plane for Laplace's equation, the stationary (time-independent) temperature distribution in a homogeneous body under Fourier's model of heat conduction, with values for the temperature specified on the boundary of the domain. The problem is physically reformulated in terms of the potential of a mass distribution. Neumann then immediately follows Green's example, giving auxiliary results and introducing the symmetric Green's function. Solutions of the problem are found for specific boundaries and the appropriate Green's functions are calculated.

1998. ———. "Zur Theorie des logarithmischen und des Newtonschen Potentials." *Mathematische Annalen* 11 (1877), 558-566.

Neumann solves the Dirichlet problem in the plane by the method of the arithmetic mean as applied to general plane domains bounded by convex curves with continuous curvature. The presentation is designedly independent of any specially chosen coordinate system. This article is a reprint of an article of the same title in the *Berichte über die Verhandlungen der mathematisch-physikalischen Classe der Königlichen Sächsischen Akademie der Wissenschaften zu Leipzig* 22 (1870), 49-56 and 264-321.

1999. Neumann, Ernst Richard Julius. *Studien über die Methoden von C. Neumann und G. Robin zur Lösung der beiden Randwertaufgaben der Potentialtheorie.* (Preisschriften gekrönt und herausgegeben von der Fürstlich Jablonowski'schen Gesellschaft, Nr 37.) Leipzig: Teubner, 1905, xxiii + 194 pp.

A prize-essay written to unify and complete the researches of the period 1870 to 1900, particularly those of Poincaré (1896) and Carl Neumann (1870). Besides these two, references are made only to Robin, Liapunov, and Schwarz. Proofs are presented for the first two boundary value problems in three dimensions (the plane case is said to be "similar"); these proofs cover convergence of the Neumann and Robin sequences of potentials. Two features stand out: Liapunov's normal derivatives and the polar function (Green's function is the sum of a series of polar functions derived from the arithmetic mean process).

2000. Ostrogradskii, Mikhail Vasil'evich. "Note sur la theorie de la chaleur" and "Deuxième note sur la theorie de la chaleur." *Mémoires de l'Académie Impériale des Sciences de Saint Pétersbourg* (1831), 129-138 and 123-126 [sic].

The material contained in these two notes dates from 1826 when it was first read to the Paris Academy. It is therefore almost simultaneous with George Green's *Essay*. The divergence theorem is explicit and for arbitrary surfaces in three-dimensional space; hence it has the same applicability as Gauss's version of 1813 and Green's formulae. Further, and maybe more importantly, the theorem is developed for rational entire functions of differential operators. These notes, especially the first, have been vastly neglected and underrated.

2001. Poincaré, Jules Henri. "La méthode de Neumann et le problème de Dirichlet." *Acta Mathematica* 20 (1896), 59-142.

First announced in the *Comptes rendus de l'Académie des Sciences de Paris* 120 (1895), 347-352, séance du 18 février 1895. This gives a rigorous proof that the Dirichlet problem has a solution for simply connected singularity-free surfaces, *convex or not*. It suggests, but only with physical arguments, that there are an infinity of potentials, the *fundamental functions*, corresponding to simple distributions on the surface, these functions being mutually orthogonal and satisfying a normal-derivative discontinuity at the surface, and providing for series expansions of all potentials.

2002. ———. "Sur les équations aux dérivées partielles de la physique mathématique." *American Journal of Mathematics* 12 (1890), 211-294.

Here is the *méthode de balayage*, the method of sweeping out mass from space onto a given surface. The method dates back to Poisson and Chasles; it gives a unique potential but not a unique mass distribution due to the selection process involved; the existence of the potential is proved rigorously. Poincaré proves by physical arguments, which he expected could be made rigorous and suggested how, that there is a sequence of fundamental functions, mutually orthogonal, which will solve a sequence of boundary value problems and so provide a series for the potential.

2003. Poisson, Siméon-Denis. "Mémoire sur la distribution de l'électricité à la surface des corps conducteurs." *Mémoires de la Classe des Sciences mathématiques et physiques de l'Institut* 12 (Première partie, année 1811) (1812), 1-92. Lu les 9 mai et 3 août 1812.

Poisson discusses spheres and surfaces not much different from spheres, both alone or in pairs. He rederives what he twice states is Laplace's result on the distribution of a thin layer of mass or charge on a surface such that the distribution exerts no force inside the surface; in making clear Laplace's argument he shows that the force suffers a jump discontinuity in the direction normal to the surface and that this jump is proportional to the density of the distribution.

2004. ———. "Second mémoire sur la distribution de l'électricité à la surface des corps conducteurs." *Mémoires de la Classe des Sciences mathématiques et physiques de l'Institut* 12 (Seconde partie, année 1811) (1814), 163-274. Lu le 6 septembre 1813.

Ostensibly about calculating charge densities or distributions on electrified surfaces (in his terms, the thickness of the charge layer), Poisson actually writes about series, the summation and convergence thereof, and about definite integrals: the Poisson integral forms a major part of this study.

2005. Riemann, Georg Friedrich Bernhard, and Karl Friedrich Wilhelm Hattendorf. *Schwere, Elektricität und Magnetismus nach den Vorlesungen von Bernhard Riemann (Göttingen 1861)*. Hannover: Carl Rümpler, 1876. Reprinted 1880, x + 358 pp.

This text dates from Riemann's lectures of 1861. References are few. The presentation breaks away from the Gauss-Dirichlet tradition since Green's work is heavily used. It is bedevilled by the German distinction between the *potential function* (attraction of a body on a mass point) and the *potential* (the internal energy of a system), a distinction stemming from a priority question on the use of the word "potential" (Gauss, Green). Green's function and its symmetry properties are used, together with Dirichlet's principle and its then standard "proof." Half the book is on potential theory, half on applications to the major interests of the Göttingen *magnetischer*

Verein and Weber: electrostatics, direct and alternating currents, magnetism, electromagnetism, electrodynamics, and the earth's magnetism.

2006. Robin, Victor Gustave. "Sur la distribution d'électricité à la surface des conducteurs fermes et des conducteurs ouverts." *Annales de l'École normale supérieure* 3 (1886, Supplément), 58 pp.

This paper is Robin's thesis, and contains both the statement of Robin's problem (distribution of charge on a regular surface so that the potential is constant on the surface) and a method for solving it.

2007. ———. *Oeuvres scientifiques, réunies et publiées sous les auspices du ministère de l'Instruction publique, par Louis Raffy*. Paris: Gauthier-Villars, 1899-1903. 3 vols: I. Théorie nouvelle des fonctions, exclusivement fondée sur l'idée de nombre (1903), vi + 215 pp.; II, première partie. *Physique mathématique* (1899), vi + 150 pp.; II, seconde partie. *Thermodynamique générale* (1901), xvi + 271 pp.; III. *Leçons de chimie physique*.

Théorie nouvelle des fonctions does calculus without "les pretendus nombres irrationels" in the tradition of Kronecker and the later work of E. Borel; these lectures were given in 1892-1893 at the Faculté des Sciences. *Physique mathématique* contains all his published papers as well as some manuscript notes on the same topics. *Thermodynamique générale* covers his course on the foundations of chemistry given in 1896 and has been fully reedited by Raffy. The third volume was never published.

2008. Schwarz, Carl Hermann Amandus. "Ueber einen Grenzübergang durch alternierendes verfahren." *Vierteljahresschrift der naturforschen Gesellschaft zu Zürich* 15 (1870), 272-286.

The existence of a solution to the Laplace equation is established by solving the boundary value problem for one domain of the plane where it can be solved and hence completing the boundary values for another domain; the new boundary value problem for the second domain is solved and this completes a new set of boundary values for the first domain. Thus one forms sequences of solutions on each domain which converge to a single solution on the union of the two domains, where both the sequences agree on the intersections in the limit, as, for example, on the nonempty intersection of a circle and a square. No asumption of continuity, curvature, or convexity on the boundary is required, but the domain must be built up as the union of domains where the problem can be solved already. This paper is reprinted on pp. 133-143 of Band 2 of his *Gesammelte mathematische Abhandlungen*.

2009. ———. *Gesammelte mathematische Abhandlungen von H.A. Schwarz*. Berlin: Springer, 1890. 2 Bände. Reprinted New York: Chelsea Publishing Co., 1972, as 2 vols. in 1, xiv + 338 pp. and vii + 370 pp.

This is by far the easiest method of access to Schwarz's work, particularly that of 1870-1872 on the integration of Laplace's equation in the plane, for circles and other amenable geometric shapes (see especially pp. 144-210 of Band 2).

2010. Sologub, Vladimir Stepanovich. *Razvitie teorii ellipticheskikh uravnenii v XVIII i XIX stoletiyakh* (Development of the theory of elliptic equations in the 18th and 19th centuries). Akademiia Nauk Ukrainskoi SSR, Sektor istorii estestvoznaniia i tekhniki, Institut istorii. Kiev: Naukova dumka, 1975, 280 pp. In Russian.

This work attempts a synthesis of the theory, motivation, and major results for elliptic equations and their boundary value problems, and to fill in the gaps left by Todhunter, Bacharach, and Burkhardt and Meyer. There are over 370 references, almost all to original papers and monographs. Despite the author's undoubted expertise and valiant efforts, there are flaws in his synthesis: major papers of Chasles and Hilbert are not mentioned, and the guiding hand of La Vallée-Poussin is sorely missed. Such faults may be ignored in an undertaking of such vast scale, as the coverage is 1750 to 1900, in detail. This work deserves to be better known.

2011. Thomson, William. "Note sur une équation aux différences partielles, qui se présente dans plusieurs questions de physique mathématique." *Journal de mathématiques pures et appliquées* 12 (1847), 493-496.

Here is the so-called Dirichlet principle in print, with variation of the function, some five to ten years before Riemann.

2012. Todhunter, Isaac. *A History of the Mathematical Theories of Attraction and the Figure of the Earth, from the Time of Newton to that of Laplace*. London: Macmillan and Co., 1873, xxxvi + 476 + 508 pp. 2 vols. Reprinted in 1 vol. New York: Dover, 1962.

This chronological list of contributions to the theory of gravitational attraction is very valuable. There are approximately 500 references; all these are analyzed, but no synthesis is attempted. Lagrange, Legendre, and Laplace are featured, and many prominent personages have at least one chapter devoted to their work. The related problem of the shape of the earth is covered from both the theoretical and practical points of view. The best coverage is for the period 1780 to 1820, but the work starts with Newton about 1680 and runs to the peak of Giovanni Plana's career around 1840. See also item 959.

PROBABILITY AND STATISTICS

General Works

2013. Biermann, K.-R. "Aus der Geschichte der Wahrscheinlichkeitsrechnung." *Wissenschaftliche Annalen* 5 (1956), 542-548.

A sketch of early developments in probability stemming from considerations of games of chance. Discusses Cardan, Pacioli, Tartaglia, Jacob Bernoulli, and Laplace.

2014. *Biometrika*. (Studies in the History of Probability and Statistics.)

An ongoing series of articles appearing regularly since 1955. The first 31 articles are reprinted in the two volumes of items 2020 and 2017. Subsequent contributions to the series can be found in issues of *Biometrika* published since 1973.

2015. Czuber, E. "Die Entwicklung der Wahrscheinlichkeitstheorie und Ihrer Anwendungen." *Jahresbericht der Deutschen Mathematiker-Vereinigung* 7 (1898), 1-279.

A dated but thorough examination of several aspects of this multifaceted subject. Discusses foundations and early phases, application to theory of repeated trials, Bayes's Theorem, applications to legal and ethical philosophy, to statistics, and to error theory.

2016. Freudenthal, H., and H.G. Steiner. "Aus der Geschichte der Wahrscheinlichkeitstheorie und der mathematischen Statistik." *Grundzüge der Mathematik, IV*. Edited by H. Behnke, G. Bertram, R. Sauer. Göttingen: Vandenhoeck & Ruprecht, 1966, 149-195.

A sketch of the development of probability and statistics from mid-17th century to the present indicating connections with other intellectual disciplines and social problems. Although the strictly mathematical developments dominate the discussion, attention is also given to game theory, economic and social theory, approximation theory, and philosophical foundations.

2017. Kendall, M.G., and R.L. Plackett, eds. *Studies in the History of Statistics and Probability*. Vol. 2. New York: Macmillan, 1977.

This book and the companion volume, item 2020, bring together a valuable collection of essays, the large majority of which appeared in the *Biometrika* series "Studies in the History of Probability and Statistics"; see item 2014.

2018. Maistrov, L.E. *Probability Theory. A Historical Sketch*. Translated and edited by S. Kotz. New York and London: Academic Press, 1974.

See item 794.

2019. Owen, D.B., ed. *On the History of Statistics and Probability.* New York and Basel: Marcel Dekker, 1976.

Proceedings of a 1974 symposium. Includes J. Neyman on the emergence of statistics in the United States, B. Harshburger on statistics in America from 1920 to 1944, W.G. Cochran on the early development of techniques in comparative experimentation, plus 18 other articles.

2020. Pearson, E.S., and M.G. Kendall, eds. *Studies in the History of Statistics and Probability.* Vol. 1. Darien, Conn.: Hafner, 1970.

This book and the companion volume, item 2017, bring together a valuable collection of essays, the large majority of which appeared in the *Biometrika* series "Studies in the History of Probability and Statistics"; see item 2014.

2021. Raymond, P. *De la combinatoire aux probabilités.* Paris: F. Maspero, 1975.

A lengthy discussion of the philosophical background, including the search for a "mathesis universalis," determinism versus free will, and early difficulties facing probability theory. There follow chapters on Pascal, Leibniz, and Jacob Bernoulli, as well as a short chapter on Huygens.

2022. Todhunter, Isaac. *A History of the Mathematical Theory of Probability from the Time of Pascal to That of Laplace.* New York: Chelsea, 1949 reprint of the 1865 edition.

Covers such figures as Cardan, Kepler, Galileo, Pascal, Fermat, Huygens, Bernoulli (James and Daniel), Montmort, de Moivre, Euler, d'Alembert, Bayes, Lagrange, Condorcet, Trembley, with 150 pages devoted to Laplace. Separate sections for combinations, mortality, and life insurance.

2023. Westergaard, H. *Contributions to the History of Statistics.* London: P.S. King & Don, 1932.

See item 796.

Pre-17th Century

2024. David, F.N. *Games, Gods and Gambling. The Origins and History of Probability and Statistical Ideas from the Earliest Times to the Newtonian Era.* London: Charles Griffin & Co., 1962.

Probably the best single account of the origins and early development of probability theory. Although the author relegates Cardano to a lesser position than does Ore (see item 2026), he receives a more sympathetic treatment than from Todhunter (see item 2022). Others considered are Galileo, Fermat, and Pascal, Huygens, Wallis, Newton, Pepys, and James Bernoulli. Also, the work of de Moivre receives strong recognition.

2025. Hacking, I. *The Emergence of Probability. A Philosophical Study of Early Ideas about Probability, Induction and Statistical Inference.* Cambridge: University Press, 1975.

See item 793.

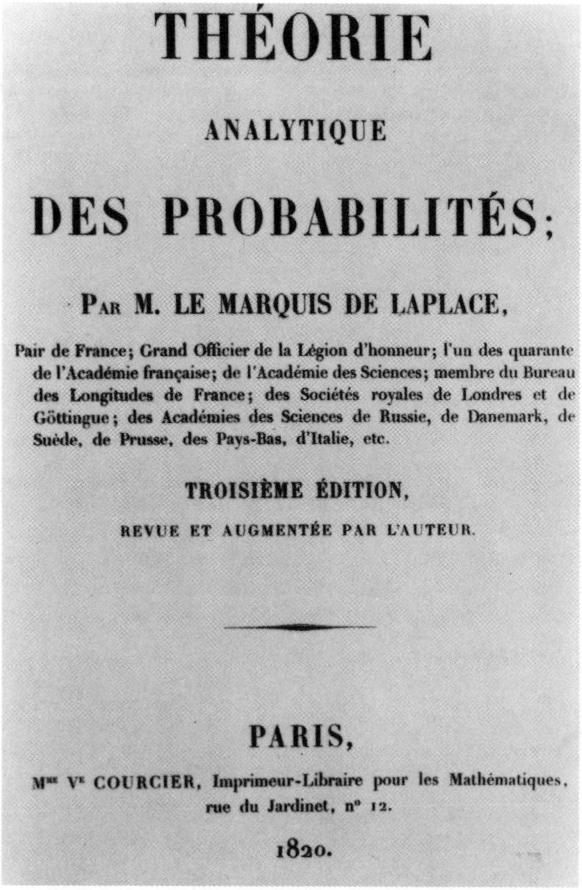

THÉORIE

ANALYTIQUE

DES PROBABILITÉS;

PAR M. LE MARQUIS DE LAPLACE,

Pair de France; Grand Officier de la Légion d'honneur; l'un des quarante de l'Académie française; de l'Académie des Sciences; membre du Bureau des Longitudes de France; des Sociétés royales de Londres et de Göttingue; des Académies des Sciences de Russie, de Danemark, de Suède, de Prusse, des Pays-Bas, d'Italie, etc.

TROISIÈME ÉDITION,

REVUE ET AUGMENTÉE PAR L'AUTEUR.

PARIS,

Mme Ve COURCIER, Imprimeur-Libraire pour les Mathématiques, rue du Jardinet, n° 12.

1820.

Figure 14. The title page from Laplace's *Théorie analytique des probabilités*, first published in 1812. This monumental work not only synthesized practically all previous work on the subject, but also anticipated many important discoveries later in the century. Laplace, like Einstein, was a determinist; he believed that probabilistic knowledge reflected the finitude of human consciousness.

2026. Ore, Øystein. *Cardano: The Gambling Scholar*. Princeton: Princeton University Press, 1953.

Besides being an entertaining biography, Ore attempts to establish Cardano as the founder of modern probability theory. Contains a translation of his *Liber de Ludo Aleae* (*Book on Games of Chance*), plus a discussion of Cardano's contribution to the subject in the last chapter, "The Science of Gambling." See also item 663.

2027. Rabinovitch, Nachum L. *Probability and Statistical Inference in Ancient and Medieval Jewish Literature*. Toronto: University of Toronto Press, 1973, 205 pp.

Argues that probabilistic thinking has deep historical roots, and in particular held a central place in ancient and medieval rabbinical literature. The author illustrates the sophistication of Jewish conceptions by the diversity of ideas presented (relative frequency, equally likely events, probabilistic logic, and foundations of a probabilistic arithmetic). See also item 438.

2028. Sheynin, O.B. "On the Prehistory of the Theory of Probability." *Archive for History of Exact Sciences* 12 (1974), 98-141.

Not confined to prehistory, but rather examines the extra-mathematical aspects of the theory: its role in jurisprudence, the fine arts, medicine, astronomy and astrology, and ancient and modern philosophy.

2029. Van Brakel, J. "Some Remarks on the Prehistory of the Concept of Statistical Probability." *Archive for History of Exact Sciences* 16 (1976), 119-136.

A philosophical essay motivated by the author's contention that recent studies (e.g., item 2028) have led to confusion by obscuring distinct notions of probability that have occurred historically. Distinguishes among the following concepts of probability: (1) epistemic notion of probable knowledge, (2) epistemic notion of chance, (3) empirical notion of permanence of statistical ratios, and (4) the conceptual notion of equipossible events.

17th Century to Present

2030. Barone, J., and A. Novikoff. "A History of the Axiomatic Formulation of Probability from Borel to Kolmogorov: Part I." *Archive for History of Exact Sciences* 18 (1978), 123-190.

Borel's contribution to probability theory is the centerpiece of Part I, which goes up to the work of Hausdorff; Part II (Hausdorff to Kolmogorov) is yet to appear. See also item 1136.

2031. Biermann, K.-R. "Ueber die Untersuchung einer speziellen Frage der Kombinatorik durch G.W. Leibniz." *Foschungen und Fortschritte* 28 (1954), 357-361.

Discusses Leibniz's technique for finding the number of combinations without repetition that contain a given element.

Illustrated with tables.

2032. Box, J.F. *R.A. Fisher. The Life of a Scientist*. New York: Wiley, 1978.

A detailed biography written by Fisher's daughter and interlaced with a good deal of biology and statistics. Individual chapters devoted to mathematical statistics, significance tests, design of experiments, scientific inference, and the biometrical movement. Evolution and eugenics are always in the foreground, but the mathematics is by no means slighted. See also item 1112.

2033. Daston, L.J. "Probabilistic Expectation and Rationality in Classical Probability Theory." *Historia Mathematica* 7 (1980), 234-260.

Argues that the concept of probabilistic expectation evolved in response to the need for rational decision making in society. Discusses economic, legalistic, psychological, and moral factors, especially during the Enlightenment period.

2034. ———. "D'Alembert's Critique of Probability Theory." *Historia Mathematica* 6 (1979), 259-279.

Seeks to understand d'Alembert's critique of conventional probability theory in terms of his larger philosophical program aimed at achieving a close correspondence among mathematics, the physical world, and psychological experience. His discussion of the St. Petersburg and Innoculation problems is reinterpreted in light of this philosophical objective. See also item 1298.

2035. Garber, D. and S. Zabell. "On the Emergence of Probability." *Archive for History of Exact Sciences* 21 (1980), 33-54.

The authors attempt to refute Hacking's thesis (item 2025) that the emergence of probability theory in the 17th century was due to the concomitant appearance of the modern notion of probability in the same era. They cite evidence that the concept of probability is much older, in works of Cicero, Quintilian, John of Salisbury, and Oresme. They also suggest that the link between probabilistic thinking and games of chance was not common until the early 17th century, and thus take the usual view that it was this connection that sparked the theory.

2036. Hacking, I. "Jacques Bernoulli's 'Art of Conjecturing.'" *British Journal for the Philosophy of Science* 22 (1971), 209-249.

A philosophical inquiry into Bernoulli's *Ars conjectandi*. Hacking first examines the conceptual framework employed by Bernoulli; he then attempts to determine the statistical value of his limit theorem as a tool for inference.

2037. Ondar, H.O., ed. *About the Theory of Probability and Mathematical Statistics*. Moscow: Nauka, 1977. In Russian.

Correspondence of A.A. Markov and A.A. Tshuprov.

2038. Pearson, E.S. *Karl Pearson: An Appreciation of Some Aspects of His Life and Work*. Cambridge: University Press, 1938.

Contains excerpts of letters, a bibliography of his works, syllabi from lecture courses, including "The Geometry of Statistics and the Laws of Chance" (1891-1894), and "Theory of Statistics" (1894-1896). Only scant reference to his mathematics.

2039. Pearson, Karl. *The History of Statistics in the 17th and 18th Centuries Against the Changing Background of Intellectual, Scientific and Religious Thought*. Edited by E.S. Pearson. New York: Macmillan, 1978.

See item 795.

2040. Schneider, Ivo. "Der Mathematiker Abraham de Moivre (1667-1754)." *Archive for History of Exact Sciences* 5 (1968-1969), 177-317.

Discusses the background to De Moivre's probabilistic studies leading to the 1718 publication *Doctrine of Chances*. Much of De Moivre's work had roots in probability theory, and he went beyond the classical application of probability to games of chance by addressing the problem of annuities and adopting Halley's conception of the "probability of life." He was the first to publish a mathematical law for decrements of life from a mortality table. See also item 890.

2041. Shafer, G. "Non-Additive Probabilities in the Work of Bernoulli and Lambert." *Archive for History of Exact Sciences* 19 (1978), 309-370.

Discusses the pioneering work of Jacob Bernoulli and Lambert in this field.

2042. Sheynin, O.B. "Newton and the Classical Theory of Probability." *Archive for History of Exact Sciences* 7 (1971), 217-243.

Newton's views on probability theory are considered together with the probabilistic aspects of his work in chronology, error theory, design of experiments, and astronomy. The influence of his probabilistic ideas on de Moivre, Arbuthnot, Bentley, and Laplace is also discussed.

2043. ———. "J.H. Lambert's Work on Probability." *Archive for History of Exact Sciences* 7 (1971), 244-256.

Discusses Lambert's work on demographical statistics and error theory. Stresses the importance of his philosophical views, and his fundamental contribution of the principle of maximum likelihood.

2044. ———. "D. Bernoulli's Work on Probability." *Rete. Strukturgeschichte der Naturwissenschaften* 1 (1972), 273-300.

The contents of eight memoirs published between 1738 and 1780 are examined. These deal with applications to demographic studies, political arithmetic and moral expectation, astronomy, and theory of errors. Bernoulli was the first to use differential equations in probability theory, and to introduce tests of statistical hypotheses. His influence on Laplace was comparable to that of de Moivre.

2045. ———. "Finite Random Sums (A Historical Essay)." *Archive for History of Exact Sciences* 9 (1973), 275-305.

Considers finite random sums in the 18th-century context of games of chance through the work of Lagrange to the early studies by Laplace. Illustrates the transition from a discrete to a continuous random variable in developments closely tied to practical problems of demography and error theory.

2046. ———. "R.J. Boscovich's Work on Probability." *Archive for History of Exact Sciences* 9 (1973), 306-324.

Boscovich's method of adjusting arc measurements is presented, as well as a manuscript which may be the first use of probability in the theory of errors. This manuscript deals with the stochastic behavior of the sum of several random variables, each with a particular discrete distribution. Another manuscript deals with the *lotto di Roma*.

2047. ———. "P.S. Laplace's Work on Probability." *Archive for History of Exact Sciences* 16 (1976), 137-187.

A detailed analysis of Laplace's *Théorie analytique des probabilités*, treating probability proper, limit theorems, and mathematical statistics. Discusses Laplace's philosophy of probability and science, concluding that his determinism was in fact compatible with his probabilistic views. Also claims that the roots of Laplace's work lay in natural science and not mathematics, and that this later proved to be an obstacle to the development of the theory, despite his numerous achievements. See also item 916.

2048. ———. "Poisson's Work in Probability." *Archive for History of Exact Sciences* 18 (1978), 245-300.

Examines Poisson's concept of probability, randomness, and distribution, as well as his work on limit theorems, the law of large numbers, mathematical statistics, and applications to jurisprudence, plus a brief summary of the contents of his probability memoirs. Poisson introduced random quantities and the cumulative distribution function, the generalized central limit theorem, and proved the law of large numbers for the case of Poisson trials.

2049. ———. "Gauss and the Theory of Errors." *Archive for History of Exact Sciences* 20 (1979), 21-72.

2050. Van der Waerden, Bartel Leendert. Historical precis to *Die Werke von Jakob Bernoulli*. Vol. III. Basel: Birkhäuser, 1975, 1-18.

2051. Yamazaki, E. "D'Alembert et Condorcet: quelques aspects de l'histoire du calcul des probabilités." *Japanese Studies in the History of Science* 10 (1971), 60-93.

A sympathetic reconsideration of d'Alembert's critique of the 18th-century theory of probability. Argues that Condorcet's views in the main supported d'Alembert's criticism.

Applications in Sociology and Biology

2052. Cullen, M.J. *The Statistical Movement in Early Victorian Britain. The Foundations of Empirical Social Research.* New York: Harvester Press/Barnes & Noble, 1975.

Covers the statistics gathered by governmental departments in Britain from 1832 to 1852. Also discusses the founding and activity of the Statistical Society of London, the Manchester Statistical Society, and other provincial societies.

2053. Pearson, Karl. *Life, Letters and Labours of F. Galton.* Vol. 2. Cambridge: University Press, 1924.

Chapter XIII discusses Galton's statistical investigations and their relationship to his anthropometric researches.

2054. Sheynin, O.B. "On the History of the Statistical Method in Biology." *Archive for History of Exact Sciences* 22 (1980), 323-371.

Concentrates on the use of the concept of randomness in evolutionary biology from Lamarck to Darwin.

2055. Walker, H.M. *Studies in the History of the Statistical Method with Special Reference to Certain Educational Problems.* Baltimore: Williams & Wilkins, 1929.

Follows the development of the normal curve, moments, percentiles, and correlation as they relate to educational statistics. The final chapter is on statistics as a subject of instruction in American universities.

Applications in Physics

2056. Brush, S.G. "Foundations of Statistical Mechanics." *Archive for History of Exact Sciences* 4 (1967), 145-183.

This paper is mostly physics, but the author also traces the mathematical background leading to the proof of the impossibility of ergodic systems by Plancherel and Rosenthal. An elementary digression on the work of Cantor, Borel, and Lebesgue helps to anchor the discussion, as does the reference to measure theory and the fundamental work of Brouwer and Baire.

2057. ———. "A History of Random Processes I. Brownian Movement from Brown to Perrin." *Archive for History of Exact Sciences* 5 (1968), 1-36.

Follows the developments from Brown's experiments and their interpretation to the theories of Einstein, Smoluchowski, and Perrin. Physics predominates over mathematics in this article.

2058. ———. "The Development of the Kinetic Theory of Gases VIII. Randomness and Irreversibility." *Archive for History of Exact Sciences* 12 (1974), 1-88.

A survey of cosmology, geology, the second law of thermodynamics, and entropy, statistics, and the kinetic theory as they came together in the 19th century. Other topics include Boltzmann's statistical theory of entropy, the recurrence paradox, and Planck's contribution to irreversible radiation processes.

2059. Schneider, Ivo. "Clausius' erste Anwendung der Wahrscheinlichkeitsrechnung im Rahmen der atmosphärischen Lichtstreuung." *Archive for History of Exact Sciences* 14 (1974), 143-158.

Seeks to establish that Clausius (in his work on meteorological optics) and not Maxwell was the first to apply probabilistic methods to physics.

2060. ———. "Rudolf Clausius' Beitrag zur Einführung wahrscheinlichkeitstheoretischer Methoden in die Physik der Gase nach 1856." *Archive for History of Exact Sciences* 14 (1975), 237-261.

Traces the entrance of probability theory in the kinetic theory of gases. Starting with ideas of Laplace and Krönig, follows Clausius's use of probability theory and the mean free path to Maxwell's use of a molecular velocity distribution.

MATHEMATICAL QUANTUM THEORY

2061. Bogolubov, N.N., A.A. Logunov, and I.T. Todorov. *Introduction to Axiomatic Quantum Field Theory*. Translated by S.A. Fulling and L.G. Popova. Reading, Mass.: W.A. Benjamin, 1975.

This monograph on the mathematical foundations of quantum theory contains numerous important works in this area with valuable commentary and supplementary remarks.

2062. Bratteli, O., and D.W. Robinson. *Operator Algebras and Quantum Statistical Mechanics 1*. New York: Springer Verlag, 1979.

The most important mathematical works related to applications in quantum theory are cited in the introduction.

2063. Dyson, F.J. "Missed Opportunities." *Bulletin of the American Mathematical Society* 78 (1972), 635-652.

With the help of many historical references, Dyson indicates in this article the connections between mathematics and physics. In particular, the mathematical foundations of quantum theory are commented upon from an historical point of view.

2064. Glimm, J. "The Mathematics of Quantum Fields." *Advances in Mathematics* 16 (1975), 221-232.

This very general article provides an overview of recent works in quantum field theory, with explanation and commentary.

2065. Jammer, M. *The Conceptual Development of Quantum Mechanics*. New York: McGraw-Hill, 1966.

The chapter entitled "Rise of Matrix Mechanics" gives a short summary of the development of matrix theory and its connections with quantum mechanics. Another chapter, "The Statistical Transformation Theory in Hilbert Space," considers Hilbert's efforts to axiomatize quantum mechanics. Finally, von Neumann's contributions to the founding of mathematical quantum mechanics are described. Along with a short description of theoretical contributions, many citations to relevant literature are also provided.

2066. Jauch, J.M. "The Mathematical Structure of Elementary Quantum Mechanics." *The Physicist's Conception of Nature*. Edited by J. Mehra. Dordrecht: D. Reidel, 1973, 300-319.

Along with a description of quantum theory from a mathematical viewpoint, a series of historical comments and bibliographic references are also given. Above all, questions of the representation of permutations and scattering theory are considered and commented upon.

2067. Kuhn, T. *Black-Body Theory and the Quantum Discontinuity, 1894-1912*. New York: Oxford University Press, 1978.

Part I deals with the classical phase of Planck's black-body theory from 1894 to 1906. Part II deals with the emergence of the quantum discontinuity between 1905 and 1912, and considers the work of Rayleigh, Jeans, Ehrenfest, Einstein, Lorentz, and Wien. An epilogue considers the decline of Planck's black-body theory, his "second theory" of radiation, its uses, and ultimate fate.

2068. Wightman, A.S. "Hilbert's Sixth Problem: Mathematical Treatment of the Axioms of Physics." *Mathematical Developments Arising from Hilbert's Problems*. Providence, R.I.: American Mathematical Society, 1976, 147-240.

A comprehensive survey article on the axiomatization of physics, especially quantum mechanics and quantum field theory. Numerous works are cited and discussed.

2069. Wintner, A. *Spektraltheorie der unendlichen Matrizen*. (Einführung in den analytischen Apparat der Quantenmechanik.) Leipzig: S. Hirzel, 1929.

A very mathematical monograph with many historical remarks and references to pertinent literature, which clarifies the status of the theory of Hilbert spaces before von Neumann. This work is itself an important primary source for mathematical quantum theory.

2070. Witt, B.S. de, and R.N. Graham. "Resource Letter IQM-1 on the Interpretation of Quantum Mechanics." *American Journal of Physics* 39 (1971), 724-738.

In addition to the most important works on quantum mechanics, a series of essential works on the mathematical foundations of quantum mechanics with remarks and commentary are to be found in Section V: "Logical Foundations."

MATHEMATICS AND RELATIVITY

2071. Abro, A. d'. *The Evolution of Scientific Thought from Newton to Einstein*. 2nd ed. New York: Dover, 1950.

This early, non-technical account gives a good introduction to the subject of relativity theory. Part I discusses mathematics relevant to Einstein's theory (manifolds, non-Euclidean geometries, electromagnetic equations). Part II treats the fundamental ideas of Special Relativity, while Part III deals with the General Theory.

2072. Borel, E. *Space and Time*. Translated by S. Rappoport and J. Dougall. London: Blackie and Son, 1926.

Examines relativity theory in relation to gravitation, measuring systems, and the role of physical space in mechanics. Other prominent topics include geometry and the shape of the earth, space and time in astronomy, abstract geometry, continuity and topology, the propagation of light, mathematical versus the physical continuum, and the finitude of the universe.

2073. Dipert, R.R. "Peirce's Theory of the Geometrical Structure of Physical Space." *Isis* 68 (1977), 404-413.

This paper examines Peirce's arguments that the geometry of physical space is non-Euclidean, and critically evaluates the methods used by Peirce to determine the curvature of space.

2074. Earman, J., and C. Glymour. "Einstein and Hilbert: Two Months in the History of General Relativity." *Archive for History of Exact Sciences* 19 (1978), 291-308.

The authors examine the evidence that Einstein's field equations were in part due to Hilbert. After presenting the relevant correspondence, they conclude that Hilbert probably had communicated his own equations to Einstein a week prior to the latter's announcement of his own. They doubt, however, that Hilbert's equations influenced Einstein, unless perhaps formally, although Hilbert may have hastened Einstein's discontent with the earlier Einstein-Grossmann theory.

2075. French, A.P., ed. *Einstein: A Centenary Volume*. Cambridge, Mass.: Harvard University Press, 1979.

A potpourri consisting of Einstein vignettes written by scholars and personal friends, excerpts from his published writings and letters, and a number of topical essays dealing with the history of relativity and quantum theory.

2076. Goldberg, S. "Henri Poincaré and Einstein's Theory of Relativity: The Role of Theory and Experiment in Poincaré's Physics." *American Journal of Physics* 35 (1967), 933-944.

The author attempts here as elsewhere to debunk Whittaker's interpretation in Chapter 2 of item 2095, "The Relativity Theory of Poincaré and Lorentz." Goldberg contends that not only did

Poincaré and Einstein have different approaches, but the problems they were attacking were fundamentally different. For Poincaré, the principal goal was to unite the Lorentz theory with classical Newtonian mechanics.

2077. ———. "The Lorentz Theory of Electrons and Einstein's Theory of Relativity." *American Journal of Physics* 37 (1969), 982-994.

Shows how Lorentz's theory differs conceptually from Einstein's treatment of the electrodynamics of moving bodies, and that Lorentz never fully rejected the ether and embraced relativity theory.

2078. ———. "In Defense of Ether: The British Response to Einstein's Special Theory of Relativity, 1905-1911." *Historical Studies in the Physical Sciences* 2 (1970), 89-126.

Contrasts the British response to relativity theory with the reactions in Germany, France, and the United States. Although the reception was mixed in Germany, and muted in France and the United States, the British physicists gave staunch opposition to what they viewed as an attack on "ether mechanics." The author traces the roots of this response to the educational training received by British physicists.

2079. ———. "Poincaré's Silence and Einstein's Relativity: The Role of Theory and Experiment in Poincaré's Physics." *British Journal for the History of Science* 5 (1970), 73-84.

Examines Poincaré's three criteria for a "good" scientific theory: simplicity, suppleness, and naturalness. On all three counts, he finds Einstein's 1905 relativity paper to be wanting. For this reason, Goldberg argues, Poincaré's silence regarding Einstein's theory should not be surprising.

2080. ———. "Max Planck's Philosophy of Nature and His Elaboration of the Special Theory of Relativity." *Historical Studies in the Physical Sciences* 7 (1976), 125-160.

Planck's receptivity to relativity theory is seen as essentially compatible with his conservative scientific outlook. Philosophical and ethical considerations are emphasized, and Planck's interpretation of Einstein's theory is shown to be consistent with Planck's view that the goal of science is to discover the absolute laws of nature.

2081. Hirosige, Tetu. "Electrodynamics before the Theory of Relativity, 1890-1905." *Japanese Studies in the History of Science* 5 (1966), 1-49.

Treats theoretical foundations and difficulties in electrodynamical theories prior to relativity. Indicates how Lorentz advanced beyond his predecessors in his 1895 *Versuch*, and that the reception to his theory was fairly wide by 1900. Cohn's work after 1900 was the first break leading to the modern view, and it was Einstein who finally rid the ether of its last mechanical property--absolute rest. This paper is a good corrective to the common view that Einstein's work represents a radical and total departure from the physics of his day. See also item 1485.

2082. ———. "Origins of Lorentz' Theory of Electrons and the Concept of the Electromagnetic Field." *Historical Studies in the Physical Sciences* 1 (1969), 151-209.

Discusses the background for Lorentz's theory under the influence of Maxwell's theory and work of Helmholtz and Hertz. Considers the optical properties of matter, the historical significance of the stationary ether hypothesis, and the relation of Lorentz's theory to Hertz's electrodynamics. Also shows how Einstein approached the subject from a different perspective. See also item 1486.

2083. ———. "The Ether Problem, the Mechanistic Worldview, and the Origins of the Theory of Relativity." *Historical Studies in the Physical Sciences* 7 (1976), 3-82.

A contribution toward understanding the distinction between Einstein's theory and those of his predecessors. The author rejects the view of Whittaker, item 2095, and Keswani, item 2085, that Einstein's theory was largely an elaboration of the work of Lorentz and Poincaré. At the same time he expands on Holton's work, item 2084, by showing that Einstein was not only uninfluenced by the ether problem and the Michelson-Morley experiment, he was motivated by altogether different considerations. Hirosige argues that Einstein's motivation was derived from Mach's philosophy of science, and in particular Mach's critique of mechanical explanations. See also item 1487.

2084. Holton, G. "Einstein, Michelson, and the 'Crucial' Experiment." *Isis* 60 (1969), 133-197.

A thorough examination of the viewpoints, accounts, and evidence linking the Michelson-Morley experiment with Einstein's own work. Holton forcefully shows that "the role of the Michelson experiment ... [was] ... so small and indirect that ... it would have made no difference to Einstein's work if the experiment had never been made at all." The larger implications he draws (public vs. private science, etc.), make this a panoramic *tour de force* based on sound argument and psychological and philosophical insight.

2085. Keswani, G.H. "Origin and Concept of Relativity." *British Journal for the Philosophy of Science* 15 (1965), 286-306; 16 (1966), 19-32 and 273-294.

Examines the extent to which Einstein was influenced by Poincaré and Lorentz. Concludes that their role in Einstein's work was considerable. The author comes short of following Whittaker's view (see item 2095), however, as Einstein was the first to unequivocally reject the ether hypothesis.

2086. Klein, F. *Vorlesungen über die Entwicklung der Mathematik im 19. Jahrhundert*. Bd. 2. Berlin: Springer Verlag, 1926-1927,

2 vols. Reprinted in 1 vol. New York: Chelsea, 1967.

Discusses relativity theory as the study of invariants of the Lorentz transformation group. Fairly technical mathematics is involved. See also items 989, 1316, 1570.

2087. Lanczos, C. *Albert Einstein and the Cosmic World Order.* New York: Wiley (Interscience), 1965.

Valuable as an elementary presentation of the mathematics underlying relativity theory. Chapters on Problem of Reference Systems, Unification of Space and Time by Einstein and Minkowski, Geometric Discoveries of Gauss, and Riemannian Geometry and Einstein's Theory of Gravitation.

2088. ———. *The Einstein Decade (1905-1915)*. New York and London: Academic Press, 1974.

A compelling study despite the liberties taken by the author. Part I is a largely biographical attempt to distill the essence of Einstein's work. Part II presents useful synopses of Einstein's published papers during the period 1905-1915.

2089. McCormmach, Russell. "Einstein, Lorentz, and the Electron Theory." *Historical Studies in the Physical Sciences* 2 (1970), 41-87.

Demonstrates the link between Lorentz's electron theory and the theoretical problems underlying special and general relativity, thus showing that Einstein's work was in fact rooted in the physics of his time. Utilizes the Lorentz-Einstein correspondence, and pays considerable attention to the historical background. See also item 1502.

2089a. Miller, Arthur I. *Albert Einstein's Special Theory of Relativity: Emergence (1905) and Early Interpretation (1905-1911)*. Reading, Mass.: Addison-Wesley, 1981, xxviii + 466 pp.

Beginning with the electrodynamical researches from 1890 to 1905 (Maxwell-Hertz and Maxwell-Lorentz theories, work of Poincaré, Kaufmann, Abraham, et al.), the author goes on to consider contemporary ideas on space, time, and simultaneity as influenced by the views of Newton, Kant, and Mach. The reception of special relativity, its experimental and theoretical implications, and the controversial issues surrounding it (e.g., Ehrenfest's Paradox) are only a few of the many aspects dealt with in this excellent study. See also item 1506.

2089b. Pais, A. *'Subtle is the Lord ...' The Science and the Life of Albert Einstein.* Oxford: Clarendon Press, 1982.

The most thoroughly documented study of Einstein's life and work to date, this biography subdivides Einstein's life primarily along lines of his dominant scientific interests: statistical physics, special relativity, general relativity, quantum theory, and unified field theories. Some of the mathematical themes that receive particular attention are relativity theory and post-Riemannian differential geometry, and the correspondence between Einstein and Hilbert that led to the gravitational field equations.

2090. Pyenson, Lewis. "La réception de la relativité généralisée: disciplinarité et institutionalisation en physique." *Revue d'Histoire des Sciences et de leurs Applications* 28 (1975), 61-73.

Among mathematical physicists, Klein and Minkowski are contrasted with theoretical physicists like Lorentz and Boltzmann. At the same time, Leyden and Vienna as centers for theoretical physics are compared with Göttingen as a center for mathematical physics. Einstein and Hilbert are contrasted in a similar way with respect to their contributions to general relativity theory.

2091. ———. "Einstein's Early Scientific Collaboration." *Historical Studies in the Physical Sciences* 7 (1976), 83-123.

An externalistic study of the influence of the Göttingen physics tradition on Einstein's work. Emphasis is on Einstein's collaboration with Laub, Ritz, and Freundlich, and the reception of his work by the Wilhelminian physical science establishment.

2092. ———. "Hermann Minkowski and Einstein's Special Theory of Relativity." *Archive for History of Exact Sciences* 17 (1977), 71-95.

Although Minkowski's physics may have been the prism through which Göttingen mathematicians viewed relativity theory, Einstein's outlook was decidedly different. Pyenson supports this thesis by comparing his outlook with those of the leading figures in mathematical physics at Göttingen: Minkowski, Hilbert, Klein, and Weyl. Einstein regarded Hilbert's work on relativity as childlike "in the sense of children who know no malice in the world." Includes an appendix reproducing an unpublished manuscript by Minkowski on complex function theory of 1907.

2093. ———. "Physics in the Shadow of Mathematics: The Göttingen Electron-Theory Seminar of 1905." *Archive for History of Exact Sciences* 21 (1979), 55-89.

The Göttingen physics seminar of 1905 studied the works of Poincaré and Lorentz. The author contrasts this mathematically oriented group and the vast array of tools and talent at its disposal with the solitary figure of Einstein. He concludes that the purist approach at Göttingen was incapable of penetrating to the core of deep, underlying problems in physics, whereas Einstein succeeded by disregarding elegant action principles and electromagnetic potential theory in favor of studying the properties of the field equations themselves. See also item 1515.

2094. Schilp, P.A., ed. *Albert Einstein: Philosopher-Scientist.* Evanston, Ill.: Library of Living Philosophers, 1949.

Contains Einstein's "Autobiographical Notes," in German and in English translation, as well as essays by numerous scientists, mathematicians, and philosophers of science dealing primarily with the general or philosophical significance of relativity theory. Among the contributors are: Sommerfeld, DeBroglie, Pauli, Born, Bohr, Franck, Reichenbach, Menger, Infeld, Laue, and Gödel. Einstein also replies to criticisms made in the volume, and there is an extensive bibliography of his work.

2095. Whittaker, Sir Edmund Taylor. *A History of the Theories of Aether and Electricity. Vol. II: The Modern Theories 1900-1926*. London: Thomas Nelson and Sons, 1953. Reprinted New York: Harper Torchbooks, 1960, 319 pp.

Chapter II, "The Relativity Theory of Poincaré and Lorentz," has been the source of considerable controversy. Few scholars seem to have accepted Whittaker's view that Poincaré's principle of relativity is no different from Einstein's theory, and that Einstein's 1905 paper "set forth the relativity theory of Poincaré and Lorentz with some amplifications...." Chapter V, "Gravitation," deals with the general theory. The book as a whole is fairly technical. See also items 1536, 1972.

SET THEORY

The history of set theory can be traced back to antiquity, insofar as the problem of continuity and infinity are related to the subject generally. The bibliography given here, however, limits itself with but few exceptions to the most modern development of set theory, beginning with the pioneering researches of Georg Cantor in the 1870s. Readers are referred to other sections of this bibliography for allied literature, especially the sections on analysis, on number (where theories of the real numbers are treated), and on topology, which has an obvious bearing on the history of set theory. References in the sections for the 19th and 20th centuries are also relevant to the history of set theory, and cross-references should help to guide readers to useful material in other parts of this bibliography.

2096. Ashworth, E.J. "An Early 15th Century Discussion of Infinite Sets." *Notre Dame Journal of Formal Logic* 18 (1977), 232-234.

Discusses the 15th-century figure John Dorp, and claims he was aware of non-denumerably infinite sets.

2097. Bunn, R. "Quantitative Relations between Infinite Sets." *Annals of Science* 34 (1977), 177-191.

Discusses Leibniz, Bolzano, and Cantor, in the context of the general history of the ordering of infinite sets from Aristotle to Cantor. See also item 2234.

2098. Cavaillès, J. *Philosophie mathématique*. Paris: Hermann, 1962.

Although not published until well after the author's death during World War II, this is a thoughtful and informative collection of observations on the history and philosophy of modern set theory. See also item 2236.

2099. Couturat, L. *De l'infini mathématique*. Paris: F. Alcan, 1896. Reprinted Paris: Blanchard, 1973.

A defense of Cantor against his adamant critic and opponent Kronecker on the subject of the infinite in mathematics. This book was widely read at the turn of the century, but is now dated.

2100. Dalen, D. van, and A.F. Monna. *Sets and Integration. An Outline of the Development*. Groningen: Wolters-Noordhoff Publishing, 1973.

Part I of this book is devoted to "Set Theory from Cantor to Cohen" and was written by van Dalen; Monna is responsible for Part II: "The Integral from Riemann to Bourbaki," which also takes up relevant aspects of set theory as it relates to theories of integration.

2101. Dauben, Joseph Warren. "C.S. Peirce's Philosophy of Infinite Sets." *Mathematics Magazine* 50 (1977), 123-135.

Compares Peirce's discovery and development of non-denumerability and his theories of infinitesimals and continuity, which arose in the context of logic, with similar interests developed by Cantor and Dedekind in the context of analysis. Stresses comparative differences, especially Cantor's opposition to infinitesimals and Peirce's adamant acceptance and development of same.

2102. ———. *Georg Cantor, His Mathematics and Philosophy of the Infinite*. Cambridge, Mass.: Harvard University Press, 1979, 361 pp.

An intellectual biography of Cantor, tracing the origins of set theory from his work on trigonometric series and rigorous definition of the real numbers, through the discovery of non-denumerable sets and the eventual development of transfinite numbers to transfinite arithmetic and the theory of transfinite ordinal and cardinal numbers and order types in general. Considers as well the social and academic context in which Cantor's work was done, as well as theological and psychological aspects of Cantor's interests. With photographs and previously unpublished material. Draws heavily on manuscripts, correspondence, and archival sources. See also items 1097, 2240.

2103. Drake, F.R. *Set Theory: An Introduction to Large Cardinals*. Amsterdam: North Holland, 1974.

Considers models of set theory, trees, reflection principles, all with historical notes. This is an advanced text, summarizing work on large cardinals primarily by Tarski and his students since 1950.

2104. Dugac, Pierre. *Richard Dedekind et les fondements des mathématiques*. Paris: Vrin, 1976.

Provides both a scientific biography of Dedekind, with numerous texts, and 58 appendixes of manuscripts and letters written by Dedekind and others (pp. 143-315). This work argues that Dedekind had a much larger role than is usually appreciated in the origins of set theory. This monograph is Number 24 in the Collection des Travaux de l'Académie international d'histoire des sciences. See also item 1098.

2105. Fraenkel, A., Y. Bar-Hillel, and A. Levy, with D. van Dalen. *Foundations of Set Theory*. 2nd rev. ed. Amsterdam: North Holland, 1973.

The revised edition, like its predecessors, is rich in historical references, although designed primarily as an advanced introduction to the subject. See also item 2243

2106. Gödel, K. "What Is Cantor's Continuum Problem?" *American Mathematical Monthly* 54 (1947), 515-525.

A popular, readable account of the essential features and significance of Cantor's Continuum Hypothesis. Gödel revised and expanded the article when it was reissued in *Philosophy of Mathematics. Selected Readings*, edited by P. Benacerraf and H. Putnam (Englewood Cliffs, N.J.: Prentice-Hall, 1964).

2107. Grattan-Guinness, Ivor. "An Unpublished Paper by Georg Cantor: 'Principien einer Theorie der Ordnungstypen. Erste Mittheilung.'" *Acta Mathematica* 124 (1970), 65-107.

This article provides an extensive introduction to and commentary upon a previously unpublished work by Cantor on the theory of order types. It was set in type and then withdrawn from *Acta Mathematica*, largely because the editor, Mittag-Leffler, regarded the developments in Cantor's paper as being of little utility and so abstruse as to cast doubt on the reputation both of Cantor's nascent set theory, as well as Mittag-Leffler's fledgling journal.

2108. ———. "Towards a Biography of Georg Cantor." *Annals of Science* 27 (1971), 345-392.

Relies upon extensive archival documents to correct many errors and falsehoods, many circulated by rumor and word-of-mouth, concerning Cantor's biography and mental illness. With photographs and transcriptions of previously unpublished materials.

2109. ———. "Some Remarks on Cantor's Published and Unpublished Work on Set Theory." *NTM-Schriftenreihe für Geschichte der Naturwissenschaften, Technik, und Medizin* 9 (1971), 1-8.

2110. ———. *Dear Russell-Dear Jourdain: A Commentary of Russell's Logic, Based on his Correspondence with Philip Jourdain*. New York: Columbia University Press; London: Duckworth, 1977, vi + 234 pp.

Discusses as well Cantor, Frege, Whitehead, Wittgenstein, raising general issues of logic, paradoxes, foundations and the Axiom of Choice. The correspondence dates from 1902 to 1919; the book includes a good selection of photographs. See also items 1651, 2248.

2111. ———. "Georg Cantor's Influence on Bertrand Russell." *History and Philosophy of Logic* 1 (1980), 61-93.

Delineates both positive and negative influences, especially on issues where Russell adopted alternative views about the nature and foundations of set theory. See also item 1145.

2112. ———, ed. *From the Calculus to Set Theory, 1630-1910. An Introductory History*. London: Duckworth, 1980, 306 pp.

Contains one article specifically devoted to the history of set theory, and three others that raise significant issues related to the development of set theory in the 19th and early 20th centuries: J. Dauben, Chapter 5: "The Development of Cantorian Set Theory," pp. 181-219; I. Grattan-Guinness, Chapter 3: "The Emergence of Mathematical Analysis and its Foundational Progress, 1780-1880," pp. 94-148; T. Hawkins, Chapter 4: "The Origins of Modern Theories of Integration," pp. 149-180; R. Bunn, Chapter 6: "Developments in the Foundations of Mathematics, 1870-1910," pp. 220-255. See also item 780.

2113. Hawkins, Thomas. *Lebesgue's Theory of Integration: Its Origins and Development*. Madison: University of Wisconsin Press, 1970. Reprinted New York: Chelsea Publishing Company, 1975.

Extensive discussion of set theory, especially from 1870 to 1900, as a background to Lebesgue's development of integration theory. See also items 1074, 1147.

2114. Jain, L.C. "Set Theory in Jaina School of Mathematics." *Indian Journal of History of Science* 8 (1973), 1-27.

Discusses Indian mathematics which the author interprets in terms of set theory and related ideas.

2115. Johnson, P.E. *A History of Set Theory*. Boston: Prindle, Weber and Schmidt, 1972.

This book is a revised version of the author's doctoral dissertation, "A History of Cantorian Set Theory," presented to the Peabody College for Teachers in 1968. It includes consideration of Georg Cantor and his pioneering contributions to set theory, developments leading to axiomatic set theory, and a lengthy bibliography. Portraits are included of Cantor, Dedekind, Kronecker, Weierstrass, Kummer, and Russell.

2116. Jourdain, Philip E.B. "The Development of the Theory of Transfinite Numbers." *Archiv der Mathematik und Physik* (Grunnert's *Archiv*), in four parts: 10 (1906), 254-281; 14 (1909), 289-311; 22 (1910), 1-21; 22 (1913), 1-21.

Part I deals with the growth of the theory of functions up to 1870; Part II is devoted primarily to Weierstrass (1840-1880); Part III studies Cantor's work on trigonometric series and his theory of real numbers, as well as other theories of irrational numbers; Part IV concludes the series with a general overview of Cantor's creation of transfinite arithmetic.

2117. ———, ed. and trans. *Contributions to the Founding of the Theory of Transfinite Numbers*, by Georg Cantor. Chicago: Open Court, 1915.

Jourdain provides his translation of Cantor's "Beiträge zur Begrundung der transfiniten Mengenlehre" (1895 and 1897) with a thorough introduction, much of it based on the articles written for the *Archiv der Mathematik und Physik*, item 2116.

2118. Juškevič, Adolf P. "Georg Cantor und Sof'ja Kovalevskaja." *Ost und West in der Geschichte des Denkens und der kulturellen Beziehungen. Festschrift für Eduard Winter zum 70. Geburtstag*. Edited by H. Mohr and C. Grau. Berlin: Akademie-Verlag, 1966.

Includes transcriptions of correspondence, as well as informative discussions of Cantor, Kovalevskaya, and Mittag-Leffler.

2119. Kartasasmita, R.B.G. "An Account of the Development of Set Theory." Dissertation, University of Illinois, 1975, 284 pp.

Discusses Cantor, Zermelo-Fraenkel set theory, von Neumann, Bernays, Robinson, and Gödel. See *Dissertation Abstracts* 36 (273 B).

2120. Manheim, Jerome H. *The Genesis of Point Set Topology*. Oxford: Pergamon Press, and New York: Macmillan, 1964.

Much of the discussion deals with set theory insofar as it relates to topology. See also items 1059, 2146.

2121. Marczewski, E. "On the Works of Wacław Sierpiński: Main Trends of His Works on Set Theory. Recollections and Reflections." *Wiadomošei Matematyczne* (Warsaw), 14 (1972), 65-72. In Polish.

Outine of Sierpiński's major works in set theory written by a former student.

2122. Medvedev, F.A. *Razvitie teorii mnozhestv v deviatnadtsatom veke*. Moscow: Nauka, 1965.

A general, detailed study of the history of set theory in the 19th century, drawing entirely upon published sources.

2123. Meschkowski, H. *Probleme des Unendlichen. Werk und Leben Georg Cantors*. Braunschweig: Vieweg, 1967.

A comprehensive study of Cantor's life and works, drawn from extensive reading in primary, secondary, and archival materials. Contains a substantial appendix of previously unpublished letters. Photographs.

2124. Moore, G.H. "Ernst Zermelo, A.E. Harward, and the Axiomatization of Set Theory." *Historia Mathematica* 3 (1976), 206-209.

In this brief note, the author shows that many of the ideas in Zermelo's famous paper of 1908 giving the first axiomatization of set theory were anticipated by the British mathematician A.E. Harward in 1905. While no claims for Harward as a precursor or an influence on Zermelo are made, the note does explore the relation between the two and why Harward's paper met no response, while Zermelo's paper three years later engendered much controversy.

2125. Murata, T. "On the Meaning of 'Virtualité' in the History of the Set Theory." *Japanese Studies in the History of Science* 5 (1966), 119-139.

The author discusses French opponents to Cantorian set theory by discussing the special use of the word "virtualité." Baire, Borel, and Lusin are featured, with emphasis given to alternative "non-atomistic" views of the continuum as being more than a simple assemblage of points.

2126. ———. "A Few Remarks on the Atomistic Way of Thinking in Mathematics." *Japanese Studies in the History of Science* 6 (1967), 47-59.

Discusses mathematicians like Borel and Brouwer who eventually had to abandon their non-atomistic conceptualizing of the continuum as a collection of points.

2127. ———. "Sur l'évolution de l'idée d''effectif' dans l'histoire de la théorie des ensembles." *Revue d'histoire des sciences* 26 (1973), 365-368.

Similar to the earlier study by Murata on "virtualité" but here emphasis is upon the concept of "effectif" in set theory at the end of the 19th century.

2128. ———. "L'évolution des principes philosophico-mathématiques de la théorie des ensembles chez Georg Cantor et leur diffusion en France jusqu'en 1905." Dissertation, University of Paris, 1974.

The author studies the reception and development of the mathematical-philosophical aspects of Georg Cantor's set theory in France during the late 19th century, finishing with the use Lebesque and others were making of Cantor's ideas in the first decade of this century.

2129. Schoenflies, A. *Entwickelung der Mengenlehre*. Leipzig: Teubner, 1900. 2nd ed., 1913.

Still a useful introduction to the subject. Particularly strong on developments at the turn of the century.

2130. ———. "Die Krisis in Cantor's mathematischem Schaffen." *Acta Mathematica* 50 (1927), 1-23.

An important study of Cantor's first major nervous breakdown, attributed in part to difficulties with Kronecker's opposition to transfinite set theory and to Cantor's own difficulties with solution of the Continuum Hypothesis. Replete with previously unpublished correspondence.

2131. Steiner, H.G. "Mengenlehre." In *Historisches Wörterbuch der Philosophie*. Vol. 5. Basel, Stuttgart: Schwabe, 1981.

Extensive article on the basic concepts and history of set theory with detailed bibliographic notes.

2132. Steur, A.J.E.M. "Georg Cantor en de leer der versamelingen." *Scientarium Historia* 15 (1973), 23-26.

Short, but with special attention to the Continuum Hypothesis.

2133. Ulam, S.M. "Infinities." In *The Heritage of Copernicus*. Edited by J. Neyman. Cambridge, Mass.: MIT Press, 1974, 378-393.

Considers Cantor, Gödel, and the problem of infinity, as well as set theory, undecidability, and notation. Ulam asserts that revolutions do occur in mathematics, and that the subject of infinities provides an example.

THERMODYNAMICS AND STATISTICAL MECHANICS

2134. Bellone, Enrico. *Aspetti dell'approccio statistico alla meccanica: 1849-1905*. Florence: G. Barbera Editore, 1972.

Surveys the theories of Laplace, Fourier, Carnot, Waterston, Joule, Rankine, Clausius, Maxwell, Boltzmann, and Planck.

2135. Brush, Stephen G. *The Kind of Motion We Call Heat: A History of the Kinetic Theory of Gases in the 19th Century*. Amsterdam and New York: North-Holland Publishing Company, 1976.

Includes a chapter "Foundations of Statistical Mechanics 1845-1915" (previously published in *Archive for History of Exact Sciences* 4 [1968], 145-183) which discusses the "ergodic hypothesis" and its relation to developments in pure mathematics (Cantor's set theory, invariance of dimensionality). There is a complete bibliography of original works on kinetic theory and statistical mechanics, 1801-1900.

2136. ———. "Proof of the Impossibility of Ergodic Systems: The 1913 Papers of Rosenthal and Plancherel." *Transport Theory and Statistical Physics* 1 (1971), 287-311.

Translations of the papers by A. Rosenthal (*Annalen der Physik* 42 [1913], 796-806), and M. Plancherel (*Annalen der Physik* [1913], 1061-1063), which proved that the original Maxwell-Boltzmann hypothesis (that the point representing the configuration of a mechanical system goes through every point in the phase space) is mathematically impossible, using the results of L.E.J. Brouwer, R. Baire, and H. Lebesgue. The introduction discusses the historical background of this important application of "pure" mathematics to theoretical physics.

2137. Cardwell, D.S.L. *From Watt to Clausius: The Rise of Thermodynamics in the Early Industrial Age*. Ithaca, N.Y.: Cornell University Press, 1971.

Development of theories of heat stressing their relations to technology.

2138. Koenig, Frederick O. "On the History of Science and of the Second Law of Thermodynamics." *Men and Moments in the History of Science*. Edited by Herbert M. Evans. Seattle: University of Washington Press, 1959, 57-111, 211-216.

Discusses the various formulations of the second law by Carnot, Clapeyron, and Clausius.

2139. Truesdell, Clifford A. "Early Kinetic Theories of Gases." *Archive for History of Exact Sciences* 15 (1975), 1-66.

Survey of theories of Euler, Herapath, Waterston, Clausius, Maxwell, and others, with a "critical bibliography of kinetic-statistical mechanics through 1866."

2140. ———. *The Tragicomical History of Thermodynamics, 1822-1854*. New York: Springer-Verlag, 1980.

A critique of theories published in this period, from a mathematical standpoint. See also item 1767.

TOPOLOGY

2141. Bollinger, Maja. "Geschichtliche Entwicklung des Homologiebegriffs." *Archive for History of Exact Sciences* 9 (2) (1972), 94-166.

A detailed history of homology theory from its beginnings to 1915. The work contains a full summary of Poincaré's fundamental contributions to the subject.

2141a. Brouwer, L.E.J. *Collected Works*. Vol. 2. *Geometry, Analysis, Topology and Mechanics*. Edited by Hans Freudenthal. Amsterdam: North-Holland Publishing Company; New York: American Elsevier Publishing Company, Inc., 1976.

Although a number of collected works of mathematicians who have made fundamental or substantial contributions to topology have been published (including those of Euler, H. Hopf, Mazurkiewicz, M. Morse, Poincaré, Riemann, Sierpiński, Uryson, and J.H.C. Whitehead), this edition of Brouwer's papers is one of the most interesting. The editor has included much historical material and many notes giving detailed coverage of Brouwer's work and his relationships with other mathematicians. The editor has succeeded in conveying to the reader something of the spirit of Brouwer, the man and his mathematics.

2142. Dehn, Max, and Poul Heegaard. "Analysis Situs." *Encyklopädie der mathematischen Wissenschaften* III.1.AB3. Leipzig: Teubner, 1907, 153-220.

An encyclopedic treatment of the subject up to 1907 written from the authors' special point of view. Although the work is not a history, it contains much material of historical interest. The references to the literature are especially valuable.

2143. Feigl, Georg. "Geschichtliche Entwicklung der Topologie." *Jahresbericht der Deutschen Mathematiker-Vereinigung* 37 (1928), 273-286.

A brief survey of the development of topology.

2143a. Hirsch, Guy. "Topologie." Chapitre X in *Abrégé d'histoire des mathématiques 1700-1900*. Edited by Jean Dieudonné. Paris: Hermann, 1978, II, 211-266.

A broad survey of recent, primarily twentieth-century, developments in topology with an emphasis on combinatorial and algebraic topology.

2144. Johnson, Dale M. "The Problem of the Invariance of Dimension in the Growth of Modern Topology, Part I." *Archive for History of Exact Sciences* 20 (1979), 97-188; Part II, 25 (1981), 85-267.

A detailed history of topological dimension theory from its beginnings to 1913 plus a survey of developments to about 1930. Part I is described in item 1058. Part II concludes Johnson's study with the work of Brouwer, the Brouwer-Lebesgue dispute, and "Glimpses of the Development of Dimension Theory after Brouwer." Substantial bibliography (pp. 251-266), and Errata to Part I, p. 267.

2145. Kline, Morris. "The Beginnings of Topology." Chapter 50 in *Mathematical Thought from Ancient to Modern Times*. New York: Oxford University Press, 1972, 1158-1181.

A general introduction to the history of topology with helpful references to the literature. See also items 82, 867, 990, 1106, 1317, 1571.

2146. Manheim, Jerome H. *The Genesis of Point Set Topology*. Oxford: Pergamon Press, and New York: Macmillan, 1964.

A sketch of the development of point set topology starting from its background in analysis. The work is a superficial treatment of the subject. See also items 1059, 2120.

2147. Pont, Jean Claude. *La topologie algébrique, des origines à Poincaré*. Paris: Presses Universitaires de France, 1974.

A broad history of algebraic topology before the fundamental work of Poincaré. It is regrettable that Poincaré's work is not covered. However, Bollinger may be consulted. See also item 1061.

2147a. Scholz, Erhard. *Geschichte des Mannigfaltigkeitsbegriffs von Riemann bis Poincaré*. Boston, Basel, Stuttgart: Birkhäuser, 1980.

A good coverage of the mathematics connected with the historical development of the topological concept of manifold, treating both the geometrical and analytical background and giving special attention to the work of Riemann and Poincaré.

2148. Tietze, Heinrich, and Leopold Vietoris. "Beziehungen zwischen den verschiedenen Zweigen der Topologie." *Encyklopädie der mathematischen Wissenschaften* III.1.AB13. Leipzig: Teubner, 1930, 141-237.

An encyclopedic treatment of the development of the various branches of topology from 1907 to 1930. The copious references are extremely valuable.

2149. Zoretti, Ludovic, and Arthur Rosenthal. "Neuere Untersuchungen über Funktionen reeller Veränderlichen, Die Punktmengen," *Encyklopädie der mathematischen Wissenschaften* II.3.C9a. Leipzig: Teubner, 1924, 855-1030.

An excellent encyclopedic treatment of point set theory and topology to 1924 including the work of Brouwer. The numerous references are extremely valuable.

VI. THE HISTORY OF MATHEMATICS: SELECTED TOPICS

This final section of guides to the history of mathematics is devoted to a number of ancillary topics, some of which have begun to receive increasing attention of late--especially the sociology of mathematics and the subject of women in mathematics. Other topics, like mathematics education and the philosophy of mathematics, have an established history of their own, and consequently the treatment given these subjects here is comparatively more substantial, more highly differentiated, and considerably more selective.

MATHEMATICS EDUCATION

The boundaries of the history of mathematics per se, the history of mathematics as an adjunct to or tool for teaching, and the history of the teaching of mathematics may seem clear, initially. However, for example, the developing curriculum is a part of the history of mathematics education which should not be separated too widely from the development of the subject. Likewise the ideas represented by such terms as mathematics "education," "teaching," "methods," "didactics," and "pedagogy" can differ from one another in the views of some. They can also be related to mathematical and educational philosophies as well as to the psychology of learning. These then are cognate areas with which the person of studying the history of mathematics education should become familiar. They are represented in the following bibliography only indirectly.

An experienced historian will also look at the periodical literature in the field of mathematics education which includes several hundred journals. Also of use are the encyclopedic works of which we have cited only one (see Brusotti, item 2152), indexing and abstracting journals (we have cited only *ZDM*, item 2163, but more general works such as the *Education Index* and the bibliographies compiled by the ERIC Center at Ohio State University can be of some aid), and even expositions of elementary mathematics and its foundations, such as those by Felix Klein, Federigo Enriques, and Augustus De Morgan.

We add to these suggestions the caveat that the following list is biased toward mathematics education in Great Britain and the United States and that in this area the most comprehensive single reference in both its coverage and bibliography is the Thirty-First Yearbook of the National Council of Teachers of Mathematics, item 2178.

Mathematics Education and the Teaching of History of Mathematics

2150. Board of Education: Reports on Special Education Subjects. *The Teaching of Mathematics in the United Kingdom*. Vol. 26. London: His Majesty's Stationery Office, 1912. 2 vols.

Valuable source material and references for the period 1830-1912.

2151. Branford, B. *A Study of Mathematical Education*. Oxford: Clarendon Press, 1908.

A fascinating study of the thesis that "the genesis of knowledge in the individual must follow the same course as the genesis of knowledge in the race." Branford was a London school inspector who combined his practical experience with a fair knowledge of the history of elementary mathematics.

2152. Brusotti, Luigi. "Questioni didattiche." *Enciclopedia delle mathematiche elementari*. Vol. III, Part 2. Milan: Ulrico Hoepli, 1950, 885-973.

Extended notes and references and the fact that nearly every topic treated includes some discussion of its development make this a valuable reference though it is not basically a history and stresses Italian problems, viewpoints, and references. Its eight sections include discussions of the goals and methods of instruction, teaching in professional schools, particular didactic questions with a special stress on geometry, and the preparation of teachers with short lists of "methods" books and mathematical-pedagogical journals. The extensive documentation of the didactic sections of the *Enciclopedia* may indirectly help the historian of mathematics education. Similar comments apply to the German and Russian encyclopedias.

2153. Butler, Charles H., and F. Lynwood Wren. *The Teaching of Secondary Mathematics*. 2nd ed. New York: McGraw-Hill Book Company, 1951, xiv + 550 pp.

Part I, pp. 1-50 of this edition, is historical and includes bibliographies. Later editions of this "methods" book used more space to discuss current reform movements and less for history. The books by Smith and Young listed below are examples of older "methods" books which contain good bibliographies and discuss older and continental views of mathematics education while presenting the needs and practices of their time.

2154. Cajori, Florian. *The Teaching and History of Mathematics in the United States*. Washington, D.C.: Government Printing Office, 1890, 400 pp.

Contains a detailed report on a questionnaire survey of curricula, libraries, texts used in the schools and colleges of the United States. See also item 987.

2155. De Morgan, A. *The Study and Difficulties of Mathematics*. Chicago: Open Court, 1943.

Popular essays originally published in the *Penny Cyclopaedia* (London: C. Knight, 1831; New York, 1910).

2156. Flegg, G. "Some Questions on the Teaching of the History of Mathematics." *Zentralblatt für Didaktik der Mathematik Analysen* 2 (1978), 67-73.

Views on the status, quality, and function of the history of mathematics, from the Open University course team leader.

2157. Grattan-Guinness, Ivor. "Not from Nowhere: History and Philosophy behind Mathematical Education." *International Journal of Mathematical Education in Science and Technology* (Chichester, England: Wiley-Interscience), 4 (3) (1973), 421-453.

Comparison of standard teaching method with the historical development of selected topics. Argues for 'history satire'--teaching based on problems as near as possible to the original, to assist the student's own understanding.

2158. *Report of the American Commissioners of the International Commission on the Teaching of Mathematics*. (United States Bureau of Education, Bulletin 1912, No. 14.) Washington, D.C.: Government Printing Office, 1918.

The International Commission, appointed in 1908, received 150 reports from 27 countries. See items 39-49 on p. 503 of the 32nd Yearbook of the N.C.T.M. for additional reports from the United States. Printed reports were also published for France, England, and Germany. See p. 298 of Cajori, *The Teaching and History of Mathematics in the United States*, for titles, item 2154 above.

2159. Rogers, L.F. "The Philosophy of Mathematics and the Methodology of Mathematics Teaching." *Zentralblatt für Didaktik der Mathematik Analysen* 2 (1978), 63-67.

Reference to ideas of Lakatos, Polanyi, Popper, and others, to emphasize the discovery process in the development of mathematics, both in history and the classroom.

2160. Tahta, D.G., and M.E. Boole. *A Boolean Anthology*. Nelson, Lancashire, Eng.: F.H. Brown and Coulton Printing Company for the Association of Teachers of Mathematics, 1972.

Selected writings of Mary Boole on mathematical education. England, nineteenth century. Useful notes on contemporary mathematical-social connections.

2161. Wilder, R.S. "History in the Mathematics Curriculum, Its Status, Quality and Function." *American Mathematical Monthly* 79 (1972), 479-495.

A rationale for the use and teaching of the history of mathematics at all levels, from school through university research department.

2162. Young, J.W.A. *The Teaching of Mathematics in the Elementary and the Secondary School*. New ed. New York: Longmans, Green, and Co., 1914, xviii + 359 pp.

Each chapter of the first (1906) edition was preceded by a bibliography ranging over time (e.g., extending back to Lagrange, 1795; Lacroix, 1816) and geography (England, France, Germany). The 1914 edition has a supplemental bibliography. Now out of print but to be found in major libraries.

2163. ZDM. *Zentralblatt für Didaktik der Mathematik Analysen*. Jahrgang 1, Heft 1, June 1969-.

The only abstracting journal which covers mathematics education at all completely. See sections A40-Biographies; History of Mathematics and D10-History of Mathematical Education.

The Use of History in the Teaching of Mathematics

2164. Cajori, Florian. *A History of Elementary Mathematics with Hints on the Methods of Teaching*. New York: The Macmillan Company, 1924, viii + 324 pp.

The "hints" are minimal, but it is a text written principally for teachers to show the relevance of the history of mathematics to the curriculum. Pages 290-309 deal with "Recent Movements in Teaching." However, there are many commentaries on teaching and its history throughout the book. See p. 323 of its index.

2165. Hallerberg, A., ed. *Historical Topics for the Mathematics Classroom*. (National Council of Teachers of Mathematics 31st Yearbook.) Washington, D.C.: 1969.

A classic in sourcebooks for the school and college teacher. Contains surveys on number, computation, algebra, geometry, trigonometry, and many short "capsules" of information on a wide variety of topics.

2166. Jones, P.S. "The History of Mathematics as a Teaching Tool." In Hallerberg, item 2165, 1-17.

2167. Popp, W. *History of Mathematics: Topics for Schools*. Cambridge, Mass.: Transworld, 1975.

Originally, *Geschichte der Mathematik im Unterricht* (Munich: Bayerischer Schulbuch-Verlag, 1968). Arithmetic, Algebra, Geometry, and Calculus. Historical background for topics in the school curriculum.

2168. Willerding, M. *Mathematical Concepts. A Historical Approach*. Boston: Prindle Weber and Schmidt, 1967.

Five volumes (series editor: Howard Eves). Short books on topics for teachers of secondary school children.

History of Mathematics Education and the Evolution of Curricula

2169. Athen, H., and H. Knule. *Proceedings of the Third International Congress on Mathematical Education. Karlsruhe, Germany, 16-21 August, 1976* (Z.D.M. Universität [West] Karlsruhe, Hertzstrasse 16, 7500 Karlsruhe, BRD).

One of the working groups was on history and pedagogy of mathematics. See also Howson, item 2175 below. Note particularly Atiyah on "Trends in Pure Mathematics."

2170. Bidwell, James K., and Robert G. Clason. *Readings in the History of Mathematics Education*. Reston, Va.: National Council of Teachers of Mathematics, 1970, xiii + 706 pp.

Reprints substantial excerpts from major documents and prominent writers spanning the period 1831-1959 in the United States. The sections are: 1. Beginnings of the Art of Teaching Mathematics (1828-1890); 2. Emergence of National Organizations as a Force in Mathematics Education (1891-1919); 3. "The 1923 Report" and Connectionism in Arithmetic (1920-1937); 4. Prewar and Postwar Reforms (1938-1959). Short commentaries point out the background and significant facets of the excerpts.

2171. Bos, Hendrik J.M., and H. Mehrtens. "The Interactions of Mathematics and Society in History. Some Exploratory Remarks." *Historia Mathematica* 4 (1977), 7-30.

Seminal paper on the "external" aspects of the history of mathematics. A landmark in the growth of social history of mathematics. See also item 2199.

2172. Duren, W.L., Jr. "CUPM, the History of an Idea." *The American Mathematical Monthly* 74, Part II (January 1975), 23-36.

Committee on the Undergraduate Program, Mathematics (CUPM) and School Mathematics Study Group (SMSG) respectively represent the collegiate and the public school mathematics teaching reform movements of the 1960s. See Wooton, item 2186 below.

2173. Grabiner, Judith V. "Mathematics in America: The First Hundred Years." *The Bicentennial Tribute to American Mathematics, 1776-1976*. Edited by Dalton Tarwater. Washington, D.C.: The Mathematical Association of America, 1977, 9-24.

Mainly concerned with collegiate education, there are some comments and references dealing with teacher training and such pedagogical-psychological questions as "mental discipline."

2174. Hollingdale, S.H. "The Teaching of Mathematics in Universities: Some Historical Notes." *Bulletin of the International Mathematics Association* 10 (1974), 312-318.

British universities, mainly 19th century. A survey with some useful references, from 13th to the 20th century.

2175. Howson, A.G., ed. *Developments in Mathematical Education. Proceedings of the International Congress of Mathematics*

Education, 1972, Exeter, England. London and New York: Cambridge University Press, 1973.

Note particularly papers by D. Hawkins, E. Leach, and R. Thom, discussing various social-historical aspects of the evolution of the mathematics curriculum. One of the working groups was on the relation between pedagogy and history of mathematics.

2176. Jones, Phillip S. "The History of Mathematical Education." *The American Mathematical Monthly* 74, Part II (January 1975), 38-55.

Surveys the pedagogy of mathematics from Socrates on--with emphasis on collegiate education in the United States and the role of the Mathematical Association of America.

2177. May, Kenneth O., ed. *The Mathematical Association of America: Its First Fifty Years*. Washington, D.C.: The Mathematical Association of America, 1972, vii + 172 pp.

The Association's major concern is with mathematics in the undergraduate colleges. Several chapters include material on mathematics education, especially Chapter I, "Historical Background and Founding of the Association," by Phillip S. Jones, and Chapter II, "The First Twenty-Five Years," by Carl B. Boyer. See also items 1167, 2195.

2178. National Council of Teachers of Mathematics. *A History of Mathematics Education in the United States and Canada*. Thirty-First Yearbook. Edited by Phillip S. Jones and Arthur Coxford, Jr. Reston, Va.: National Council of Teachers of Mathematics, 1970, xx + 557 pp.

Emphasizes curricular and methodological changes in the elementary and secondary schools. Issues and forces causing the changes are stressed. Philosophical and psychological trends are noted but not elaborated. Extensive bibliography. Lists of persons serving on study and planning committees and a section on the education of teachers are included.

2179. Siddons, A.W. "Fifty Years of Change." *Mathematical Gazette* 40 (1956), 161-169.

History of the Mathematical Association and its influence on school curricula in Britain. A number of useful references, including the origins of the Association for the Improvement of Geometrical Teaching, and some Mathematical Association Teaching Committee Reports.

2180. Smith, David Eugene. *The Teaching of Elementary Mathematics*. New York: The Macmillan Company, 1900. Reprinted 1908, xv + 312 pp.

Written by one of the most influential mathematics educators of the period. Chapter XIII, "The Teacher's Book-Shelf," is short, but references to the literature in footnotes are plentiful. The author wrote "There are many books on the teaching of mathematics, some of them quite pretentious in

Figure 15. The Mathematics Club of Göttingen, as photographed in 1902. Seated at the table is Felix Klein; to the left is David Hilbert; to the right, Karl Schwarzschild and Grace Chisholm Young. On the far right is Ernst Zermelo. Just behind Zermelo is Felix Bernstein.

their claims, a few published in America, a few in England and France, and a large number in Germany." This book represents the first serious attempt to examine the historical evolution of the mathematical curriculum.

2181. ———, and Jekuthiel Ginsburg. *A History of Mathematics in America before 1900*. (The Carus Mathematical Monographs, No. 5.) Chicago: The Mathematical Association of America, and Open Court Publishing Co., 1934, x + 209 pp.

Includes discussions of colleges, their mathematics curricula, texts, and prominent teachers in the United States. The chapters are: I. The Sixteenth and Seventeenth Centuries; II. The Eighteenth Century; III. The Nineteenth Century, General Survey; IV. The Period 1875-1900. The last includes the beginning of advanced study and research. See also item 993.

2182. ———, and Charles Goldhizer. *Bibliography of the Teaching of Mathematics, 1900-1912*. (United States Bureau of Education, Bulletin 1912, no. 29.) Washington, D.C.: Government Printing Office, 1912.

Prepared in connection with work for the International Commission on the Teaching of Mathematics.

2183. Swetz, F. *Mathematics Education in China: Its Growth and Development*. Cambridge, Mass.: MIT Press, 1974.

Discusses the position of mathematics in the traditional education system, the effects of the system on mathematical thinking and instruction, and the reforms from 1870 to 1970.

2184. Wilder, R.L. "The Role of the Axiomatic Method." *American Mathematical Monthly* 74 (1967), 115-127.

A history of the axiomatic method and its relation to the teaching of mathematics.

2185. Wilson, D.K. *The History of Mathematical Teaching in Scotland to the End of the 18th Century*. London: University of London Press, 1935.

Particular to Scotland, and establishing different traditions to English education.

2186. Wooton, William. *SMSG. The Making of a Curriculum*. New Haven: Yale University Press, 1965, ix + 182 pp.

A history of the School Mathematics Study Group (SMSG), the major school curriculum reform project of the 1960s. See Duren, item 2172 above, for a history of the major collegiate reform project.

2187. Yeldham, F.H. *The Teaching of Arithmetic through Four Hundred Years. 1535-1935*. London: Harrap, 1936.

Content, methods, and style of popular arithmetic texts from Tonstall (1522) through the nineteenth century. Particular to England and works in English.

HISTORY OF INSTITUTIONS

2188. Archibald, Raymond C. *A Semicentennial History of the American Mathematical Society 1888-1938*. New York: American Mathematical Society, 1938, xi + 262 pp.

Includes chapters on the history of the Society, its financial affairs, meetings, Gibbs Lectureship, library, Council and Board of Trustees, Chicago section, *Bulletin*, *Transactions*, and *Colloquium Lectures* and *Publications*. Over half of the volume is devoted to biographical and bibliographical notes on the Society's chief secretaries and past presidents.

2189. Ball, Walter W. Rouse. *A History of the Study of Mathematics at Cambridge*. Cambridge: Cambridge University Press, 1889, xvii + 264 pp.

Notes on eminent Cambridge mathematicians, in particular, the Newtonian and Analytical schools, from the 14th century to about 1859. Valuable for its detail on the structure of Cambridge studies and on the very important role of mathematics in those studies, especially through the mathematical tripos. See also item 981.

2190. Biermann, Kurt-R. *Die Mathematik und ihre Dozenten an der Berliner Universität, 1810-1920. Stationen auf dem Wege eines mathematischen Zentrums von Weltgeltung*. Berlin: Akademie-Verlag, 1973, viii + 265 pp.

Surveys the work of such Berlin mathematicians as Dirichlet, Steiner, Jacobi, Kummer, Weierstrass, Kronecker, and Frobenius. Includes much information on the German Academy of Sciences, lists of *dozenten*, dissertations, *habilitationen*, prize problems, samples of the curriculum, and many informative documents. See also item 1099.

2191. Crosland, Maurice. *The Society of Arcueil. A View of French Science at the Time of Napoleon I*. London: Heinemann, 1967, xx + 514 pp.

Explores the organization and patronage of science in early nineteenth-century France. Laplace, Biot, Arago, and Poisson were among the members of the society. Its leaders, Laplace and Berthollet, promoted a research program which stressed the mathematization of the experimental sciences.

2192. Enros, Philip C. "The Analytical Society: Mathematics at Cambridge University in the Early Nineteenth Century." Dissertation, University of Toronto, 1979. *Dissertation Abstracts International* A 40, no. 12, pt. 1 (June 1980), p. 6396-A.

A detailed study of a group of Cambridge undergraduates (1812-1813), including Babbage, Peacock, and Herschel, interested in promoting analytical mathematics. Establishes an intellectual and social framework for the Cambridge mathematical revival movement.

2193. Gericke, H. "Aus der Chronik der Deutschen Mathematiker-Vereinigung." *Jahresbericht der Deutschen Mathematiker-Vereinigung* 68 (1966), 46-74.

Relies upon previously unpublished documents to provide a thorough study of the Union. See also Dauben, item 2102, 159-165, and Gutzmer, item 2194.

2194. Gutzmer, A. "Geschichte der Deutschen Mathematiker-Vereinigung." *Jahresbericht der Deutschen Mathematiker-Vereinigung* 10 (1909), 1-49.

Gutzmer's report, made originally to the Third International Congress of Mathematicians in Heidelberg, 1903, was printed separately and in full by Teubner in 1904. For a more recent history of the German Mathematicians Union, see Gericke, item 2193, and Dauben, item 2102, especially pp. 159-165.

2195. May, Kenneth O., ed. *The Mathematical Association of America: Its First Fifty Years*. Washington, D.C.: The Mathematical Association of America, 1972, vii + 172 pp.

Six essays present the background and founding of the Association, the periods 1915-1940, World War II, 1946-1965, as well as the sections and financial history of the Association. Appendices list the constitution and by-laws, officers, committees, publications, films, Hedrick Lectures, national meetings, membership, awards, and competitions, as well as finances. See also items 1167, 2177.

2196. Rees, M.S. "Mathematics and the Government: The Post-War Years as Augury of the Future." *The Bicentennial Tribute to American Mathematics 1776-1976*. Edited by Dalton Tarwater. Washington, D.C.: The Mathematical Association of America, 1977, 101-116.

Sketches the efforts of the Office of Naval Research to support students of mathematics as well as basic research in mathematics during 1946-1950. The article views these efforts from the perspective of government support for research.

SOCIOLOGY OF MATHEMATICS

2197. Bloor, David. *Knowledge and Social Imagery*. London: Routledge and Kegan Paul, 1976, 156 pp.

Discussion of the "strong programme in the sociology of knowledge"; application to mathematics: J.S. Mill vs. Frege, numbers, "negotiation" of logic and proof, Azande logic. For a critique of the general approach see S. Woolgar, "Interests and Explanation in the Social Study of Science," *Social Studies of Science* 11 (1981), 365-394.

2198. ———. "Wittgenstein and Mannheim on the Sociology of Mathematics." *Studies in the History and Philosophy of Science* 4 (1973), 173-191.

Takes Wittgenstein's philosophy of mathematics to defeat mathematical realism as a major obstacle to a sociological analysis of mathematics.

2199. Bos, Hendrik J.M., and H. Mehrtens. "The Interactions of Mathematics and Society in History. Some Exploratory Remarks." *Historia Mathematica* 4 (1977), 7-30.

Survey; discussion of social forms of mathematical practice, bibliography. See also item 2171.

2200. Crane, Diana. *Invisible Colleges: Diffusion of Knowledge in Scientific Communities*. Chicago: The University of Chicago Press, 1972, 213 pp.

Structural and quantitative analysis of scientific communication in two specialities, one being the theory of finite groups; no concern for the mathematical contents of the theory.

2201. Fang, J., and K.P. Takayama. *Sociology of Mathematics. A Prolegomenon*. Happauge: Paideia Press, 1975, 364 pp.

Written by Fang, not a sociologist, to provide an interdisciplinary "meeting place" for studies on mathematics. A (very) preliminary survey of problems of method and theory in social aspects of the history and philosophy of mathematics, presented on the background of a personal philosophy of the "working mathematician." Special topics: Greek mathematics, critique of Fisher, items 2202, 2203, Japanese mathematics (Wasan), non-Euclidean geometry.

2202. Fisher, Charles S. "The Death of a Mathematical Theory." *Archive for History of Exact Sciences* 3 (1966), 137-159.

2203. ———. "The Last Invariant Theorists." *Archives Européennes de Sociologie* 8 (1967), 216-244.

Study of the fate of a mathematical theory, "theory" being taken as a social entity.

2204. ———. "Some Social Characteristics of Mathematicians and Their Work." *American Journal of Sociology* 78 (1972/1973), 1094-1118.

Case study of types of professional behavior in problem-solving along with the history of the Poincaré conjecture.

2205. Folta, Jaroslav. "Social Conditions and the Founding of Scientific Schools. An Attempt at an Analysis on the Example of the Czech Geometric School." *Acta historiae rerum naturalium necnon technicarum. Czechoslovak Studies in the History Science*, Special Issue 10. Prague: Czechoslovak Academy of Sciences, 1977, 81-179.

Starting from a general discussion of mathematical "schools," the development of Czech mathematics is shown to be conditioned by the social and economic conditions generated by the onset of the industrial revolution in this region.

2206. Forman, Paul. "Weimar Culture, Causality, and Quantum Theory, 1918-1927. Adaption by German Physicists and Mathematicians to a Hostile Intellectual Environment." *Historical Studies in the Physical Sciences* 3 (1971), 1-115.

Analysis of the reaction of scientists to "environmental pressure" from the antirationalist intellectual milieu of Weimar Germany as a source of receptivity for a new, acausal theory.

2207. Hessen, Boris. "The Social and Economic Roots of Newton's Principia." In *Science at the Cross Roads*. Edited by N.I. Bukharin et al. London: Kniga Ltd., 1931. New ed. London: F. Cass & Co., 1971, 149-212. Separate ed. New York, 1971.

The classical externalist interpretation of Newton's achievements from an orthodox Soviet-Marxist point of view, relating science directly to class struggle and the development of means and modes of production; not specific for mathematics.

2208. MacKenzie, Donald A. *Statistics in Britain 1865-1930. The Social Construction of Scientific Knowledge*. Edinburgh: Edinburgh University Press, 1981, 306 pp.

Detailed and well-based account on the rise of statistical theory in Britain (Galton, Pearson, Fisher), arguing for the social construction of scientific contents on the background of social interests connected with the contemporary eugenics movement.

2209. MacLeod, Roy. "Changing Perspectives in the Social History of Science." *Science, Technology and Society: A Cross-Disciplinary Perspective*. Edited by I. Spiegel-Rösing and D. de Solla-Price. London: Sage, 1977, 149-195.

General survey; bibliography.

2210. Mehrtens, Herbert, H. Bos, and I. Schneider, eds. *Social History of Nineteenth Century Mathematics*. Boston: Birkhäuser, 1981, 301 pp.

Thirteen papers of varying length, method, and aim from a workshop, Berlin 1979. Three parts: "aspects of a fundamental change--the early nineteenth century," "the professionalization of mathematics and its educational context," "individual achievements in social context," and an appendix, including a general paper and select bibliography.

2211. Needham, Joseph. "Mathematics and Science in China and the West." *Science and Society* 20 (1956), 320-343. Reprinted in *Sociology of Science; Selected Readings*. Edited by B. Barnes. Harmondsworth: Penguin Books, 1972, 21-44.

Comparative analysis of the sources of the "scientific revolution" of the Renaissance: "Apparently a mercantile culture alone was able to do what agrarian bureaucratic civilisation could not--bring to fusion point the formerly separated disciplines of mathematics and nature knowledge."

2212. Pedersen, Olaf. "The 'Philomaths' of 18th Century England." *Centaurus* 8 (1963), 238-262. Wallis, Peter J. "British Philomaths--Mid-Eighteenth Century and Earlier." *Centaurus* 17 (1972/1973), 301-314. Wallis, Peter and Ruth. "Female Philomaths." *Historia Mathematica* 7 (1980), 57-64.

Studies of a group of minor mathematicians and mathematical laymen important for the dissemination of mathematical knowledge. See also item 2383.

2213. Schneider, Ivo. "Der Einfluss der Praxis auf die Entwicklung der Mathematik vom 17. bis zum 19. Jahrhundert." *Zentralblatt für Didaktik der Mathematik* 9 (1977), 195-205.

Presents examples of the influence of social practice on the development of mathematical knowledge from Galilei to the *Ecole Polytechnique*. See also item 800.

2214. *Social Studies of Science*. 8, No. 1 (February 1978): *Special Theme Issue: Sociology of Mathematics*.

D. MacKenzie and B. Norton on Karl Pearson and statistics in intellectual and social context; B. Martin on Game Theory; N. Stern on age and achievement.

2215. Struik, Dirk J. "On the Sociology of Mathematics." *Science and Society* 6 (1942), 58-70.

"Classical" introduction to the interaction of mathematics and society in history.

2216. Taylor, Eva G.R. *The Mathematical Practitioners of Tudor and Stuart England, 1485-1714*. Cambridge: Cambridge University Press, 1954, xi + 443 pp., 12 pls.

See items 630, 801, 1809.

2217. ———. *The Mathematical Practitioners of Hanoverian England, 1714-1840*. Cambridge: Cambridge University Press, 1966, 503 pp.

History of a field of mathematical practice with extensive biographical information; little sociological analysis. See also items 995, 1810.

2218. Zilsel, Edgar. "The Sociological Roots of Science." *American Journal of Sociology* 47 (1942), 245-279. German translation, *Die sozialen Ursprünge der neuzeitlichen Wissenschaft*. Essays translated from the English and edited by W. Krohn. Frankfurt/M.: Suhrkamp, 1976, 49-65.

Argues for comparative and sociological analysis in the history of science, and finds the roots of the scientific revolution in the breakdown of the social barrier between intellectual and manual labor during the Renaissance. The German edition contains eight papers by Zilsel and a competent introduction by the editor on the sociological interpretation of modern science.

PHILOSOPHY OF MATHEMATICS

The scope of philosophy of mathematics, so far as it is covered in the present bibliography, comprises: criticism and clarification of basic mathematical concepts, assumptions, and methods of proof; discussions of mathematical objects, truth, certainty, and objectivity; views concerning mathematical infinity; views about what the nature of mathematics is or should be (e.g., logicism and intuitionism), leading to systematic reconstructions of mathematics. The literature cited here includes works on the ancient philosophy of mathematics as represented by Plato and Aristotle, conceptual and methodological issues involved in the development of the calculus and extensions of the number system, the pre-Cantorian paradoxes of the infinite, Kant's philosophy of mathematics, and the philosophy of geometry. The period most extensively covered is the one extending from the 1870s to the 1920s--a time of remarkable advances by Frege, Dedekind, Cantor, Russell, Hilbert, and Brouwer, as well as of vigorous controversies concerning such matters as the antinomies of set theory and the relation between mathematics and logic. Some references also deal with main positions in the philosophy of mathematics which were formulated somewhat more recently.

2219. *Dictionary of the History of Ideas. Studies of Selected Pivotal Ideas*. Edited by P.P. Wiener. New York: Charles Scribner's Sons, 1974. 4 vols. plus an index vol.

Articles by R. Blanché on "Axiomatization," by S. Bochner on "Continuity and Discontinuity in Nature and Knowledge" and on "Infinity," by R.L. Wilder on "Relativity of Standards of Mathematical Rigor." Bibliographies.

2220. *The Encyclopedia of Philosophy*. Edited by P. Edwards. New York: Macmillan, 1967. 8 vols.

High quality articles, often quite substantial and always with bibliographies, on: "Zeno of Elea" by G. Vlastos, "Frege" by M. Dummett, "Russell, Logic and Mathematics" by A.N. Prior, "Hilbert" by P. Bernays, "Brouwer" by C. Parsons, "Antinomies" by J. van Heijenoort, "Continuum Problem" by R. Smullyan, "Foundations of Mathematics" by C. Parsons, "Geometry" by S.F. Barker, "Gödel's Theorem" by J. van Heijenoort, "Infinity in Mathematics and Logic" by J. Thomson, "Types" by Y.A. Bar-Hillel. The long article "Logic, History of," edited by A.N. Prior, has parts which are relevant to the philosophy of mathematics.

2221. *The Journal of Symbolic Logic*.

Volume 1 (1936) contains a bibliography by A. Church which was "intended to be a complete bibliography of symbolic logic for the period 1666-1935 inclusive." Many items belonging to the philosophy of mathematics are included. Additions, corrections, and indices to this bibliography are given in volume 3, and it is extended past 1935 by listings in each volume. A large number of entries are reviewed, but beginning with vol. 41, no. 2 (June 1976) only books are reviewed.

2222. *The Philosopher's Index. An International Index to Philosophical Periodicals*. Bowling Green, Ohio: Philosophical Documentation Center, Bowling Green State University, 1969-.

Coverage is from 1967 on. Abstracts are included in the 1969 and later editions. This index has been supplemented by *The Philosopher's Index. A Retrospective Index to U.S. Publications from 1940* (Bowling Green, Ohio: Philosophical Documentation Center, Bowling Green State University, 1978. 3 vols.). "Includes original philosophy books published in the United States between 1940 and 1976, and articles published in philosophy journals in the United States between 1940 and 1966."

2223. Barker, S.F. "Logical Positivism and the Philosophy of Mathematics." *The Legacy of Logical Positivism. Studies in the Philosophy of Science*. Edited by P. Achinstein and S.F. Barker. Baltimore: The Johns Hopkins Press, 1969, 229-257.

Gives an explanation of some basic positivist views relating to philosophy of mathematics and then attempts to account for an asymmetry in the positivist treatments of arithmetic and geometry: the formulas of arithmetic, but not those of geometry, are granted "a standard prevailing interpretation ... under which they become analytic truths." See also A. Hausmann, "Non-Euclidean Geometry and Relative Consistency Proofs," *Motion and Time, Space and Matter. Interrelations in the History of Philosophy and Science*, edited by P.K. Machamer and R.G. Turnbull (Columbus: Ohio State University Press, 1976), 418-435.

2224. Becker, Oscar. *Grundlagen der Mathematik in geschichtlicher Entwicklung*. Freiburg: Alber, 1954.

With historical thoroughness, the author concentrates on Greek, seventeenth-century, nineteenth-century, and twentieth-century foundations of mathematics. In particular, he discusses the first from the Eleatics to Proclus, the second in terms of the calculus, the third as it concerns the foundations of geometry, and the fourth vis-à-vis logicism, intuitionism, and formalism up to Lorenzen's work. See also item 63.

2225. Benacerraf, P., and H. Putnam, eds. *Philosophy of Mathematics, Selected Readings*. Englewood Cliffs, N.J.: Prentice-Hall, 1964.

Contains selections from Frege, Russell, Poincaré, Brouwer, and Hilbert. Also included are the 1931 symposium on the foundations of mathematics in which Carnap, Heyting, and von Neumann participated; two papers by Gödel, "Russell's Mathematical Logic" and "What Is Cantor's Continuum Problem?," which are classic expressions of mathematical realism; Bernays's "On Platonism in Mathematics," which is very rich in content, insightfully discussing much more than is indicated by the title. (For further papers by Bernays, see his *Abhandlungen zur Philosophie der Mathematik* [Darmstadt: Wissenschaftliche Buchgesellschaft, 1976]). "Hilbert's Program" by G. Kreisel remarks briefly on the development of Hilbert's views and the opposition between his approach and that of Brouwer, and then proceeds to a reconstruction of Hilbert's program. There are some very influential papers by W.V. Quine on ontology, conventional truth, and the analytic-synthetic distinction. C.G. Hempel's classic exposition of logicism is also reprinted here. The large section devoted to selections from and critical commentary on Wittgenstein's *Remarks on the Foundations of Mathematics*, edited by G.H. von Wright et al., translated by G.E.M. Anscombe (Cambridge, Mass.: MIT Press, 1967), is out of all proportion to its influence in the philosophy of mathematics. Bibliography.

2226. Bernays, P. "Hilberts Untersuchungen über die Grundlagen der Arithmetik." In D. Hilbert, *Gesammelte Abhandlungen*. Vol. 3. Berlin: Springer, 1935, 196-216.

Describes the development of Hilbert's views on foundations, beginning with "Uber den Zahlbegriff" (1900); also discusses papers bearing on Hilbert's program by Ackermann, von Neumann, and Gödel.

2227. Beth, E.W. *The Foundations of Mathematics, A Study in the Philosophy of Science*. Amsterdam: North-Holland, 1959. Rev. ed., 1964.

Though the greater part of this large volume presents the fundamentals of many mathematical and logical theories of modern foundational studies, it begins with two chapters on ancient philosophy; for the author thinks "we must go back at least to Aristotle if we want to grasp the ultimate roots of the doctrinal divergences which have arisen from the results of modern research into the foundations of mathematics." The chapter on Aristotle's theory of science has sections on the theories of science of Nieuwentyt, a 17th-century critic of Leibniz's calculus, and of Kant. There are many brief historical and philosophical passages providing background for an exposition of technical results. Also included are chapters on logicism, Cantorism, intuitionism, and nominalism, as well as a very comprehensive chapter on the various mathematical paradoxes (17 are listed) and the most common methods of avoiding them.

2228. Black, M. *The Nature of Mathematics*. Paterson, N.J.: Littlefield, Adams & Co., 1933.

This is a philosopher's detailed critical analysis of logicism and, to a lesser extent, of formalism and intuitionism. The account of logicism includes not only Russell's and Whitehead's *Principia Mathematica*, but also the views of F. Ramsey, L. Chwistek, and L. Wittgenstein. The author emphasizes the need for mutual interaction among the three schools rather than simply within each of them.

2229. Bolzano, B. *Paradoxien des Unendlichen*. Edited by F. Prihonský. English translation with historical introduction by D. Steele: *Paradoxes of the Infinite*. London: Routledge and Kegan Paul, 1950.

In preparation for a discussion of the paradoxes or *apparent* contradictions, Bolzano tries to give precise explanations or definitions of such concepts as *set*, *sum*, *quantity*, *series*, *finite*, and *infinite*. An infinite set is defined as one which is not finite, but his definition of *finite* is not satisfactory. He proceeds to criticize definitions of infinite given by others, to refute those denying the possibility of anything infinite, and to offer a proof of the existence of infinite sets. The most interesting of the paradoxes discussed is the fact that an infinite set is reflexive. He resolves this by denying that one-one correspondence is the criterion of quantitative equality for infinite sets, but is then left with an unexplained concept, "equality in multiplicity." Bolzano also discusses the fundamentals of differential calculus and the concepts of a continuum and of dimension.

2230. Boyer, Carl B. *The Concepts of the Calculus, a Critical and Historical Discussion of the Derivative and the Integral*. New York: Columbia University Press, 1939. Reprinted as *The History of the Calculus and Its Conceptual Development*. New York: Dover, 1959.

An account is given of ancient and medieval thought concerning number, continuity, indivisibles and infinitesimals, the infinite, and motion; over half the book is devoted to the conceptual developments preceding Newton and Leibniz. There is also an account of the various criticisms of the unclear Newtonian and Leibnizian "foundations" by Nieuwentyt, Berkeley, and the French Academy of Sciences, as well as of 18th-century attempts to improve the conceptual foundations of the calculus. Both favorable aspects and deficiencies are indicated; there is attention to the influences of philosophical, geometrical, mechanical, and formalistic ideas. The last chapter contains a description of the aspects of the full rigorous formulation contributed by Bolzano, Cauchy, Weierstrass, Dedekind, and Cantor. See also F. Cajori, *A History of the Conceptions of Limits and Fluctions in Great Britain from Newton to Woodhouse* (items 927 and 1259 above)--nearly a whole volume, centering on the criticisms in Berkeley's *Analyst*, of quotations of explanations and criticisms relating to the concepts of fluctions and limits. See also items 777, 860.

2231. Brittan, G.G., Jr. *Kant's Theory of Science*. Princeton: Princeton University Press, 1978.

Chapter 2, "Kant's Philosophy of Mathematics," discusses the main recent interpretations or reconstructions of the meaning and justification of Kant's thesis that mathematical propositions are synthetic a priori truths; references are given to the relevant works of Beth, Hintikka, and Parsons, in which many further references are to be found. In chapter 3, "Geometry, Euclidean and Non-Euclidean," an attempt is made to separate Kant's arguments for the theses that Euclidean geometry is synthetic and that it is a priori. Views in opposition to the first thesis are evaluated.

2232. Brouwer, L.E.J. *Collected Works*. Vol. 1. Philosophy and Foundations of Mathematics. Edited by A. Heyting. Amsterdam: North-Holland, 1975.

Contains a translation of Brouwer's 1907 thesis, "On the Foundations of Mathematics"--from the historical point of view, the most interesting of his philosophical writings. Here he advances his view that all legitimate mathematics is constructed from a basic intuition: "a unity of continuity and discreteness, a possibility of thinking together several entities, connected by a 'between,' which is never exhausted by the insertion of new entities," "the intuition of many-oneness." It is maintained that mathematical construction is separate from language and logical reasoning. There are observations on the antinomies and criticisms of Dedekind's system of arithmetic, the logistic systems of Russell and Peano, Cantor's theory of the transfinite, and Hilbert's 1904 paper. The thesis does not reject the law of excluded middle; doubts are first expressed about it in a 1908 paper appearing here in an English translation, "The Unreliability of the Logical Principles." Numerous other papers on philosophy and intuitionistic mathematics are also included in this volume.

2233. Brunschvicg, L. *Les étapes de la philosophie mathêmatique.* Paris: Alcan, 1929.

A distinctively French philosophical analysis of the entire history of European mathematics from the Pythagoreans to the First World War. Although the author mentions logicism, he treats the modern period briefly. By contrast, he dwells on Descartes, Leibniz, Kant, and Comte.

2234. Bunn, R. "Quantitative Relations Between Infinite Sets." *Annals of Science* 34 (1977), 177-191.

Examines some medieval arguments concerning the once widespread view that one infinite cannot be greater than another. The different treatments by Galileo, Leibniz, and Bolzano of the paradox of the infinite, involving the reflexivity of infinite sets, are analyzed and the logical error common to their arguments is pointed out. The relation of Cantor's theory to this matter is also discussed. See also item 2097.

2235. Cantor, Georg. *Gesammelte Abhandlungen mathematischen und philosophischen Inhalts.* Edited by E. Zermelo. Berlin: J. Springer, 1932. Reprinted Hildesheim: Olms, 1962, and Berlin: Springer-Verlag, 1980.

Contains Cantor's attempts to defend his theory of the transfinite from traditional objections to the infinite, many of which go back to Aristotle, and his opposition to the restrictions on mathematical concepts and theories backed by his contemporary critic, Leopold Kronecker. Also included are letters to Dedekind in which are stated certain of the antinomies which Cantor was the first to discover, and his means of reformulating the foundations of set theory in light of these antinomies. See also I. Grattan-Guinness, "The Rediscovery of the Cantor-Dedekind Correspondence," *Jahresbericht der Deutschen Mathematiker-Vereinigung* 76 (1974), 104-139, and "The Correspondence Between Georg Cantor and Philip Jourdain," *Jaresbericht der Deutschen Mathematiker-Vereinigung* 73 (1971), 112-130. The

1980 reprint also contains a list of works by Cantor not included in the 1932 *Gesammelte Abhandlungen*, as well as publications containing letters and previously unpublished materials by Cantor.

2236. Cavaillès, J. *Philosophie mathématique*. Paris: Hermann, 1962.

This volume consists of three parts: a historical study of set theory and its philosophical presuppositions from Cantor to Zermelo; a French translation of the Cantor-Dedekind correspondence; and an essay on the researches of Gentzen and Gödel vis-à-vis consistency. See also the author's *Méthode axiomatique et formalisme* (Paris: Hermann, 1938), in three volumes. Here three essays are included: "Le problème du fondement des mathématiques," "Axiomatique et système formel," and "La non-contradiction de l'arithmétique." See also item 2098.

2237. Chihara, C.S. *Ontology and the Vicious-Circle Principle*. London, Ithaca, N.Y.: Cornell University Press, 1973.

Not intended as a history, though it deals with the views of Russell, Gödel, Quine, and Poincaré. The chapter, "Poincaré's Philosophy of Mathematics," is the most valuable as history. There the author stresses "the difficulties involved in reconstructing a coherent view of mathematics from the many paradoxical claims [Poincaré] made"; his treatment, however, is most sympathetic. The chapter on Russell is also of some use in the difficult project of trying to understand his thought on the solution of the antinomies, and related matters.

2238. Church, Alonzo. *Introduction to Mathematical Logic*. Princeton, N.J.: Princeton University Press, 1956, x + 378 pp.

In addition to historical notes (pp. 155-166, 288-294), this volume contains a lengthy quasi-historical introduction to the notions of proof, syntax, semantics, and the logistic method, by the mathematician who successfully established the study of mathematical logic in the United States. See also item 1649.

2239. Curry, H.B. *Outlines of a Formalist Philosophy of Mathematics*. Amsterdam: North-Holland, 1951.

Although not a history, this book attempts to present a philosophy of mathematics from the standpoint of a working mathematician. To do so, he treats the notion of "formal system" in depth. He argues for a formalist definition of mathematics as the science of formal systems.

2240. Dauben, Joseph Warren. *Georg Cantor, His Mathematics and Philosophy of the Infinite*. Cambridge, Mass.: Harvard University Press, 1979, 361 pp., Chapters 6, 10, 11.

Along with much else relating to Cantor's philosophy, an account is given of Cantor's answers to traditional criticisms of the actual infinite by philosophers and theologians; his view that though transfinite numbers have a sort of metaphysical reality, only their consistent and distinct conception is necessary for mathematical admissability; Frege's critique of Cantor's use of abstraction and his definitions of such concepts as *power* of a set and *finite* set; Cantor's discovery, use, and probable interpretation of the antinomies. Various post-Cantorian developments concerning the antinomies and axiomatics are also described; in particular, there is a detailed relation of the opposition by Borel and Lebesgue to Zermelo's proof of the well-ordering theorem and its defense by Hadamard. There are copious references to Cantor's correspondence as well as to his publications, and a large bibliography. See also items 1097, 2102.

2241. Fraenkel, A. *Einleitung in die Mengenlehre*. 3rd ed. Berlin: J. Springer, 1928. Reprinted New York: Dover, 1946.

Not just a text on set theory; a work of considerable historical interest, long a principal reference on set theory and foundations of mathematics. Attention is given to conceptual matters throughout, and the last two chapters contain extensive discussions of the antinomies, intuitionism, impredicative definitions, logicism, the axioms of set theory and the axiomatic method in general, and Hilbert's program. A scholarly work giving a large number of references. Substantial bibliography.

2242. ———. *Abstract Set Theory*. Amsterdam: North-Holland, 1953.

Has a bibliography section of 137 pages containing many references belonging to the philosophy of mathematics. More recent editions contain much shorter bibliographies.

2243. ———, Y.A. Bar-Hillel, and A. Levy, with D. van Dalen. *The Foundations of Set Theory*. 2nd rev. ed. Amsterdam: North-Holland, 1973.

Not on the whole a historical treatment of the foundations of set theory, but contains a number of valuable historical sections and references. In particular, there are historical introductions on the antinomies, on intuitionism, and on Hilbert's program. There is also a section on attitudes taken to the axiom of choice and a section on the logistic thesis. The last section of the book is a survey of some of the main views on the question of the reality of mathematical objects as these were formulated in the 1940s and 1950s. The 1958 edition contains a more extensive bibliography of older works and a large bibliography on the antinomies. See also item 2105.

2244. Frege, Gottlob. *Grundgesetze der Arithmetik, begriffsschriftlich abgeleitet*. Vol. 1, 1893; vol. 2, 1903. Reprinted Hildesheim: Olms, 1962. Vol. 1 is partially translated by Montgomery Furth under the title *The Basic Laws of Arithmetic*. Berkeley: University of California Press, 1964. Selections from vol. 2 are translated by P. Geach and M. Black in *Translations from the Philosophical Writings of Gottlob Frege*. 2nd ed. Oxford: Blackwell, 1960.

This is Frege's main work in logic and foundations of mathematics. The introduction to volume 1 contains an exposition of his objective point of view mainly by contrasting it with that of the idealistic logicians of his time. It seems safe to say that nobody so vigorously opposed any intrusion of psychology into mathematics as Frege. The amount of critical examination of others is a prominent feature of Frege's writings. About one-third (sections 55-164) of the second volume is devoted to a discussion of matters relating to the foundations of the theory of real numbers; most of this is remarkably detailed criticism of the methods of Cantor, Dedekind, Weierstrass, Hankel, and the formalists of the time represented mainly by Heine and Thomae. (Sections not translated by Geach and Black are: 55, 68-85, on Cantor; 138, 145, 148-164, on Weierstrass and on Frege's way of conceiving the real numbers.) Frege's criticism of formalism had the approval of Hilbert and is apparently still topical: in a paper commenting on "popular discussions of foundations," G. Kreisel says it is "quite wrong not to use the penetrating and detailed analysis by the pioneers of modern logic [such as Frege]."

2245. ———. *Grundlagen der Arithmetik: Eine logisch mathematische Untersuchung über den Begriff der Zahl*. Breslau: Koebner, 1884. English translation by J.L. Austin. *The Foundations of Arithmetic: A Logico-Mathematical Enquiry into the Concept of Number*. 2nd ed. Oxford: Blackwell, 1953. Reprinted Evanston: Northwestern University Press, 1968.

This great classic critically examines previous views on the basic concepts of arithmetic and addresses the question whether the propositions of arithmetic are synthetic a priori as Kant maintained, or analytic in the sense of being derivable by logical principles from definitions. Frege tried to show the latter by defining the concept "finite number" and deriving the principle of mathematical induction, "which is ordinarily held to be peculiar to mathematics, but is really based on the universal principles of logic." His definitions were formulated as they were, not because they were most "natural," but because they could serve as the foundation of proofs. While nobody was more insistent on the necessity of proving existence and uniqueness theorems justifying definitions than Frege, these justifying demonstrations have to come from logic if the theorems derived from the definitions are to be analytic. But principles from which existential propositions follow are prime candidates for the designation "peculiar to mathematics" or even "synthetic," rather than "universal principles of inference." See Charles Parsons, "Frege's Theory of Number" in *Philosophy in America*, edited by M. Black (Ithaca: Cornell University Press, 1965), 180-203.

2246. ———. *Philosophical and Mathematical Correspondence*. Chicago: University of Chicago Press, 1980.

A translation of the German edition (item 1604), this book contains correspondence on the foundations of mathematics (especially geometry and logic) by Frege and Hilbert, Husserl, Peano, Russell, and G. Vailati, among others. See also item 1603--Frege's *Nachgelassene Schriften und wissenschaftlicher Briefwechsel*, edited by H.H.F. Kambartel and F. Kaulbach (Hamburg: Meiner, 1969-1976), in 2 vols.: posthumous writings appear in vol. 1; correspondence in vol. 2.

2247. Gauthier, Y. *Fondements des mathématiques: Introduction à une philosophie constructiviste*. Montreal: Université de Montréal, 1976.

The author--a philosopher well versed in recent technical results from the foundations of mathematics--analyzes such results à propos of set theory, intuitionism, and category theory. In conclusion, he uses this analysis to discuss what mathematical logic has contributed to the philosophy of mathematics.

2248. Grattan-Guinness, Ivor. *Dear Russell-Dear Jourdain: A Commentary on Russell's Logic, Based on His Correspondence with Philip Jourdain*. New York: Columbia University Press; London: Duckworth, 1977, vi + 234 pp.

Russell's correspondence with Jourdain reveals much regarding his struggle from 1904 to 1907 to find a solution to the antinomies and formulate a satisfactory system of basic concepts and assumptions for the logical development of mathematics. There is material on Russell's thought about the existence or, better, the nonexistence of classes and about the axioms of choice and infinity. The author provides background information and some criticism, as well as a translation of Russell's 1911 "Sur les axiomes de l'infini et du transfini." Large bibliography. See also items 1651, 2110, 2248.

2249. Hankins, Thomas L. "Algebra as Pure Time: William Rowan Hamilton and the Foundations of Algebra." *Motion and Time. Space and Matter. Interrelations in the History of Philosophy and Science*. Edited by P.K. Machamer and R.G. Turnbull. Columbus: Ohio State University Press, 1976, 327-359.

Using correspondence and manuscripts, the author gives an account of the various influences on and the development of Hamilton's thinking on foundations and its relation to Kant. For a reconstruction of Hamilton's efforts, an analysis of defects, and a discussion of the lack of influence of his ideas, see J. Mathews, "William Rowan Hamilton's Paper of 1837 on the Arithmetization of Analysis," items 1015 and 1330 above. See also item 1040.

2250. Heath, T. *Mathematics in Aristotle*. Oxford: Oxford University Press, 1949.

A translation, with commentary, of those parts of the various works of Aristotle which in any way concern mathematics. See also H.G. Apostle, *Aristotle's Philosophy of Mathematics* (Chicago: The University of Chicago Press, 1952); and J. Hintikka, "Aristotelian Infinity," *The Philosophical Review* 75 (1966), 197-218. See also item 217.

2251. Heyting, A. *Intuitionism: An Introduction*. 2nd rev. ed. Amsterdam: North-Holland, 1966.

Written by Brouwer's chief disciple, this introduction to intuitionism contains a dialogue contrasting intuitionist and other views on the philosophy of mathematics, as well as a detailed exposition of intuitionistic concepts applied to algebra, geometry, analysis, and mathematical logic.

2252. Hilbert, David. "Neubegründung der Mathematik." *Abhandlungen aus dem mathematischen Seminar der Hamburgischen Universität* 1 (1922), 157-177. Reprinted in *Gesammelte Abhandlungen*. Vol. 3. Berlin: Springer, 1935, 157-177, item 1116 above.

Hilbert's first presentation of his thoughts on foundations appeared in 1904, but it was not until 1922 that a much improved formulation of his ideas was published. The principle stimulus for returning to foundations in 1922 would seem to have been the constructivistic writings of Weyl, especially the 1921 paper "Über die neue Grundlagenkrise ..." (see annotation on Weyl, item 2274 below). According to Hilbert, Weyl did not detect a vicious circle in the usual methods of concept formation in analysis; rather, he showed that a circle would occur if the usual methods were employed from a constructivistic basis like Weyl's (see also Hilbert and Ackermann, *Grundzuge der theoretischen Logik* [Berlin: Springer, 1928], chapter 4, sections 5 and, especially, 9--this material does not occur in later editions). "The vicious circle is artificially brought into analysis by Weyl." Hilbert vigorously emphasized that analysis was not uncertain and its foundations were not unstable; Weyl's representations did not correspond to the real facts. Moreover, it was Hilbert's opinion that the constructive tendency had been misdirected by Brouwer and Weyl. A constructive mathematics should not replace the classical; the constructive tendency should be applied to attain incontestible consistency proofs for the axiomatized classical systems, which would restore the reputation for absolute certainty enjoyed by mathematics before the antinomies and put an end to controversy once and for all. Hilbert's 1925 paper, "On the Infinite," included in the volumes edited by Benacerraf and Putnam, item 2225, and by van Heijenoort, items 1619, 2272, is also of considerable interest for the philosophy of mathematics.

2253. Hintikka, J. *The Philosophy of Mathematics*. Oxford: Oxford University Press, 1969.

In this selection of eleven essays, the author argues that a true understanding of the philosophy of mathematics can only be based on an adequate understanding of mathematical logic. The essays themselves are by L. Henkin (completeness of first-order and higher-order logics), R. Smullyan (self-reference), G. Kreisel (informal rigor), S. Feferman (predicative analysis), K. Gödel (intuitionistic logic), H. Rogers (computability), A. Robinson (metaphysics of the calculus), and A. Tarski (first-order geometry).

2254. Lakatos, I., ed. *Problems in the Philosophy of Mathematics*. Amsterdam: North-Holland, 1972.

This collection of essays, the proceedings of an international colloquium, offers a diversity of approaches to the history of the philosophy of mathematics. A. Szabo discusses the Eleatics vis-à-vis the rise of Greek axiomatics, while A. Robinson considers the history of the calculus from the standpoint of non-standard analysis. F. Somers analyzes Frege's use of quantification, S. Körner ponders the effect on philosophy of twentieth-century mathematical logic, and G. Kreisel argues for the value of informal rigor. Last, P. Bernays and A. Mostowski discuss how P. Cohen's results have affected the foundations of mathematics. Each essay has several commentaries, together with the essayist's reply.

2255. ———. *Proofs and Refutations: The Logic of Mathematical Discovery*. Cambridge: Cambridge University Press, 1976.

Influenced by G. Polya and K. Popper, this book argues for a dialectical approach to the history of proofs in mathematics. Two case studies provide evidence for the thesis that mathematics grows by proof, counterexample, then revised proof.

2256. Lewis, C.I. *A Survey of Symbolic Logic*. Berkeley: University of California Press, 1918.

This book is at once a history of symbolic logic--discussing the Boole-Schröder algebra of logic on the one hand and the Peano-Russell logistic on the other--and a summary of the author's results on modal logic. It concludes with the most comprehensive bibliography of research in symbolic logic up to 1918. See also item 1668.

2257. Mooij, J.J.A. *La philosophie des mathématiques de Henri Poincaré*. Paris: Gauthier-Villars, 1966.

A comprehensive historical treatment of the background, context, and content of Poincaré's philosophy of mathematics, including his philosophy of geometry. In particular, the relation of Poincaré's philosophy to Kant's is discussed and the views of the logicists are expounded; a detailed coverage of the controversy between Russell and Poincaré is given, and there are chapters on Poincaré's criticisms of Hilbert and of Zermelo's set theory. Large bibliography.

2258. Moore, Gregory H. "The Origins of Zermelo's Axiomatization of Set Theory." *Journal of Philosophical Logic* 7 (1978), 307-329.

Describes the controversy provoked by E. Zermelo's proof of the well-ordering theorem in 1904. Investigates the factors motivating Zermelo to axiomatize set theory in 1907. The chief factor was his desire to secure his proof of 1904 against its numerous critics and, in particular, to preserve his axiom of choice.

2259. Mostowski, Andrzej. *Thirty Years of Foundational Studies*. New York: Barnes & Noble, 1966.

See items 1151, 1663.

2260. Nagel, Ernest. "'Impossible Numbers': A Chapter in the History of Modern Logic." *Studies in the History of Ideas*. Vol. 3. New York: Columbia University Press, 1935, 429-474. Reprinted in E. Nagel. *Teleology Revisited and Other Essays in the Philosophy and History of Science*. New York: Columbia University Press, 1979, 166-194.

Describes attempts to interpret various extended number systems in the transition in which "mathematics has grown from having been the science of quantity to becoming the science which explores the most abstract properties of any subject matter whatsoever." Ideas of Euler, Wallis, Playfair, Peacock, D. Gregory, De Morgan, Hamilton, and Boole are discussed. See also items 1043, 1214.

2261. ———. "The Formation of Modern Conceptions of Formal Logic in the Development of Geometry." *Osiris* 7 (1939), 142-224. Reprinted in *Teleology Revisited*, item 2260, 195-259.

Discusses philosophically interesting mathematical developments in geometry during the 19th century. Nagel briefly examines the reasons which emerged for rejecting the component of the traditional conception of geometry according to which it is "an inherently quantitative science." He then proceeds to a more thorough presentation of matters bearing on the thesis that geometry is "the science of extension or space." Also considered are "the consequent reorientations provoked by this material to the traditional claim that geometry is an a priori science." Among the ideas discussed are those of Gergonne, Poncelet, Grassmann, von Staudt, Chasles, Plücker, Helmholtz, Pasch, Klein, Hilbert, and Poincaré. Numerous references.

2262. Poincaré, Henri. *The Foundations of Science*. Translated by G.B. Halsted. Lancaster, Pa.: The Science Press, 1913.

The original version (1893) of the chapter, "The Nature of Mathematical Reasoning," written well before the controversy with the logicists and Cantorists, apparently in ignorance of the work of Dedekind and Frege on the foundations of arithmetic, presents the view that the natural numbers are indefinable and the principle of induction is synthetic a priori. The question of the nature of mathematics is pursued in terms of the opposing views of Leibniz and Kant, with Poincaré defending the position of Kant. Later he would combat the new Leibnizians, the logicists, and continue to espouse a Kantian view. Thus, in "Mathematics and Logic," the original version of which appeared in 1905, the question raised is "whether ... once the principles of logic are admitted we can ... demonstrate all mathematical truths without making a fresh appeal to intuition"; his answer is no. This drew replies from Couturat and Russell. An account of the circumstances of the controversy is given by P. Jourdain in *The Monist* 23 (1912), 481-483, and a translation of Couturat's paper "For Logistic" appears there, pp. 483-523. The paper by Russell is included in *Essays in Analysis* (see annotation for item 2267 below). Poincaré was also critical of Hilbert for apparently having to use complete induction to establish the consistency of a system containing an axiom of complete induction. There is a reply by Hilbert in his "Neubegründung der Mathematik" (1922), see item 2252 above; see also the annotations to the van Heijenoort volume, items 1619 and 2272. Though Poincaré was a forerunner of intuitionism, he held that in mathematics *existence* means "exemption from contradiction." For this, he was reprimanded by Brouwer in 1907; see item 2232.

2263. ———. *Dernières pensées*. Paris: E. Flammarion, 1913. English translation by J. Bolduc, *Mathematics and Science, Last Essays*. New York: Dover, 1963.

Contains Poincaré's later essays "The Logic of Infinity" and "Mathematics and Logic." The former is critical of the 1908 systems of Russell and of Zermelo; his objection to Russell is that he seems to employ the theory of numbers in describing the system of logical types and that the axiom of reducibility is unclear and lacks justification, and to Zermelo, that his axioms are arbitrary. In "Mathematics and Logic," Poincaré discusses the views he calls "pragmatism" (his own) and "Cantorism." Pragmatists consider only objects which are finitely definable and recognize as meaningful theorems only those which are "verifiable."

2264. ———. *Science and Method*. New York: Dover, 1952.

This collection of provocative essays includes his three polemics against the logics of Peano, Russell, and Hilbert, and well as his seminal essay on the psychology of mathematical creation.

2265. Resnik, M.D. "The Frege-Hilbert Controversy." *Philosophy and Phenomenological Research* 34 (1974), 386-403.

The controversy concerned the nature of the axiomatic method, and, in particular, its application in foundations of geometry. One main issue was the sense (if any) in which axiom systems can constitute definitions. The author provides a clear presentation of the issues and of the strengths and weaknesses of Frege's criticisms of Hilbert. The sources involved--the correspondence, the articles on foundations of geometry by Frege, as well as an article in defense of Hilbert by A. Korselt--are translated by E.W. Kluge in *Gottlob Frege on the Foundations of Geometry and Formal Theories of Arithmetic* (New Haven: Yale University Press, 1971).

2266. ———. *Frege and the Philosophy of Mathematics*. Ithaca: Cornell University Press, 1980.

This full-length study of Frege includes but supersedes Resnik's essay on the Frege-Hilbert controversy (see item 2265). Here the author also discusses Frege's criticisms of psychologism, formalism, and empiricism, and analyzes Frege's larger philosophy of mathematics.

2267. Russell, Bertrand. *Essays in Analysis*. Edited by D. Lackey. London: George Allen & Unwin, 1973.

Contains fifteen of Russell's papers from the period 1904-1913, dealing with his views on classes, the antinomies and various ways of avoiding or solving them, the axiom of choice, the axiom of infinity, and other matters relating to the philosophy of mathematics. Some previously unpublished manuscripts are also included. Russell's articles of 1906 and 1910 replying to Poincaré appear here in English versions; the latter is almost identical to the second chapter of the introduction to *Principia Mathematica*. The article, "The Axiom of Infinity" (1904), was written before Russell had fixed on a method of avoiding the antinomies which required a special hypothesis of infinity; it maintains that the existence of an infinite class is provable. At the time Russell favored some form of the "zig-zag" theory, which is described as a possibility of avoiding the antinomies, in "Some Difficulties in the Theory of Transfinite Numbers and Order Types" (1906), also reprinted in this collection. Historical introductions to main sections of the book are provided by the editor. An appendix contains a bibliography of all Russell's published and unpublished writings on logic. *The Collected Papers of Bertrand Russell*, presently being prepared by the Russell Editorial Project at McMaster University, will eventually supersede this volume.

2268. ———. *The Principles of Mathematics*. 1st ed. Cambridge: Cambridge University Press, 1903. 2nd ed. London: Allen, 1937.

Contains Russell's most extensive philosophical presentation of logicism, discussions of the work of many of his predecessors, an explanation of his famous antinomy, attempts to deal with the antinomies, including an appendix on the theory of types, his analysis of the various number concepts, an exposition of his logic of relations, and much else. It is a work in which theories are expounded to a certain point, dropped, then, later, something else is tried; it is not an exposition of finished thoughts, but is exploratory; experimenting with ideas, Russell is struggling with problems new and difficult, and *Principles* is the record of his thoughts up to late 1902. Russell thought he was laying mathematical foundations for mathematics in place of the nonsense of idealistic philosophers and refuting the Kantian philosophy of mathematics. "Philosophy asks of Mathematics: What does it mean? Mathematics in the past was unable to answer, and Philosophy answered by introducing the totally irrelevant notion of mind. But now Mathematics is able to answer, so far at least as to reduce the whole of its propositions to certain fundamental notions of logic." The view that "mathematical reasoning is not strictly formal, but always uses intuitions" was the Kantian doctrine which Russell believed "capable of a final and irrevocable refutation." He did not, at the time of *Principles*, oppose the doctrine that mathematics is in some sense synthetic.

2269. Skolem, Thoralf. *Selected Works in Logic*. Edited by Jens Erik Fenstad. Oslo: Universitetforlaget, 1970, 732 pp.

This volume contains, in addition to papers on logic and the philosophy of mathematics, a biography of Skolem by J. Fenstad and a detailed historical study by H. Wang.

2270. Tarski, Alfred. *Logic, Semantics, Metamathematics: Papers from 1923 to 1938*. Edited by J.H. Woodger. Oxford: Clarendon Press, 1956, xiv + 472 pp. Rev. ed. by J. Corcoran. Indianapolis: Hackett, 1980.

Not a history, this work contains the translation into English of many of the most significant articles by one of this century's best logicians. Particularly useful is the translation of "The Concept of Truth in Formalized Languages."

2271. Torretti, R. *Philosophy of Geometry from Riemann to Poincaré*. Dordrecht: Reidel, 1978.

The four chapters of this substantial volume deal with "background information about the history of science and philosophy ... the development of non-Euclidean geometries until ... Klein's papers 'On the So-called Non-Euclidean Geometry' in 1871-73 ... 19th-century research into the foundations of geometry ... philosophical views about the nature of geometrical knowledge from John Stuart Mill to Henri Poincaré." The background information given includes such matters as Aristotle's theory of science and the nature of the Euclidean assumptions, which the author doubts were considered by Euclid to be self-evident truths. In addition to the account of philosophical thought, there is a good deal of mathematical exposition. Numerous notes and a large bibliography. See also item 1589.

2272. Van Heijenoort, Jean. *From Frege to Gödel. A Source Book in Mathematical Logic, 1879-1931*. Cambridge, Mass.: Harvard University Press, 1967, xi + 660 pp.

Many of the papers collected here are of some relevance to philosophy of mathematics. Among these are Zermelo's 1908 "A New Proof ...," in which he defends his 1904 proof of the well-ordering theorem against criticisms of, among other things, its use of the axiom of choice and impredicative definition; Russell's 1908 paper on type theory; Hilbert's "On the Infinite" (1925); Skolem's 1922 "Some Remarks ...," in which he asserts that "axiomatizing set theory leads to a relativity of set-theoretic notions, and this relativity is inseparably bound up with every thoroughgoing axiomatization"; Weyl's 1927 "Comments...." Commenting on a paper by Hilbert, Weyl says "a few words in defense of intuitionism," which include a vindication of Poincaré's view on mathematical induction. The editor's introduction to this selection provides an excellent historical and critical treatment of the issues involved in this controversial subject. See also item 1620.

2273. Wedberg, A. *Plato's Philosophy of Mathematics*. Stockholm: Almqvist & Wiksell, 1955.

Reconstructs Plato's philosophy of geometry and arithmetic in a way which is in agreement with Aristotle's description of Plato's doctrine. Wedberg compares and contrasts Plato's views with others common in his time as well as those common in recent times. In the case of arithmetic, from before Plato's time and long after, numbers were usually said to be pluralities of units; this is the definition used by Euclid and repeated for centuries. In postulating numbers as ideas, "Plato took, within the framework of his idealistic system, essentially the same step which ... Frege took ... when he rejected the still current definitions of positive integers as sets of units...."

2274. Weyl, H. *Das Kontinuum. Kritische Untersuchungen über die Grundlagen der Analysis*. Leipzig: Veit, 1918.

Weyl's initial work in foundations (1910) aimed at making Zermelo's axiomatization of set theory more precise, but unsatisfactory attempts to eliminate the concept *finite number* from his explication of Zermelo's concept of *definite property* combined with his philosophical reflections led him to the conviction that Poincaré was right in holding iteration to be the ultimate foundation of mathematics. He also became convinced that concept formation in classical analysis was infected with a sort of vicious circularity. The "semi-intuitionistic" system of *Das Kontinuum* avoids this circularity by accepting the natural numbers as an infinite totality on the basis of which further sets and relations are predicatively constructed. A number of papers on philosophy of mathematics are contained in his *Gesammelte Abhandlungen*, edited by K. Chandrasekharan (Berlin: Springer, 1968, 4 vols.). Of particular note are "Der circulus vitious in der heutigen Begründung der Analysis" (1919), in which he tries to explain in the most direct way the nature of the vicious circle in analysis claimed in *Das Kontinuum*; "Über die neue Grundlagenkrise der Mathematik" (1921), which begins with an account of his 1918 system but proceeds to a lengthy explanation of Brouwer's intuitionism; and "Die heutige Erkenntnislage in der Mathematik" (1925), which contains a section on Hilbert. See also "Mathematics and Logic, A Brief Survey serving as Preface to a Review of *The Philosophy of Bertrand Russell*," *American Mathematical Monthly* 53 (1946), 2-13; the review appeared in the same volume, pp. 208-214.

REGIONAL STUDIES

AFRICAN MATHEMATICS

This bibliography deals with the "sociomathematics" of Africa, mainly in the region south of the Sahara. Emphasis is given to applications of mathematics in the lives of the hundreds of diverse African ethnic groups, and, conversely, the influence of African institutions upon the evolution of their mathematics. Topics include numeration systems and accompanying gesture counting, mystical beliefs involving numbers, monetary systems, time reckoning, weights and measures, record keeping, geometry in art and architecture, and games of strategy and of chance.

Source Materials

2275. *Human Relations Area Files*.

Collection of basic source materials of about 300 mainly non-Western cultures. Reproductions of pages from published books, articles, and reports, filed by culture code and then by appropriate subject codes, such as numbers and measures, numeration, games, weights and measures.

2276. *Peabody Museum, Harvard University.*

Published catalogs of books and articles in anthropology, by area and subject.

General Surveys

2277. Raum, O.F. *Arithmetic in Africa.* London: Evans, 1938.

Raised in Tanganyika, Raum used both his own experiences and secondary sources to analyze numeration systems of many African peoples, and to describe applications of arithmetic in the marketplace, in games, in home construction, and other aspects of daily life.

2278. Zaslavsky, Claudia. "Black African Traditional Mathematics." *Mathematics Teacher* 63 (1970), 345-356.

Brief survey of number systems, applications of numbers, and games in several different cultures.

2279. ———. "Mathematics of the Yoruba People and of Their Neighbors in Southern Nigeria." *The Two-Year College Mathematics Journal* 1 (1970), 76-99.

Yoruba numeration is based upon groupings by twenties. Discusses symbolism of the numbers four and two hundred; the four-day market week, and cowrie shell currency groupings.

2280. ———. *Africa Counts: Number and Pattern in African Culture.* Boston: Prindle, Weber & Schmidt, 1973; paperback, Westport, Conn.: Lawrence Hill, 1979, 328 pp.

Most complete on the subject to date. Examines the mathematical contributions of many African peoples living south of the Sahara, in the context of their social and economic development. Using both primary and secondary sources, the author discusses numeration systems, mystical attributes of numbers, applications to time reckoning, currency, and measures, the geometry of African art and architecture, and mathematical games. Includes section by D.W. Crowe on geometric symmetries in African art and artifacts. In-depth studies of southwestern Nigeria and East Africa follow. Photographs, diagrams, tables, maps; 200 references.

Architecture

2281. Denyer, Susan. *African Traditional Architecture.* New York: Africana Publishing Co., 1978.

General survey of architecture and architectural decoration, and their relation to ecological and historic factors. Over 300 illustrations; detailed line drawings of the numerous styles and forms; maps; bibliography.

2282. Garlake, P.S. *Great Zimbabwe.* London: Thames & Hudson, 1973.

Research into the history of the ruins of the "great stone house," the legendary palace and fortified complex of buildings constructed centuries ago in the country now called Zimbabwe. Europeans had attributed this feat of construction to King Solomon or other non-Africans. Discussion of shapes and dimensions of the many buildings and walls.

2283. Oliver, Paul, ed. *Shelter in Africa*. New York: Praeger, 1971.

Detailed articles on the use of space and shape in construction, as well as geometric patterns in house decoration.

2284. Prussin, Labelle. *Architecture in Northern Ghana*. Berkeley and Los Angeles: University of California Press, 1969.

Prussin has done extensive fieldwork in Africa. She stresses the influence of historical and cultural factors in the use of space.

2285. ———. "Sudanese Architecture and the Manding." *African Arts* 3 (4) (1970), 13-18, 64-67.

Islam introduced new building technology and the square shape to a region that had known only round houses.

2286. ———. "Fulani-Hausa Architecture." *African Arts* 10 (1) (1976), 8-19, 97 (notes).

Effect of the 1804 Islamic revolution on the use of space in homes and public buildings in the region now known as northern Nigeria and southern Niger.

Art

2287. *African Arts*. Periodical. University of California at Los Angeles, since 1967.

See articles on architecture, patterns in art, and related topics.

2288. Crowe, D.W. "The Geometry of African Art I. Bakuba Art." *Journal of Geometry* 1 (1971), 169-182.

2289. ———. "The Geometry of African Art II. A Catalog of Benin Patterns." *Historia Mathematica* 2 (1975), 253-271.

Group theoretical analysis of repeated patterns, lavishly illustrated by examples from the Bakuba (Zaire) and Benin City (Nigeria).

Games

Mancala is the generic Arabic name of the game played throughout Africa, based entirely on mathematical principles.

2290. Avedon, Elliott M., and B. Sutton-Smith. *The Study of Games*. New York: John Wiley, 1971.

Includes Stewart Culin, "Mancala, the National Game of Africa," originally published by U.S. Government Printing Office in 1894, and other references to this universal African game, as well as to other games of chance and of strategy. African bibliography on pp. 132-135 and 255.

2291. Béart, Charles. *Jeux et jouets de l'Ouest Africain*. Dakar: IFAN, 1955. 2 vols.

Magic squares, mathematical puzzles and riddles, mancala, games of chance, etc.

2292. Bell, R.C. *Board and Table Games from Many Civilizations*. London: Oxford University Press, 1960. 2 vols. Vol. I in paperback.

Versions of mancala and other African games, their origin and history. See also Bell's article on the game of mangola, "Mangola and Mancala Boards," in *Games and Puzzles* 46 (1976).

2293. Centner, T. *L'enfant africain et ses jeux*. Elizabethville, Belgian Congo; Lubumbashi, Zaire: Centre d'Etude des Problèmes Sociaux Indigènes. Collection mémoires 17, 1963.

Networks, counting games, mancala, etc.

2294. Murray, H.J.R. *A History of Board Games Other than Chess*. Oxford: Clarendon Press, 1952.

Most complete collection available. Origins, history, and strategy of mancala and other games going back to ancient Egypt. Hundreds of references, good organization, illustrations, tables.

2295. Nsimbi, M.B. *Omweso, a Game People Play in Uganda*. (Occasional Paper 6.) African Studies Center, University of California at Los Angeles, 1968, 48 pp.

Another version of the universal African game of mancala; history, customs, rules of play, strategy.

2296. Pankhurst, Richard. "Gabata and Related Board-Games of Ethiopia and the Horn of Africa." *Ethiopia Observer* 14 (1971), 154-206.

Well-documented description and history of many versions of mancala.

2297. Russ, Laurence. *Mancala Games*. Algonac, Mich.: Reference Publications, 1984.

Most comprehensive work on the many versions in Africa, as well as in other regions of the world. Illustrations, bibliography.

Markets and Currency

2298. Einzig, Paul. *Primitive Money*. 2nd ed. Oxford: Pergamon Press, 1966.

Part 3 is devoted to Africa. Many references. Deals with such currency items as gold dust in Ghana, cattle in eastern and southern Africa, beads and cloth in various regions.

2299. Johnson, Marion. "The Cowrie Currencies of West Africa." 2 parts. *Journal of African History* 11 (1970), 17-49; 331-353.

Cowrie currency arithmetic, value of cowrie shells from the 14th century to the present, and depreciation of this currency as a factor in the expansion of African numeration systems. Many references.

2300. Kirk-Greene, A.H.M. "The Major Currencies in Nigerian History." *Journal of the Historical Society of Nigeria* 2 (1960), 132-150.

Currencies used in the last five centuries included copper bracelets and bars, beads, cloth, salt, iron bars, and, most popular of all, cowrie shells. Only guns and alcohol maintained stable exchange rates in trade between Europeans and Africans in the 19th century.

2301. Pankhurst, Richard. "'Primitive Money' in Ethiopia." *Journal de la Société des Africanistes* 32 (1962), 213-247.

The use as currency of bars of salt, pieces of cloth, and bars of iron, common in most parts of Africa, is unusually well documented in Ethiopia by historical records covering a period of many centuries. Measures of value used before standardized state currencies.

Numbers and Numeration Systems

2302. Armstrong, Robert G. *Yoruba Numerals*. Ibadan: Oxford University Press, 1962, 36 pp.

History and explanation of Yoruba numeration system, based on grouping by twenties. The operation of subtraction predominates in constructing the higher numerals.

2303. Delafosse, Maurice. "La numération chez les Nègres." *Africa* 1 (1928), 387-390.

Construction of numeration systems in many languages.

2304. Dieterlen, Germaine. *Essai sur la religion Bambara*. Paris: Presses Universitaires de France, 1951.

Includes discussion of the religious symbolism of numbers among the Bambara of Mali, based on fieldwork.

2305. Ganay, Solange de. "Graphies Bambara des nombres." *Journal de la Société de Africanistes* 20 (1950), 295-305.

Written symbols for numbers developed by the Bambara of Mali.

2306. Monteil, Charles. "Considérations générales sur le nombre et la numération chez les Mandés." *L'Anthropologie* 16 (1905), 485-502.

Number systems of some West African peoples.

2307. Schmidl, Marianne. "Zahl und Zählen in Afrika." *Mitteilungen der anthropologischen Gesellschaft in Wien* 45 (1915), 165-209.

Extensive discussion of the numbers developed by hundreds of African peoples, based on 260 references; numeration systems, finger gestures, symbolism, and beliefs about numbers, organized by regions.

2308. Seidenberg, A. "The Diffusion of Counting Practices." *University of California Publications in Mathematics*, Vol. 3. Berkeley and Los Angeles: University of California Press, 1960, 215-299.

Traces the different types of counting systems throughout the world from centers of diffusion. Tables, maps.

Time Reckoning

2309. Nilsson, M.P. *Primitive Time Reckoning*. Lund: C.W.K. Gleerup, 1920.

Discusses societies throughout the world, including African. Calendars based on seasons and on lunar cycles. Market week. Many references.

Weights and Measures

2310. Pankhurst, Richard. "A Preliminary History of Ethiopian Measures, Weights and Values." *Journal of Ethiopian Studies* 7 (1969), 31-54 and 99-164.

Well-documented research.

2311. Paulme, Denise. "Systèmes pondéraux et monetaires en Afrique noire." *Revue Scientifique* 80 (1942), 219-226.

Arithmetic relationships among weights and measures; currency values.

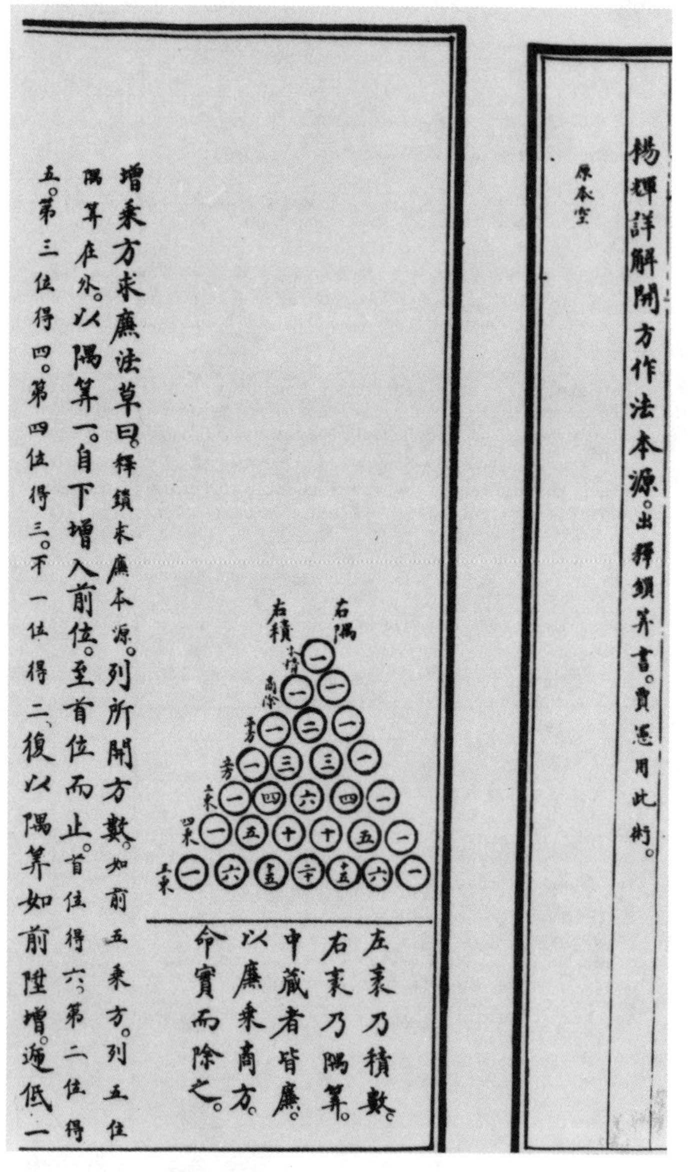

楊輝詳解開方作法本源。出釋鎖算書。賈憲用此術。

原本空

一
一 一
一 二 一
一 三 三 一
一 四 六 四 一
一 五 十 十 五 一
一 六 十五 二十 十五 六 一

左袤乃積數
右袤乃隅算
中藏者皆廉
以廉乘商方
命實而除之。

增乘方求廉法草曰。釋鎖求廉本源。列所開方數。如前五乘方。列五位
隅算在外。以隅算一。自下增入前位。至首位而止。首位得六。第二位得
五。第三位得四。第四位得三。下一位得二。復以隅算如前陞增。遞低一

Figure 16. The "Pascal" Triangle and a description of the formation in Yang Hui's *Hsiang-chieh suan-fa* (item 1261). The previous page states that the method was used by Chia Hsien (fl. ca. 1100). Taken from *Yung-lo ta-tien*, Ch. 16344.

EAST ASIAN MATHEMATICS

History of Chinese Mathematics

In this bibliography more emphasis is placed on algebra and the period most extensively covered is from the thirteenth to the early fourteenth century. There is a selection of essential works on the history of Chinese mathematics up to the Ch'ing dynasty. The recent publications are mainly monographs on important works in Chinese mathematics.

2312. Ch'ien Pao-tsung. *Chung-kuo sku-hsüeh shih* (History of Chinese mathematics). Peking: Science Publishing Co., 1964.

This book traces the historical development of Chinese mathematics in chronological order from the pre-Ch'in dynasty (before 211 B.C.) through over two millennia to the end of the Ch'ing or Manchu dynasty.

2313. ——— et al. *Sung Yüan shu-hsüeh shih lun-wen-chi* (Collected essays on Sung and Yuan mathematics). Peking: Science Publishing Co., 1966.

In this collection of essays by Ch'ien Pao-tsung, Mei Yung-chao, Yen Tun-chieh, Tu Shih-jan, and Pai Shang-shu, there is a wealth of information on thirteenth-century and early fourteenth-century Chinese mathematics. However, concentration is on the works of the four mathematicians, Ch'in Chiu-shao, Li Chih, Yang Hui, and Chu Shih-chieh.

2314. Ho Peng-Yoke. "Ch'in Chiu-shao" (III, 249-256), "Chu Shih-chieh" (III, 265-271), "Li Chih" (VIII, 313-320), "Liu Hui" (VIII, 418-425), "Yang Hui" (XIV, 538-546). *Dictionary of Scientific Biography*. New York: Scribner's, 1970-1976.

The biographical articles on the Chinese mathematicians Ch'in Chiu-shao (ca. 1202-1261), Chu Shih-chieh (fl. 1280-1303), Li Chih (1192-1279), Liu Hui (fl. ca. A.D. 250), and Yang Hui (fl. 1261-1275) are excellent for reference purposes. See also item 10 above.

2315. Hoe, John. *Les systèmes d'équations polynômes dans le Siyuan Yujian (1303)*. Paris: Collège de France, 1977, 341 pp.

This volume concentrates on the first four problems of Chu Shih-chieh's book *Ssu-yüan yü-chien* (The jade mirror of the four unknowns). These present intricate methods of solutions of systems of equations involving up to four unknowns. There is general discussion of other subjects in the book. The actual translation of the text is in a second volume which has not yet been published. See also item 1173.

2316. Hsü Ch'un-fang. *Chung suan-chia ti tai-shu-hsüeh yen-chiu* (Study of algebra by Chinese mathematicians). Peking: Chung-kuo Ch'ing-nien Pub., 1954.

A book on the history of Chinese algebra which includes topics such as simultaneous linear equations, progressions and series, indeterminate analysis, the Chinese remainder theorem, polynomial equations, and systems of equations of higher degree.

2317. Lam Lay-Yong. *A Critical Study of the Yang Hui Suan Fa. A Thirteenth-century Chinese Mathematical Treatise*. Singapore: Singapore University Press, 1977.

This book contains a detailed and analytical study of one of the best-known Chinese mathematical treatises of the thirteenth century, the *Yang Hui Suan Fa* (Yang Hui's methods of computation). The first part of the book gives a complete translation of this work.

2318. Li Yen. *Chung-kuo suan-hsüeh shih* (History of Chinese mathematics). Shanghai: Commercial Press, 1937. Reprinted 1955.

In this compact volume, the history of Chinese mathematics is divided into five main periods: (1) From the Huang-ti to the Chou and Ch'in dynasties (2700 B.C. to 200 B.C.), (2) from the Han to the Sui dynasty (200 B.C. to 600 A.D.); (3) from the T'ang to the Sung and Yuan dynasties (600 to 1367); (4) from the Ming to the beginning of the Ch'ing dynasty (1367-1750); and (5) from middle to the end of the Ch'ing dynasty (1750-1912).

2319. ———. *Chung-suan shih lun-ts'ung* (Collected essays on the history of Chinese mathematics). Peking: Science Publishing Co., 1954-1955 (second series). 5 vols.

This large collection of essays ranges over a wide selection of topics. It also includes bibliographical works and excerpts of earlier mathematical texts, which are out of print. Comparisons and reference to mathematical works outside China are made and numerous medieval methods are explained with modern notations.

2320. ——— and Tu Shih-jan. *Chung-kuo ku-tai shu-hsüeh chien-shih* (A Short History of Ancient Chinese Mathematics). Peking, 1963-1964. Reprinted Hong Kong: Commercial Press, 1976. 2 vols.

This is an improved historical account of Chinese mathematics, with separate chapters on the abacus and the infusion of Western mathematics into China.

2321. Libbrecht, Ulrich. *Chinese Mathematics in the Thirteenth Century. The Shu-shu chiu-chang of Ch'in Chiu-shao*. Cambridge, Mass.: MIT Press, 1973.

A useful book not only for Chinese mathematics in the thirteenth century but also for an insight into Chinese thought and life. It gives a detailed study of Ch'in Chiu-shao's work *Shu-shu chiu-chang* (Mathematical treatise in nine sections).

2322. Mikami, Yoshio. *The Development of Mathematics in China and Japan*. 2nd ed. New York: Chelsea Publishing Co., 1974.

The first part of the book traces the development of Chinese mathematics from the earliest periods to the Ming dynasty. It also includes special chapters on the solution of equations, values of π, and circle measurement. The second part is on Japanese mathematics. See also item 2336.

2323. Needham, J. *Science and Civilisation in China*. Vol. 3. Cambridge: Cambridge University Press, 1959.

Volume 3 of this massive project is on Mathematics and the Sciences of the Heavens and the Earth. In the Mathematics section, a chronological survey is first presented, followed by a detailed coverage of the various topics. It is indisputably essential reading for anyone interested in the history of Chinese mathematics.

2324. Sarton, George. *Introduction to the History of Science*. Baltimore: Williams and Wilkins Co., 1927-1947. 3 vols.

The biographical notes on Chinese mathematicians in these volumes provide useful guidance for general and research purposes. See also items 348, 469.

2325. Vogel, K. *Neun Bücher arithmetischer Technik*. Braunschweig: Friedr. Vieweg und Sohn, 1968.

This is a translation of the *Chiu-chang suan-shu* (Nine chapters on the mathematical art) which is the backbone of the early Chinese mathematical texts. There is also an exposition of the mathematical topics accompanied by explanations on the methods.

2326. Wang Ling and J. Needham. "Horner's Method in Chinese Mathematics: Its Origins in the Root-Extraction Procedures of the Han Dynasty." *T'oung Pao* 43 (1955), 345-401.

The article examines, analyzes, and translates the relevant passages in the *Chiu-chang suan-shu* (Nine chapters on the mathematical art) on root extraction procedures. The method of root extraction is compared with Horner's method and the similarities are shown. An explanation of how the method was discovered in China is presented.

2327. Wylie, A. *Chinese Researches*. Shanghai: n.p., 1897. Reprinted London: K. Paul, Trench, Trubner & Co., 1937.

Although this is a nineteenth-century book, the section "Jottings on the Science of the Chinese Arithmetic" still provides interesting and informative reading.

History of Japanese Mathematics

2328. Endo, Toshisada. *Zoshu Nippon Sugakushi* (The enlarged and revised history of Japanese mathematics). Tokyo: Koseisha-koseikaku, 1960; 1st ed., 1897.

Compiled for the first time in 1897; revised after the death of the author by Y. Mikami in 1918, and again, revised and enlarged in 1960 by A. Hirayama.

2329. Hayashi, Tsuruichi. *Hayashi Tsuruichi Hakushi Wasan Kenkyu Shuroku* (Collected papers on the Wasan by T. Hayashi). Tokyo: Tokyo Kaiseikan, 1937. 2 vols.

Almost all works by the author on the history of Japanese mathematics, including not a few English papers.

2330. Hirayama, Akira, et al., eds. *Ajima Naonobu Zenshu* (Collected works of Naonobu Ajima). Tokyo: Fuji Junior College Press, 1966.

Naonobu Ajima wrote on the theory of integers, binomial theorem, and double integral; with an English summary.

2331. ———. *Seki Kowa Zenshu* (Takakazu Seki's collected works, edited with explanations). Osaka: Osaka Kyoiku Tosho, 1974.

All extant works of Kowa or Takakazu Seki, who wrote much about the theory of numbers, extrapolations, determinants, and so on, with a useful analysis and English translation.

2332. Hosoi, So. *Wasan Shiso no Tokushitsu* (Characteristics of the thought of the Wasan). Tokyo: Kyoritsu Shuppan, 1941.

The author, whose grandfather and great grandfather were traditional mathematicians of the Bakumatsu period, regards himself as a scholar belonging to the old tradition.

2333. The Japan Academy. *Meijizen Nippon Sugakushi* (The history of Japanese mathematics of the pre-Meiji period). Tokyo: Iwanami Shoten, 1954-1960. 5 vols.

The most detailed study of Japanese mathematics, discussing technical parts of the history.

2334. Kato, Heizaemon. *Wasan no Kenkyu: Gyoretsu Shiki oyobi Enri* (A study of the Wasan: The determinant and the circle theorem). Tokyo: Kaiseikan, 1944.

2335. Matsuzaki, Toshio. *Edo Jidai no Sokuryo-jyutsu* (The art of surveying in the Edo period). Tokyo: Sogokagaku Shuppan, 1979.

With many illustrations.

2336. Mikami, Yoshio. *The Development of Mathematics in China and Japan*. New York: G.E. Stechert & Co., 1913. New York: Chelsea Publishing Co., 1961.

The first monograph written in English on the history of Japanese and Chinese mathematics. See also item 2322.

2337. ———, and D.E. Smith. *A History of Japanese Mathematics*. Chicago: The Open Court Publishing Co., 1914.

Written in English and read by researchers around the world. But the book has many mistakes and should be revised.

2338. ———. "Seki Kowa no Gyoseki to Keihan no Sanka narabini Shina no Sanpo tono Kankei oyobi Hikaku (Kowa Seki's achievement and the Keihan mathematicians; and their relation and comparison to the method of calculation in China)." *Sugakushi Kenkyu* (Journal for the history of mathematics), 22 and 23 (1964), first published in the *Toyo Gakuho* from 1932 to 1935.

One of the most important treatises of Mikami's, discussing the influence of Chinese mathematics upon Seki and analyzing works of mathematicians of the Keihan region.

2339. ———. *Bunkashi-jyo kara Mitaru Nippon no Sugaku* (Japanese mathematics viewed from cultural history). Tokyo: Sogensha, 1947.

An epoch-making work on Japanese mathematics from the viewpoint of cultural history, written in 1921.

2340. Ogura, Kinnosuke. *Nippon no Sugaku* (Japanese mathematics). Tokyo: Iwanami Shoten, 1940.

A brief and good introduction to Japanese mathematics as broadcast by the author, but with several mistakes.

2341. ———. *Kindai Nippon no Sugaku* (The mathematics of modern Japan). Tokyo: Shinjusha, 1956.

An excellent study on Japanese mathematics after the Meiji restoration.

2342. ———. *Chosakushu* (Collected works). Tokyo: Keisho Shobo, 1973-1975. 8 vols.

Important papers on mathematical education and the history of mathematics in China and Japan.

2343. Ohya, Shinichi, et al. *Chugoku no Kagaku* (Chinese science). Tokyo: Chuokoron, 1975.

Includes Ohya's Japanese translation of the *Chiu Chang Suan Shu* (Nine chapters on the mathematical art) and explanations of it.

2344. ———. "Wasan Izen (Before the Wasan)." *Shizen* (August 1977-June 1978).

Sheds light on Japanese mathematics before the sixteenth century, which has not been studied well.

2345. Sawada, Goichi. *Naracho-jidai Minsei Keizai no Suteki-kenkyu* (A numerical study of the public welfare and economy of the Nara period). Tokyo: Kashiwa Shobo, 1972. 1st ed., 1927.

A laborious work on the population and industry of the Nara period.

2346. Shimizu, Tatsuo, et al., eds. *Surikagaku* (Mathematical sciences). Vol. 12 of *Nippon Kagaku Gijyutsushi Taikei* (An outline of the history of science and technology in Japan). Tokyo: Daiichi-hoki Shuppan, 1969.

A brief survey of mathematics of the pre-Meiji period and a detailed study of mathematics after the Meiji restoration.

2347. Shimodaira, Kazuo. *Nipponjin no Sugaku* (The mathematics made by the Japanese). Tokyo: Kawade-shobo Shinsha, 1972.

A well-written general survey of the field.

2348. Yamazaki, Yowemon. *Shuzan Sanpo no Rekishi* (The history of the calculation by the abacus and counting rod). Tokyo: Morikita Shuppan, 1958.

Interprets the method of calculation by the counting rod (*sangi*) and the abacus (*soroban*) in China and Japan.

WOMEN IN MATHEMATICS

General Reference Works

2349. *Association for Women in Mathematics Newsletter*. Women's Research Center, Wellesley College, 828 Washington Street, Wellesley, Mass. 02181; published since 1971.

Contains many biographical articles on women mathematicians and other articles on issues which concern women and mathematics. Specific historical articles appear starting in 1976. Many lack references. Normally issued six times a year.

2350. Campbell, Paul J., and Louise S. Grinstein. "Women in Mathematics: A Preliminary Selected Bibliography." *Philosophia Mathematica* 13/14 (1976-1977), 171-203.

A listing of 70 women mathematicians, providing birth and death dates, nationality, areas of interest, and related reference material. There follows an extensive index of references including biographical dictionaries, encyclopedias, books, and periodical literature. Has many valuable annotations. Errata available from Paul J. Campbell, Beloit College, Wis. 53511.

2351. Chinn, Phyllis Zweig. *Women in Science and Mathematics Bibliography*. Distributed by the American Association for the Advancement of Science, Washington, D.C., June 1979, 38 pp.

Broad coverage of "topics about women and science or mathematics, rather than biographies of women scientists." Many articles in psychology, sociology, and on sex differences in learning.

2352. *Directory of Women Mathematicians: 1973-1974*. Providence, R.I.: American Mathematical Society, 1973, 49 pp. With annual supplements through 1977-1978.

Each listing contains information supplied by the individual on professional position, Ph.D. degree date and institution, thesis title and supervisor, fields of interest, and a partial bibliography.

2353. Høyrup, Else. *Women and Mathematics, Science and Engineering, a Partially Annotated Bibliography with Emphasis on Mathematics and with References on Related Topics*. Roskilde, Denmark: Roskilde University Library, 1978, 62 pp.

Includes historical, sociological, and psychological sources arranged by subject, with an author index.

2354. King, Amy C., with Rosemary McCroskey. "Woman Ph.D.'s in Mathematics in USA and Canada: 1886-1973." *Philosophia Mathematica* 13/14 (1976-1977), 79-129.

A listing of 926 (of the more than 1200) women who earned Ph.D.s in the period 1886-1973. Year of degree, degree-granting institution, and area of specialization are included for each woman listed. Tables presenting summaries of annual totals of doctoral degrees and number of degrees from the degree-granting institutions are of limited value since the data on which they are based are not comprehensive. Many typographical errors.

2355. Valentin, G. "Die Frauen in den exakten Wissenschaften." *Bibliotheca Mathematica* 9 (1895), 65-76.

Bibliography of works by 71 women in mathematics, astronomy, and physics. In some cases includes biographical references. Corrections and additions appear in G. Eneström, "Note bibliographique sur les femmes dans les sciences exactes," *Bibliotheca Mathematica* 10 (1896), 73-76.

Source Materials

2356. Cambridge, Mass. Radcliffe College. Schlesinger Library. Helen Brewster Owens papers.

Includes reprints of dissertations and articles by women mathematicians, lists of American women with Ph.D.s in mathematics and science, questionnaires sent to women mathematicians, and correspondence about women in mathematics.

2357. London. British Museum. Manuscript Department. Charles Babbage correspondence.

Includes correspondence between Babbage and Ada Lovelace covering the period before, during, and after their work together.

2358. Washington, D.C. Smithsonian Institution. National Museum of American History. Division of Mathematics. Data base on American mathematicians active before 1940.

Computer files include biographical and bibliographical information on over 300 women.

Articles and Monographs

Readers should note that in addition to the works listed below, the bibliography by Campbell and Grinstein, item 2350, contains an extensive list of works about individual women mathematicians. Similarly, one should consult the *Dictionary of Scientific Biography*, item 10, for biographies and bibliographies devoted to Agnesi, du Châtelet, Hypatia, Germain, Somerville, Kovalevskaya, and Noether.

2359. Archibald, Raymond Clare. "Women as Mathematicians and Astronomers." *American Mathematical Monthly* 25 (1918), 136-139.

The first of a large number of articles featuring short biographies of women mathematicians. Included are Hypatia, Agnesi, Germain, Kovalevskaya, and Herschel. Early biographical and bibliographical references.

2360. Brewer, James W., and Martha K. Smith, eds. *Emmy Noether: A Tribute to Her Life and Work*. New York: Marcel Dekker, 1981, x + 180 pp.

Includes a lengthy biographical essay by Clark Kimberling; personal reminiscences by Saunders Mac Lane and Olga Taussky; obituaries by B.L. van der Waerden and P.S. Alexandrov; technical articles by Richard G. Swan, E.J. McShane, Robert Gilmer, T.Y. Lam, and A. Fröhlich; a translation of Noether's address to the 1932 International Congress of Mathematicians; and a list of her publications.

2361. Bucciarelli, Louis L., and Nancy Dworsky. *Sophie Germain: An Essay in the History of the Theory of Elasticity*. Dordrecht and Boston: D.R. Reidel Publishing Company, 1980, xi + 147 pp.

Discusses briefly Germain's correspondence with Gauss and her contributions to number theory, as well as her better-known accomplishments in mathematical physics, for which she won the *prix extraordinaire* of the Institut de France in 1816. The gold medal was awarded to Germain, despite technical difficulties with her mathematical analysis of the problem of elasticity. What little is known of Germain's biography is also presented, along with nine figures and illustrations. See also item 1772.

2362. Channell, Ruth E. "A Compendium: The Women of Mathematics." Master's thesis, Emporia State University, Emporia, Kansas, 1977, 82 pp.

Contains a general survey, brief biographical sketches (based on sometimes unreliable secondary sources) of 33 women associated with mathematics, a miscellaneous list of 59 other women in mathematics, and general resource information. Valuable chiefly for its 399-item bibliography (not annotated).

2363. Coolidge, Julian L. "Six Female Mathematicians." *Scripta Mathematica* 17 (1951), 20-31.

Biographies of Hypatis, Agnesi, du Châtelet, Somerville, Germain, and Kovalevskaya. Few references.

2364. Dick, Auguste. *Emmy Noether, 1882-1935*. Translated by H.I. Blocher. Boston: Birkhäuser, 1981, xiv + 193 pp.

Updated English translation of a 1970 biography of Emmy Noether. It also includes obituaries by B.L. van der Waerden, Hermann Weyl, and P.S. Alexandrov, the latter newly added for this edition, as well as lists of Noether's publications, dissertations completed under her direction, and obituaries. See also item 1110.

2365. Dubreil-Jacotin, Marie-Louise. "Women Mathematicians." *Great Currents of Mathematical Thought I*. Edited by F. LeLionnais. New York: Dover, 1971, 268-280. A new English translation of the 1962 enlarged edition of *Les grands courants de la pensée mathématique*. Paris: A. Blanchard, 1962.

Short biographies of Agnesi, Germain, Somerville, Kovalevskaya, and Noether, who are described as "those who ... can ... cut an honorable figure among the great mathematicians." Few references.

2366. Eells, Walter Crosby. "American Doctoral Dissertations on Mathematics and Astronomy written by Women in the Nineteenth Century." *The Mathematics Teacher* 50 (1957), 374-376.

Lists eleven women (plus Christine Ladd-Franklin who earned a degree in 1882 which was not awarded until 1926), their dates, degree institution, dissertation title, date, and publication source. Missing from this list is Agnes Sime Baxter, Cornell, 1895.

2367. Grattan-Guinness, Ivor. "A Mathematical Union: William Henry and Grace Chishom Young." *Annals of Science* 29 (1972), 105-184.

Based mainly on family documents and letters, as well as much autobiographical material by Grace Young.

2368. Grinstein, Louise S. "Some 'Forgotten' Women of Mathematics: A Who Was Who." *Philosophia Mathematica* 13/14 (1976-1977), 73-78.

Short professional profiles of "a dozen women [who] have been selected arbitrarily as a representative group" of contributors to mathematics. References for each. These American and British women are Louise Duffield Cummings, Ellen Amanda Hayes, Marie Litzinger, Sheila Scott MacIntyre, Isabel Maddison, Helen Abbot Merrill, Mary Winston Newson, Charlotte Angas Scott, Lao Genevra Simons, Clara Eliza Smith, Anna Johnson Pell Wheeler, and Grace Chisholm Young.

2369. Kimberling, Clark H. "Emmy Noether." *American Mathematical Monthly* 79 (1972), 136-149.

Many references to source materials within body of article.

2370. Kochina, P.Y. *Sof'ya Vasil'ievna Kovalevskaya*. Moscow: Izdatel'stvo 'Nauka,' 1981, 312 pp.

Makes extensive use of correspondence and the Kovalevskaya archives of the Soviet Academy of Sciences. Forty-four photographs of people and places significant in Kovalevskaya's life. For the English translation, see item 2380.

2371. Kovalevskaya, Sofya. *A Russian Childhood.* Translated, edited, and introduced by Beatrice Stillman. With an Analysis of Kovalevskaya's Mathematics by P.Y. Kochina, USSR Academy of Sciences. New York: Springer-Verlag, 1978, xiii + 250 pp.

New English translation of a recent Russian edition based on the original manuscript. First published edition appeared in novel form in Swedish in 1889. Volume also contains "An Autobiographical Sketch," many notes and references, and a list of Kovalevskaya's scientific works. The Kochina article concentrates on the Cauchy-Kovalevskaya theorem and the problem of a heavy rigid body rotating around a fixed point.

2372. Kramer, Edna E. "Six More Female Mathematicians." *Scripta Mathematica* 23 (1957), 83-95.

Written to show that Coolidge's small number of women mathematicians attaining "real eminence" can be greatly expanded if the period subsequent to 1900 is considered. Includes short biographies (based primarily on interviews with the mathematicians themselves) of a sample of six twentieth-century Europeans: Hanna Neumann, Maria Pastori, Maria Cinquini-Cibrario, Jacqueline Lelong-Ferrand, Paulette Libermann, and Sophie Piccard. A biography of Florence Nightingale David is added in a revised version in Kramer, *The Nature and Growth of Modern Mathematics* (New York: Hawthorn Books Inc., 1970), 703-714.

2373. Loria, Gino. "Les femmes mathématiciennes." *Revue Scientifique* Series 4, 20 (1903), 385-392.

Example of a mathematician's attempt to discredit the contributions of women mathematicians. Criticized by J. Joteyko in "À propos des femmes mathématiciennes," *Revue Scientifique* Series 5, 1 (1904), 12-15. Reply to criticism by Loria in "Encore les femmes mathématiciennes," *Revue Scientifique* Series 5, 1 (1904), 338-340.

2374. Mozans, H.J. *Woman in Science.* 1913. Reprinted Cambridge, Mass.: The MIT Press, 1974, xvii + 452 pp.

Contains a 21-page chapter on women in mathematics featuring the same women as Rebière, item 2381. Naive historical judgments on mathematics. Several biographical references.

2375. Osen, Lynn M. *Women in Mathematics.* Cambridge, Mass.: The MIT Press, 1974, xii + 185 pp.

Biographies of Hypatia, Agnesi, du Châtelet, Herschel, Germain, Somerville, Kovalevskaya, and Noether. Brief comments on contributions by contemporary women mathematicians. References. Not written for mathematicians.

2376. Owens, Helen B. "Early Scientific Work of Women and Women in Mathematics." Mimeographed. State College, Pa.: Pennsylvania State University, University Park, 1940?, 10 pp.

Text of a talk detailing highlights of women in science from "primitive" times to the present. Information on some American women mathematicians who received Ph.D.s before 1900 and statistics on American women who received Ph.D.s by the late 1930s are included. No references.

2377. Patterson, Elizabeth. "Mary Somerville." *British Journal for History of Science* 4 (1969), 311-339.

Paper sketches Somerville's career, its 19th-century English science setting, and the contents of the Somerville papers.

2378. Perl, Teri. *Math Equals*. Reading Mass.: Addison-Wesley Publishing Company, 1978.

Biographies of nine women mathematicians. Includes elementary mathematics activities related to the areas in which they worked. Final chapter details founding of Association for Women in Mathematics. Referenced footnotes for each biography.

2379. ———. "The *Ladies' Diary* or *Woman's Almanack*, 1704-1841." *Historia Mathematica* 6 (1979), 36-53.

Analyzes patterns of participation by female contributors to a popular women's magazine which was largely devoted to mathematical problems and puzzles. An earlier version appears as "The Ladies' Diary ... Circa 1700," *The Mathematics Teacher* 70 (1977), 354-358.

2380. Polubarinova-Kochina, P. *Sophia Vasilyevna Kovalevskaya, Her Life and Work*. Moscow: Foreign Languages Publishing House, 1957.

The point of view is different from that found in other sources. Kovalevskaya is presented as a political, radical individual. No references appear in the English version of this work. For the Russian original, see item 2370.

2381. Rebière, A. *Les femmes dans la science*. 2nd ed. Paris: Libraire Nony et Cie, 1897. 1st ed. 1894, with subtitle, *Conférence faite au Cercle Saint-Simon le 24 février 1894.*

Probably the first extensive collection of biographies of women in science. First edition an 87-page brochure. Second edition includes 285 pages of biography and two appendices: "Si la femme est capable de science--opinions diverses" and "Menus propos sur les femmes et les sciences--notules diverses." Includes substantial entries on Agnesi, du Châtelet, Germain, Hypatia, Kovalevskaya, and Somerville, each with bibliographies and references. Also contains short entries on other women mathematicians and on women who influenced male mathematicians.

2382. Tee, G.J. "Sof'ya Vasil'yevna Kovalevskaya." *Mathematical Chronicle* 5 (1977), 113-139.

References include many Russian sources as well as German and French material. Several photographs not previously published. Journal published in New Zealand.

2383. Wallis, Ruth, and Peter Wallis. "Female Philomaths." *Historia Mathematica* 7 (1980), 57-64.

"The evidence from the Ladies' Diary ... is extended, and other sources of information, particularly subscription lists, are discussed" in order to determine "the extent of women's interest in and practice of mathematics in Britain in the 18th century."

2384. Weyl, Hermann. "Emmy Noether." *Scripta Mathematica* 8 (1935), 201-220.

Memorial address, Bryn Mawr College, April 26, 1935.

AUTHOR INDEX

Includes authors, editors, translators.

SUBJECT INDEX

Lists names which appear in titles and annotations.